11th International Spring Seminar on Nuclear Physics

Shell Model and Nuclear Structure – Achievements of the Past Two Decades

Journal of Physics: Conference Series
Volume 580

Ischia, Italy
12 - 16 May 2014

ISBN: 978-1-5108-0075-5
ISSN: 1742-6588

Printed from e-media with permission by:

Curran Associates, Inc.
57 Morehouse Lane
Red Hook, NY 12571

Some format issues inherent in the e-media version may also appear in this print version.

Printed by Curran Associates, Inc. (2015)

For permission requests, please contact the Institute of Physics
at the address below.

Institute of Physics
Dirac House, Temple Back
Bristol BS1 6BE UK

Phone: 44 1 17 929 7481
Fax: 44 1 17 920 0979

techtracking@iop.org

Additional copies of this publication are available from:

Curran Associates, Inc.
57 Morehouse Lane
Red Hook, NY 12571 USA
Phone: 845-758-0400
Fax: 845-758-2634
Email: curran@proceedings.com
Web: www.proceedings.com

Table of contents

Volume 580

11th International Spring Seminar on Nuclear Physics: Shell Model and Nuclear Structure – achievements of the past two decades
12–16 May 2014, Ischia, Italy

Accepted papers received: 9 December 2014
Published online: 9 February 2015

Preface

Papers

Journal of Physics: Conference Series **580** (2015) 011001 doi:10.1088/1742-6596/580/1/011001

Foreword

The 11th International Seminar on Nuclear Physics was held in Ischia from May 12 to May 16, 2014. This Seminar was dedicated to Aldo Covello, who has been the promoter of this series of meetings, which started in Sorrento in 1986 and continued with meetings held every two or three years in the Naples area.

Aldo's idea was to offer to a group of researchers, actively working in selected fields of Nuclear Physics, the opportunity to confront their points of view in a lively and informal way. The choice for the period of the year, Spring, as well as the sites chosen reflected this intent.

The first meeting was of a purely theoretical nature, but it was immediately clear that the scope of these conferences needed to be enlarged calling into play the experimental community. Then, starting from the second meeting, all the following ones have been characterized by fruitful discussion between theoretical and experimental researchers on current achievements and future developments of nuclear structure. This may be read, in fact, as one of the motivating factors for Aldo's election as Fellow of the American Physical Society in 2008 "... *for his outstanding contributions to the international nuclear physics community by providing, for over two decades, a venue for theorists and experimentalists to share their latest ideas."*

The present meeting, organized by Aldo's former students and with the benefit of his suggestions, has maintained this tradition. The title "Shell model and nuclear structure: achievements of the past two decades" recalls that of the 2nd International Spring Seminar "Shell Model and Nuclear Structure: where do we stand?". The main aim of this 11th Seminar was, in fact, to discuss the changes of the past two decades on our view of nuclei in terms of shell structure as well as the perspectives of the shell model, which has been one of the key points in Aldo's research. This point is well accounted by the Opening Speech of Igal Talmi, one of the fathers of the shell model. Then, as usual, the program of the meeting consisted of general talks and more specialized contributions, which covered five main topics: i) From nuclear forces to nuclear structure; ii) Exploring nuclear structure toward the drip line; iii) Role of the shell model in the study of exotic nuclei; iv) Nuclear structure aspects outside the shell model; and v) Special topics. The main conclusions were drawn in two keynote talks given by Amand Faessler and Franco Iachello.

The Conference had about 90 participants from some 20 countries [please see the list of participants]. This is well in line with the tradition of these meetings, as is the fact that more than 50% of the present participants attended one or more of the previous Seminars.

We received 58 manuscripts out of the 73 invited papers and contributions presented at the Seminar. All of these have been peer reviewed and are collected in this volume. We would like to thank all the colleagues who have acted as referees to assess the suitability of the various articles for publication in the *Journal of Physics: Conference Series*. We are confident that the high quality of both invited and contributed papers contained in these Proceedings will be appreciated by the nuclear physics community.

We gratefully acknowledge the members of the Advisory Committee for their valuable cooperation in preparing the scientific program as well as the financial support of the Istituto Nazionale di Fisica Nucleare, the University of Naples Federico II, and the Dipartimento di Fisica who helped make the Seminar possible.

Angela Gargano
Luigi Coraggio
Nunzio Itaco
Editors

LOCAL ORGANIZING COMMITTEE

A Gargano
L Coraggio
N Itaco
F Andreozzi
N Lo Iudice
A. Porrino

INTERNATIONAL ADVISORY COMMITTEE

F Azaiez *(Orsay)*
B R Barrett *(Tucson)*
N Benczer-Koller *(Rutgers)*
Y Blumenfeld *(IPN Orsay)*
A Bracco *(Milano)*
B A Brown *(Michigan)*
G de Angelis *(Legnaro)*
J Dobaczewski *(Warsaw)*
A Faessler *(Tubingen)*
K Gelbke *(Michigan)*
M Gorska *(GSI)*
J H Hamilton *(Vanderbilt)*
M Hjorth-Jensen *(Michigan and Oslo)*
F Iachello *(Yale)*
T T S Kuo *(Stony Brook)*
G La Rana *(Napoli)*
R Machleidt *(Moscow)*
L Moretto *(Berkeley)*
T Otsuka *(Tokyo)*
N Pietralla *(Darmstadt)*
A Poves *(Madrid)*
A Schwenk *(Darmstadt)*
O Sorlin *(Ganil)*
I Talmi *(Rehovot)*
P Van Duppen *(Leuven)*
A Vitturi *(Padova)*
V Voronov *(Dubna)*

SPONSORS OF THE SEMINAR

Dipartimento di Fisica, Università di Napoli "Federico II"
Istituto Nazionale di Fisica Nucleare
Università di Napoli Federico II

11th INTERNATIONAL SPRING SEMINAR ON NUCLEAR PHYSICS

SHELL MODEL AND NUCLEAR STRUCTURE: achievements of the past two decades
in honor of Aldo Covello

ISCHIA, MAY 12–16, 2014

List of participants

Y. Alhassid	Center for Theoretical Physics, Sloane Physics Laboratory, Yale University, New Haven, Connecticut 06520, USA *yoram.alhassid@yale.edu*
F. Andreozzi	Dipartimento di Fisica, Università di Napoli Federico II, Complesso Universitario di Monte S. Angelo, Via Cintia, 80126 Napoli, Italy *andreozzi@na.infn.it*
F. Azaiez	IPN, F-91406 Orsay, France *azaiez@ipno.in2p3.fr*
B. R. Barrett	Department of Physics, PO Box 21008, University of Arizona, Tucson, Arizona 85721-0081, USA *bbarrett@Physics.arizona.edu*
N. Benczer-Koller	Department of Physics and Astronomy, Rutgers University, New Brunswick, NJ 08816, USA *nkoller@physics.rutgers.edu*
M. A. Bentley	Department of Physics, University of York, Heslington, York, YO10 5DD, United Kingdom *michael.bentley@york.ac.uk*
A. N. Bezbakh	Joint Institute for Nuclear Research, Dubna 141980, Russia, *bezbakh@ther.jinr.ru*
A. Bonaccorso	Istituto Nazionale di Fisica Nucleare, Sezione di Pisa, Largo B. Pontecorvo 3, 56127 Pisa, Italy *bonac@df.unipi.it*
S. Bönig	Institut für Kernphysik, Technische Universität Darmstadt, Darmstadt, Germany *sboenig@ikp.tu-darmstadt.de*
M. J. Garcia Borge	ISOLDE-PH, CERN, 1121 Genève 23, Switzerland *mgb@cern.ch*
M. Bouhelal	Laboratoire de Physique Appliquée et Théorique, Université de Tébessa, Route de Constantine 12002, Tébessa, Algeria m_bouhelal@yahoo.fr

M. Böyükata
Department of Elementary Science Education, Faculty of Education,
Çanakkale Onsekiz Mart University,
TR-17100 Çanakkale, Turkey
boyukata@comu.edu.tr

D. Bozik
Institute of Particle and Nuclear Physics MFF UK, Charles University in Prague,
V Holesovickach 2, 180 00, Prague, Czech Republic
daniel.bozik@gmail.com

R. Broda
Niewodniczański Institute of Nuclear Physics PAN,
PL-31-342 Kraków, Poland
rafal.broda@ifj.edu.pl

B. A. Brown
Department of Physics and Astronomy, and National Superconducting Cyclotron Laboratory, Michigan State University,
East Lansing, Michigan 48824-1321, USA
brown@nscl.msu.edu

F. Catara
Dipartimento di Fisica ed Astronomia, Università di Catania,
Via S. Sofia 64, 95123 Catania, Italy
francesco.catara@ct.infn.it

L. Coraggio
Istituto Nazionale di Fisica Nucleare, Sezione di Napoli,
Complesso Universitario di Monte S. Angelo, Via Cintia, 80126 Napoli, Italy
coraggio@na.infn.it

A. Covello
Dipartimento di Fisica, Università di Napoli Federico II,
Complesso Universitario di Monte S. Angelo, Via Cintia, 80126 Napoli, Italy
covello@na.infn.it

J. Cseh
MTA ATOMKI ,
Debrecen, Bem ter 18/C, Hungary 4026
cseh@atomki.mta.hu

J. Darai
Department of Experimental Physics, University of Debrecen,
Debrecen, Bem ter 18/A, Hungary 4026
darai@atomki.mta.hu

C. H. Dasso
Departamento de Física Atómica, Molecular y Nuclear, Universidad de Sevilla,
Apartado 1065, E-41080 Sevilla, Spain
dasso@us.es

G. de Angelis
Istituto Nazionale di Fisica Nucleare, Laboratori Nazionali di Legnaro,
Viale dell'Università 2, 35020 Legnaro (Pd), Italy
deangelis@lnl.infn.it

G. de Gregorio
Dipartimento di Fisica, Università di Napoli Federico II,
Complesso Universitario di Monte S. Angelo, Via Cintia, 80126 Napoli, Italy
degregorio@na.infn.it

N. Dinh Dang
Theoretical Nuclear Physics Laboratory,
Nishina Center for Accelerator-Based Science, RIKEN ,
2-1 Hirosawa, Wako city, 351-0198 Saitama, Japan
dang@riken.jp

G. Dong	The School of Physics, Peking University, Beijing 100871, China *gxdongpku@gmail.com*
J. P. Draayer	Department of Physics and Astronomy, Louisiana State University, Baton Rouge, LA 70803, USA *draayer@sura.org*
A. Etile	CSNSM, Université Paris-Sud, CNRS/IN2P, Orsay, France *etile@csnsm.in2p3.fr*
A. Faessler	Institut für Theoretische Physik, Universität Tübingen, Auf der Morgenstelle 14, D-72076 Tübingen, Germany *faessler@uni-tuebingen.de*
L. S. Ferreira	Centro de Fiísica das Interacções Fundamentais, and Departmento de Física, Instituto Superior Técnico, Universidade de Lisboa, Av Rovisco Pais, 1049 001, Lisboa, Portugal *flidia@ist.utl.pt*
S. Franchoo	IPN, F-91406 Orsay, France *franchoo.ipno.in2p3.fr*
A. Gade	National Superconducting Cyclotron Laboratory, Michigan State University, 640 South Shaw Lane, East Lansing, Michigan 48824, USA *gade@nscl.msu.edu*
M. Gai	LNS at Avery Point, University of Connecticut 1084 Shennecossett Road, Groton, CT 06340-6097, USA *moshe.gai@yale.edu*
D. Gambacurta	GANIL, CEA/DSM-CNRS/IN2P3, Bvd Henri Becquerel, 14076 Caen, France *gambacurta@ganil.fr*
U. Garg	Physics Department, University of Notre Dame, Notre Dame, IN 46556, USA *Umesh.Garg.1@nd.edu*
A. Gargano	Istituto Nazionale di Fisica Nucleare, Sezione di Napoli, Complesso Universitario di Monte S. Angelo, Via Cintia, 80126 Napoli, Italy *gargano@na.infn.it*
P. Garrett	University of Guelph, Department of Physics, Guelph, Ontario NIG 2WI, Canada *pgarrett@physics.uoguelph.ca*
J. N. Ginocchio	Theoretical Division, Los Alamos National Laboratory, MS B283 Los Alamos, NM 87545, USA *gino@lanl.gov*

J. M. G. Gómez	Departamento de Física Atómica, Molecular y Nuclear, Universidad Complutense de Madrid, E-28040 Madrid, Spain *gomezk@nuc1.fis.ucm.es*
F. Gramegna	Istituto Nazionale di Fisica Nucleare, Laboratori Nazionali di Legnaro, Viale dell'Università 2, 35020 Legnaro (Pd), Italy *gramegna@lnl.infn.it*
J. H. Hamilton	Department of Physics and Astronomy, Vanderbilt University, Nashville, Tennessee 37235, USA *j.h.hamilton@vanderbilt.edu*
A. Hennig	Institute for Nuclear Physics, University of Cologne, Zülpicher Straße 77, D-50937 Köln, Germany *hennig@ikp.uni-koeln.de*
A. Heusler	Gustav-Kirchhoff-Strasse 7/1, D-69120 Heidelberg, Germany *A.Heusler@mpi-hd.mpg.de*
F. Iachello	Center for Theoretical Physics, Yale University, New Haven, Connecticut 06520-8120, USA *Francesco.Iachello@yale.edu*
L. W. Iskra	Institute of Nuclear Physics, Polish Academy of Sciences, PL-31342 Kraków, Poland *lukasz.iskra@ifj.edu.pl*
N. Itaco	Dipartimento di Fisica, Università di Napoli Federico II Complesso Universitario di Monte S. Angelo, Via Cintia, 80126 Napoli, Italy *itaco@na.infn.it*
D. G. Jenkins	Department of Physics, University of York, York YO10 5DD, UK *david.jenkins@york.ac.uk*
M. W. Kirson	Department of Particle Physics and Astrophysics, Weizmann Institute of Science, 76100 Rehovot, Israel *michael.kirson@weizmann.ac.il*
F. Knapp	Institute of Particle and Nuclear Physics MFF UK, Charles University in Prague, V Holesovickach 2, 180 00, Prague, Czech Republic *knapp@ipnp.troja.mff.cuni.cz*
T. T. S. Kuo	Department of Physics and Astronomy, Stony Brook University, Stony Brook, New York 11794, USA *kuo@tonic.physics.sunysb.edu*
J. Kvasil	Institute of Particle and Nuclear Physics MFF UK, Charles University in Prague, V Holesovickach 2, 180 00, Prague, Czech Republic *kvasil@ipnp.troja.mff.cuni.cz*

K. Langanke
GSI Helmholtzzentrum für Schwerionenforschung,
D-64291 Darmstadt, Germany
k.langanke@gsi.de

S. M. Lenzi
Dipartimento di Fisica e Astronomia, Università di Padova,
Via Marzolo 8, 35131 Padova, Italy
lenzi@pd.infn.it

N. Lo Iudice
Dipartimento di Scienze Fisiche, Università di Napoli Federico II,
Complesso Universitario di Monte S. Angelo, Via Cintia, 80126 Napoli, Italy
loiudice@na.infn.it

S. Lunardi
Dipartimento di Fisica e Astronomia, Università di Padova,
Via Marzolo 8, 35131 Padova, Italy
lunardi@pd.infn.it

R. Machleidt
Department of Physics, University of Idaho
Moscow, ID 83844-0903, USA
machleid@uidaho.edu

D. Mancusi
CEA, Centre de Saclay, IRFU/Service de Physique Nucléaire,
F-91191 Gif-sur-Yvette, France
davide.mancusi@cea.fr

T. Marchi
Istituto Nazionale di Fisica Nucleare, Laboratori Nazionali di Legnaro,
Viale dell'Università 2, 35020 Legnaro (Pd), Italy
Tommaso.Marchi@lnl.infn.it

L. E. Marcucci
Dipartimento di Fisica E. Fermi, Università di Pisa,
Largo B. Pontecorvo 3, 56127 Pisa, Italy
marcucci@pi.infn.it

R. Massarczyk
Helmholtz-Zentrum Dresden-Rossendorf,
01328 Dresden, Germany
massarczyk@hzdr.de

B. Melon
Dipartimento di Fisica e Astronomia, Università di Firenze,
Via G. Sansone 1, 50125 Sesto Fiorentino (Fi), Italy
melon@fi.infn.it

L. G. Moretto
LBL and UCB Berkeley,
CA USA
lgmoretto@lbl.gov

R. Moro
Dipartimento di Fisica, Università di Napoli Federico II,
Complesso Universitario di Monte S. Angelo, Via Cintia, 80126 Napoli, Italy
moro@na.infn.it

H. Naïdya
Institut Pluridisciplinaire Hubert Curien (IPHC)
23 rue du Loess -BP 28, F-67037, Strasbourg, Cedex 2, France
houda.naidja@iphc.cnrs.fr

S. Nishimura
RIKEN Nishina Center,
Wako city, 351-0198 Saitama, Japan
nishimu@ribf.riken.jp

T. Otsuka	Department of Physics, University of Tokyo, Hongo, Bunkyo-ku, Tokyo 113-0033, Japan otsuka@phys.s.u-tokyo.ac.jp
L. Pellegri	Istituto Nazionale di Fisica Nucleare, Sezione di Milano, Via Celoria 16, 20133 Milano, Italy, luna.pellegri@mi.infn.it
C. M. Petrache	Centre de Sciences Nucléaires et de Sciences de la Matière, Université Paris-Sud and CNRS/IN2P3, 91405 Orsay, France petrache@csnsm.in2p3.fr
N. Pietralla	Institut für Kernphysik, TU Darmstadt, Schlossgartenstraße 9, 64289 Darmstadt, Germany pietralla@ikp.tu-darmstadt.de
S. Pirrone	Istituto Nazionale di Fisica Nucleare, Sezione di Catania, Via Santa Sofia 64, 95123 Catania, Italy pirrone@ct.infn.it
Z. Podolyák	Department of Physics, University of Surrey, Guildford GU2 7XH, UK z.podolyak@surrey.ac.uk
A. Poves	Departamento de Física Teórica e IFT-UAM/CSIC, Universidad Autónoma de Madrid, E-28049 Madrid, Spain alfredo.poves@uam.es
A. V. Ramayya	Department of Physics and Astronomy, Vanderbilt University, Nashville, Tennessee 37235, USA a.v.ramayya@vanderbilt.edu
A. Repko	Institute of Particle and Nuclear Physics MFF UK, Charles University in Prague, V Holesovickach 2, 180 00, Prague, Czech Republic anton@a-repko.sk
P. Ring	Physik Department, Technische Universität München, D-85747 Garching, Germany ring@ph.tum.de
R. N. Sagaidak	Flerov Laboratory of Nuclear Reactions, Joint Institute for Nuclear Research, 141980, Dubna, Moscow region, Russian Federation sagaidak@nrmail.jinr.ru
F. Sammarruca	Physics Department, University of Idaho Moscow, Idaho 83844-0903, USA fsammarr@uidaho.edu
E. Santopinto	Istituto Nazionale di Fisica Nucleare, Sezione di Genova, via Dodecaneso 33, 16146 Genova, Italy elena.santopinto@ge.infn.it

R. Schwengner	Institute of Radiation Physics, Helmholtz-Zentrum Dresden-Rossendorf Bautzner Landstr. 400, 01328 Dresden Germany *r.schwengner@hzdr.de*
S. Stoica	Horia Hulubei Foundation, P.O. Box MG12, 407, Atomistilor street, 077125 Bucharest-Magurele, Ilfov, Romania *stoica@theory.nipne.ro*
T. Suzuki	Department of Physics, College of Humanities and Sciences, Nihon University, Sakurajosui 3, Setagaya-ku, Tokyo 156-8550, Japan *suzuki@phys.chs.nihon-u.ac.jp*
I. Talmi	Department of Physics, The Weizmann Institute of Science, Rehovot 76100, Israel *igal.talmi@weizmann.ac.il*
J. J. Valiente Dobón	Istituto Nazionale di Fisica Nucleare, Laboratori Nazionali di Legnaro, Viale dell'Università 2, 35020 Legnaro (Pd), Italy *valiente@lnl.infn.it*
N. Van Giai	Institut de Physique Nucléaire, IN2P3-CNRS, Université Paris-Sud, F-91406 Orsay Cedex, France nguyen@ipno.in2p3.fr
P. Veselý	Institute of Nuclear Physics, Academy of Sciences of the Czech Republic, V Holesovickach 2, Řež, Czech Republic *p.vesely@ujf.cas.cz*
E. Vigezzi	Istituto Nazionale di Fisica Nucleare, Sezione di Milano, Via Celoria 16, 20133 Milano, Italy *vigezzi@mi.infn.it*
A. Vitturi	Dipartimento di Fisica e Astronomia Università di Padova, Via Marzolo 8, 35131 Padova, Italy *vitturi@pd.infn.it*
A. A. Voinov	Flerov Laboratory of Nuclear Reaction, Joint Institute for Nuclear Research, RU-141980 Dubna, Russian Federation *voinov_2000@mail.ru*
V. V. Voronov	Bogoliubov Laboratory of Theoretical Physics, Joint Institute for Nuclear Research, 141980 Dubna, Moscow region, Russia *voronov@theor.jinr.ru*
X. Wang	Build. 3, Room 203, No.1 Xueshi Road, Huzhou city, Zhejiang Province, People's Republic of China *wangxiaobaopku@gmail.com*
O. Wieland	Istituto Nazionale di Fisica Nucleare, Sezione di Milano, Via Celoria 16, I-20129 Milano, Italy *Wieland@mi.infn.it*
A. Zilges	Institut für Kernphysik, University of Cologne, 50937 Köln, Germany *zilges@ikp.uni-koeln.de*

Journal of Physics: Conference Series **580** (2015) 011003 doi:10.1088/1742-6596/580/1/011003

Seminar Photo

Peer review statement

All papers published in this volume of *Journal of Physics: Conference Series* have been peer reviewed through processes administered by the proceedings Editors. Reviews were conducted by expert referees to the professional and scientific standards expected of a proceedings journal published by IOP Publishing.

Journal of Physics: Conference Series **580** (2015) 012001 doi:10.1088/1742-6596/580/1/012001

The Shell Model – Simplicity from Complexity

Igal Talmi

The Weizmann Institute of Science, Rehovot, Israel

Email: igal.talmi@weizmann.ac.il

Abstract. In order to apply the shell model to most nuclei, an effective ("residual") interaction between nucleons should be added to the single nucleon Hamiltonian. It cannot be the bare interaction which would introduce strong short range correlations into the shell model wave functions of nucleons moving independently in a potential well. For many years, many authors have been developing methods for calculating from the bare interaction, an effective interaction which may be used in the shell model. Only recently, the results seem to agree with experiment. During these years, successful shell model calculations have been carried out by using effective interactions determined consistently from measured energies of nuclei. Some old and some new results exhibit the success of this approach. In principle, shell model wave functions may be determined by a renormalized Hamiltonian which contains the effective interactions rather than the bare ones. Shell model wave functions may be used to calculate other observables if matrix elements are taken of renormalized operators. Such a program could succeed irrespective of the relation between shell model wave functions and the real ones. This would certainly be the case for ab initio calculations where a central potential well is not assumed and derivation of nuclear structure is attempted by using many body theory of nucleons interacting by the bare interaction. Still, there are observables which have been successfully calculated by shell model wave functions from real, un-renormalized operators. Several examples are briefly described below. This fact may be taken as some evidence that the "wounds" inflicted on shell model wave functions by the short range correlations occupy only a small volume and that there is a considerable overlap between shell model wave functions and the real ones. The calculated overlap between shell model and real wave functions may be obtained from derivations of the effective interaction. It is more difficult to see how the shell model may emerge as a good approximation from ab initio calculations. The question is whether the simplicity of the shell model as a good approximation will emerge from those complex calculations.

1. Introduction

It is a great honor and pleasure to open the 11th Spring Seminar on Nuclear Physics. It is intended to pay tribute to our friend and colleague Aldo Covello who established a series of superb conferences which have been highly interesting, informative and stimulating. We wish him and Renata, many years of good health, fruitful activity and much happiness. The present conference was organized by his younger colleagues and it followed very successfully the great tradition which he established. We are very grateful to them for their efforts.

The subject of the conference this year is the Shell Model and Nuclear Structure. There are several approaches to the shell model, like the situation of the proverbial elephant and the blind men. One approach is the simple shell model with (two-body) effective interactions determined from experiment. Other efforts aim to derive the effective interaction to be used in the shell model, from the interaction between free nucleons. In a more ambitious approach no shell structure is adopted. It begins *ab initio*, from the "bare" interaction between free nucleons and attempts to obtain good approximations to the real wave functions of nuclei. In the following, a brief description of these approaches is given. In the shell model, nucleons are taken to move independently in a central potential well. Magic nuclei contain closed shells of protons and neutrons. There is only one shell model wave function of each of these nuclei and it is well defined. This is the case also for nuclei in which there is one (valence) nucleon outside closed shells or one nucleon missing from them (one hole state). If there are several valence nucleons or holes, there are several states of the system and they are all degenerate in the central potential well. If an interaction is included in the shell model Hamiltonian, the eigenvalues are no longer degenerate and the various eigenstates have definite wave functions with well defined values of total spin J and projection M.

In the early days, the rather mild potentials used in that time to describe the interaction between free nucleons, were used in the shell model. The agreement between calculated energies and experimental ones was at best only qualitative. Later, it was established by experiments, that the bare interaction is too strong and singular at short distances and hence, cannot be used in the shell model. Thus, it turned out that shell model wave functions may be used only with a strongly renormalized interaction. Matrix elements of the bare interaction between the real wave functions could be hopefully replaced by matrix elements of the renormalized, or effective interaction between shell model wave functions. More than 50 years ago, Brueckner introduced the G-matrix and he was followed by many authors who developed methods of many-body theory to be applied to actual (finite) nuclei. Some of those authors who made important contributions to this approach are among the participants in this conference. This is a very difficult problem and only recently some reliable results were obtained, like those of Covello et al. [1]

It should be realized that in addition to the computational difficulties, there is an additional problem which is more serious in the *ab initio* approach. There is now extensive and detailed experimental information about the interaction between free nucleons. Most of it is obtained from scattering experiments which yield information about "on-shell" features of the interaction. There are many theoretical models which reproduce rather well the experimental data. There is no theory, however, from which this interaction could be derived. The QCD theory which gives excellent results for high energy phenomena is not very helpful in the low energy, non perturbative domain. This subject will be discussed in this conference.

There is a great promise but there are very great difficulties in the *ab initio* approach. Its success depends on the knowledge of the bare interaction and the accuracy of the necessary approximations of the nuclear many-body problem. Reliable information on the (real) nuclear wave functions would enable the calculation of nuclear states and their energies, also of nuclei which are inaccessible experimentally. It will also enable the calculation of various moments and transitions, electromagnetic and weak ones including neutrino-less beta decay. There will be a talk about this approach by one of its originators. An intriguing question about this (no-core shell model) approach is whether the simplicity of the shell model will emerge from the very complicated calculations. This simplicity is demonstrated in the following by a few examples, some old and some new ones.

2. The simple shell model

Without reliable theoretical derivation, matrix elements of the effective interaction were extracted from energies of nuclear states in a consistent way. Restriction to two-body interactions leads to matrix elements between n-nucleon states which are linear combinations of two-nucleon matrix elements. Thus, energies of nuclear states can be expressed in terms of a smaller set of two-nucleon matrix elements. If there are sufficient measured energies, these matrix elements may be determined. If nuclear energies, calculated with these matrix elements are in good agreement with the measured ones, this procedure is consistent. The two-nucleon matrix elements thus determined, may be used in calculating and predicting energies which have not yet been measured.

This method has been extensively used in many cases. In order to obtain more detailed and accurate information, more configurations have been included. In recent years, many millions of nuclear states are included in large scale shell model calculations. They need very many two-nucleon matrix elements and single nucleon energies. There are not enough measured energies to determine all of them but the important ones are determined from experiment [2]. In the following, however, only simple calculations will be presented, demonstrating how the very simple shell model may be used in many cases.

Using simple shell model configurations, only a few matrix elements could be determined. Still, some general features of the effective interaction were extracted from simple cases. As has been well known, the $T = 1$ interaction is attractive and strong in $J = 0$ states. In other states, however, it is weak on the average and its average is repulsive in jj'-configurations. In j^n-configurations it leads to eigenstates with good *seniority* [3]. The average $T = 0$ interaction which acts between protons and neutrons, is strong and attractive. It breaks seniority in a major way leading to rotational spectra in many nuclei with both valence protons and neutrons.

A direct consequence of these features is that the potential well of the shell model is created by the attractive proton-neutron interaction. It determines its depth and its shape. Hence, energies of proton orbits in the well (single nucleon energies) are determined by the occupation numbers of neutrons and *vice versa*. A clear demonstration of this conclusion is offered by the prediction of the ^{11}Be ground state spin [4]. Some examples presented below, demonstrate these features.

3. Some shell model applications to nuclei

Some properties of the $T = 1$ interaction may be seen in j^n-configurations of identical nucleons. Ground states have maximum $J = 0$ pairing and hence, lowest seniorities, $v = 0$ ($J = 0$) for even n values and $v = 1$ ($J = j$) for odd ones. There are many two-body interactions which are diagonal in seniority and all of them yield a simple expression for ground state energies. Binding energies are given by the following expression.

$$BE(j^n) = BE(n = 0) + nC + \alpha n(n-1)/2 + \beta(n/2) \ . \tag{1}$$

Equation (1) includes the binding energy of the nucleus with closed shells, without j-nucleons, and n single nucleon energies. The contribution of the mutual interaction of the j-nucleons is given by the quadratic term in n and the pairing term. The coefficients of these terms are linear combinations of two-nucleon matrix elements

$$\alpha = \frac{[(2j+2)\bar{V}_2 - V_0]}{2j+1} \qquad \beta = (2j+2)\frac{V_0 - \bar{V}_2}{2j+1} \tag{2}$$

where

$$V_o = \langle j^2\, v = 0, J = 0|V|j^2\, v = 0, J = 0\rangle$$

$$\bar{V}_2 = \Sigma(2J+1)\langle j^2\, v = 2, J|V|j^2\, v = 2, J\rangle / \Sigma(2J+1) \ . \tag{3}$$

The summations in (3) are on all j^2 states with even values of $J > 0$. The expression (1) of binding energies yields a very simple result for single nucleon separation energies. They should lie on two straight and parallel lines, one for odd and the other for even numbers n. In figure 1, measured single neutron separation energies of calcium isotopes are plotted as a function of n. In these nuclei, valence neutrons are taken to occupy the $1f_{7/2}$ orbit

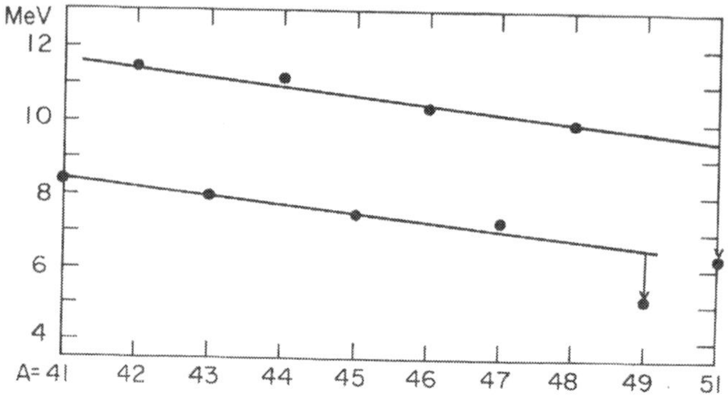

Figure 1. Neutron separation energies from some calcium isotopes.

The agreement between measured energies and the expression (1) is rather good. The coefficient of the pairing term is large and attractive whereas the coefficient of the quadratic term is small and repulsive. From figure 1 follows that the magic nucleus ^{48}Ca is not "tightly bound". In fact, it is less tightly bound than its preceding even-even neighbours. Magic nuclei have extra stability due to nuclei following them being less bound. This fact has been known for many years but references are still published to "an enhanced binding" of magic nuclei.

An important question may be raised in view of figure 1. The experimental data agree rather well with the formula which is valid for two-nucleon effective interaction between the valence nucleons. *Where are the three-body forces*? Such forces are used by many authors who start from the bare interaction. Also, if there are no such forces in the bare interaction, they may well be introduced into the effective interaction by the renormalization. There is no clear evidence for three-body interactions between valence nucleons also in other cases, including the large scale shell model calculations. Still, three-body interactions could contribute implicitly. Single nucleon energies, usually taken form experiment, could include contributions from such interaction between two core nucleons with each valence nucleon. The two-nucleon matrix elements extracted from experiment could include contributions from three-nucleon interactions between a core nucleon and two valence ones. Three-nucleon terms, proportional to n^3, as well as n and n^2 terms, could simply arise from polarization of the core by the valence nucleons. The first term in (1), $\mathrm{BE}(n = 0)$ may not be constant as the valence orbit is being filled. It may depend on n and yield n^3 terms in addition to terms proportional to n and n^2. Before reliable calculations will be available, there is no way to determine such possible contributions. Still, the fact that three-body interactions between valence nucleons do not appear, may indicate that they are rather weak and do not depend strongly on the nuclear states.

In the shell model, magic nuclei have closed (fully occupied) shells. Shells may include one j-orbit or several orbits whose energies are close and hence, the effective interaction causes mixing of configurations. In either case, the first excited state of magic nuclei is due to a nucleon elevated to a higher orbit and hence, it is much higher than in other nuclei. Another feature of a magic nucleus is that if a nucleon is added to it, it must occupy a higher shell. Hence, its separation energy is lower than in preceding nuclei. This effect is clearly demonstrated in figure 1 for ^{49}Ca and ^{50}Ca. The magicity of ^{48}Ca is seen also in its high first excited state compared to ^{46}Ca as shown in figure 2.

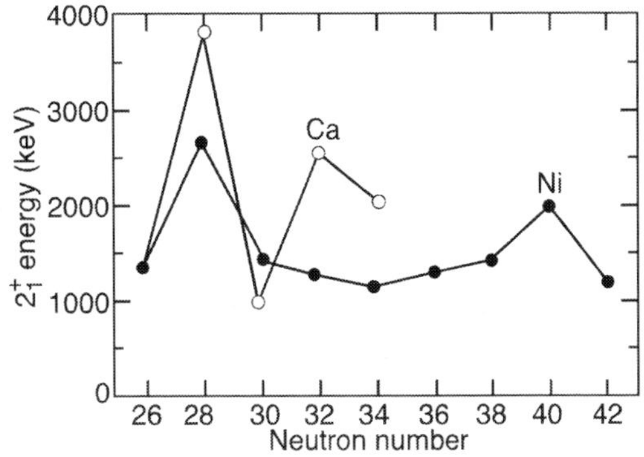

Figure 2. Positions of first excited $J = 2$ states in Ca and Ni isotopes.

The data in figure 1 are rather old but recently, it became possible to investigate experimentally calcium isotopes beyond ^{48}Ca [5]. To assign shell model configurations of those nuclei it is useful to look at the single neutron energies of the orbits higher than $1f_{7/2}$. The lowest orbits may be seen in figure 3, in which levels of ^{49}Ca are shown.

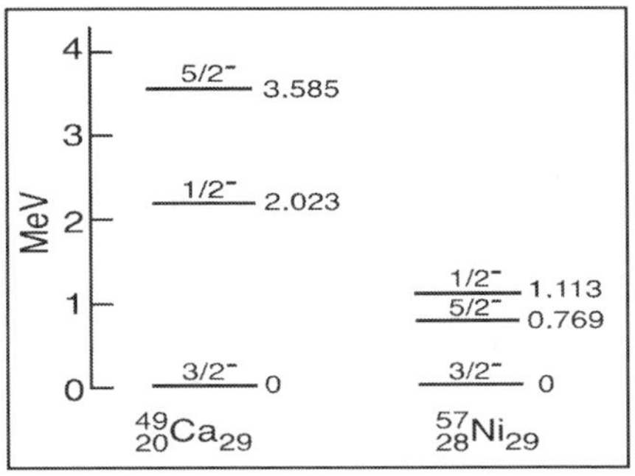

Figure 3. Single neutron energies of $2p_{3/2}$, $2p_{1/2}$ and $1f_{5/2}$ orbits.

Thus, the expected orbit to be occupied by valence neutrons is $2p_{3/2}$. Non-diagonal matrix elements between this orbit and the $2p_{1/2}$ and $1f_{5/2}$ orbits were determined many years ago from energies of nickel isotopes [6]. They are not expected to cause appreciable mixing of higher configurations into the $(1p_{3/2})^n$configuration. To see whether this expectation is correct, we look at the neutron separation energies of calcium isotopes which were recently measured (figure 4).

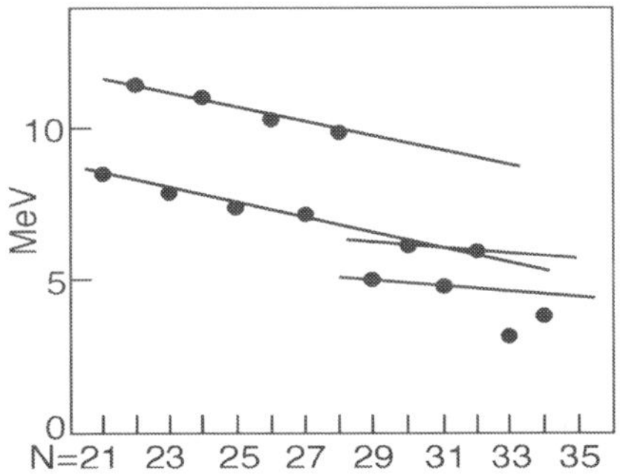

Figure 4. Recently measured neutron separation energies from Ca isotopes.

The neutron separation energies in figure 4 agree well with valence neutrons occupying a $j = 3/2$ orbit which is full in ^{52}Ca. The first excited $J = 2$ level in that nucleus is considerably higher than in ^{50}Ca. In fact, the position of the $J = 2$ level in ^{50}Ca is consistent with the measured binding energies. According to (2), it is equal to

$$V_2 - V_0 = \frac{4\beta}{5} = 1.36\times0.8 = 1.088 \text{ MeV} \;.$$

This value is in fair agreement with the measured 1.027 MeV.

4. Generalized seniority

The behavior of first excited $J = 2$ states in figure 2 of nickel isotopes is very different from that in calcium isotopes. There is no trace of the "magic number" $n = 34$ in ^{52}Ca. It seems that in the case of nickel, there is mixing between configurations of the various orbits which are filled at $n = 40$ where the first excited $J = 2$ is higher than in lighter nuclei. The different behaviour can be traced to the difference in single nucleon levels between ^{49}Ca and ^{57}Ni shown in figure 3. This difference is due to the attractive interaction between valence neutrons and the 8 protons occupying the$1f_{7/2}$ orbit in nickel.

The neutron separation energy from ^{49}Ca is 5.146 MeV whereas from ^{57}Ni it is 10.250 MeV. Hence, the average interaction between a $1f_{7/2}$ proton and a $2p_{3/2}$ neutron is 0.638 MeV. The average attraction between a $1f_{7/2}$ and a $1f_{5/2}$ neutron is stronger, which accounts for the lowering of the 5/2 state in ^{57}Ni. It is equal to {(10.250-5.146)+(3.585-0.769)}/8=0.990 MeV. This attraction is close to the average attraction, not considered here, between a $1f_{7/2}$ proton and a $1f_{7/2}$ neutron, 1.068 MeV. Similarly, the average attraction between a $1f_{7/2}$ proton and a $2p_{1/2}$ neutron is calculated to be 0.752 MeV, a little stronger than with its spin orbit partner, $2p_{3/2}$.

In view of the single nucleon energies in figure 3, it is expected that configurations will be strongly admixed in nickel isotopes heavier than ^{57}Ni. There are some similarities in the energies of nickel isotopes heavier than ^{57}Ni and those in the seniority scheme. Some are related to binding energies and some to positions of first excited $J = 2$ states. The spacings of the latter and the $J = 0$ ground states are certainly not equal but they do not vary much compared to levels of odd Ni isotopes of which some are presented in figure 5.

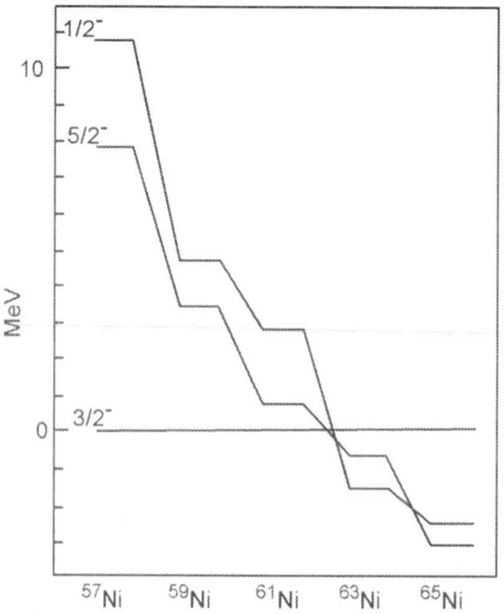

Figure 5. Low lying levels of odd isotopes of Nickel.

States of j^n-configurations of identical nucleons have well defined seniorities. Where such configurations are mixed, some of their states may be analogous to those of seniority. In a single j-orbit, the seniority scheme is defined by using the pair creation operator

$$S_j^+ = \frac{1}{2}\Sigma(-1)^{j-m} a_{jm}^+ a_{j-m}^+ \ . \tag{4}$$

In *generalized seniority* this role is played by a linear combination

$$S^+ = \Sigma\, \alpha_j\, S_j^+ \tag{5}$$

which creates an eigenestate of the shell model Hamiltonian, $HS^+|0> = VS^+|0>$. If also $(S^+)^2|0>$ is an eigenstate, whose eigenvalue may be expressed as $2V + W$, then the following relation is obeyed

$$\left[[H, S^+], S^+\right] = W(S^+)^2 \ . \tag{6}$$

If this is the case, all analogs of seniority $v = 0$ states are eigenstates

$$H(S^+)^N|0> = \left\{NV + \frac{1}{2}WN(N-1)\right\}\ (S^+)^N|0> \ . \tag{7}$$

From (7) follows that nucleon *pair* separation energies as a function of N lie on a straight line. In the case of nickel isotopes this is fairly the case as plotted in figure 6.

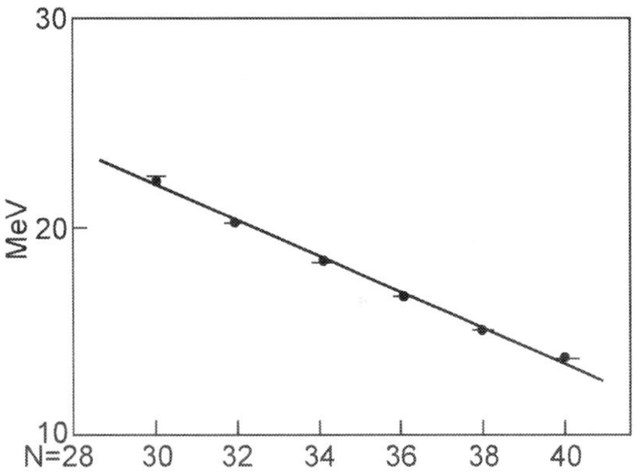

Figure 6. Neutron pair separation energies from *even* Ni isotopes.

In the seniority scheme in a single j-orbit, spacings between levels with seniorities $v \leq n$ are constant, independent of n. In the case of generalized seniority, this is the case only if all coefficients in (5) are equal. Still, also in the case of unequal coefficients, there may be analogs of $v = 2$ states which are eigenstates. Consider an eigenstate with $J = 2$ in the two nucleon configuration. It is created by an operator

$$D_M^+ = \Sigma\, \beta_{jj'}\, D_{Mjj'}^+ = \Sigma\, \beta_{jj'}\left(1 + \delta_{jj'}\right)^{-\frac{1}{2}} (jmj'm'|jj'J = 2, M) a_{jm}^+ a_{j'm'}^+ \ , \tag{8}$$

i.e $HD_M^+|0>= V_2D_M^+|0>$. If also $HS^+D_M^+|0>$ is an eigenstate, than the condition

$$[[H,S^+],D_M^+] = WS^+D_M^+ \tag{9}$$

is obeyed. In this case, the analogs of $v = 2, J = 2$ states are eigenstates. The eigenvalues are simply given by

$$H(S^+)^{N-1}D_M^+|0>= \left\{(N-1)V + V_2 + \frac{1}{2}WN(N-1)\right\} (S^+)^{N-1}D_M^+|0> . \tag{10}$$

The difference between the eigenvalues in (10) and those in (7) for the same value of N, is constant, independent of N. Thus, the positions of the special eigenstates in (10) above the ground states in (7) are constant and equal to $V_2 - V$. This is only approximately seen in nickel isotopes. There are more impressive examples which will not be considered here.

The examples above exhibit the fact that the simple shell model may be very useful in some cases. It seems to be a very useful approximation even for some nuclei away from the valley of stability.

5. Shell model wave function

In view of the complication in deriving the effective interaction, not much attention has been paid to the emerging wave functions. Shell model wave functions have been considered just as model wave functions. Since they do not include the important short range correlations, they could be very different from the real wave functions of nuclei. The short range correlation introduced by the strong and short range bare interaction, were described as "wounds" inflicted on the wave functions of nucleons moving independently in a central potential well. Still, such wounds may be limited to short distances and not cause too big changes in the shell model wave functions. The correlations are important for "short range" observables, but perhaps not for other ("long range") ones. There is evidence, some of it mentioned below, that there are observables which have been successfully calculated by using shell model wave functions with un-normalized operators.

It is interesting to have some idea about the possible extension of the wounds. The change in the wave function of two nucleons is more appreciable when they are close. It is possible to find out how many encounters of two nucleons occur at zero range by calculating the expectation value of the zero range δ-potential in given nuclear states. This may be carried out in some simple cases.

Consider n identical nucleons in a single j-orbit. As mentioned above, eigenstates of this system are states of the seniority scheme. The δ-potential interaction is diagonal in this scheme. Its eigenvalues are given by (1) without the quadratic term. Hence, the eigenvalues in ground states are equal to $nV_0/2$. Thus, the number of zero range encounters is linear in n, considerably smaller than the total number, $n(n-1)/2$.

In the case of a harmonic oscillator potential well, the calculation above has a more precise meaning. There, any wave function of two nucleons may be expressed as a linear combination with a finite number of terms, of products of two oscillator wave functions. One is a function of their center-of-mass, $(r_1+r_2)/2$ and the other, a function of their distance $r = r_2 - r_1$ [7]. The matrix elements of $\delta(r)$ between wave functions with $l > 0$ vanish since these wave functions vanish at $r = 0$. The only contribution to the eigenvalue $nV_0/2$ is due to s-states wave functions of r. The state $j^2J = 0$ has in its expansion s-states but their weight is always smaller than 1 due to higher l-states in the expansion. Hence, the eigenvalue $nV_0/2$ is an upper bound of the number of s-states which suffer the most severe wounds.

Another simple case is offered by protons and neutrons occupying a single j-orbit. In this case, no simple prescription is available for the structure of such a system. There is, however, only one wave function when both proton and neutron j-orbits are completely filled. This wave function may be expressed by the seniority scheme where for a simple δ-potential, the eigenvalue is calculated to be equal to $5(2j+1)V_0/2$. It is much smaller than the total number of matrix elements which is $(4j+2)(4j+1)/2$. The eigenvalue includes the contribution of the proton-proton and neutron-neutron interactions, $2\times(2j+1)V_0/2$. The proton-neutron interactions contribute, in this case, $3(2j+1)V_0/2$.

Some indications that shell model wave functions are real to a considerable extent, are listed below. The analysis of pick-up and stripping direct reactions has always been based on shell model wave functions. They will, however, not be considered here.

Some direct evidence is offered by the rather large extension ("halo") of the ^{11}Be nucleus. It is due to the $2s_{1/2}$ wave function of the valence neutron in the $1/2^+$ ground state. It has a rather small separation energy and no centrifugal barrier. Hence, the extension of the wave function is indeed rather large.

Striking evidence is offered by measurements of charge distributions due to valence protons. Precise measurements were carried out by Frois *et al* [8] of the difference between the charge distributions of the ^{206}Pb and ^{205}Tl nuclei. The result has a remarkable similarity to the probability distribution of a proton in a $3s_{1/2}$ orbit. This is the last orbit to be filled in the 50 - 82 shell. It is interesting to see if such a result could be obtained by many body theories.

Some evidence for the reality of shell model wave functions is offered by energy differences of mirror states in nuclei. Differences of electrostatic energies have been successfully calculated using shell model wave functions and the un-normalized Coulomb potential. Here, the dependence of Coulomb energy differences on the wave functions of valence nucleons is considered. A simple example, the difference between mirror levels of ^{17}O and ^{17}F nuclei, is shown in figure 7.

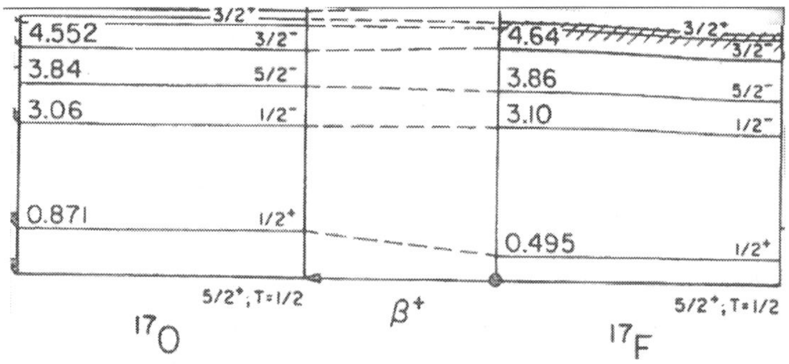

Figure 7. Differences between mirror states of ^{17}O and ^{17}F nuclei.

At a first glance at figure 7 it may seem that the Coulomb energy difference of the $1/2^+$ states is larger than that of the $5/2^+$ground states. Actually, the opposite is true. The wrong impression is due to the fact that in figure 7, the ground states positons were plotted at the same energy. The energy of the ^{17}F ground state is higher by 3.54 MeV than the ^{17}O binding energy. The energy difference between the $1/2^+$ states is *smaller* by $0.871 - 0.495 = 0.376$ MeV than the difference between ground state energies. This difference is due to the $2s_{1/2}$ wave function of the valence nucleon being more extended than its $1d_{5/2}$ wave function in the ground states.

The dependence of Coulomb energy differences on the shell model wave functions of the state considered, is evident in many cases. An example is shown in figure 8 where mirror levels of ^{14}C and ^{14}N are shown. Also in this case, the ground state of ^{14}C is plotted at the same energy as the corresponding $T = 1, J = 0$ state of ^{14}N which lies 2.31 MeV above the $T = 0, J = 1$ ground state.

IOP Publishing
doi:10.1088/1742-6596/580/1/012001

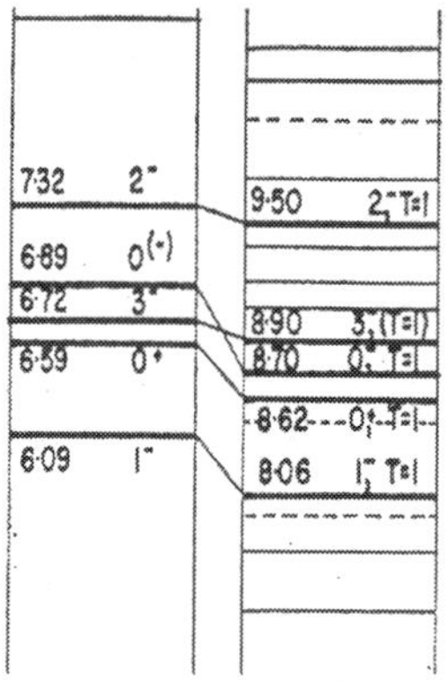

Figure 8. Some mirror levels of ^{14}C (left) and ^{14}N (right).

According to the simple shell model, the negative parity states in figure 8 are due to coupling of a $2p_{1/2}$ nucleon to a nucleon in the $2s_{1/2}$ or the $1d_{5/2}$ orbit. The states with $J = 2$ and $J = 3$ are $p_{1/2}d_{5/2}$ states. As in the preceding example, the differences in Coulomb energy of these states are smaller than that of the ground states which mainly contain *p*-shell nucleons. The $J = 0$ and $J = 1$ states are $p_{1/2}s_{1/2}$ states and their Coulomb energies differences are smaller than those of the ground states. They are also smaller than the states with $d_{5/2}$ nucleons, as is seen in the case of $A = 17$ nuclei.

6. Conclusion – will simplicity arise from complexity?

Shell model wave functions, of independent motion of individual nucleons, cannot be the real wave functions of the nucleons. They do not include short range correlations which are due to the strong bare interaction. The simple shell model, however, has considerable predictive power and its wave functions seem to have some reality. It is a very good approximation of nuclear structure and may be used for many aspects. Examples were given above as a reminder of its power and simplicity.

In *ab initio* calculations the starting point is the bare interaction leading to admixtures of many states. In one of these approaches, the No Core Shell Model, harmonic oscillator wave functions are used just as a convenient scheme. States from higher and higher oscillator shells are included, their number limited only by computational difficulties.

Will the simplicity of the shell model as a good approximation, emerge from the complexity of such calculations? This should be a challenge to every many-body theory of nuclear structure. A reliable many-body theory should explain why the simple shell model works so well, which interactions lead to it and where it becomes useless.

References

[1] Coraggio L, Covello A, Gargano A, Itaco N and Kuo T T S 2009 *Prog. Part. Nucl. Phys.* **62** 135
[2] See *e.g.* Yamada Y, Otsuka T, Shimitzu N, Honma M, and Otsuno Y 2014 *Phys. Rev. C* 89 031301(R)
[3] Seniority and generalized seniority are described in detail in Talmi I 1993 Simple models of complex nuclei, Harwood
[4] Talmi I and Unna I 1960 *Phys. Rev. Lett.* **4** 469
[5] Kreim S, Blaum K, Dilling J and Litvinov Y A 2013 *Nucl. Phys. News* **23** 18 and references therein
[6] Auerbach N 1966 *Nucl. Phys.* **76** 321
[7] Talmi I 1952 *Helv. Phys. Acta* **25** 185
[8] Cavedon J *et al* 1982 *Phys. Rev. Lett.* **49** 978

The explosion of chiral many-body forces: How to deal with it?

R. Machleidt

Department of Physics, University of Idaho, Moscow, Idaho 83844, USA

E-mail: machleid@uidaho.edu

Abstract. During the past two decades, it has been demonstrated that chiral effective field theory represents a powerful tool to deal with nuclear forces in a systematic and model-independent way. Two-, three-, and four-nucleon forces have been derived up to next-to-next-to-next-to-leading order (N^3LO) and (partially) applied in nuclear few- and many-body systems—with, in general, a good deal of success. But in spite of these achievements, we are still faced with some great challenges. Among them is the problem of a proper renormalization of the two-nucleon potential. Another issue are the subleading many-body forces, where the "explosion" of the number of terms with increasing order and the order-by-order convergence are reasons for concern. In this talk, I will mainly focus on the latter topic.

1. Introduction

The problem of a proper derivation of nuclear forces is as old as nuclear physics itself, namely, almost 80 years [1]. The modern view is that, since the nuclear force is a manifestation of strong interactions, any serious derivation has to start from quantum chromodynamics (QCD). However, the well-known problem with QCD is that it is non-perturbative in the low-energy regime characteristic for nuclear physics. For many years this fact was perceived as the great obstacle for a derivation of nuclear forces from QCD—impossible to overcome except by lattice QCD.

The effective field theory (EFT) concept has shown the way out of this dilemma. For the development of an EFT, it is crucial to identify a separation of scales. In the hadron spectrum, a large gap between the masses of the pions and the masses of the vector mesons, like $\rho(770)$ and $\omega(782)$, can clearly be identified. Thus, it is natural to assume that the pion mass sets the soft scale, $Q \sim m_\pi$, and the rho mass the hard scale, $\Lambda_\chi \sim m_\rho \sim 1$ GeV, also known as the chiral-symmetry breaking scale. This is suggestive of considering a low-energy expansion arranged in terms of the soft scale over the hard scale, $(Q/\Lambda_\chi)^\nu$, where Q is generic for an external momentum (nucleon three-momentum or pion four-momentum) or a pion mass. The appropriate degrees of freedom are, obviously, pions and nucleons, and not quark and gluons. To make sure that this EFT is not just another phenomenology, it must have a firm link with QCD. The link is established by having the EFT observe all relevant symmetries of the underlying theory, in particular, the broken chiral symmetry of low-energy QCD [2].

The early applications of chiral perturbation theory (ChPT) focused on systems like $\pi\pi$ and πN, where the Goldstone-boson character of the pion guarantees that the expansion converges. But the past 15 years have also seen great progress in applying ChPT to nuclear forces (see

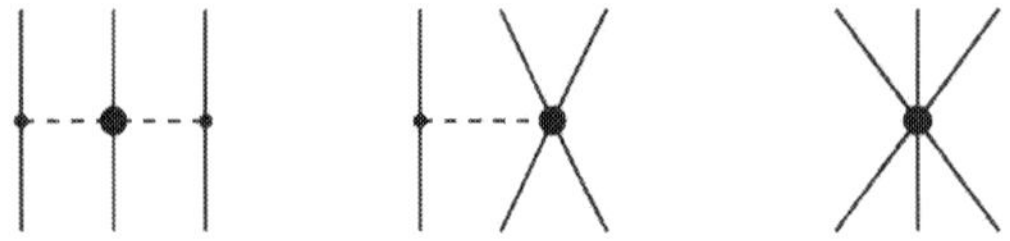

Figure 1. The three-nucleon force at NNLO. From left to right: 2PE, 1PE, and contact diagrams. Solid lines represent nucleons and dashed lines pions. Small solid dots denote vertices of index $\Delta_i = 0$ and large solid dots are $\Delta_i = 1$.

Refs. [3, 4] for recent reviews and find comprehensive lists of references therein). As a result, nucleon-nucleon (NN) potentials of high precision have been constructed, which are based upon ChPT carried to next-to-next-to-next-to-leading order ($\mathrm{N^3LO}$) [5, 3], and applied in nuclear structure calculations with great success.

However, in spite of this progress, we are not done. Due to the complexity of the nuclear force issue, there are still many subtle and not so subtle open problems. We will not list and discuss all of them, but just mention two, which we perceive as the most important ones:

- The proper renormalization of chiral nuclear potentials and
- Subleading chiral few-nucleon forces.

This talk is mainly focused on the latter issue.

2. The chiral NN potential

In terms of naive dimensional analysis or "Weinberg counting", the various orders of the low-energy expansion, which define the chiral NN potential, are given by:

$$V_{\mathrm{LO}} = V_{\mathrm{ct}}^{(0)} + V_{1\pi}^{(0)} \tag{1}$$

$$V_{\mathrm{NLO}} = V_{\mathrm{LO}} + V_{\mathrm{ct}}^{(2)} + V_{1\pi}^{(2)} + V_{2\pi}^{(2)} \tag{2}$$

$$V_{\mathrm{NNLO}} = V_{\mathrm{NLO}} + V_{1\pi}^{(3)} + V_{2\pi}^{(3)} \tag{3}$$

$$V_{\mathrm{N^3LO}} = V_{\mathrm{NNLO}} + V_{\mathrm{ct}}^{(4)} + V_{1\pi}^{(4)} + V_{2\pi}^{(4)} + V_{3\pi}^{(4)} \tag{4}$$

where the superscript denotes the order ν of the expansion. LO stands for leading order, NLO for next-to-leading order, etc.. Contact potentials carry the subscript "ct" and pion-exchange potentials can be identified by an obvious subscript.

NN potentials have been constructed at all of the above orders, and it has been shown [5] that at $\mathrm{N^3LO}$ the precision is finally achieved, which is necessary and sufficient for reliable applications in *ab initio* nuclear structure calculations. Thus, the NN problem appears to be under control, at least for the time being (and see Ref. [6] for the renormalization issue).

3. Nuclear many-body forces

The chiral two-nucleon force (2NF) at $\mathrm{N^3LO}$ has been applied in microscopic calculations of nuclear structure with, in general, a great deal of success. However, from high-precision studies conducted in the 1990s, it is well-known that certain few-nucleon reactions and nuclear structure issues require three-nucleon forces (3NFs) for their microscopic explanation. Outstanding examples are the A_y puzzle of N-d scattering and the ground state of $^{10}\mathrm{B}$. *An important advantage of the EFT approach to nuclear forces is that it creates two- and many-nucleon forces on an equal footing.* In this section, we will now explain in some detail those chiral three- and four-nucleon forces.

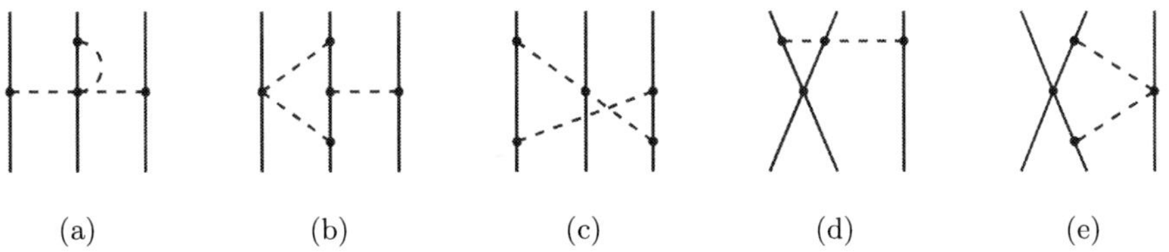

Figure 2. Leading one-loop 3NF diagrams at N^3LO. We show one representative example for each of five topologies, which are: (a) 2PE, (b) 1PE-2PE, (c) ring, (d) contact-1PE, (e) contact-2PE. Notation as in Fig. 1.

3.1. Three-nucleon forces

The order of a 3NF is given by

$$\nu = 2 + 2L + \sum_i \Delta_i \,, \tag{5}$$

where L denotes the number of loops and Δ_i is the vertex index. We will use this formula to analyze 3NF contributions order by order.

3.1.1. Next-to-leading order. The lowest possible power is obviously $\nu = 2$ (NLO), which is obtained for no loops ($L = 0$) and only leading vertices ($\sum_i \Delta_i = 0$). As it turns out, the contribution from these NLO diagrams vanishes. So, the bottom line is that there is no genuine 3NF at NLO. The first non-vanishing 3NF appears at NNLO.

3.1.2. Next-to-next-to-leading order. The power $\nu = 3$ (NNLO) is obtained when there are no loops ($L = 0$) and $\sum_i \Delta_i = 1$, i.e., $\Delta_i = 1$ for one vertex while $\Delta_i = 0$ for all other vertices. There are three topologies which fulfill this condition, known as the two-pion exchange (2PE), one-pion exchange (1PE), and contact graphs (Fig. 1).

The 1PE and contact 3NF terms involve each a new parameter, which are commonly denoted by D and E and which do not appear in the 2N problem. There are many ways to pin these two parameters down. The triton binding energy and the nd doublet scattering length $^2a_{nd}$ have been used for this purpose. But one may also choose the binding energies of ^{3}H and ^{4}He, an optimal over-all fit of the properties of light nuclei, or electroweak processes like the tritium β decay. Once D and E are fixed, the results for other 3N, 4N, etc. observables are predictions.

The 3NF at NNLO has been applied in calculations of few-nucleon reactions, structure of light- and medium-mass nuclei [7, 8], and nuclear and neutron matter [9, 10, 11] with a good deal of success. Yet, the famous 'A_y puzzle' of nucleon-deuteron scattering is not resolved. When only 2NFs are applied, the analyzing power in p-^{3}He scattering is even more underpredicted than in p-d. However, when the NNLO 3NF is added, the p-^{3}He A_y substantially improves (more than in p-d) [12]—but a discrepancy remains. Furthermore, the spectra of light nuclei leave room for improvement.

To summarize, the 3NF at NNLO is a remarkable contribution: It represents the leading many-body force within the scheme of ChPT; it includes terms that were advocated already some 50 years ago; and it produces noticeable improvements in few-nucleon reactions and the structure of light nuclei. But unresolved problems remain. Moreover, in the case of the 2NF, we have pointed out that one has to proceed to N^3LO to achieve sufficient accuracy. Therefore, the 3NF at subleading order is needed for at least two reasons: for consistency with the 2NF and to hopefully resolve outstanding problems in microscopic nuclear structure and reactions.

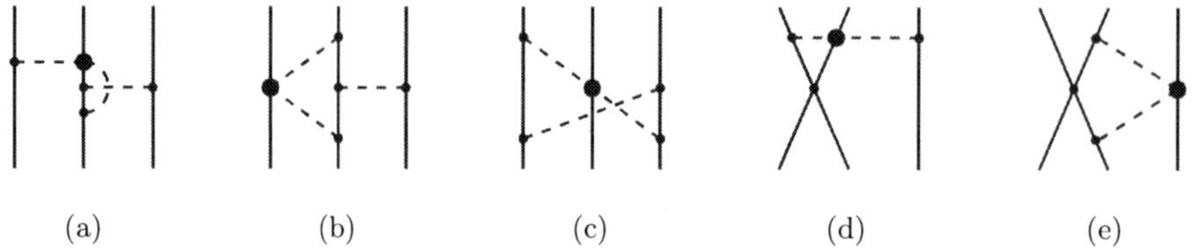

Figure 3. Sub-leading one-loop 3NF diagrams which appear at N^{4}LO. We show one representative example for each of five topologies, which are: (a) 2PE, (b) 1PE-2PE, (c) ring, (d) contact-1PE, (e) contact-2PE. Notation as in Fig. 1.

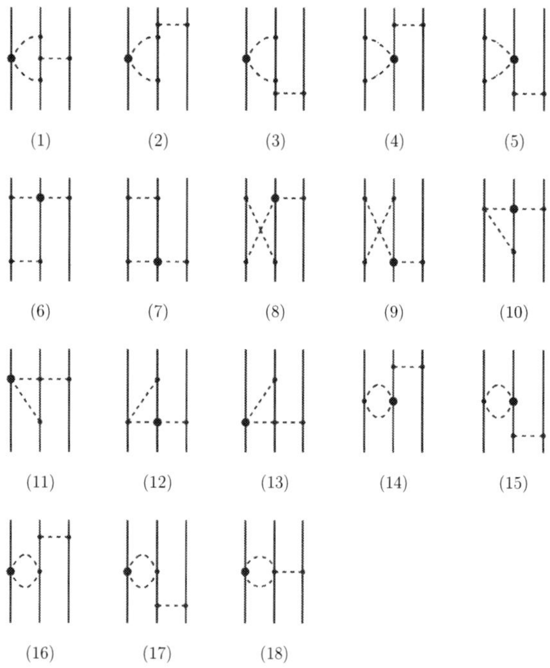

Figure 4. The topology (b) of Fig. 3 (1PE-2PE) in detail. Notation as in Fig. 1.

3.1.3. Next-to-next-to-next-to-leading order. At N^{3}LO, there are loop and tree diagrams. For the loops (Fig. 2), we have $L = 1$ and, therefore, all Δ_i have to be zero to ensure $\nu = 4$. Thus, these one-loop 3NF diagrams can include only leading order vertices, the parameters of which are fixed from πN and NN analysis. The long-range part of the chiral N^{3}LO 3NF has been tested in the triton [13] and in three-nucleon scattering [14] yielding only moderate effects and no improvement of the A_y puzzle. The long- and short-range parts of this force have been used in neutron matter calculations (together with the N^{3}LO 4NF) producing surprisingly large contributions from the 3NF [15]. Thus, the ultimate assessment of the N^{3}LO 3NF is still outstanding and will require more few- and many-body applications. But we expect that, overall, the 3NF at N^{3}LO is small and will most likely not solve the outstanding problems.

3.1.4. The 3NF at N^{4}LO. Because the 3NF at N^{3}LO is presumably small, it is necessary to move on to the next order of 3NFs, which is N^{4}LO or $\nu = 5$ (of the Δ-less theory which we

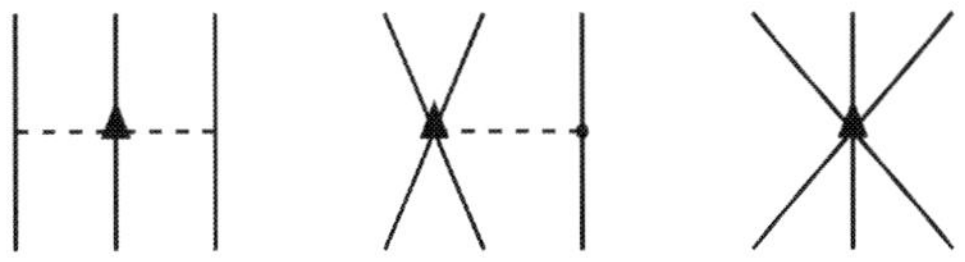

(a) (b) (c)

Figure 5. 3NF tree graphs at N^4LO ($\nu = 5$) denoted by: (a) 2PE, (b) 1PE-contact, and (c) contact. Solid triangles represent vertices of index $\Delta_i = 3$. Other notation as in Fig. 1.

have silently assumed so far). The loop contributions that occur at this order are obtained by replacing in the N^3LO loops one vertex by a $\Delta_i = 1$ vertex (with LEC c_i), Fig. 3, which is why these loops may be more sizable than the N^3LO loops. The 2PE, 1PE-2PE, and ring topologies have been evaluated [16]. Note that each diagram in Fig. 3 stands symbolically for a group of diagrams. We demostrate this for the 1PE-2PE topology, for which we display in Fig. 4 all diagrams for that topology. This applies to each topology and, thus, provides an idea of the "explosion" of 3NF contributions at subleading orders.

In addition to the loops, we have at N^4LO three 'tree' topologies (Fig. 5), which include a new set of 3N contact interactions, which have recently been derived by the Pisa group [17]. Contact terms are typically simple (as compared to loop diagrams) and their coefficients are essentially free. Therefore, it would be an attractive project to test some terms (in particular, the spin-orbit terms) of the N^4LO contact 3NF [17] in calculations of few-body reactions (specifically, the p-d and p-^{3}He A_y) and spectra of light nuclei.

3.2. Four-nucleon forces

For four-nucleon forces (4NFs), the power is given by

$$\nu = 4 + 2L + \sum_i \Delta_i \,. \tag{6}$$

Therefore, a 4NF appears for the first time at $\nu = 4$ (N^3LO), with no loops and only leading vertices, Fig. 6. This 4NF includes no new parameters and does not vanish [18]. It has been applied in a calculation of the ^{4}He binding energy, where it was found to contribute a few 100 keV [19]. It should be noted that this preliminary calculation involves many approximations, but it provides an idea of the order of magnitude of the 4NF, which is indeed small as compared to the full ^{4}He binding energy of 28.3 MeV.

4. Conclusions

The past 15 years have seen great progress in our understanding of nuclear forces in terms of low-energy QCD. Key to this development was the realization that low-energy QCD is equivalent to an effective field theory (EFT) which allows for a perturbative expansion that has become known as chiral perturbation theory (ChPT). In this framework, two- and many-body forces emerge on an equal footing and the empirical fact that nuclear many-body forces are substantially weaker than the two-nucleon force is explained naturally.

In this talk, I have focused mainly on nuclear many-body forces based upon chiral EFT. The 3NF at NNLO has been known for a while and applied in few-nucleon reactions, structure of light- and medium-mass nuclei, and nuclear and neutron matter with some success. However, the famous 'A_y puzzle' of nucleon-deuteron scattering is not resolved by the 3NF at NNLO. Thus, one important open issue are the few-nucleon forces beyond NNLO ("sub-leading few-nucleon forces") which, besides the A_y puzzle, may also resolve some important outstanding

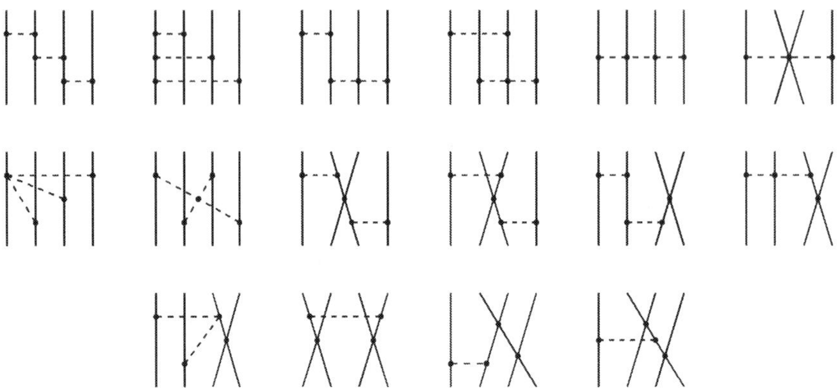

Figure 6. Leading four-nucleon force at N^3LO.

nuclear structure problems. As explained, this may require going even beyond N^3LO. However, as demonstrated, with each higher order, the number of diagrams increases enormously. Thus, practitioners are faced with the problem of how to deal with this explosion of 3NF contributions. My advice is that, for a while, one should not aim at complete calculations at given higher orders. Rather one will have to be selective and try to identify the more important 3NF terms in the "forrest" of diagrams. The N^4LO 3NF contact terms [Fig. 5(c)] [17] are a promising and manageable starting point.

Finally, let me note that, because of lack of space, I have discussed here only the so-called Δ-less version of ChPT. There is also the Δ-full version (see Ref. [3] for details), in which the number of diagrams is even larger.

Acknowledgments

Work supported by the US Department of Energy under Grant No. DE-FG02-03ER41270.

References

[1] Machleidt R 1989 *Adv. Nucl. Phys.* **19** 189
[2] Weinberg S 1979 *Physica* **96A** 327
[3] Machleidt R and Entem D R 2011 *Phys. Rep.* **503** 1
[4] Epelbaum E, Hammer H-W and Meißner Ulf-G 2009 *Rev. Mod. Phys.* **81** 1773
[5] Entem D R and Machleidt R 2003 *Phys. Rev.* C **68** 041001
[6] Marji E, Canul A, MacPherson Q, Winzer R, Zeoli Ch, Entem D R and Machleidt R 2013 *Phys. Rev.* C **88** 054002
[7] Hagen G, Hjorth-Jensen M, Jansen G R, Machleidt R, and Papenbrock T 2012 *Phys. Rev. Lett.* **108** 242501,
[8] Hagen G, Hjorth-Jensen M, Jansen G R, Machleidt R, and Papenbrock T 2012 *Phys. Rev. Lett. 109* 032502
[9] Sammarruca F, Chen F B, Coraggio L, Itaco N and Machleidt R 2012 *Phys. Rev.* C **86** 054317
[10] Coraggio L, Holt J W, Itaco N, Machleidt R and Sammarruca F 2013 *Phys. Rev.* C **87** 014322
[11] Coraggio L, Holt J W, Itaco N, Machleidt R, Marcucci L E and Sammarruca F 2014 *Phys. Rev.* C **89** 044321
[12] Viviani M, Giarlanda L, Kievsky A and Marcucci L E 2013 *Phys. Rev. Lett.* **111** 172302
[13] Skibinski R *et al* 2011 *Phys. Rev.* C **84** 054005
[14] Witala H *et al* 2013 *Few-Body Syst.* **54** 897
[15] Tews I, Krüger T, Hebeler K and Schwenk A 2013 *Phys. Rev. Lett.* **110** 032504
[16] Krebs H, Gasparyan A and Epelbaum E 2012 *Phys. Rev.* C **85** 054006
Krebs H, Gasparyan A and Epelbaum E 2013 *Phys. Rev.* C **87** 054007
[17] Girlanda L, Kievsky A and Viviani M 2011 *Phys. Rev.* C **84** 014001
[18] Epelbaum E 2007 *Eur. Phys. J.* **A34** 197
[19] Rozpedzik D, Golak J, Skibinsk, R, Witala H, Glöckle W, Epelbaum E, Nogga A and Kamada H 2006 *Acta Phys. Polon.* **B37** 2889

Journal of Physics: Conference Series **580** (2015) 012003 doi:10.1088/1742-6596/580/1/012003

Deriving the nuclear shell model from first principles

B R Barrett[1], E Dikmen[1,2], A F Lisetskiy[3], P Maris[4], A M Shirokov[4,5,6] and J P Vary[4]

[1] Department of Physics, PO Box 210081, University of Arizona, Tucson, AZ 85721-0081, USA
[2] Departmant of Physics Suleyman Demirel University, Isparta, Turkey
[3] Mintec, Inc., Tucson, AZ, USA
[4] Department of Physics and Astronomy, Iowa State University, Ames, IA 50011 USA
[5] Skobeltsyn Institute of Nuclear Physics. Lomonosov Moscow State University, Moscow 119991, Russia
[6] Pacific National University, 136 Tikhookeanskaya st., Khabarovsk 680035, Russia

E-mail: bbarrett@physics.arizona.edu

Abstract. A procedure for calculating microscopically the input for standard shell-model calculations, *i.e.*, the core and single-particle energies plus the two-body effective model-space interactions, is presented and applied to nuclei at the start of the *sd*-shell. Calculations with the JISP16 and Idaho χEFT N^{3}LO nucleon-nucleon interactions are performed and yield consistent results, which also are similar to phenomenological results in the *sd*-shell as well as with other theoretical calculations, utilizing other techniques. All results show only a weak A-dependence.

1. Introduction

Tremendous progress has been made in the last 15 to 20 years in the constructing of a unified microscopic formalism capable of describing the properties of nuclei and nuclear reactions. This has been achieved: 1.) by a better understanding of the fundamental interactions among the nucleons, mainly through the use of Chiral Effective Field Theory (χEFT) [1, 2, 3] and 2.) by the development of new approaches or by the improvement of existing methods, for solving the nuclear many-body problem, such as, the Greens Function Monte Carlo (GFMC) [4, 5], No-Core Shell Model (NCSM) [6, 7, 8, 9] and Coupled Cluster (CC) [10, 11, 12, 13] approaches, among others. The success of these approaches is well-documented, especially for light nuclei, *e.g.*, $A \leq 16$ (see the above references).

However, the main problem facing most of these techniques is how to extend them to heavier mass nuclei, because of the rapid growth of the model spaces required to obtain a converged result or other computational problems related to the increase in A. Consequently, there is currently much interest in developing a method for computing microscopically the input for Standard Shell Model (SSM) calculations, *i.e.*, the inert core energy, the single-particle (s.p.) energies and the effective two-body interactions in the valence space, using one of the successful, existing many-body methods, *e.g.*, those mentioned above. Such a procedure was put forward by Navrátil *et al.* in 1998 [14] and applied successfully to a calculation of the effective charges and gyromagnetic ratios for $0p$-shell nuclei. This approach was expanded by Lisetskiy *et al.* [15, 16] to all nuclei in the $0p$-shell. Their results exhibited a strong A-dependence, which arose from specific choices made in the definition of the core and s.p. energies. We note that there is considerable freedom in the definitions of the contributions to the effective valence-space Hamiltonian.

This contribution presents the approach with another choice of the core and valence-space energies, along with its application to nuclei at the start of the *sd*-shell. This modified method exhibits only a very weak A-dependence and produces results, which are consistent with two recent, but different theoretical approaches to this problem [17, 18, 19].

2. Formalism

The basic formalism is well-described in Refs. [15, 16] and the reader is referred to these publications for the details of this method. The procedure can be outlined as follows:

(i) Start with a nucleus, such as, ^{18}F, which has the SSM structure of an *inert* core, (*e.g.*, ^{16}O) with two valence nucleons in the first major shell outside of the core, in this case, the *sd*-shell.

(ii) Pick some existing many-body method, such as the NCSM or the CC approach, to calculate the eigenvalues and eigenfunctions for this nucleus. In the case of the NCSM the nucleon-nucleon (NN) interaction used in performing the numerical calculations is truncated, *i.e.*, renormalized, into a tractable two-nucleon model space using the Okubo-Lee-Suzuki (OLS) transformation [20, 21, 22]. This procedure is referred to as determining the two-nucleon (2N) cluster in a finite model (*i.e.*, basis) space.

(iii) Use this 2N-cluster to construct the A-nucleon Hamiltonian in the model space, which is then diagonalized in the model space to obtain the A-nucleon eigenenergies and eigenfunctions.

(iv) Perform a second OLS transformation of these converged eigenenergies and eigenfunctions into the $0\hbar\Omega$ space, which is, by definition, the space of the inert core plus two-valence nucleons in the first major shell outside of the core. Because this space has zero excitation energy available to it, the nucleons in the core are *energetically frozen* and, hence, do indeed behave like an inert core. Thus, all of the A-body correlations obtained by solving the A-nucleon problem in the large model space are now fully contained in the matrix elements of the two-nucleons in the valence space.

(v) Now separate the obtained two-body matrix elements into a core energy, s.p. energies and effective two-body interactions, as described in Refs. [15, 16], except that the A-dependence for the kinetic energy is taken to be A_{core} instead of A, when calculating the core energy, and $A_{core} + 1$ for the s.p. energies.

(vi) These core and s.p. energies and two-body effective interactions can now be used for performing SSM calculations for the rest of the nuclei in that major shell, *e.g.*, the *sd*-shell.

3. Applications

We have performed NCSM calculations in the *sd*-shell for nuclei with $A = 18$ and 19 using the JISP16 [23] and Idaho χEFT N^{3}LO [24] NN interactions. The maximum model-space size that we could handle in these calculations was $N_{max} = 4$, *i.e.*, all configurations up to energies of $4\hbar\Omega$, where $\hbar\Omega$ is the harmonic-oscillator energy.

The eigenenergies and eigenfunctions obtained in our calculations for the ground and excited states of $A = 18$ and $A = 19$ nuclei were then projected into the $0\hbar\Omega$ space, utilizing the OLS transformation. For example, for $A = 18$ this yields 18-body matrix elements for four nucleons in the $0s$-shell, 12 nucleons in the $0p$-shell and two nucleons in the *sd*-shell. Because of the restriction to a total excitation energy of $0\hbar\Omega$ the lowest-lying 16 nucleons form an ^{16}O core, because they are *energetically frozen* against excitations into the *sd*-shell. Thus, all the 18-body correlations of the large model-space results are now contained within the two-body matrix elements (TBMEs) of the two nucleons in the *sd*-shell, *i.e.*, the valence nucleons.

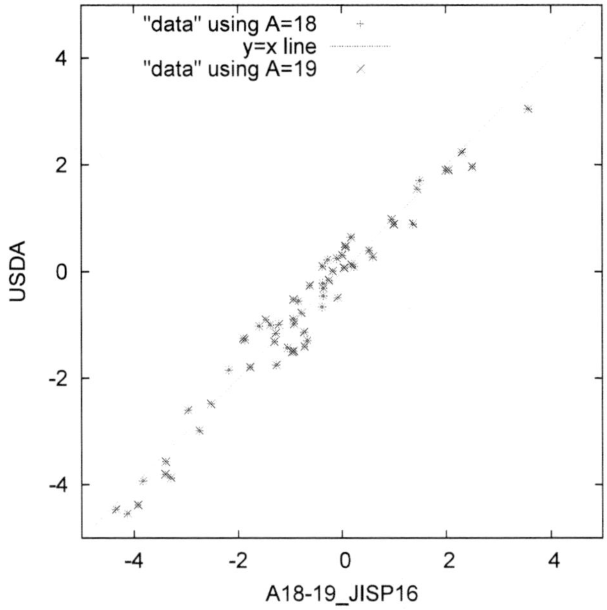

Figure 1. Comparison of our two-body effective shell-model interactions, *i.e.,* after the OLS transformation into the $0\hbar\Omega$ space, for A=18 (red +) and A=19 (blue x) nuclei with the phenomenological USDA two-body matrix elements of Brown and Richter [26].

These highly correlated TBMEs can now be separated into an inert ^{16}O core energy, s.p. energies for each of the two nucleons in the *sd*-shell and residual effective two-body shell-model interactions. This is done by first calculating the g.s. energy of the ^{16}O using the TBMEs generated for the original $A = 18$ calculation, *i.e.,* this is an ^{16}O core *inside* of ^{18}O or ^{18}F. Note that the relative kinetic-energy operator used in the ^{16}O calculations should be for only 16 nucleons. The *sd*-shell s.p. energies are then obtained by calculating the low-lying spectra of ^{17}O and ^{17}F and subtracting the previously calculated ^{16}O core energy. Finally, the two-body effective shell-model interactions are found by subtracting the ^{16}O core energy plus the appropriate s.p. energies for the two *sd*-shell valence nucleons from the original, highly correlated TBMEs obtained by the second OLS transformation. Thus, we now have all the input necessary for performing SSM calculations for other nuclei within the *sd*-shell.

To study the A-dependence of our results, we can repeat the above calculations for the $A = 18$-cluster inside of nuclei with $A = 19, A = 20$, *etc.*

4. Results

Figure 1 compares our two-body effective shell-model interactions for $A = 18$ and $A = 19$ with the phenomenological two-body effective *sd*-shell matrix elements of Brown and Richter [26], for their so-called USDA interaction. First of all, one notes how closely these three sets of TBMEs are correlated. Secondly, one observes that our $A = 18$ and $A = 19$ results are essentially the same, *i.e.*, they are only weakly A-dependent.

Table 1 gives the ^{16}O core energies and the $A = 17$ s.p. energies for the $A = 18$ and $A = 19$ systems, calculated with the JISP16 NN interaction [23]. Again one notices the weak A-dependence. Calculations have also been carried out using the Idaho χEFT N^{3}LO

Table 1. ^{16}O core energies (E_{core}) and proton and neutron single-particle energies for the JISP16 NN interactions for the $A = 18$ and $A = 19$ systems.

	$A = 18$ ($E_{core} = -115.529$)			$A = 19$ ($E_{core} = -115.319$)		
a	$\frac{1}{2}$	$\frac{3}{2}$	$\frac{5}{2}$	$\frac{1}{2}$	$\frac{3}{2}$	$\frac{5}{2}$
ϵ_a^n	-3.068	6.262	-2.270	-3.044	6.289	-2.248
ϵ_a^p	0.603	9.748	1.398	0.627	9.774	1.419

Table 2. ^{16}O core energies (E_{core}) and proton and neutron single-particle energies for the chiral N3LO NN interactions for the $A = 18$ and $A = 19$ systems.

	$A = 18$ ($E_{core} = -118.469$)			$A = 19$ ($E_{core} = -118.306$)		
a	$\frac{1}{2}$	$\frac{3}{2}$	$\frac{5}{2}$	$\frac{1}{2}$	$\frac{3}{2}$	$\frac{5}{2}$
ϵ_a^n	-3.638	3.763	-3.042	-3.625	3.770	-3.031
ϵ_a^p	0.044	7.299	0.690	0.057	7.307	0.700

NN interaction [24]. These results are shown in Table 2. The results for both interactions are similar and again show only a weak A-dependence.

In a recent paper, Jansen *et al.* [17] report on investigations similar to ours but using the Coupled Cluster approach for performing their large model-space calculations and including the Idaho χEFT 3NF at N^{2}LO with the χEFT NN interaction at N^{3}LO [24]. They find $E_{^{16}O,A=18} =$ -130.462 MeV and $E_{^{16}O,A=19} =$ -130.056 MeV, for the ^{16}O core energies for $A = 18$ and $A = 19$, respectively. One observes a slightly lower ^{16}O core energy, by about 10%, due to the inclusion of the 3NF, and a weak A-dependence, similar to ours. The input for SSM calculations can also be obtained using the In-Medium Similarity Renormalization Group (IM-SRG) approach [18, 19]. Preliminary results, obtained using this method [25], are consistent with our results given here.

Finally, columns 1 and 2 of Fig. 2 compare, respectively, the SSM calculation for $A = 18$ with the exact NCSM calculation for $A = 18$. By construction these two results must be the same. Columns 3 and 4 compare, respectively, the results of SSM calculations performed for $A = 19$ using the $A = 18$ input and the $A = 19$ input. These are compared with the full NCSM results for $A = 19$ in column 5. The strong similarity of these results again demonstrates the weak A-dependence of the SSM input. The close agreement with the full NCSM results suggests that the effective three-body interactions in the $A = 19$ systems are small. However, this point needs to be tested by performing calculations for larger A systems in the sd-shell.

5. Conclusions

We have performed *ab initio* NCSM calculations for $A = 18$ and $A = 19$ nuclei, which were then projected into the sd-shell, using the OLS transformation. These projected results were then used to determine the ^{16}O core energy, sd-shell s.p. energies and two-body effective sd-shell interactions. Our results correlate well with both phenomenological results and theoretical results obtained using the CC and IM-SRG methods. The results are found to be weakly A-dependent, so that they should be appropriate for performing SSM calculations for nuclei throughout the sd-shell.

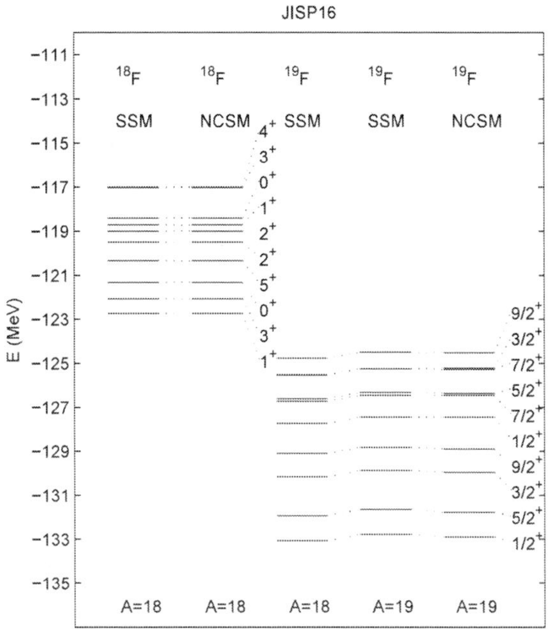

Figure 2. The ground and low-lying excited-state energies of ^{18}F and ^{19}F obtained by SSM and NCSM calculations, using the shell-model input for ^{18}F and ^{19}F, respectively, and generated as described in the text. All calculations are based on the JISP16 NN interaction [23].

Acknowledgments

This work was supported in part by the U.S. National Science Foundation under Grant Nos. PHYS-0845912 and 0904782, the U.S. Department of Energy under Grant Nos. DESC0008485 (SciDAC/NUCLEI) and DE-FG02-87ER40371, and the Higher Education council of Turkey (YOK) and The Scientific and Technological Research Council of Turkey (TUBITAK). Computational resources were provided by the National Energy Research Supercomputer Center (NERSC), which is supported by the Office of Science of the U.S. Department of Energy under Contract No. DE-AC02-05CH11231.

References

[1] Weinberg S 1992 *Nucl. Phys.* B **363** 3, and references therein.
[2] Ordonez C, Ray L and van Kolck U 1996 *Phys. Rev.* C **53** 2086
[3] van Kolck U 1999 *Prog. Part. Nucl. Phys.* **43** 337
[4] Pieper S C, Pandharipande V R, Wiringa R B and Carlson J 2001 *Phys. Rev.* C **64** 014001
[5] Pieper S C and Wiringa R B 2001 *Annu. Rev. Nucl. Part. Sci.* **51** 53
[6] Navrátil P, Vary J P and Barrett B R 2000 *Phys. Rev. Lett.* **84** 57288
[7] Navrátil P, Vary J P and Barrett B R 2000 *Phys. Rev.* C **62** 054311
[8] Navrátil P, Quaglioni S, Stetcu I and Barrett B R 2009 *J. Phys.* G **36** 083101
[9] Barrett B R, Navrátil P and Vary J P 2013 *Prog. Part. Nucl. Phys.* **69** 131
[10] Coester F 1958 *Nucl. Phys.* **7** 421
[11] Coester F and Kümmel H 1960 *Nucl. Phys.* **17** 477
[12] Hagen G, Papenbrock T, Dean D J and Hjorth-Jensen M 2010 *Phys. Rev.* C **82** 034330

[13] Hagen G, Papenbrock T, Hjorth-Jensen M and Dean D J 2014 *Rep. Prog. Phys.* **77** 096302
[14] Navrátil P, Thoresen M and Barrett B R 1997 *Phys. Rev.* C **55** R573
[15] Lisetskiy A F, Barrett B R, Kruse M K G, Navrátil P, Stetcu I and Vary J P 2008 *Phys. Rev.* C **78** 044302
[16] Lisetskiy A F, Kruse M K G, Barrett B R, Navrátil P, Stetcu I and Vary J P 2009 *Phys. Rev.* C **80** 024315
[17] Jansen G R, Engel J, Hagen G, Navrátil P and Signoracci A 2014 *Phys. Rev. Lett.* **113** 142502
[18] Tsukiyama K, Bogner S K and Schwenk A 2012 *Phys. Rev.* C **85** 061304
[19] Bogner S K, Hergert H, Holt J D, Schwenk A, Binder S, Calci A, Langhammer J and Roth R, *Phys. Rev. Lett.* **113** 142501
[20] Okubo 1954 *Prog. Theor. Phys.* **12** 603
[21] Suzuki K and Lee S Y 1980 *Prog. Theor. Phys.* **64** 2091
[22] Suzuki K 1982 *Prog. Theor. Phys.* **68** 246
[23] Shirokov A M, Vary J P, Mazur A I and Weber T A 2007 *Phys. Lett.* B **644** 33
[24] Machleidt R and Entem D 2011*Phys. Rep.* **503** 1
[25] H. Hergert, private communication 2014 *to be published*
[26] Brown B A and Richter W A 2006 *Phys. Rev.* C **74** 034315

The isospin dependence of the nuclear force and its impact on the many-body system

F. Sammarruca, L. White, B. Chen

Physics Department, University of Idaho, Moscow, ID, 83844, USA

E-mail: fsammarr@uidaho.edu

Abstract. A major goal of contemporary nuclear physics is to improve our knowledge of the nuclear matter equation of state. In particular, the equation of state of isospin-asymmetric matter (that is, with unequal concentrations of protons and neutrons), is not well understood, mostly due to our limited knowledge of the symmetry energy. The latter reduces the binding energy in a nucleus with unequal number of protons and neutrons, and is crucial for understanding nuclear stability. We review experimental, phenomenological, and theoretical facts about the symmetry energy. We emphasize the importance of adopting a microscopic approach towards a better understanding of this important quantity.

1. Introduction

Nearly eight decades ago, Bethe and Weizäcker [1, 2] proposed the "liquid drop model" for the nuclear binding energies which, in spite of its simplicity, turned out to be remarkably successful. The model is essentially a mass formula which provides the total binding energy per nucleon as the sum of several contributions,

$$\frac{B}{A} = a_V - \frac{a_s}{A^{1/3}} - \frac{a_C Z(Z-1)}{A^{4/3}} - \frac{a_{sym}(A-2Z)^2}{A^2} - \frac{\Delta E}{A} \,. \tag{1}$$

The first four contributions are known as the volume, the surface, the Coulomb, and the symmetry term, respectively. The last term contains smaller contributions, originating from effects such as pairing. The (attractive) volume term is the largest contribution and represents the energy per nucleon of an infinite, charge-neutral, isospin-symmetric system. Such idealized system is known as symmetric nuclear matter, and the value of the a_V coefficient is the binding energy per nucleon of infinite symmetric matter, approximately 15 MeV. Clearly, the (repulsive) Coulomb term favors a nucleus with a large neutron excess. On the other hand, the distabilizing effect of the symmetry term increases with larger values of $N - Z = A - 2Z$. Fits to empirical nuclear binding energies determine the coefficents of the mass formula, resulting in a value of about 23 MeV for a_{sym}.

Although this simple model captures the main features of the nuclear binding energies, a deep understanding of nuclear structure and nuclear stability, all the way to the (yet unknown) drip lines, requires a microscopic approach to the interaction of hadrons in nuclei. This is still a very challenging problem that goes to the very core of nuclear physics. In fact, our present knowledge of the nuclear force in free space is, in itself, the result of decades of struggle [3]. The nature of the nuclear force in the medium is of course a much more complex problem, as it involves aspects of the force that cannot be constrained through free-space nucleon-nucleon (NN) scattering. Predictions of nuclear properties are the ultimate test for many-body theories.

Nuclear matter, which was defined above, is an alternative and convenient theoretical laboratory for many-body theories. Being an infinite (and thus translationally invariant) system of nucleons subjected only to their mutual strong interactions, it is theoretically simpler than an actual nucleus.

Adopting what is known as "local density approximation," one can relate the energy per particle in nuclear matter to the one in finite systems, through an appropriate energy-density functional. *Vice versa*, empirical information on nuclei can be extrapolated to an infinite system to provide constraints for theoretical models of nuclear matter.

For the case of (cold) isospin-asymmetric nuclear matter, where proton and neutron densities are different, the energy per particle is a function of both the total density and the relative concentrations of protons and neutrons. Naturally, the latter dependence is closely related to the symmetry term in the Bethe-Weizäcker formula, Eq. (1).

In the *ab initio* approach to the many-body problem, one starts with fundamental few-nucleon forces. Recently, chiral effective theories of the nuclear force, originally proposed by Weinberg [4], have become popular as a mean to respect the symmetries of Quantum Chromo Dynamics (QCD) while retaining the basic degrees of freedom typical of low-energy nuclear physics, pions and nucleons. Chiral effective theories provide a well-defined scheme to determine the appropriate many-body diagrams to be included at each order of the perturbation. However, being based on a low-momentum expansion, interactions derived from chiral perturbation theory are not suitable for applications in dense nuclear/neutron matter, where high Fermi momenta are involved. We will discuss this issue further in Sec. 3.

Although predictions of symmetric nuclear matter properties from different models are reasonably convergent, predictions of the symmetry energy can vary dramatically, particularly so for its density dependence. It is the purpose of this article to explore and discuss the present status of our knowledge of the symmetry energy and, more broadly, the equation of state (EoS) of asymmetric matter. One of our main goals is to demonstrate the importance of pursuing a microscopic approach towards a better understanding of the properties of dense many-body systems.

2. The Symmetry Energy

2.1. Some phenomenological features

Isospin-asymmetric nuclear matter is characterized by the neutron density, ρ_n, and the proton density, ρ_p. It is more convenient to refer to the total density $\rho = \rho_n + \rho_p$ and the asymmetry (or neutron excess) parameter $\alpha = \frac{\rho_n - \rho_p}{\rho}$. Clearly, α=0 corresponds to symmetric matter and α=1 to neutron matter.

Expanding the energy/particle in isospin asymmetric matter with respect to the asymmetry parameter yields

$$e(\rho, \alpha) = e_0(\rho) + \frac{1}{2}\Big(\frac{\partial^2 e(\rho, \alpha)}{\partial \alpha^2}\Big)_{\alpha=0} \alpha^2 + \mathcal{O}(\alpha^4) \,. \quad (2)$$

To a very good degree of approximation, the energy per particle in isospin asymmetric matter can be written as

$$e(\rho, \alpha) \approx e_0(\rho) + e_{sym}(\rho)\alpha^2 \,, \quad (3)$$

which is to be compared with the sum of the first and fourth terms on the right-hand side of Eq. (1), with α replacing $\frac{N-Z}{A}$. A typical value for e_{sym} at nuclear matter saturation density (ρ_0) is 30 MeV, with theoretical predictions spreading approximately between 26 and 35 MeV.

A measure for the density dependence of the symmetry energy is the parameter defined as

$$L = 3\rho_0\Big(\frac{\partial e_{sym}(\rho)}{\partial\rho}\Big)_{\rho_0} \approx 3\rho_0\Big(\frac{\partial e_{n.m.}(\rho)}{\partial\rho}\Big)_{\rho_0}, \tag{4}$$

where we have used Eq. (3) with α=1. Thus, L is sensitive to the gradient of the energy per particle in neutron matter ($e_{n.m.}$), that is, the neutron matter pressure. As to be expected on physical grounds, the neutron skin,

$$S = \sqrt{< r_n^2 >} - \sqrt{< r_p^2 >}\,, \tag{5}$$

where $\sqrt{< r_n^2 >}$ and $\sqrt{< r_p^2 >}$ are the *r.m.s.* radii of the neutron and proton distributions, is highly sensitive to the same energy gradient.

Predictions of L by phenomenological models show a very large spreading. Values ranging from -50 to +100 MeV are found from the numerous parametrizations of Skyrme interactions [5], all chosen to fit the binding energies and the charge radii of a large number of nuclei.

The definition of L in Eq. (4) originates from an expansion of the symmetry energy in powers of the density (relative to saturation density):

$$e_{sym}(\rho) \approx e_{sym}(\rho_0) + \frac{L}{3}\Big(\frac{\rho-\rho_0}{\rho_0}\Big) + \frac{K_{sym}}{18}\Big(\frac{\rho-\rho_0}{\rho_0}\Big)^2, \tag{6}$$

where

$$K_{sym} = 9\rho_0^2\Big(\frac{\partial^2 e_{sym}(\rho)}{\partial^2\rho}\Big)_{\rho=\rho_0}. \tag{7}$$

2.2. Experimental constraints on the symmetry energy

In this section we review some recent measurements to constrain the symmetry energy. We refer the reader to Ref. [6] and references therein for more details.

Several types of experiment can be exploited to constrain the density dependence of the symmetry energy. They include:

Heavy ion collisions.

The EoS is an important part of the input of transport models describing heavy ion collisions, and thus can be constrained through measurements of selected observables in ion-ion scattering. Sensitivity to the EoS can be attained by varying the number of protons and neutrons of the colliding nuclei. Emission of particles from neutron-rich or proton-rich systems is sensitive to the symmetry potential, which is repulsive on neutrons and attractive on protons (in a neutron-rich environment). Neutron-proton spectral ratios as well as neutron, hydrogen, and fragment flows are sensitive to the density dependence of the symmetry energy and have been used for that purpose. Also, in a collision where target and projectile have different values of N/Z, the system tends towards a state of constant value of the asymmetry parameter α, a phenomenon known as "isospin diffusion", the rate of which is controlled by the strength of the symmetry potential. Neutron and proton spectra from central collisions of ^{124}Sn + ^{124}Sn and ^{112}Sn + ^{112}Sn at 50 MeV per nucleon have been measured, as well as transverse collective flows of hydrogen, helium isotopes, and other fragments at 35 MeV per nucleon in ^{70}Zn + ^{70}Zn, ^{64}Zn + ^{64}Zn, and ^{64}Ni + ^{64}Ni reactions.

Nuclear binding energies.

This method relies on the liquid droplet model we referred to in Section 1. Because the symmetry energy contribution to the total binding energy is small, a reliable determination of both the symmetry energy and its slope can be difficult. All other contributions must be known with very high accuracy. With improved droplet models which allow to reproduce binding energies within

0.1%, it was possible to determine the symmetry energy within 1 MeV and its slope within 30 MeV [6].

Neutron skin measurements with hadronic or electroweak probes.

The relation between the thickness of the neutron skin and the slope of the symmetry energy can be understood in terms of the pressure gradient (due to the symmetry potential) that pushes neutrons outwards (and acts in a similar manner in the interior of neutron stars). Experiments aimed at extracting the neutron skin as a measure of the slope of the symmetry energy use either electroweak or hadronic probes. Particularly for the latter, theoretical uncertainties in the models used to analyse tha data are hard to quantify.

Parity-violating electron scattering experiments are now a realistic option to determine neutron distributions with unprecedented accuracy. The neutron radius of ^{208}Pb is expected to be re-measured at the Jefferson Laboratory in the PREXII experiment planned for the near future. Parity-violating electron scattering at low momentum transfer is especially suitable to probe neutron densities, as the Z^0 boson couples primarily to neutrons. A much higher level of accuracy can be achieved with electroweak probes than with hadronic scattering. With the success of this program, reliable empirical information on neutron skins will be able to provide more stringent constraints on the density dependence of the symmetry energy.

Measurements of the neutron skin of ^{48}Ca have also been proposed and are expected to take place in the near future with the CREX experiment [7]. Being much lighter than ^{208}Pb, ^{48}Ca is considerably more sensitive to surface and shell effects, which makes it possible to extract useful structure information. Furthermore, *ab initio* calculations are feasible for Calcium isotopes, thus CREX experiments have the potential to test models of three-nucleon forces and density functional theories.

Brief summary of present constraints.

In summary, the authors of Ref. [6] conclude that consistent constraints can be obtained from structure measurements and heavy-ion collisions. Their analyses, which include densities between $0.3\rho_0$ and ρ_0, yield a constraint centered at approximately (32.5, 70) MeV for the symmetry energy and the L parameter, respectively. The extracted value for the neutron skin of ^{208}Pb is 0.18 ± 0.027 fm, which is found to be consistent with the symmetry energy constraints.

3. Our microscopic approach to isospin-asymmetric nuclear matter

After the development of QCD and the understanding of its symmetries, chiral effective theories [4] were developed as a way to respect the symmetries of QCD while keeping the degrees of freedom (nucleons and pions) typical of low-energy nuclear physics. However, chiral perturbation theory (ChPT) has definite limitations as far as the range of allowed momenta is concerned. For the purpose of applications in dense matter, where higher and higher momenta become involved with increasing Fermi momentum, NN potentials based on ChPT are unsuitable. On the other hand, relativistic meson theory is an appropriate framework to deal with the high momenta encountered in dense matter. In particular, the one-boson-exchange (OBE) model has proven very successful in describing NN elastic data in free space up to high energies, and has a good theoretical foundation. We seek a momentum-space potential developed within a relativistic scattering equation. Furthermore, we require a potential that uses the pseudovector coupling for the interaction of nucleons with pseudoscalar mesons. With these constraints in mind, as well as the requirement of a good description of the NN data, Bonn B [3] is a reasonable choice.

Concerning our approach to nuclear matter, we adopt the Dirac-Brueckner-Hartree-Fock (DBHF) method. The main strength of the DBHF approach is its inherent ability to account for important three-body forces through its density dependence. More specifically, we refer to three-body forces arising from excitation of a virtual nucleon-antinucleon pair. For a detailed review of our application of the DBHF method to symmetric and asymmetric matter, we refer the reader to Ref. [8].

4. The symmetry energy and the role of the isovector mesons

Recenlty we explored the role of all isovector channels for the symmetry energy from the point of view of an *ab initio* model [9]. The main point of the *ab initio* approach is that mesons are tightly constrained by the free-space data and their parameters are never readjusted in the medium (this is what we mean by "parameter-free"). Furthermore, the contributions from the various mesons are fully iterated, thus giving rise to correlation effects. The corresponding predictions can be dramatically different than those which may be produced in first-order calculations.

In Fig. 1 we show the density dependence of the symmetry energy obtained with the Bonn A, B, and C potentials. The potential model dependence comes almost entirely from differences among predictions of the energy in symmetric nuclear matter. With the three sets of predictions, we mean to estimate the uncertainty to be expected when using different parametrizations for the isovector mesons, while describing the free-space NN data quantitatively.

In contrast, we show in Fig. 2 predictions of the symmetry energy with the numerous parametrizations of the Skyrme model [5]. Clearly, the constraint from the free-space data reduces dramatically the spreading among theoretical predictions.

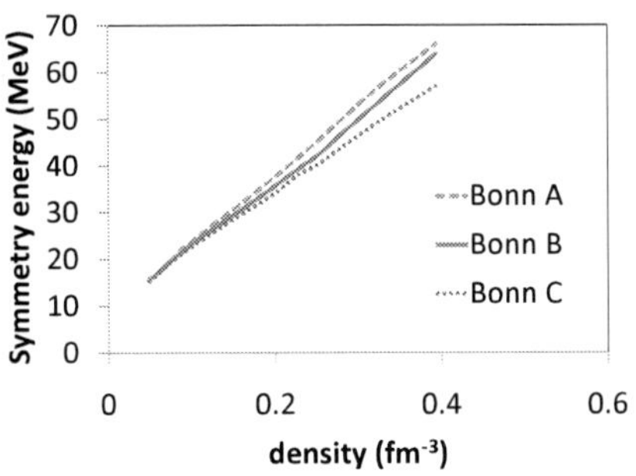

Figure 1. The symmetry energy as predicted with Bonn A, B, and C.

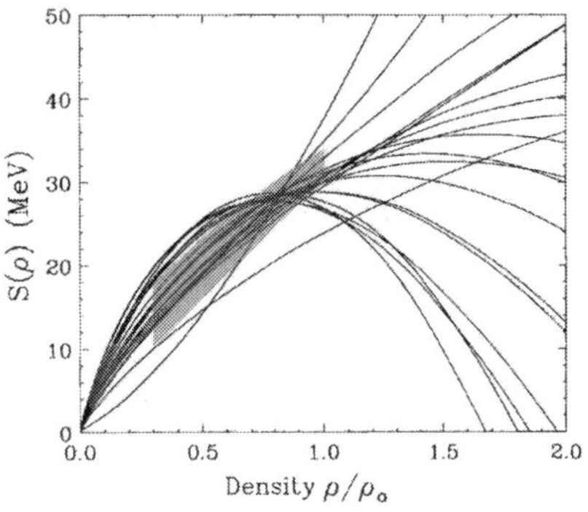

Figure 2. Predictions of the symmetry energy *vs.* density for various parametrizations of the Skyrme model.

We have examined the effect of the isovector mesons (π, ρ, and δ) on the difference between the potential energies of pure neutron matter and symmetric matter [9]. Our findings are easily understood in terms of the contributions of each meson to the appropriate component of the nuclear force and the isospin dependence naturally generated by isovector mesons. We find that the pion gives the largest contribution to this difference. The contribution of the pion is often overlooked, possibly because this meson is missing from some mean field models, which are popular among users of equations of state. It is our opinion that conclusions regarding the interplay of ρ and δ in phenomenological models [10] should be taken with caution.

We emphasize the fundamental differences between our approach and the one of mean field models, particularly pionless QHD theories. First, these differences are of conceptual relevance, since free-space NN scattering and the NN bound state are essentially determined by pion physics. Furthermore, they can impact in a considerable way conclusions with regard to isospin dependent systems/phenomena. In order to have a fundamental basis, a microscopic theory of the nuclear many-body problem has to start from the bare NN interaction with all its components.

5. Conclusions and outlook

The symmetry energy plays a crucial role for the binding energies of nuclei and thus nuclear stability. In the idealized system known as infinite nuclear matter, it represents the reduction of binding energy per particle in the presence of isospin asymmetry.

The details of the density dependence of the symmetry energy [i. e. its derivative(s)] are still quite controversial. And yet, important observables depend crucially on those details, the most egregious example being the neutron skin thickness. Furthermore, the potential energy part of the symmetry energy controls the isospin-dependent dynamics in heavy-ion reactions. We have reviewed the rich and diverse effort which is presently going on to improve the available constraints on the symmetry energy and, more generally, the energy and pressure in asymmetric matter.

In reviewing our theoretical approach to symmetric and asymmetric matter, we argued that relativistic nuclear physics is a valid paradigm, particularly when high momenta (and/or high Fermi momenta) are involved. The ability of the DBHF approach to effectively incorporate important three-body effects is what makes this method attractive and convenient.

We then concentrated on what builds the symmetry energy in a relativistic meson-theoretic model. On simple physical grounds, one can expect that the isovector mesons would be the important building blocks. This is in fact the case. Of those, we identified the pion as by far the most important. This brings the tensor force into the spot light, not only as a powerful saturation mechanism, but also as the main nuclear force component for the symmetry energy. We observed that the spreading of symmetry energy predictions obtained with different NN potentials (constrained by free-space data) is moderate, in contrast with phenomenological models.

In closing, we point out that our microscopic effective interactions (that is, the Dirac-Brueckner-Hartree-Fock G-matrix) can be useful for applications to nuclear reactions [11, 12]. Most recent work of our group includes the solution of the Bethe-Goldstone equation in three-dimensional space (so that standard angle-average approximations could be removed from the Pauli blocking operator) [13], and the developement of microscopic in-medium NN cross sections suitable for applications in nucleus-nucleus collisions [12].

Acknowledgments

This work was supported by the U.S. Department of Energy, Office of Science, Office of Basic Energy Sciences, under Award No. DE-FG02-03ER41270.

References

[1] Bethe H A and Bacher R F 1936 *Rev. Mod. Phys.* **8** 82
[2] von Weizäcker C F and Fur Z 1935 *Physik* **96** 431
[3] Machleidt R 1989 *Adv. Nucl. Phys.* **19** 189
[4] Weinberg S 1979 *Physica* **96A** 327; 1990 *Phys. Lett.* **B251** 288
[5] Brown B A 2000 *Phys. Rev. Lett.* **85** 5296
[6] Tsang M B *et al.* 2012 *Phys. Rev. C* **86** 015803
[7] *hallaweb.jlab.org/parity/prex; http://www.jlab.org/conferences/crex/.*
[8] Sammarruca F 2010 *Int. J. Mod. Phys. E* **19** 1259
[9] Sammarruca F 2011 *Phys. Rev. C* **84** 044307
[10] Baran V, Colonna M, Greco V and Di Toro M 2005 *Phys. Rep.* **410** 335
[11] Sammarruca F and White L 2011 *Phys. Rev. C* **83** 064602
[12] Chen B, Sammarruca F and Bertulani C 2013 *Phys. Rev. C* **87** 054616
[13] White L and Sammarruca F 2013 *Phys. Rev. C* **88** 054619

Journal of Physics: Conference Series **580** (2015) 012005 doi:10.1088/1742-6596/580/1/012005

On the way to a microscopic derivation of covariant density functional theory

P. Ring

Abstract. Several problems encountered on the way to a microscopic derivation of covariant density functionals are discussed. The starting point are semi-microscopic functionals adjusted to ab-initio calculations in nuclear matter. Since it is impossible to derive in this way the strength of tensor terms in such functionals we propose a solution of the relativistic Brueckner-Hartree-Fock equations in finite nuclei and show first results.

Fakultät für Physik, Technische Universität München, D-85748 Garching, Germany

E-mail: ring@ph.tum.de

1. Introduction

In recent years considerable progress has been achieved in ab-initio derivations of nuclear properties from bare nucleon-nucleon forces. For light nuclei it is possible to solve the exact nuclear many-body problem on the computer. For heavier nuclei approximate methods have been applied. The calculation of nuclear properties for the majority of nuclear systems, however, is left to density functional methods. Non-relativistic [1, 2] and relativistic [3, 4, 5, 6, 7] versions enable an effective description of the nuclear many-body problem not only for bulk properties, such as binding energies and radii, but also for collective excitations such as rotations and giant resonances, and, by going beyond mean field, complicated configurations [8, 9, 10, 11, 12] and sophisticated low-lying spectra in transitional nuclei [13, 14].

At present most of these functionals are purely phenomenological. Of course, one of the main goals in nuclear physics is to build a universal density functional theory based on microscopic calculations [15, 16]. This functional should be able to explain as many as possible measured data within the same parameter set and to provide reliable predictions for properties of nuclei far from stability not yet or never accessible to experiments in the laboratory. It should be derived in a fully microscopic way from the interactions between bare nucleons. At present, however, attempts to derive such a density functional provide only qualitative results for two reasons: first, the three-body term of the bare interaction is not known well enough and, second, the methods to derive such a functional are not precise enough to achieve the required accuracy. Note that a 1 per mille error in the binding energy per particle of symmetric nuclear matter leads to an error of several MeV in the binding energy of heavy nuclei, an error which is an order of magnitude larger than required by astrophysical applications. From such considerations it is evident, that successful ab-inito functionals in nuclear physics, at least in foreseeable future will always depend on a very small number phenomenological parameters for fine tuning. It is the number of final parameters which counts.

One of the underlying symmetries of QCD is Lorentz invariance and therefore covariant density functionals are of particular interest in nuclear physics. This symmetry not only allows

Journal of Physics: Conference Series **580** (2015) 012005 doi:10.1088/1742-6596/580/1/012005

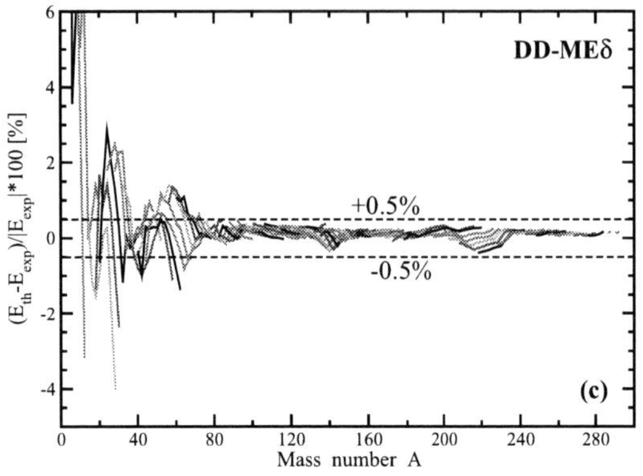

Figure 1. The relative difference between theoretical and experimental masses of 835 even-even nuclei investigated in relativistic Hartree Bogoliubov (RHB) calculations with the set DD-MEδ If $E_{th} - E_{exp} < 0$, the nucleus is more bound in the calculations than in experiment. Dashed lines show the ± 0.5 % limit (from Ref. [17]).

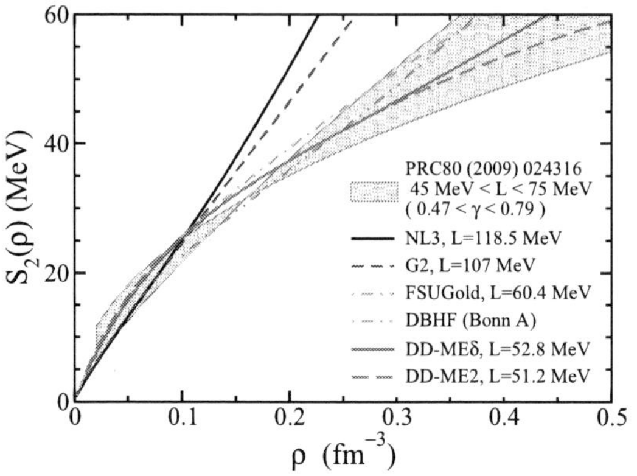

Figure 2. The symmetry energy as a function of the density. The results for DD-MEδ are compared with those of the density functionals NL3 [18], G2 [19], FSUGold [20], and microscopic DBHF calculations [21]. The shaded area represents the empirical region suggested by the available constraints on the L parameter (from Ref. [22]).

to describe the spin–orbit coupling, which has an essential influence on the underlying shell structure, in a consistent way, but it also puts stringent restrictions on the number of parameters in the corresponding functionals without reducing the quality of the agreement with experimental data. Therefore we restrict ourselves in the following to covariant density functionals.

2. Semi-microscopic functionals

In Coulombic system an essential input for the derivation of microscopic functionals is the exact numerical solution of the homogeneous electron gas at various densities. Starting from this energy functional $E[\rho]$ additional gradient terms and many other corrections have been added with great success. Therefore it seems to be reasonable to apply a similar concept in nuclear physics. Of course, at present there are no exact solutions of the nuclear matter problem available producing at least saturation at reasonable densities.

Therefore we followed the spirit of the Barcelona group [23], who used modern Brueckner-Hartree-Fock calculations in nuclear matter as a starting point for a semi-phenomenological non-relativistic potential, and derived in a similar way a relativistic functional DD-MEδ [22] with density dependent meson couplings. In contrast to the phenomenological density functionals with meson exchange such as NL3 [18] which contain usually only the σ-meson in the isoscalar scalar channel, the ω-meson in the isoscalar vector channel, and the ρ-meson in the isovector vector channel, microscopic Dirac-Brueckner-Hartree-Fock (DBHF) calculations show very clearly that there is also isospin-dependence in the scalar channel described by the δ-meson (a_0-meson). The density dependence of the meson-nucleon coupling vertices in the new functional DD-MEδ [22] is derived from *ab-initio* DBHF calculations in nuclear matter. The starting point forms the Equation of State (EoS) of symmetric nuclear matter and of neutron matter derived by Baldo et al. [24] in a state-of-the-art non-relativistic Brueckner-Hartree-Fock calculation including relativistic corrections and three-body forces. In addition the isovector part of the effective Dirac mass $m_p^* - m_n^*$ derived by the Tübingen group [25] in DBHF theory determines the coupling constant of the δ-meson and its density dependence. Only four additional parameters,

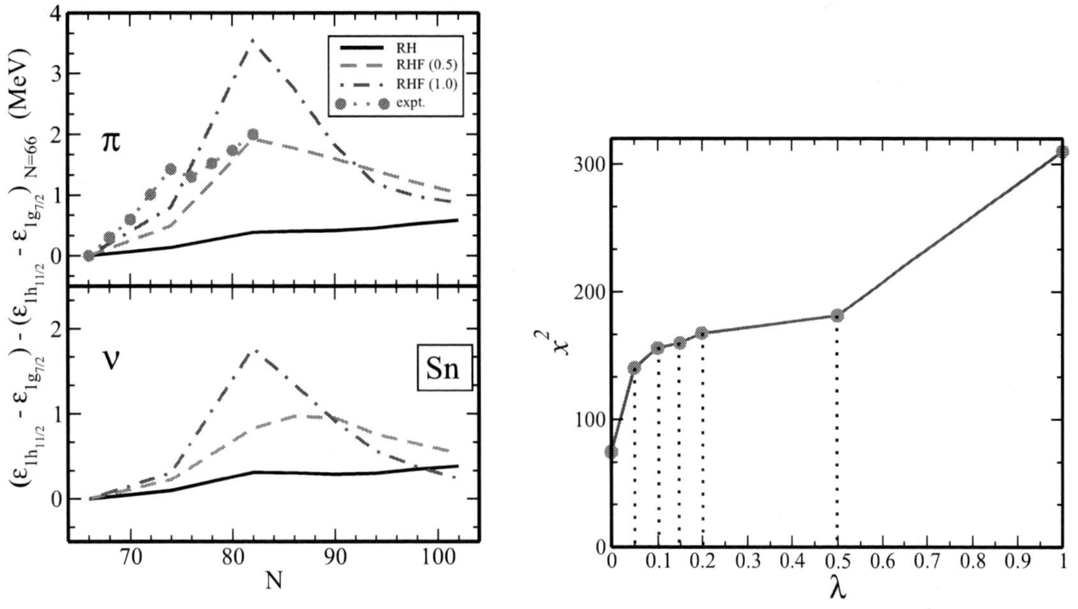

Figure 3. (left panel) The evolution of the $1h_{11/2}$ and $1g_{7/2}$ neutron and proton energy gap. The s.p levels of ^{116}Sn are taken as reference point. (right panel) χ^2 of the fit as function of the pion coupling constant in terms of the free pion coupling constant.(from Ref. [27]).

the coupling constants of the σ- the ω- and the ρ-meson at saturation density and the mass of the σ-meson are used for a fine-tuning of binding energies.

Extended applications of this functional on a global scale [17] have shown that it is comparable with all the modern relativistic high precision functionals containing considerably more phenomenological parameters. Fig. 1 shows that the deviations between theoretical and experimental binding energies are well below 0.5 %, with some exceptions for light nuclei. Since it is derived to a large extent from microscopic calculations, this functional contains a δ-meson. Its contributions in the iso-vector channel are not small. However, they are opposite in sign to those of the ρ-meson and for all the densities important in finite nuclei they are more or less compensated. The final iso-vector properties in this region are identical to high precision covariant density functionals without δ-meson. This is shown in Fig. 2 for the symmetry energy.

We can conclude that, at present, we are able to derive very successful semi-microscopic covariant density functionals [22, 26] from ab-initio results for nuclear matter and neutron matter, containing only few remaining parameters for a fine tuning.

3. Single particle energies and tensor terms

There are, however, additional problems, which cannot be treated with such simple Lagrangians. It is an open question, whether one needs further extensions. Famous examples are the tensor terms. In particular the momentum dependent tensor terms $\sim (\boldsymbol{\sigma}_1 \boldsymbol{k}_1)(\boldsymbol{\sigma}_2 \boldsymbol{k}_2)$ contained in the exchange terms of the σ- and ρ-meson. It leads to very specific properties of single particle energies, which cannot be reproduced in any of the simple mean field calculations. A famous case is the splitting of $1h_{11/2}$ and $1g_{7/2}$ proton configurations in the Sn-region. As it is shown in the left panel of Fig. 3, the experimental data can be reproduced by including the exchange term of a one-pion exchange using a strength parameter roughly one half of the experimental pion-nucleon coupling in the vacuum. The right panel of Fig. 3 shows that it is impossible to determine the size of this parameter by fitting to bulk properties, such a masses and radii, because the best fit is obtained without pion contributions. Similar results have also been found

Journal of Physics: Conference Series **580** (2015) 012005 doi:10.1088/1742-6596/580/1/012005

in density dependent Relativistic Hartree-Fock (RHF) theory [28].

Of course, the results of the left panel of Fig. 3 would suggest to proceed on the phenomenological way and to fit to single particle energies. There are two reasons against such a procedure, first, according to the principle of the Kohn-Sham method in density functional theory, the single particle energies are only mathematical auxiliary objects and no measurable quantities, and second, there exists particle-vibrational coupling (PVC), which leads to a considerable fragmentation of the single particle levels. At present, it is not only conceptually, but also technically impossible to carry out such a fit. The experimental knowledge on the spectroscopic factors is mostly very limited and PVC calculations in deformed nuclei are extremely complicated. Most of the very few investigations with PVC are based on simple separable forces. In fact, recent investigations [29] show, that PVC plays also a considerable role in the Sn-isotopes shown in the left panel of Fig. 3.

4. Relativistic Brueckner-Hartree-Fock theory in finite nuclei

The concept to derive density functionals from properties of infinite nuclear matter, is very successful for general properties of these functionals. However, it cannot teach us too much when it comes to properties of the functionals, which do not show up in nuclear matter as for instance the tensor term in spin non-saturated systems. Nuclear matter is usually spin-saturated and therefore the contributions of the tensor term in first order are small. In order to study such effects from ab-inito calculations, we have to carry out microscopic calculations in finite nuclei, in particular also in heavy nuclei, which are not spin-saturated.

An obvious way to carry out such investigations is Brueckner theory, the mother of modern density functional theory in nuclei [30, 31]. The advantage of relativistic Brueckner theory is that one does not need three-body force to get saturation in nuclear matter. Therefore Dirac-Brueckner-Hartree-Fock (DBHF) theory has been used to describe also finite nuclei [32, 33, 34, 35]. To our knowledge in all these applications the local density approximation has been used, i.e. in a first step the self-energies are calculated in nuclear matter of various densities. In a second step the relativistic Hartree-Fock equations are solved in finite nuclei with density dependent coupling constants adjusted to the results of nuclear matter calculations at the corresponding density.

Although such calculations are successful, they cannot reach the accuracy of present day phenomenological covariant density functionals. It is also clear that this concept is relatively useless for a microscopic study of the influence of the tensor force in finite non spin-saturated nuclei, because, as discussed above, nuclear matter is spin-saturated.

By this reason we started a new project to solve the DBHF equations in finite systems. The relativistic Hartree-Fock equations will be solved in a large oscillator basis with N_f major shells [36, 27] by iteration. The interaction will be the G-matrix and its matrix elements $\langle ab|G(\omega)|cd\rangle$ will be determined, in each step of this iteration, by the solution of the Bethe-Goldstone (BG) equation

$$G(\omega) = \bar{V}^N + \bar{V}^N Q_F \frac{1}{\omega - H_0} Q_F G(\omega). \tag{1}$$

in the same relativistic oscillator basis (for details see Ref. [36]). Here $\bar{V}^N$ are the anti-symmetrized matrix elements the relativistic bare force *Bonn A* [37], H_0 are the self-consistent two-particle energies $\epsilon_m + \epsilon_{m'}$ of the RHF-operator at this step of the iteration, and Q_F is the Pauli operator

$$Q_F = \sum_{m<m'} |mm'\rangle\langle mm'|. \tag{2}$$

summing over intermediate states m, m' above the Fermi surface. Of course, since the single particle energies ϵ_m and $\epsilon_{m'}$ refer to the eigenvalues of the RHF-operator, the Pauli operator

(2) is defined in the corresponding relativistic Hartree-Fock basis. Therefore the full solution of this problem requires the transformation of the Pauli-operator from the RHF basis to the oscillator basis and this is the final goal of such calculations. In a first step we plan the so-called *oscillator* approximation for the Pauli operator, i.e. we use in Eq. (2) oscillator states for the wave functions $|mm'\rangle$. On the other hand self-consistent RHF-energies $\epsilon_m + \epsilon_{m'}$ with the same number of radial nodes will be used for H_0. For light nuclei this seems to be a reasonable approximation.

The corresponding code has been developed and first preliminary results show the possibility of such a concept. However further tests are necessary and therefore we show here only first results in the local density approximation, where the mean field equations are solved for a density dependent interaction determined by the solution of the DBHF-equtions in nuclear matter [38].

As an application we consider the nucleus ^{16}O. We use the realistic NN interaction Bonn A which has been adjusted to the NN scattering data in Ref. [37].

Table 1. Ground state properties of ^{16}O: total binding energy (MeV), charge radius (fm), and spin-orbit splitting (MeV) for the $1p$ proton shell. Experiments are compared with calculations discussed in the text. The theoretical values are given by their deviations from the experiment (in %). Negative (positive) values for the binding energies mean under- (over-) binding

	EXP. [39, 40]	PKO1 [28]	BHF [41]	DDBH	DDBHF	CC-NN [42]	CC-NNN [43]
E	-127.62	−0.6 %	−17.8 %	−15.8 %	−8.0 %	−5.3 %	+6.6 %
r_c	2.74	−1.6 %	−16.3 %	−4.9 %	−3.8 %	—	—
ΔE^{ls}_{1p}	6.3	+1.6 %	+19.0 %	−17.5 %	−23.8 %	—	—

The ground state properties, total binding energy, charge radius and the proton spin-orbit splitting for the $1p$ shell of ^{16}O are listed in Table 1. The experimental results are compared with a RHF calculation with the phenomenological set PKO1 [28], a non-relativistic Brueckner calculation BHF [41] with the potential *Bonn A*, two calculations using the local density approximations on the Hartree (DDBH) and on the Hartree-Fock (DDBHF) level. It is seen that the ground state properties in the Brueckner calculations in finite nuclei are improved considerably as compared to the non-relativistic results even in the local density approximation. The deviation from the experimental values have been decreased from 18% to 8% in the case of the energy and from 16 % to 4 % in the case of the charge radius, which is consistent with the conclusions in the infinite nuclear matter [44]. For the binding energy of ^{16}O the same quality is reached as in the coupled-cluster calculations based on chiral interactions at N^{3}L order using only two-body forces (CC-NN) [42] and in addition three-body forces (CC-NNN) [45].

In summary, a code has been developed for the solution of the Dirac-Brueckner-Hartree-Fock (DBHF) equations for finite nuclei in an oscillator basis. We hope that we can improve in this way the results of the local density apprimation shown in Table 1. The only input is the bare NN-force Bonn A adjusted to the scattering phase shifts in Ref. [37]. No other parameter is used.

Of course there are many open problems: The oscillator approximation for the Pauli operator should be avoided. Heavier nuclei have to be investigated, in particular those with non spin-saturated configurations, as for instance ^{132}Sn or ^{208}Pb. This will give us information about the importance of the tensor term in finite nuclei. Pairing correlations and deformations should be included for the description of open shell nuclei.

Acknowledgments
I would like to express my deep gratitude to all my collaborators, who contributed to the results presented in this contribution, in particular to Jinniu Hu, Xavier Roca-Maza, Milena Serra†, M. Centelles, George Lalazissis, Jie Meng, Takaharu Otsuka, Xavier Vinas, and Peter Schuck. This work was partly supported by the DFG cluster of excellence "Origin and Structure of the Universe" (www.universe-cluster.de).

References
[1] Vautherin D and Brink D M 1972 *Phys. Rev. C* **5** 626
[2] Dechargé J and Gogny D 1980 *Phys. Rev. C* **21** 1568
[3] Serot B D and Walecka J D 1986 *Adv. Nucl. Phys.* **16** 1
[4] Reinhard P G 1989 *Rep. Prog. Phys.* **52** 439
[5] Ring P 1996 *Prog. Part. Nucl. Phys.* **37** 193
[6] Vretenar D, Afanasjev A V, Lalazissis G A and Ring P 2005 *Phys. Rep.* **409** 101
[7] Meng J, Toki H, Zhou S G, Zhang S Q, Long W H and Geng L S 2006 *Prog. Part. Nucl. Phys.* **57** 470 (2006).
[8] Litvinova E and Ring P 2006 *Phys. Rev. C* **73** 044328
[9] Litvinova E, Ring P and Tselyaev V I 2007 *Phys. Rev. C* **75** 064308
[10] Litvinova E, Ring P and Tselyaev V I 2008 *Phys. Rev. C* **78** 014312
[11] Litvinova E, Ring P and Tselyaev V I 2010 *Phys. Rev. Lett.* **105** 022502
[12] Ring P and Litvinova E 2009 *Phys. At. Nucl.* **72** 1285
[13] Niksic T, Vretenar D and Ring P 2011 *Prog. Part. Nucl. Phys.* **66** 519
[14] Yao J M, Hagino K, Li Z P, Meng J and Ring P 2014 *Phys. Rev. C* **89** 054306
[15] *Lecture Notes in Physics*, Eds. Lalazissis G A, Ring P and Vretenar D 2004 (Springer-Verlag, Heidelberg) Vol. 641
[16] Drut J, Furnstahl R and Platter L 2010 *Prog. Part. Nucl. Phys.* **64** 120
[17] Agbemava S E, Afanasjev A V, Ray D and Ring P 2014 *Phys. Rev. C* **89** 054320
[18] Lalazissis G A, König J and Ring P 1997 *Phys. Rev. C* **55** 540
[19] Furnstahl R J, Serot B D and Tang H B 1997 *Nucl. Phys. A* **615** 441
[20] Todd-Rutel B G and Piekarewicz J 2005 *Phys. Rev. Lett.* **95** 122501
[21] Warda M, Viñas X, Roca-Maza X and Centelles M 2009 *Phys. Rev. C* **80** 024316
[22] Roca-Maza X, Viñas X, Centelles M, Ring P and Schuck P 2011 *Phys. Rev. C* **84** 054309
[23] Baldo M, Robledo L M, Schuck P and Viñas X 2010 *J. Phys. G* **37** 064015
[24] Baldo M, Maieron C, Schuck P and Viñas X 2004 *Nucl. Phys. A* **736** 241
[25] van Dalen E N E, Fuchs C and Faessler A 2007 *Eur. Phys. J. A* **31** 29
[26] Niksic T, Vretenar D and Ring P 2008 *Phys. Rev. C* **78** 034318
[27] Lalazissis G A, Karatzikos S, Serra M, Otsuka T and Ring P 2009 *Phys. Rev. C* **80** 041301
[28] Long W H, Van Giai N and Meng J 2006 *Phys. Lett. B* **640** 150
[29] Afanasjev A V and Litvinova E 2014 *arXiv* [nucl-th]1409.4855
[30] Negele J W 1970 *Phys. Rev. C* **1** 1260
[31] Negele J W and Vautherin D 1972 *Phys. Rev. C* **5** 1472
[32] Brockmann R and Toki H 1992 *Phys. Rev. Lett.* **68** 3408
[33] Fritz R, Müther H and Machleidt R 1993 *Phys. Rev. Lett.* **71** 46
[34] Fritz R and Müther H 1994 *Phys. Rev. C* **49** 633
[35] Shi H L, Chen B Q and Ma Z Y 1995 *Phys. Rev. C* **52** 144
[36] Gambhir Y K, Ring P and Thimet A 1990 *Ann. Phys. (N.Y.)* **198** 132
[37] Machleidt R 1989 *Adv. Nucl. Phys.* **19** 189
[38] Hu J N 2012 *Phd thesis, Osaka University, Osaka 567-0047, Japan*
[39] Wang M, Audi G, Wapstra A H, Kondev F G, MacCormick M, Xu X and Pfeiffer B 2012 *Chin. Phys. C* **36** 1603
[40] Angeli I and Marinova K 2013 *At. Data Nucl. Data Tab.* **99** 69
[41] Müther H, Machleidt R and Brockmann R 1990 *Phys. Rev. C* **42** 1981
[42] Hagen G, Papenbrock T, Dean D J, Hjorth-Jensen M and Asokan B V 2009 *Phys. Rev. C* **80** 021306
[43] Hagen G, Hjorth-Jensen M, Jansen G R, Machleidt R and Papenbrock T 2012 *Phys. Rev. Lett.* **108** 242501
[44] Brockmann R and Machleidt R 1990 *Phys. Rev. C* **42** 1965
[45] Hagen G, Hjorth-Jensen M, Jansen G R, Machleidt R and Papenbrock T 2012 *Phys. Rev. Lett.* **109** 032502

Journal of Physics: Conference Series **580** (2015) 012006 doi:10.1088/1742-6596/580/1/012006

Relativistic Hartree-Fock-Bogoliubov predictions of superheavy magic nuclei

J J Li[1,2], **W H Long**[1], **J Margueron**[3] **and N Van Giai**[2]

[1] School of Nuclear Science and Technology, Lanzhou University, Lanzhou 730000, China
[2] Institut de Physique Nucléaire, IN2P3-CNRS, Université Paris-Sud, F-91406 Orsay Cedex, France
[3] Institut de Physique Nucléaire de Lyon, IN2P3-CNRS, Université de Lyon, F-69622 Villeurbanne Cedex, France

E-mail: nguyen@ipno.in2p3.fr

Abstract. The occurrence of spherical shell closures in the superheavy nuclei region is explored in the framework of the relativistic Hartree-Fock-Bogoliubov theory (RHFB), and also with the simpler version of relativistic mean field with quasiparticles, in other words the relativistic Hartree-Bogoliubov (RHB) approximation. Both theoretical frameworks are often used for nuclei in the known regions of the nuclear chart. Shell closures characterized in terms of two-nucleon gaps indicate the magic regions. The results depend slightly on the effective Lagrangians used, but the magic numbers beyond ^{208}Pb are generally predicted to be $Z = 120$, 138 for protons and $N = 172$, 184, 228, 258 for neutrons. Shell effects are sensitive to various terms of the mean-field, such as the spin-orbit coupling, the scalar and effective masses. This explains the (relatively small) variations in the predictions, depending on the effective Lagrangians employed. The most complete model (RHFB) favors the nuclide ^{304}120 as the next spherical doubly-magic nucleus beyond ^{208}Pb.

1. Introduction

In modern nuclear physics, a challenging frontier remains in exploring the mass and charge limits of the superheavy nuclei (SHN) with proton numbers $Z \geqslant 104$. In the past years, many efforts have been devoted to identifying the position of the so-called stability island of the SHN, which depends on the balance of various contributions to the nuclear binding energy [1]. Experimentally, superheavy elements (SHE) have been synthesized up to $Z = 118$ [2]. The increasing survival probabilities of the measured SHE from $Z = 114$ to 118 seem to indicate enhanced shell effects with increasing Z and therefore a possible proton magic shell in the superheavy region. This would correspond to the position of the superheavy island, in the region $Z \geqslant 120$ [3].

It is thus interesting to explore with various microscopic models the doubly closed-shell systems within a spherical symmetry assumption. The shells are essentially determined by the spin-orbit (SO) splittings, and by the nucleon effective masses. On the other hand, the occurrence of shell closures could be somehow connected with approximate degeneracies of pseudo-spin (PS) partners [4, 5]. In the non-relativistic self-consistent mean field theory [6, 7], the SO splittings depend directly on an extra SO parameter in the energy density functional. In the covariant density functional (CDF) theory, such as the relativistic Hartree-Bogoliubov

(RHB) [8, 9] or the relativistic Hartree-Fock-Bogoliubov (RHFB) [10] approaches, the SO potential is naturally brought about by the same Lorentz scalar and vector fields which create the central potential. The number of independent degrees of freedom is thus reduced, and this could be an advantage for exploring unknown regions. Furthermore, in the more complete RHFB version of the CDF theory where the Lorentz $\rho - N$ tensor coupling can be included, the SO effects are found to be somewhat enhanced by this coupling [5], a feature which is not present in the simple RHB. This is one of the main motivations for undertaking the present study within the RHFB frame.

Aiming at the possible magic shells in the superheavy region, we investigate the superheavy nuclides covering $Z = 110 - 140$ and we compare the predictions of RHFB with those of RHB results where Fock terms are not explicitly considered. This short contribution is based on a recently published paper [11] where a more detailed discussion can be found. This contribution is organized as follows. In Sec. 2 we briefly introduce the RHFB theory. The occurrence of superheavy magic shells as well as the evolutions of the predicted shell closures are discussed in Sec. 3. A brief summary is presented in Sec. 4.

2. The Relativistic Hartree-Fock-Bogoliubov model

We briefly sketch the theoretical framework used here to describe the ground states of nuclei, and we apply it to spherically symmetric systems. The model consists of a Lagrangian $\mathscr{L}$ containing two parts. The first part $\mathscr{L}_{free}$ is a sum of terms describing the free nucleon field, free meson and photon fields. The second part $\mathscr{L}_I$ represents the nucleon-nucleon interaction mediated by the meson-nucleon and photon-nucleon couplings. This interaction Lagrangian can be expressed as:

$$\begin{aligned}\mathscr{L}_I =&\bar{\psi}\Big(- g_\sigma\sigma - g_\omega\gamma^\mu\omega_\mu - g_\rho\gamma^\mu\vec{\rho}_\mu\cdot\vec{\tau} + \frac{f_\rho}{2M}\sigma_{\mu\nu}\partial^\nu\vec{\rho}^\mu\cdot\vec{\tau}\\ &- \frac{f_\pi}{m_\pi}\gamma_5\gamma_\mu\partial^\mu\vec{\pi}\cdot\vec{\tau} - e\gamma^\mu A_\mu\frac{1-\tau_3}{2}\Big)\psi,\end{aligned} \tag{1}$$

where ψ denotes the nucleon field whereas σ, ω_μ, $\vec{\rho}_\mu$, $\vec{\pi}$, and A_μ stand respectively for the isoscalar-scalar, isoscalar-vector, isovector-vector, isovector-pseudovector and photon fields. It can be noticed from Eq.(1) that, in this model the meson field $\vec{\rho}_\mu$ can be coupled to the nucleonic field ψ in two different ways, by vector-coupling with a coupling strength g_ρ, and by tensor-coupling with a coupling strength f_ρ. This Lagrangian is the most complete relativistic model to date. All meson-nucleon couplings g_i anf f_i are density-dependent, and their values have been adjusted in previous studies [5].

Following the standard variational procedure, one can obtain the field equations of the nucleon, mesons and photon, namely the Dirac, Klein-Gordon and Proca equations. The corresponding Hamiltonian H in nucleon space can be obtained through a Legendre transformation, as it is done, e.g., in Ref.[12]. This allows one to express H in terms of nucleonic operators only:

$$H = \sum_{\alpha\beta} c_\alpha^\dagger c_\beta T_{\alpha\beta} + \frac{1}{2}\sum_{\alpha\alpha'\beta\beta'} c_\alpha^\dagger c_\beta^\dagger c_{\beta'} c_{\alpha'} \sum_\phi V^\phi_{\alpha\beta\beta'\alpha'}, \tag{2}$$

where $T_{\alpha\beta}$ is the nucleon kinetic energy operator, and $V^\phi_{\alpha\beta\beta'\alpha'}$ stands for different types of two-body interactions mediated by the mesons (photon):

$$T_{\alpha\beta} = \int d\boldsymbol{r}\bar{\psi}_\alpha(\boldsymbol{r})\left(-i\boldsymbol{\gamma}\cdot\boldsymbol{\nabla} + M\right)\psi_\beta(\boldsymbol{r}), \tag{3}$$

$$V^\phi_{\alpha\beta\beta'\alpha'} = \int d\boldsymbol{r}d\boldsymbol{r}'\bar{\psi}_\alpha(\boldsymbol{r})\bar{\psi}_\beta(\boldsymbol{r}')\Gamma_\phi D_\phi(\boldsymbol{r},\boldsymbol{r}')\psi_{\beta'}(\boldsymbol{r}')\psi_{\alpha'}(\boldsymbol{r}), \tag{4}$$

and ϕ denotes the various meson (photon)-nucleon coupling channels, namely the σ-S, ω-V, ρ-V, ρ-T and ρ-vector-tensor (ρ-VT), π-PV and A-V couplings. The expressions of the Γ_ϕ and D_ϕ can be found in Ref.[12].

The pairing correlations are treated self-consistently by the RHFB method (see, e.g., Ref.[10]).

In this work, the integro-differential RHFB equations are solved by using a Dirac Woods-Saxon basis [13] with a radial cutoff $R = 28$ fm. The numbers of positive and negative energy states in the basis expansion for each single-particle (s.p.) angular momentum (l, j) are chosen to be 44 and 12, respectively.

3. Results and Discussion

The numerical applications have been carried out in RHFB approach with the parameter sets PKA1 [5] and the PKOi series (i=1, 2, 3) [14, 15] for the RHFB calculations, and PKDD [16] and DD-ME2 [17] Lagrangians for the RHB calculations. The DD-ME2 model is one of the most frequently used RMF Lagrangians, and we choose to compare with the PKDD parametrization to have an indication of the model dependence of the predictions. The PKOi parameter sets are the first density-dependent RHFB models in the literature whereas PKA1 is an improved version containing also a Lorentz tensor ρ-nucleon coupling.

The pairing fields are calculated with the finite-range Gogny force D1S [18] with a rescaling strength factor f, which is introduced to compensate for level-density differences among various mean field approaches. It is indeed known that pairing related quantities, such as odd-even mass differences and moments of inertia, are systematically overestimated in the RHFB calculations of heavy nuclei with the original Gogny pairing force [19]. Therefore, a renormalization factor $f = 0.9$ is applied to reproduce the odd-even mass differences in Pb isotopes.

In ordinary nuclei, the magicity is generally well marked, e.g., by a sizable shell gap along the isotopic or isotonic chains. In SHN, the shell effects are not so clearly visible. To identify the occurrence of shell closures in the superheavy region, we use the so-called two-nucleon gaps - δ_{2p} for protons and δ_{2n} for neutrons - namely the differences of two-nucleon separation energies of the neighboring isotopes or isotones which provide us with an efficient evaluation of the shell effects [20, 21]:

$$\delta_{2p}(N, Z) = S_{2p}(N, Z) - S_{2p}(N, Z + 2), \tag{5}$$

$$\delta_{2n}(N, Z) = S_{2n}(N, Z) - S_{2n}(N + 2, Z). \tag{6}$$

The peak values of δ_{2p} or δ_{2n} are essentially determined by the sudden jump of S_{2p} or S_{2n}, respectively. This constitutes a good indication of the emergence of closed shells.

To have further understanding on the model deviations, we compare the s.p. spectra obtained by the selected mean field functionals. Taking the predicted doubly magic SHN $^{304}120_{184}$ as an example, Fig. 1 shows the proton (left panel) and neutron (right panel) canonical s.p. spectra. It is found that PKA1 provides the most evident magicity at $Z = 120$ and $N = 184$, respectively, although these shell closures are relatively weaker than in ordinary nuclei. For the neutron shell $N = 184$, it is essentially determined by the degeneracy of two PS partners $\{2h_{11/2}, 1j_{13/2}\}$ and $\{4s_{1/2}, 3d_{3/2}\}$, respectively above and below the shell. For the latter, the PS partners are predicted to be almost degenerate by all the models considered, while for the former, the PS partners have high angular momentum and some differences among models are observed: PKA1 predicts a weak PS splitting, at variance with the predictions of the other Lagrangians.

It is interesting to discuss the structure of the s.p. levels for the proton shell closure $Z = 120$. As shown in the left panel of Fig. 1 the proton shell closure coincides with a large PS splitting, $\{3p_{3/2}, 2f_{5/2}\}$, whereas the SO doublet $\{3p_{1/2}, 3p_{3/2}\}$ above the shell is almost degenerate. The shell gap at $Z = 120$ can therefore be interpreted as a manifestation of a large PS splitting and a weak SO splitting. In fact as shown in Fig. 2, the evolution of proton (left plot) and neutron

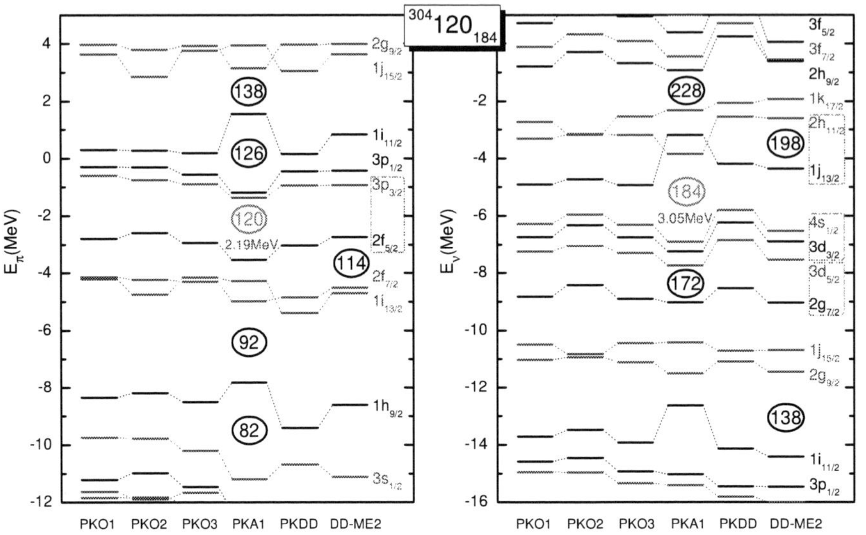

Figure 1. Proton (left panel) and neutron (right panel) single-particle spectra in the $^{304}120$ nuclide. The results with PKOi and PKA1 correspond to RHFB calculations, those with PKDD and DD-ME2 are obtained in RHB approximation.

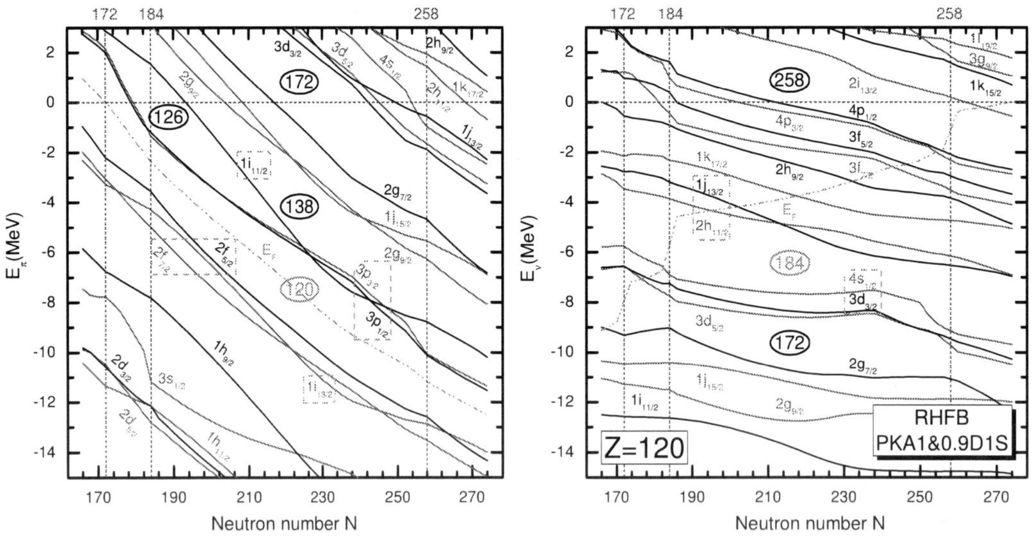

Figure 2. Proton (left plot) and neutron (right plot) single-particle energies along the isotopic chain of $Z = 120$. The results are extracted from the calculations of RHFB with the effective Lagrangian PKA1 and the D1S pairing force rescaled by a factor 0.9.

(right plot) s.p. energies along the $Z = 120$ isotopic chain determined by PKA1-RHFB model, it is also found that the spin partners $3p$ are almost degenerated and the SO splittings between the $2f$ states are not so distinct as generally expected. Coincident with the relation between the SO effects and PS symmetry conservation pointed out in Ref. [22], the reduced SO effects may lead to enlarged splitting between PS partners (i.e., $3p_{3/2}$ and $2f_{5/2}$).

4. Summary

In summary, the occurrence of spherical shell closures for SHN and the physics therein have been investigated using the RHFB theory with density-dependent meson-nucleon couplings, in comparison with the predictions of some RHB models. The shell effects are quantified in terms of two-nucleon gaps $\delta_{2n(p)}$. The results indicate that the nuclide $^{304}120_{184}$ could be the next spherically doubly magic nuclide beyond ^{208}Pb. It is also found that the shell effects in SHN are sensitive to the values of both scalar mass and effective mass, which essentially determine the spin-orbit effects and level density, respectively, and consistently the splittings between PS partners are essentially affects, which is also found to be tightly related with the emergence of superheavy magic shells. Additionally the analysis on the shell evolution as well as the density profile indicate that the emergence or disappearance of shell closure is tied up with the evolution of the central and spin-orbit mean fields, a feature that covariant mean field models may describe in a more unified way as compared to non-relativistic energy density functional (EDF) approaches. A further advantage of the RHFB framework is that exchange (Fock) terms are explicitly treated rather than approximately included by readjusted direct (Hartree) contributions as it is done in RHB (this is particularly true for the Coulomb exchange energy which is basically absent in RHB).

In this work, we have discussed the shell closure effects only at the mean field level, with the aim of comparing the predictions of different relativistic models. A more complete study would require, of course, to examine the changes brought about by the effects of particle-vibration coupling (see, e.g., Ref.[23]).

Experimental measurement of Q_α for at least one isotope of $Z = 120$ nucleus would help to set a proper constraint in determining the shell effects of SHN and to test further the reliability of the models as well. One also has to admit that for a more extensive exploration one needs to take into account the deformation effects.

Acknowledgements

This work is partly supported by the National Natural Science Foundation of China under Grant No. 11375076, and the Specialized Research Fund for the Doctral Program of Higher Education under Grant No. 20130211110005. J Li is supported in part by the China Scholarship Council (CSC) under grant No. 201206180063.

References

[1] Block M et al. 2010 *Nature* **463** 785
[2] Oganessian Y T et al. 2010 *Phys. Rev. Lett.* **104** 142502
[3] Adamian G G, Antonenko N V and Sargsyan V V 2009 *Phys. Rev. C* **79** 054608
[4] Ginocchio J N 2005 *Phys. Rep.* **414** 165
[5] Long W H, Sagawa H, Van Giai N and Meng J 2007 *Phys. Rev. C* **76** 034314
[6] Bender M, Heenen P-H and Reinhard P-G 2003 *Rev. Mod. Phys.* **75** 121
[7] Stone J R and Reinhard P-G 2007 *Prog. Part. Nucl. Phys.* **58** 587
[8] Vretenar D, Afanasjev A V, Lalazissis G A and Ring P 2005 *Phys. Rep.* **409** 101
[9] Meng J, Toki H, Zhou S G, Zhang S Q, Long W H and Geng L S 2006 *Prog. Part. Nucl. Phys.* **57** 470
[10] Long W H, Ring P, Van Giai N and Meng J 2010 *Phys. Rev. C* **81** 024308
[11] Li J J, Long W H, Margueron J and Van Giai N 2014 *Phys. Lett. B* **732** 169
[12] Bouyssy A, Mathiot J-F, Van Giai N and Marcos S 1987 *Phys. Rev. C* **36** 380
[13] Zhou S G, Meng J and Ring P 2003 *Phys. Rev. C* **68** 034323
[14] Long W H, Van Giai N and Meng J 2006 *Phys. Lett. B* **640** 150
[15] Long W, Sagawa H, Meng J and Van Giai N 2008 *Europhys. Lett* **82** 12001
[16] Long W, Meng J, Van Giai N and Zhou S G 2004 *Phys. Rev. C* **69** 034319
[17] Lalazissis G A, Nikšiś T, D. Vretenar D and Ring P 2005 *Phys. Rev. C* **71** 24312

[18] Berger J F, Girod M and Gogny D 1984 *Nucl. Phys. A* **428** 23
[19] Wang L J, Sun B Y, Dong J M and Long W H 2013 *Phys. Rev. C* **87** 054331
[20] Rutz K, Bender M, Bürvenich T, Schilling T, Reinhard P-G, Maruhn J and Greiner W 1997 *Phys. Rev. C* **56** 238
[21] Zhang W, Meng J, Zhang S Q, Geng L S and Toki H 2005 *Nucl. Phys. A* **753** 106
[22] Shen S, Liang H, Zhao P, Zhang S and Meng J 2013 *Phys. Rev. C* **88** 024311
[23] Litvinova E V and Afanasjev A V 2011 *Phys. Rev. C* **84** 014305

Journal of Physics: Conference Series **580** (2015) 012007 doi:10.1088/1742-6596/580/1/012007

From N=2Z in ^{60}Ca to N=Z in ^{80}Zr: Connecting the driplines

A. Poves[1], F. Nowacki[2], K. Sieja[2], A. P. Zuker[2], and S. M. Lenzi[3]

[1]Departamento de Física Teórica and IFT, UAM-CSIC, Universidad Auónoma de Madrid, Spain and Isolde, CERN, Geneve, Switzerland

[2]Université de Strasbourg and CNRS, IPHC, 23 rue du Loess 67037 Strasbourg, France

[3]Dipartimento di Fisica dell'Universitá and INFN, Sezione di Padova, I-35131 Padova, Italy

E-mail: alfredo.poves@uam.es

Abstract. We present Large Scale Shell Model calculations using the effective interaction LNPS, which reproduce nicely the experimental data available in the N=40 island of deformation. We deal as well with the triple (shape?) coexistence in the low energy spectrum of ^{68}Ni, showing that the highly deformed band based upon the excited 0^+ state at 2627 keV is in fact the doorway to the island. The onset of prolate quadrupole collectivity and shape coexistence in the heavy N=Z nuclei, is explained by means of the realizations of the pseudo plus quasi SU3 symmetry, with special emphasis on the shape mixing in the ground state of ^{72}Kr.

1. Introduction

Why do the quadrupole correlations thrive in the nucleus? The fact that the spherical nuclear mean field is close to the harmonic oscillator has profound consequences, because its dynamical symmetry, responsible for the accidental degeneracies of its eigenvalues, is SU(3), which has among its generators the quadrupole operator. When valence protons and neutrons occupy the degenerate orbits of a major oscillator shell, and for an attractive Q·Q interaction, the many body problem has an analytical solution in which the ground state of the nucleus is maximally deformed (Elliott's model) [1]. Where else do the quadrupole correlations thrive? In cases when both valence neutrons and protons occupy quasi-degenerate orbits with $j_r - j_s$=2 and $l_r - l_s$=2, (quasi-SU3) [2], or quasi-spin multiplets (pseudo-SU3) [3, 4]. For example, $0f_{7/2}$ and $1p_{3/2}$, or $0g_{9/2}$, $1d_{5/2}$ and $2s_{1/2}$ form quasi-SU3 multiplets and $0f_{5/2}$, $1p_{3/2}$ and $1p_{1/2}$ a pseudo-SU3 triplet.

2. The minimal valence space for the N=40 isotones

In this region of the chart of nuclei, the valence space should ideally include the full pf shell for neutrons and protons as well as (at least) the quasi-SU3 sequence $0g_{9/2}$, $1d_{5/2}$, $2s_{1/2}$. However, the dimensions of this space are out of reach for the moment. Thus we shall enforce the following restrictions which are physically sound: for the very neutron rich isotopes we adopt a ^{48}Ca core and exclude the gds proton orbits and the $2s_{1/2}$ neutron orbit. For the proton rich isotopes beyond ^{68}Se, we take a ^{56}Ni core and exclude the $2s_{1/2}$ orbit both for protons and neutrons. We use the interaction LNPS [5], whose effective single particle energies in the neutron rich sector are plotted in Figure 1.

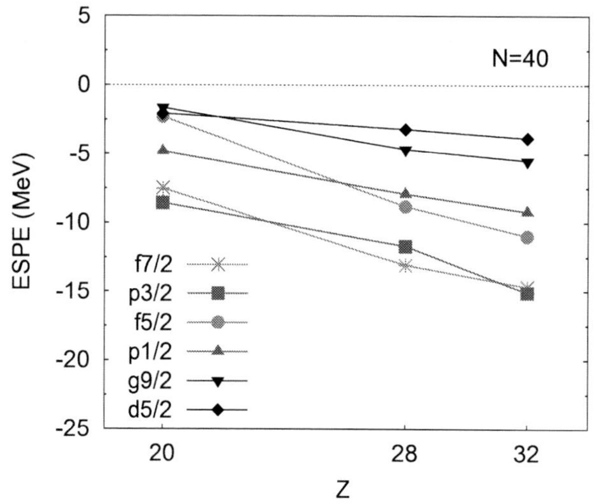

Figure 1. Effective neutron single particle energies of the N=40 isotopes from ^{60}Ca to ^{72}Ge.

These ESPE's are the phenomenological input of the Large Scale Shell Model (LSSM) calculations and are dictated by the experiment. They should provide a meaningful meeting point with the "ab initio" calculations. Notice the unorthodox spherical mean field at Z=20 with the orbits $0g_{9/2}$, $1d_{5/2}$ and $0f_{5/2}$ nearly degenerated. Similar situations are found for the N=20 and N=28 isotones toward Z=8, and for the N=8 isotopes approaching Z=2.

3. Triple coexistence in ^{68}Ni

In a first approximation, the ground state of ^{68}Ni behaves as double magic. But its low energy spectrum is much more complex, as three coexisting 0^+ states appear between 0 and ~2.5 MeV [6]. This is clearly seen in Figure 2. The excitation energy of the 0_2^+ state has been recently remeasured and now the state is placed at 1.6 MeV [7, 8]. This state is dominated by 2p-2h neutron excitations and is mildly oblate according to the calculations. The 0_3^+ is predicted by the LSSM calculations to be the band head of a very low-lying superdeformed band of 6p-6h nature. These calculations describe transition rates ranging over more than two orders of magnitude. One can understand the lowering of the 0_3^+ state in terms of the variants of SU(3) already mentioned. The neutron 4p-4h configuration takes full advantage of the quadruple correlations (quasi-SU3 for the particles and pseudo-SU3 for the holes) if the corresponding proton configuration is a 2p-2h across Z=28, as it is the case in the calculations, because the two protons excited are in a pseudo-SU3 configuration as well.

4. The island of deformation south of ^{68}Ni

For Z<28, the protons in the $0f_{7/2}$ orbit contribute efficiently to the build-up of the quadrupole collectivity of the configurations with np-nh neutron excitations across N=40, which take advantage of the quasi-SU3 coherence of the doublet $0g_{9/2}$- $1d_{5/2}$. Our LSSM calculations in the valence space of the full *pf*-shell for the protons and the $0f_{5/2}$ $1p_{3/2}$ $1p_{1/2}$ $0g_{9/2}$ and $1d_{5/2}$ orbits for the neutrons, predict another region of deformation centered in ^{64}Cr.

The mechanism at play is similar to the one that produces the other "islands of inversion" at N=20 and N=28. On the one side, the N=40 gap disappears at Z=20 (as it does at Z=8 for the N=20 isotopes). In addition, when the proton *pf*-shell is not completely filled, the quadrupole

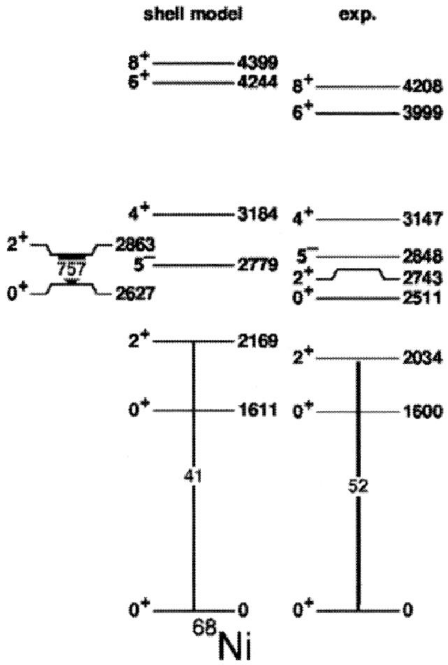

Figure 2. Level scheme of ^{68}Ni: Experiment [6, 7] compared to the results of the LSSM calculations with the LNPS interaction [9].

energy gains of the intruder configurations with several neutrons excited across N=40 is very large. The final result is that configurations alike to that of the third 0^+ state in ^{68}Ni become the ground states of ^{66}Fe, ^{64}Cr and ^{62}Ti. This can be seen in the behavior of the 2^+ excitation energies and B(E2) values depicted in Figure 3. In this sense we can submit that the 0_3^+ state in ^{68}Ni is the precursor of the N=40 island of deformation.

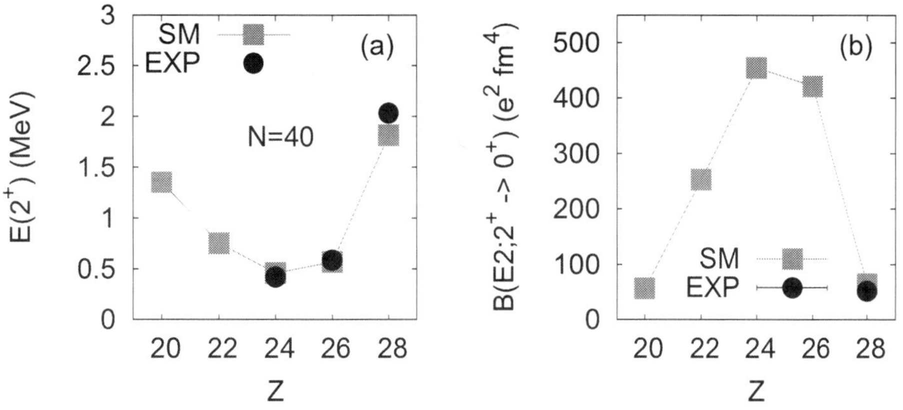

Figure 3. Excitation energies and B(E2) values for the N=40 isotopes, LNPS results *vs* experiment [10].

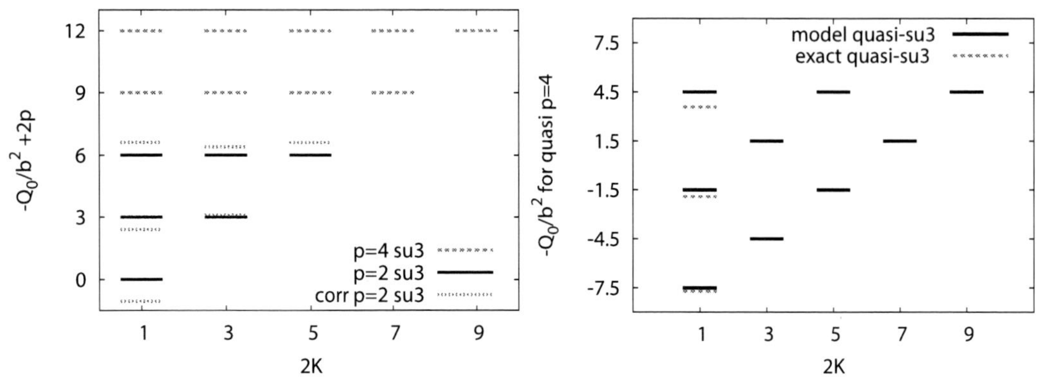

Figure 4. The intrinsic orbits of SU3 for p=2 and p=4 (p is the principal quantum number of the HO and b its length parameter) and the pseudo-SU3 (p=2) ones for the $1p_{3/2}$-$1p_{1/2}$-$0f_{5/2}$ (left panel). The quasi-SU3 orbits for the *gds* space ($0g_{9/2}$, $1d_{5/2}$ and $2s_{1/2}$) (right panel).

5. Deformation in heavy N=Z nuclei; The Quasi+Pseudo SU3 framework

The ground states of the N=Z nuclei above ^{68}Se and below ^{88}Ru, (and their neighbors) are dominated by configurations with np-nh jumps across N=Z=40. Why? Because the **n** particles sit in quasi-SU3 orbits and the **n** holes in pseudo-SU3, thus maximizing their quadrupole moments and, a fortiori, their quadrupole correlation energy. The gains in correlation energy overwhelm the monopole energy cost of crossing the gap between the upper *pf*-shell orbits and the intruder $0g_{9/2}$ and $1d_{5/2}$. This is clearly seen in the diagrams of the intrinsic SU3-like states in Figure 4. Filling the lowest quasi and pseudo SU3 intrinsic orbits we can determine for each nucleus which are the most favorable configurations from the quadrupole point of view. They are the following (b is the harmonic oscillator length):

- ^{72}Kr
  - 4p-4h; Q_0= 60 b^2
  - 8p-8h; Q_0= 73 b^2
  - 12p-12h; Q_0= 74 b^2
- ^{76}Sr
  - 4p-4h; Q_0= 51 b^2
  - 8p-8h; Q_0= 77 b^2
  - 12p-12h; Q_0= 79 b^2
- ^{80}Zr
  - 8p-8h; Q_0= 72 b^2
  - 12p-12h; Q_0= 83 b^2
  - 16p-16h; Q_0= 85 b^2

For instance the quadrupole moment of the 8p-8h configuration in ^{80}Zr is obtained from the figure as follows; put 8 particles in Quasi-SU3 to get Q_0= 4 x 7.5 + 4 x 4.5 = 48 b^2 and 16 in Pseudo-SU3 Q_0=5.06 x 4 + 1.41 x 4 + 1.08 x 4 + (−2.35) x 4 = 20.72 b^2 for a total Q_0= 48 + 21 + 3 = 72 b^2 (the extra three is an SU3 technicality).

We have recently shown that these quadrupole moments are very resilient to departures of the single particle energies from their degenerated limit [11]. The key point here is that the quadrupole energy gains grow with the square of the quadrupole moment whereas the monopole losses are at most proportional to the number of particle-hole jumps (modulo the N=40 gap). In ^{76}Sr and ^{80}Zr the deformed (prolate) configurations, 8p-8h and 12p-12h, win comfortably. In ^{72}Kr the 4p-4h prolate and oblate solutions (oblate corresponding to $(0g_{9/2})^4$ instead of $(gds)^4$) are degenerated as we shall discuss next. From the Q_0's one can deduce the B(E2)'s. The $2^+ \to 0^+$ is equal to $Q_0^2/50.3$ and the $4^+ \to 2^+$ a factor 1.43 larger. To compare with experiment we set b^2=4.5 fm^2 and obtain the following B(E2) values:

- ^{72}Kr; $2^+ \to 0^+$; 1470 e^{2}fm^4; $4^+ \to 2^+$; 2100 e^{2}fm^4
- ^{76}Sr; $2^+ \to 0^+$; 2380 e^{2}fm^4; $4^+ \to 2^+$; 3410 e^{2}fm^4
- ^{80}Zr; $2^+ \to 0^+$; 2800 e^{2}fm^4; $4^+ \to 2^+$; 4000 e^{2}fm^4

The available experimental results are:

- ^{72}Kr; $2^+ \to 0^+$; 810(150) e^{2}fm^4; $4^+ \to 2^+$; 2720(550) e^{2}fm^4 [12, 13]
- ^{76}Sr; $2^+ \to 0^+$; 2200(270) e^{2}fm^4 [14]
- ^{80}Zr; no data yet.

Notice that the agreement is excellent except for the B(E2) $2^+ \to 0^+$ of ^{72}Kr. But this is a blessing in disguise because it may led us to understand better the prolate oblate coexistence in this nucleus.

6. ^{72}Kr, a case of full prolate oblate mixing

It is common lore to speak of prolate-oblate or prolate-spherical coexistence when an excited 0^+ state appears at very low energy. This is the case in ^{72}Kr, whose first excited state is a 0^+ at 671 keV followed by a 2^+ at 710 keV. The very large B(E2) of the transition $4^+ \to 2^+$ strongly suggest that the 2^+ belongs to a prolate band which extends up to J=16^+ [15]. But, if it is so, where is the band-head? If we follow the J(J+1) sequence from the upper part of the band we should expect it to be 250 keV below the 2^+, which is very close to the experimental excitation energies of the 2^+'s in ^{76}Sr and ^{80}Zr [16, 17]. Obviously the distortion must be due to the mixing of the prolate band-head with a close-lying oblate state. This is borne out by a recent Coulex measurement of the spectroscopic quadrupole moment of the 2^+ state, made at Isolde [18], which finds it prolate.

We can thus propose a very simple model to explain the prolate oblate mixing in ^{72}Kr. The first element to take into account is that the oblate and prolate 4p-4h states do not mix directly; *i.e.* $\langle p|H|o\rangle \approx 0$. The mixing should then proceed through 2p-2h or 6p-6h states. Lets take these to be represented by an auxiliary state $|I\rangle$, and further assume that it lies at about ΔE=4 MeV (as our calculations indicate) and that its coupling to both prolate and oblate states is equal to δ. Taking the prolate and oblate states degenerated for simplicity, the mixing matrix reads:

$$\begin{pmatrix} 0 & 0 & \delta \\ 0 & 0 & \delta \\ \delta & \delta & \Delta E \end{pmatrix}$$

The mixing can proceed as well through a cloud of N states. In this, more realistic, case the matrix would be:

$$\begin{pmatrix} 0 & 0 & \beta & \beta & \beta & \dots \\ 0 & 0 & \beta & \beta & \beta & \dots \\ \beta & \beta & \Delta E & 0 & 0 & \dots \\ \beta & \beta & 0 & \Delta E & 0 & \dots \\ \beta & \beta & 0 & 0 & \Delta E & \dots \\ \dots & \dots & \dots & \dots & \dots & \dots \end{pmatrix}$$

Which has the same two lowest eigenvalues and eigenvectors than the previous one, provided $\delta = \sqrt{N}\ \beta$. Taking $\delta \sim 1$ MeV, or β=0.4 and N=6, as our calculations suggest, the eigenvalues are: −0.5 MeV, 0.0 MeV and +4.5 MeV. They fit nicely the experimental energies. The eigenstates corresponding to the two lower eigenvalues are:

$$|0^+_1\rangle = 43\%\ |p\rangle + 43\%\ |o\rangle + 14\%\ |I\rangle \text{ and } |0^+_2\rangle = 50\%\ |p\rangle + 50\%\ |o\rangle$$

The mixing of the oblate and prolate 2^+'s is bound to be much smaller, and we disregard it. Therefore, the B(E2)($2^+ \rightarrow 0^+_1$) will be approximately one half of the expected value for the prolate band in full accord with the experimental data.

How to name the shape of an object which is an even mixture of prolate and oblate? What is the nature of this mixing of shapes? Or should we rather speak of a shape entangled state?

7. Conclusions

The onset of the different modes of quadrupole collectivity depend on the structure of the spherical mean field. In the valence space comprising the pf shell and the $0g_{9/2}$ and $1d_{5/2}$ orbits we have explained the appearance of large prolate deformation at N=Z in terms of different realizations of SU3. Similar arguments explain the onset of deformation in the new island of inversion around ^{64}Cr and its precursor, the superdeformed band of ^{68}Ni.

Acknowledgments

This work is partly supported by a grant of the MICINN (Spain) (FPA2011-29854), by the Nupnet network SARFEN (PRI-PIMMNUP-2011-1361) and by MINECO's (Spain) Programme "Centros de Excelencia Severo Ochoa" grant SEV-2012-0249.

References

[1] Elliott J P 1956 *Proc. R. Soc. London, Ser. A* **245** 128
[2] Zuker A P, Retamosa J, Poves A, Caurier E 1995 *Phys. Rev. C* **52** 1741(R)
[3] Arima A, Harvey M and Shimizu K 1969 *Phys. Lett. B* **30** 517
[4] Hecht K and Adler A 1969 *Nucl. Phys. A* **137** 129
[5] Lenzi S M, Nowacki F, Poves A and Sieja K 2010 *Phys. Rev. C* **82** 054301
[6] Dijon A *et al.* 2012 *Phys. Rev. C* **85** 031301.
[7] Recchia F *et al.* 2013 *Phys. Rev. C* **88** 041302(R)
[8] Suchyta S *et al.* 2014 *Phys. Rev. C* **89** 021301(R)
[9] Nowacki F private communication
[10] `http://www.nndc.bnl.gov/ensdf`
[11] Zuker A P, Poves A, Nowacki F and Lenzi S M arXiv:1404.0224
[12] Gade *et al.* 2005 *Phys. Rev. Lett.* **95** 022502
[13] Iwasaki H *et al.* 2014 *Phys. Rev. Lett.* **112** 142502
[14] Lemasson A *et al.* 2012 *Phys. Rev. C* **85** 041303
[15] de Angelis G. *et al.* 1997 *Phys. Lett. B* **415** 217
[16] Lister C J *et al.* 1994 *Phys. Rev. C* **42** 1191(R)
[17] Lister C J *et al.* 1987 *Phys. Rev. Lett.* **59** 1270
[18] Nara Singh B S private communication.

Journal of Physics: Conference Series **580** (2015) 012008 doi:10.1088/1742-6596/580/1/012008

Nuclear structure studies far east and out west on the nuclear chart

Alexandra Gade

National Superconducting Cyclotron Laboratory and Department of Physics and Astronomy, Michigan State University, East Lansing, MI 48824, USA

E-mail: gade@msu.edu

Abstract. The nuclear potential and resulting shell structure are well established for the valley of stability, however, dramatic modifications to the familiar ordering of single-particle orbitals in rare isotopes have been found: new shell gaps emerge, some conventional magic numbers are no longer valid, and theoretical models that extrapolate from stability to describe observables in exotic nuclei can fail. Current efforts in nuclear structure physics are aimed at unraveling the driving forces behind this structural evolution, which was found most dramatic in neutron-rich species. This manuscript will outline some of the recent efforts at NSCL aimed at shedding light on the evolution of nuclear structure around neutron number $N = 40$ in neutron-rich Ti isotopes and in neutron-deficient Sn nuclei approaching ^{100}Sn.

1. Introduction

One important goal in nuclear physics is the development of a predictive model of the atomic nucleus. The nuclear shell model is one of the cornerstones for a comprehensive understanding of nuclei. Concerted experimental and theoretical efforts are aimed at unfolding the driving forces behind departures from the well-established, traditional shell structure, which have been observed mostly in the shortest-lived nuclei beyond the valley of stability [1, 2, 3].

Among the robust driving forces of shell evolution identified in neutron-rich nuclei so far are the spin-isospin parts of the nucleon-nucleon NN interaction [4], in particular the monopole part of the tensor force [5]. The important roles of $3N$ forces and the central part have been pointed out recently as well [6, 7]. Toward the nucleon driplines, the density dependence of the spin-orbit force and couplings to the particle continuum were shown to become important [8]. Examples for the shell evolution in neutron-rich nuclei driven by the tensor force are the breakdown of $N = 20$ and $N = 28$ as a magic number in neutron-rich nuclei in the vicinity of ^{32}Mg and ^{42}Si, respectively, and the emergence of new shell gaps at $N = 32, 34$ in the neutron-rich fp shell [2, 3, 4, 5].

One important experimental task is the quantification of changes in the nuclear structure from the measurement of experimental observables that are calculable and that ultimately allow to discriminate between different theoretical approaches. The manuscript outlines how the results from in-beam γ-ray spectroscopy experiments performed at the National Superconducting Cyclotron Laboratory (NSCL) [9] on neutron-rich Ti isotopes [10] and neutron-deficient Sn nuclei informed large-scale nuclear shell-model calculations [11].

2. In-beam γ-ray spectroscopy at NSCL

Experimental approaches aimed at tracking the changes in nuclear shell structure are manifold. One way of probing nuclear structure quantitatively is the application of nuclear reactions that selectively probe specific degrees of freedom, e.g. single-particle or collective degrees of freedom. Coulomb excitation of nuclei has long been used to investigate collectivity. Intriguing possibilities arise along isotonic and isotopic chains of nuclei where the onset of collectivity or the migration of single-particle levels can be tracked consistently as function of the neutron number. One- and two-nucleon knockout reactions probe the single-particle degree of freedom with measured cross sections sensitive to the wavefunction overlaps between the projectile ground state and the knockout residue's final states.

At in-flight facilities, exotic medium-mass and light nuclei away from stability can be efficiently produced by fragmentation of stable, primary beams impinging upon stable targets at high beam velocities. The resulting secondary beams of rare isotopes are then available for experiments at velocities typically exceeding 30% of the speed of light. Well-established experimental techniques used for many decades to study the structure of stable nuclei are typically not applicable at the low beam intensities encountered for the shortest-lived rare isotopes. Powerful new experimental techniques have been developed to enable in-beam spectroscopy studies of fast rare-isotope beams with intensities several orders of magnitude less than what is required for measurements with traditional low-energy techniques. The intensities of rare-isotope beams are typically lower than stable-beam rates by orders of magnitude. The experimental technique of in-beam γ-ray spectroscopy compensates for the reduced intensities by employing thick reaction targets, taking advantage of the high beam velocity, and enabling experiments at luminosities comparable to stable-beam experiments but with projectile rates of up to factors of thousands less.

At NSCL, reactions with fast rare-isotope beams, for example intermediate-energy Coulomb excitation and nucleon knockout are induced by targets of several hundred mg/cm^2 thickness. The detection of γ rays is used for the identification of the reaction residue's final state in knockout, or to tag the inelastic process in Coulomb excitation measurements [3]. Since the γ-ray emission from the reaction residues occurs *in flight*, the γ-ray detection arrays have to be as granular as possible to enable an accurate angle-dependent, event-by-event reconstruction of the Doppler-shifted γ-ray transition energies into the rest frame of the emitter. The γ-ray spectroscopy of ^{60}Ti was performed with the advanced γ-ray tracking array GRETINA [12] and the Coulomb excitation of ^{104}Sn was executed with the high-efficiency, 192-element, CsI(Na) scintillator array CAESAR [13].

The choice of the target material is dictated by the desired reaction channel: Direct one- and two-nucleon knockout reactions [14] are typically induced by light targets, for example ^{9}Be or ^{12}C, while intermediate-energy Coulomb excitation [15, 16] requires high-Z targets like gold. For the measurements described in this manuscript, the reacted beam emerging from the target was identified with the large acceptance S800 magnetic spectrograph [17] to uniquely select the reaction residue of interest in the one-proton knockout leading to ^{60}Ti while a fixed opening-angle phoswich detector was used for particle identification in the Coulomb excitation of ^{104}Sn.

3. Towards ^{60}Ca – Nuclear structure of ^{60}Ti

Data for chains of magic isotopes and regions of rapid shell evolution both offer challenging tests of nuclear models, allowing changes in nuclear structure to be tracked as a function of isospin, and providing crucial benchmarks for calculations incorporating new structural effects. The Ca isotopes and the region of neutron-rich nuclei near $N = 40$, which are subject to rapid shell and shape evolution [18, 19, 20, 21], meet at the very neutron rich ^{60}Ca ($N = 2Z$). While even the existence of ^{60}Ca has not been established yet, spectroscopy of ^{60}Ti was in reach at NSCL. The measurements were enabled by the luminosity inherent to the use of fast beams from fragmentation [3] and the efficiency and spectral quality provided by GRETINA [12]. Figure 1 shows where ^{60}Ti and ^{60}Ca are located on the nuclear chart.

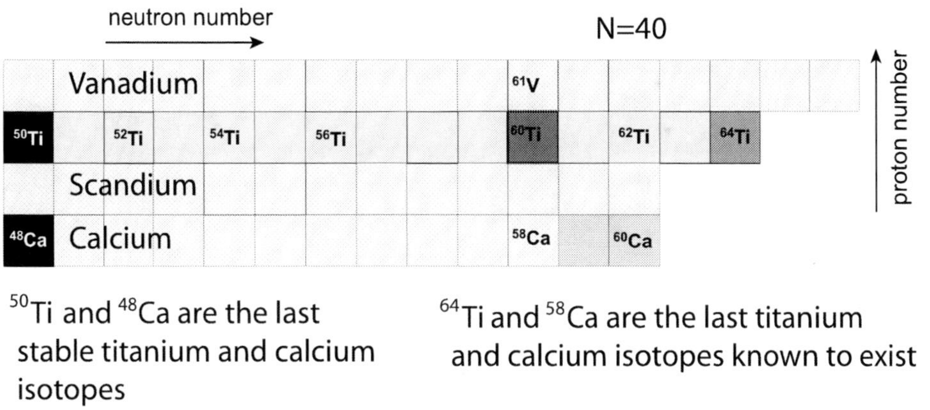

Figure 1. (Color online) The region of interest on the nuclear chart.

Excited states in the neutron-rich Ti isotopes were populated in the $^{9}\mathrm{Be}(^{61}\mathrm{V}, ^{58,60}\mathrm{Ti}+\gamma)X$ nucleon removal reactions at 90.0 MeV/u at NSCL. Typical on-target rates of 15 ^{61}V/s were obtained following the fragmentation of a ^{82}Se primary beam.

In the projectile-like reaction residue ^{60}Ti, a peak at about 860 keV is observed on top of very little background. One-proton knockout is a direct reaction with sensitivity to the details of the nuclear wavefunction and it quantifies the overlap in structure between the projectile ground state and the final states populated in the knockout residue [14]. The partial cross section to an excited final state is measured from the efficiency-corrected peak area relative to the number of knockout residues. Assuming that the peak structure in ^{60}Ti corresponds to a single transition then implies that 111(12)% of the knockout proceeds to the state depopulated by this 860-keV transition and that there is essentially no population of any other bound final state in ^{60}Ti. For a nucleus bound by more than 5 MeV, this scenario is highly unlikely. Indeed, the asymmetric peak shape of the structure at 860 keV supports the presence of a doublet of two transitions. Analysis as a doublet implies the presence of two coincident γ rays at 850(5) and 866(5) keV (see figure 2, inset), presumably corresponding to the $2_1^+ \to 0_1^+$ and $4_1^+ \to 2_1^+$ transitions in ^{60}Ti. GRETINA's $\gamma\gamma$ coincidence efficiency and the extremely low background in the spectra affords a $\gamma\gamma$ coincidence analysis consistent with a doublet of two coincident transitions (see [10] for details).

The partial cross sections to the 2_1^+ and 4_1^+ states were deduced from the feeding-corrected γ-ray intensities and the 0_1^+ population results from subtraction. The so-determined strength constitutes 29(12)%, 30(11)%, and 40(10)% population of the 0_1^+, 2_1^+, 4_1^+ states, which potentially includes unobserved feeding from higher-lying states that could not be observed due to limited statistics. The measured partial cross sections to final states in ^{60}Ti were compared to calculations using spectroscopic factors from GXPF1A [22] and LNPS [23] shell-model effective interactions, respectively. The comparison demonstrates that the neutron $g_{9/2}$ and $d_{5/2}$ orbitals that are included in the model space of LNPS and that are absent in GXPF1A are crucial for the description of the very neutron-rich ^{61}V ground state and ^{60}Ti final states populated in the knockout [10].

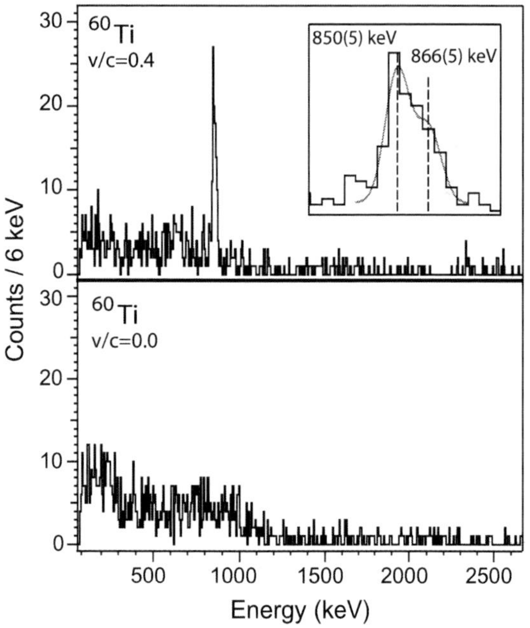

Figure 2. (Color online) Gamma-ray spectra of ^{60}Ti Doppler reconstructed into the rest frame of the emitter using the online signal decomposition of GRETINA (upper panel) and as detected in the laboratory frame (lower panel). The inset shows the doublet structure in more detail.

4. Collectivity in neutron-deficient Sn isotopes

Of particular importance for the development of nuclear models is experimental data that consistently tracks structural changes as function of isospin and varied binding, for example. The chain of Sn isotopes has been an important testing ground for nuclear theory as some spectroscopic data is available from $N = Z = 50$ ^{100}Sn [24] to ^{134}Sn [25]. In even-even nuclei, the electromagnetic $B(E2\uparrow) = B(E2; 0_1^+ \rightarrow 2_1^+)$ excitation strength is a measure of quadrupole collectivity, sensitive to the presence of shell gaps and nuclear deformation, among others. In the Sn isotopes, this transition strength has been reported from ^{104}Sn to ^{130}Sn. The trend is pronounced asymmetric with respect to mid-shell and not even the largest-scale shell-model calculations have been able to describe the evolution of transition strength across the isotopic chain without varying effective charges [11].

At NSCL, the experimental study of quadrupole collectivity in the neutron-deficient nucleus ^{104}Sn using intermediate-energy Coulomb excitation was possible. The $B(E2; 0_1^+ \rightarrow 2_1^+)$ value for the excitation of the first 2^+ state in ^{104}Sn was measured to be 0.180(37) e^2b^2 relative to the well-known $B(E2)$ value of ^{102}Cd. This result disagrees beyond 1-σ with another published measurement [26]. Contrary to the conclusions of [26], our result indicates that the most modern many-body calculations in the shell-model framework remain unable to describe the enhanced collectivity below mid-shell in Sn approaching $N = Z = 50$. In this picture, we attribute the enhanced collectivity to proton particle-hole configurations beyond the necessarily limited shell-model spaces and suggest the asymmetry of the $B(E2)$-value trend around mid-shell to originate from enhanced proton excitations across $Z = 50$ as $N = Z$ is approached [11].

The measurement was performed at the NSCL at Michigan State University. The secondary projectile beam containing ^{104}Sn and ^{102}Cd was produced by fragmentation of a 140 MeV/u ^{124}Xe stable beam on a 240 mg/cm^2 ^{9}Be production target and separated using a 150 mg/cm^2 Al wedge degrader in the A1900 fragment separator [27] and NSCL's Radio Frequency Fragment Separator (RFFS) [28].

With the known value of $B(E2\uparrow) = 0.28(3)$ e^2b^2 for ^{102}Cd [29], a transition strength of $B(E2\uparrow) = 0.180(37)$ e^2b^2 for ^{104}Sn was deduced. This normalization eliminates a variety of systematic uncertainties and absolute efficiencies in general. the result compares well with the work by Doornenbal [30] and disagrees within 1-σ with the recently published value from a relativistic Coulomb excitation measurement performed at GSI [26].

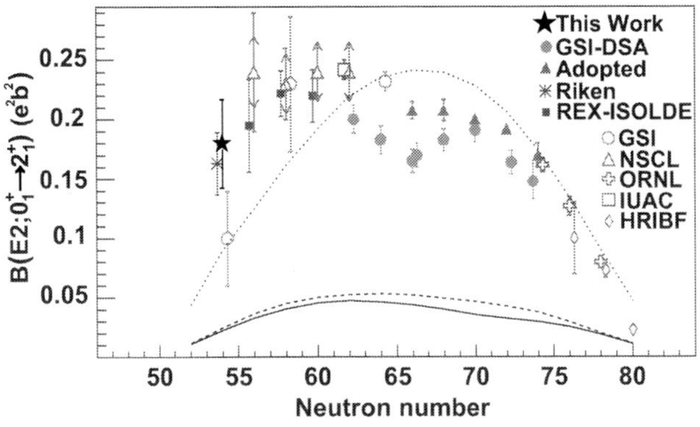

Figure 3. Measured $B(E2)$ values for the chain of Sn isotopes [31, 32, 33, 34, 35, 26, 36, 37, 38, 39, 40, 30]. Solid line: Large-scale shell-model calculations with NNLO interaction [41]. Dashed line: Large-scale shell-model calculations with N^{3}LO interaction [42]. The dotted line reproduces the *gds* model-space calculation shown in [34]. (Figure adapted from [11])

First, large-scale shell-model calculations using an inert $Z = 50$ core were performed to describe the data across the Sn isotopic chain. In figure 3, the results for the newly optimized NNLO interaction [41], together with those obtained with the N^{3}LO interaction [42] are presented. The latter interaction also gives the overall best reproduction of the excited states and binding energies. Thus, unless one adopts a phenomenological adjustment of the effective neutron charges, see for example Ref. [43], theory based on an inert $Z = 50$ core fails to describe the $B(E2\uparrow)$ strengths. If one increases the neutron effective charge to $1.0e$, then the $B(E2)$ values are increased by a factor of four and come closer to the data but do not describe the asymmetry toward neutron-deficient Sn [11]. In [34], the model space was increased to allow up to four protons to be excited from the $0g_{9/2}$ orbital to $1d$, $2s_{1/2}$ and $0g_{7/2}$. The results from this calculation, with standard effective charges $1.5e$ for protons and $0.5e$ for neutrons, are also shown in figure 3. Overall, the data is much better described, except that the extended calculation is symmetric around the middle, whereas experiment shows an asymmetry with an enhancement at the neutron-deficient end. This indicates that proton excitations across $Z = 50$ are critical for the description of collectivity. It is argued, as was in [36], that the increased $B(E2)$ strength toward the neutron-deficient nuclei is due to α correlations toward $N = Z$ outside of the necessarily truncated shell-model spaces [11].

5. Summary

First nuclear structure information on ^{60}Ti was obtained at NSCL thanks to the spectral quality of GRETINA and the ^{61}V projectile beam produced by fragmentation of the unique ^{82}Se primary beam. The first 2^+ state of ^{60}Ti, at an energy of 850(5) keV, is almost twice as high as the excitation energy of the corresponding 2^+ level in the $N = 38$ isotone ^{62}Cr, indicating a steep decrease in collectivity. The data on ^{60}Ti are consistent with a shell-model prediction using the LNPS effective interaction which allows for the largest neutron model space yet, while they disagree with calculations restricted to the neutron fp shell. ^{60}Ti represents an important benchmark, being one of the most neutron-rich systems from which to extrapolate towards ^{60}Ca, a nucleus with its structure closely tied to the location of the neutron drip-line in the semi-magic Ca isotopic chain [10].

On the neutron-deficient side of the nuclear chart, we have determined the ^{104}Sn $B(E2;0^+ \to 2^+_1)$ strength from intermediate-energy Coulomb excitation. Unlike the conclusion of [26], the departure from large-scale shell-model calculations persists for the neutron-deficient Sn isotopes approaching ^{100}Sn. It is proposed that this deviation from shell model to originates from the interplay of proton particle-hole configurations beyond the necessarily limited shell-model spaces and it is argued that their effect on quadrupole collectivity is a common phenomenon along proton-magic isotope chains [11].

Acknowledgments
GRETINA was funded by the DOE, Office of Science. Operation of the array at NSCL was supported by NSF under Cooperative Agreement PHY-1102511 (NSCL) and DOE under Grant DE-AC02-05CH11231 (LBNL). Thank you, Aldo, for inspiration and your contagious passion for nuclear science.

References
[1] Brown BA 2001 *Prog. Part. Nucl. Phys.* **47** 517
[2] Sorlin O and Porquet MG 2008 *Prog. Part. Nucl. Phys.* **61** 602
[3] Gade A and Glasmacher T 2008 *Prog. Part. Nucl. Phys.* **60** 161
[4] Otsuka T et al. 2001 *Phys. Rev. Lett.* **87** 082502
[5] Otsuka T et al. 2005 *Phys. Rev. Lett.* **95** 232502
[6] Otsuka T et al. 2010 *Phys. Rev. Lett.* **104** 012501
[7] Otsuka T et al. 2010 *Phys. Rev. Lett.* **105** 032501
[8] Dobaczewski J et al. 2007 *Prog. Part. and Nucl. Phys.* **59** 432
[9] Gade A and Gelbke CK 2010 *Scholarpedia* **5(1)** 9651
[10] Gade A et al. 2014 *Phys. Rev. Lett.* **112** 112503
[11] Bader VM et al. 2013 *Phys. Rev. C* **88** 051301(R)
[12] Paschalis S et al. 2013 *Nucl. Instrum. Methods Phys. Res., Sect. A* **709** 44
[13] Weisshaar D et al. 2010 *Nucl. Instrum. Methods Phys. Res., Sect. A* **624** 615
[14] Hansen PG and Tostevin JA 2003 *Annu. Rev. Nucl. Part. Sci.* **53** 221
[15] Glasmacher T 1998 *Annu. Rev. Nucl. Part. Sci.* **48** 1
[16] Motobayashi T et al. 1995 *Phys. Lett.* B **346** 9
[17] Bazin D et al. 2003 *Nucl. Instrum. Methods Phys. Res., Sect. B* **204** 629
[18] Sorlin O et al. 2003 *Eur. Phys. J. A* **16** 55
[19] Adrich P et al. 2008 *Phys. Rev. C* **77** 054306
[20] Gade A et al. 2010 *Phys. Rev. C* **81** 051304(R)
[21] Carpenter M et al. 2013 *Phys. Rev. C* **87** 041305(R)
[22] Honma M et al. 2005 *Eur. Phys. J. A* **25** 499
[23] Lenzi S et al. 2010 *Phys. Rev. C* **82** 054301
[24] Hinke CB et al. 2012 *Nature* **486** 341
[25] Korgul A et al. 2000 *Eur. Phys. J. A* **7** 167
[26] Guastalla G et al. 2013 *Phys. Rev. Lett.* **110** 172501
[27] Morrissey DJ et al. 2003 *Nucl. Instrum. Methods Phys. Res., Sect. B* **204** 90
[28] Bazin D et al. 2009 *Nucl. Instrum. Methods Phys. Res., Sect. A* **606** 314
[29] Ekström A et al. 2009 *Phys. Rev. C* **80** 054302
[30] Doornenbal P et al. 2013 arXiv:1305.2877[nucl-ex]
[31] Cederkäll J et al. 2007 *Phys. Rev. Lett.* **98** 17
[32] Ekström A et al. 2008 *Phys. Rev. Lett.* **101** 1
[33] Jungclaus A et al. 2011 *Phys. Lett. B* **695** 110
[34] Banu A et al. 2005 *Phys. Rev. C* **72** 061305
[35] Doornenbal P et al. 2008 *Phys. Rev. C* **78** 031303
[36] Vaman C et al. 2007 *Phys. Rev. Lett.* **99** 162501
[37] BNL Evaluated Nuclear Structure Data File (ENSDF), http://www.nndc.bnl.gov
[38] Allmond JM et al. 2011 *Phys. Rev. C* **84** 061303
[39] Kumar R et al. 2010 *Phys. Rev. C* **81** 024306
[40] Radford DC et al. 2005 *Nucl. Phys. A* **757** 264
[41] Ekström A et al. 2013 *Phys. Rev. Lett.* **110** 192502
[42] Entem RD and Machleidt R 2003 *Phys. Rev. C* **68** 041001
[43] Bäck T et al. 2013 *Phys. Rev. C* **87** 031306(R)

Recent Advances in the Application of the Shell Model Monte Carlo Approach to Nuclei

Y. Alhassid[1], M. Bonett-Matiz[1], A. Mukherjee[2], H. Nakada[3] and C. Özen[4]

[1]Center for Theoretical Physics, Sloane Physics Laboratory,Yale University, New Haven, CT 06520, USA
[2]ECT*, Villa Tambosi, I-38123 Villazzano, Trento, Italy
[3]Department of Physics, Graduate School of Science, Chiba University, Inage, Chiba 263-8522, Japan
[4]Faculty of Engineering and Natural Sciences, Kadir Has University, Istanbul 34083, Turkey

E-mail: yoram.alhassid@yale.edu

Abstract. The shell model Monte Carlo (SMMC) method is a powerful technique for calculating the statistical and collective properties of nuclei in the presence of correlations in model spaces that are many orders of magnitude larger than those that can be treated by conventional diagonalization methods. We review recent advances in the development and application of SMMC to mid-mass and heavy nuclei.

1. Introduction

Most microscopic studies of heavier nuclei are based on mean-field methods such as density functional theory. However, important correlation effects beyond the mean field can be missed. The configuration-interaction (CI) shell model approach accounts for correlations and shell effects but conventional diagonalization methods are limited by the size of the many-particle model space. The auxiliary-field Monte Carlo (AFMC) method enables calculation of thermal observables and ground state properties in model spaces that are many orders of magnitude larger than model spaces that can be treated by conventional diagonalization methods. In nuclear physics this method is also known as the shell model Monte Carlo (SMMC) method [1, 2]. SMMC has proven to be a powerful method in calculating statistical and collective properties of nuclei [3, 4, 5, 6, 7].

Here we review recent advances in the development and application of SMMC to nuclei. In particular, we describe a method to circumvent the sign problem that arises from the projection on an odd number of particles and its application to calculate pairing gaps in mid-mass nuclei and level densities of odd-mass isotopes. We also discuss the application of SMMC to describe the microscopic emergence of collectivity in heavy rare-earth nuclei and present results for the collective enhancement factors of level densities in such nuclei.

2. Auxiliary-field Monte Carlo (AFMC) method

The AFMC method utilizes the Hubbard-Stratonovich transformation [8], in which the imaginary-time propagator $e^{-\beta H}$ of a nucleus that is described by a Hamiltonian H at inverse temperature $\beta = 1/T$ is represented by a functional integral of one-body propagators of non-interacting nucleons over external auxiliary fields

$$e^{-\beta H} = \int \mathcal{D}[\sigma] G_\sigma U_\sigma \,. \tag{1}$$

Here G_σ is a Gaussian weight and U_σ is the propagator of non-interacting nucleons moving in time-dependent auxiliary fields $\sigma(\tau)$. The thermal expectation value of an observable O is

$$\langle O \rangle_T = \frac{\mathrm{Tr}(e^{-\beta H} O)}{\mathrm{Tr} e^{-\beta H}} = \frac{\int D[\sigma] W_\sigma \Phi_\sigma \langle O \rangle_\sigma}{\int D[\sigma] W_\sigma \Phi_\sigma} \,, \tag{2}$$

where $W_\sigma = G_\sigma |\,\mathrm{Tr}\, U_\sigma|$ is a positive-definite weight function and $\Phi_\sigma = \mathrm{Tr}\, U_\sigma / |\,\mathrm{Tr}\, U_\sigma|$ is the Monte Carlo sign function. The quantity $\langle O \rangle_\sigma = \mathrm{Tr}\,(O U_\sigma)/\mathrm{Tr}\, U_\sigma$ is the thermal expectation value of the observable for a given configuration σ of the auxiliary fields. In SMMC, we choose samples σ_k distributed according to W_σ and estimate the observables from $\langle O \rangle_T \approx \sum_k \langle O \rangle_{\sigma_k} \Phi_{\sigma_k} / \sum_k \Phi_{\sigma_k}$. AFMC often suffers from a sign problem at low temperatures. However, there are classes of good-sign interactions that are free of the sign problem and are composed of the dominant collective components of effective nuclear interactions [9]. Small bad-sign components can be treated by the extrapolation method of Ref. [2].

3. Circumventing the odd-particle sign problem

Applications of SMMC to odd-even and odd-odd nuclei have been hampered by a sign problem that originates from the projection on a odd number of particles. Such a sign problem occurs at low temperatures even for good-sign interactions and makes it difficult to estimate accurately the ground-state energy of an odd-particle system. A breakthrough was a method we introduced to calculate accurately the ground-state energy of an odd-particle system that circumvents this sign problem [10]. It is based on the scalar imaginary-time single-particle Green's functions

$$G_\nu(\tau) = \sum_m \langle \mathcal{T} a_{\nu m}(\tau) a^\dagger_{\nu m}(0) \rangle \,. \tag{3}$$

Here $\mathcal{T}$ denotes time ordering, $\nu = (nlj)$ denotes the radial quantum number n, orbital angular momentum l and total spin j of the single-particle levels, and m is the magnetic quantum number of j. In an asymptotic regime of sufficiently large τ but $\tau \ll \beta$, the Green's functions of an even-even nucleus A have the form $G_\nu(\tau) \sim e^{-\beta[E_j(A\pm1) - E_{\mathrm{g.s.}}(A)]|\tau|}$, where $E_j(A \pm 1)$ is the lowest energy of spin $J = j$ in the neighboring even-odd nuclei $A \pm 1$ and $E_{\mathrm{g.s.}}$ is the ground-state energy of the nucleus A. Thus $E_j(A \pm 1)$ can be determined from the slope of the calculated $\ln G_\nu$ and the ground-state energy of the even-even nucleus. The ground-state energy of the even-odd nucleus is then found by minimizing $E_j(A \pm 1)$ over j.

We used this method to calculate neutron pairing gaps in mid-mass nuclei from the second-order energy difference in neutron number $\Delta_n = (-)^N [E(Z, N+1) + E(Z, N-1) - 2E(Z, N)]/2$, where $E(Z, N)$ is the ground-state energy of the nucleus with Z protons and N neutrons. Fig. 1 shows neutron pairing gaps for mid-mass isotopes. The calculations were carried out in the complete $fpg_{9/2}$ shell with the interaction of Ref. [3]. The SMMC results (solid circles) are in overall good agreement with the experimental values (open circles).

4. Level densities

SMMC has been successful in calculating nuclear state densities $\rho(E_x)$ versus excitation energy E_x in the presence of correlations. The density calculated in SMMC is the *state* density, in which each level with spin J is counted $2J + 1$ times. However, often the density measured in the experiments is the *level* density, in which each level with spin J is counted once.

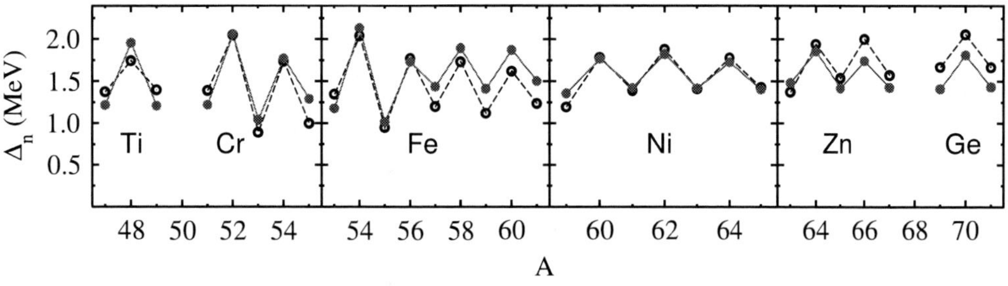

Figure 1. Neutron pairing gaps Δ_n as a function of mass number A in families of $fpg_{9/2}$-shell isotopes. The gaps calculated in SMMC (solid circles) are compared with the experimental gaps (open circles). Adapted from Ref. [10].

We recently introduced a method to calculate the level density directly in SMMC [11]. Denoting by ρ_M the level density at given value M of the angular momentum component J_z, the level density $\tilde{\rho}$ is given by $\tilde{\rho} = \rho_{M=0}$ for even-mass nuclei and by $\tilde{\rho} = \rho_{M=1/2}$ for odd-mass nuclei. J_z projection can be carried out exactly in SMMC as discussed in Ref. [5].

Level densities were recently measured by proton evaporation spectra in a family of nickel isotopes $^{59-64}$Ni [12]. We calculated these level densities [13] within the $fpg_{9/2}$ shell using the Hamiltonian of Ref. [3]. The calculation of the ground-state energies of the odd-mass nickel isotopes, and hence their excitation energies, was made possible by the method discussed in Sec. 3. Fig. 2 shows the level densities of $^{59-64}$Ni. The SMMC results are in close agreement with the level densities determined from various experimental data sets.

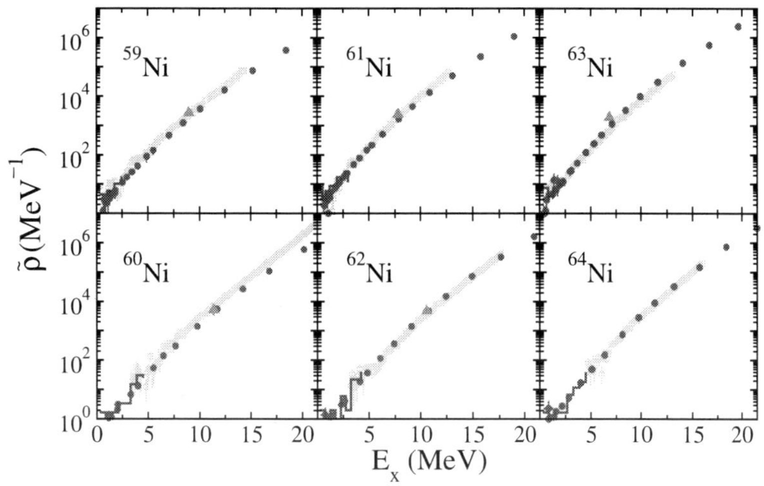

Figure 2. Level densities $\tilde{\rho}$ in a family of nickel isotopes. The SMMC level densities (solid circles) are compared with level densities determined from various experimental data sets: proton evaporation spectra (open squares and quasi-continuous lines), level counting data (histograms) and neutron resonance data when available (triangles). Taken from Ref. [13].

5. Microscopic emergence of collectivity in heavy nuclei

Heavy nuclei are known to exhibit various types of collectivity. Typically, nuclei near shell closure are spherical and display vibrational collectivity, while mid-shell nuclei are strongly deformed and display rotational collectivity. Different types of collectivity are well described by empirical models but much less is understood microscopically. Here we use SMMC to describe the emergence of collectivity in heavy nuclei within the framework of the CI shell model approach.

5.1. Thermal signatures of collectivity

A given type of collectivity is usually identified by its associated spectrum. While SMMC is a powerful method for calculating thermal and ground-state properties in very large model spaces, it does not provide detailed spectroscopic information. To overcome this difficulty, we have identified a thermal observable whose low-temperature behavior is sensitive to the type of collectivity. Such an observable is $\langle \mathbf{J}^2 \rangle_T$, where $\mathbf{J}$ is the total angular momentum. Assuming an even-even nucleus with a rotational or a vibrational ground-state band and an excitation energy E_{2^+} of the first 2^+ state, we have [6, 7]

$$\langle \mathbf{J}^2 \rangle_T \approx \begin{cases} 30\dfrac{e^{-E_{2^+}/T}}{\left(1-e^{-E_{2^+}/T}\right)^2} & \text{vibrational band} \\ \dfrac{6}{E_{2^+}}T & \text{rotational band} \end{cases} . \tag{4}$$

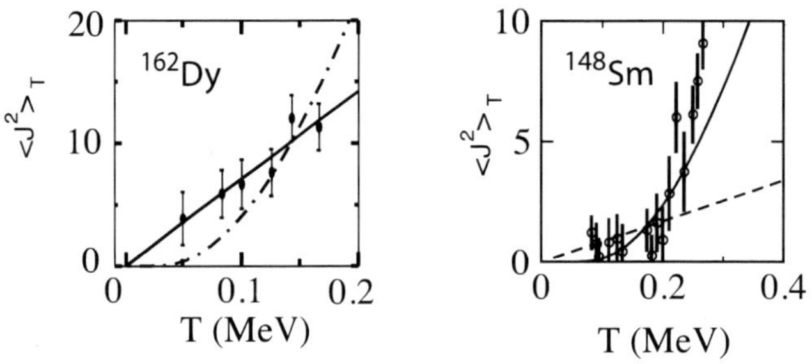

Figure 3. Left: $\langle \mathbf{J}^2 \rangle_T$ in ^{162}Dy. The SMMC results (solid circles) are compared with fits to the rotational model (solid line) and vibrational model (dashed-dotted line). Adapted from Ref. [6]. Right: $\langle \mathbf{J}^2 \rangle_T$ in ^{148}Sm. The SMMC results (open circles) are compared with fits to the vibrational model (solid line) and rotational model (dashed line).

We carried out calculations for rare-earth nuclei using the $50-82$ shell plus $1f_{7/2}$ for protons, and the $82-126$ shell plus $0h_{11/2}, 1g_{9/2}$ for neutrons. The Hamiltonians we used are given in Refs. [6, 7]. The left panel of Fig. 3 shows $\langle \mathbf{J}^2 \rangle_T$ for the deformed rare-earth nucleus ^{162}Dy. The SMMC results are compared with fits to both the rotational and vibrational models. The agreement of the SMMC results with the rotational model confirms that we are able to describe the rotational character of ^{162}Dy in the framework of a truncated spherical shell model approach. The fit to the rotational formula gives $E_{2^+} = 84.5 \pm 8.9$ keV, in agreement with the experimental result of $E_{2^+} = 80.6$ keV. The right panel of Fig. 3 shows $\langle \mathbf{J}^2 \rangle_T$ for the spherical rare-earth nucleus ^{148}Sm. Here the SMMC results follow closely the vibrational formula fit with $E_{2^+} = 538 \pm 31$ keV, in agreement with the experimental value of $E_{2^+} = 550$ keV.

5.2. The crossover from vibrational to rotational collectivity

Heavy nuclei are known to exhibit a crossover from vibrational to rotational character as the number of neutrons increases from shell closure towards mid-shell. In a mean-field approximation, this is described by a phase transition from spherical to deformed nuclei. We have studied two such families of rare-earth isotopes, $^{148-154}$Sm and and $^{144-152}$Nd [7].

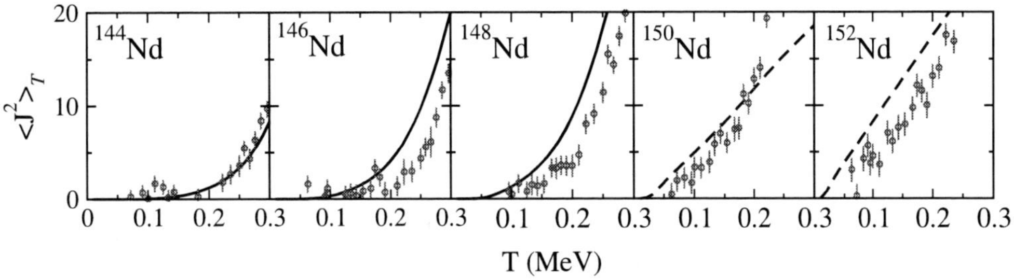

Figure 4. $\langle \mathbf{J}^2 \rangle_T$ as a function of temperature T for a family of even neodymium isotopes. The solid lines describe the experimental values extracted using Eq. (5). For $^{150-152}$Nd we only use the discrete levels in Eq. (5) (dashed lines). Adapted from Ref. [7].

Fig. 4 shows $\langle \mathbf{J}^2 \rangle_T$ for a family of even neodymium isotopes. The open circles are the SMMC results. We observe a crossover from a soft response to temperature in $^{144-146}$Nd, typical of a vibrational nucleus, to a rigid response in ^{152}Nd, typical of a rotational nucleus. The solid lines are extracted from experimental results using

$$\langle \mathbf{J}^2 \rangle_T = \frac{1}{Z(T)} \left(\sum_i^N J_i(J_i+1)(2J_i+1)e^{-E_i/T} + \int_{E_N}^{\infty} dE_x \, \rho(E_x) \, \langle \mathbf{J}^2 \rangle_{E_x} \, e^{-E_x/T} \right) , \quad (5)$$

where $Z(T) = \sum_i^N (2J_i+1)e^{-E_i/T} + \int_{E_N}^{\infty} dE_x \rho(E_x) e^{-E_x/T}$ is the corresponding experimental partition function. Here E_i are the measured low-lying energy levels with spin J_i, which we assume to be complete up to to certain threshold energy E_N. Above the threshold energy, we use a backshifted Bethe formula for the level density $\rho(E_x)$ whose parameters are determined from a fit to level counting data at low energy and neutron resonance data at the neutron resonance threshold. We use in Eq. (5) the spin cutoff model estimate $\langle \mathbf{J}^2 \rangle_{E_x} = 3\sigma^2$, where $\sigma^2 = IT/\hbar^2$ is calculated assuming a rigid-body moment of inertia I. We observe an overall good agreement between the SMMC values of $\langle \mathbf{J}^2 \rangle_T$ and the experimentally determined values.

6. Collective enhancement factors of level densities in heavy nuclei

SMMC state densities in rare-earth nuclei were found to be in good agreement with experimental data [14]. Collective states in these nuclei lead to enhancement of their state density, which is described by a collective enhancement factor K. The decay of K with excitation energy is one of the least understood topics in the modeling of level densities [15], and it is often parameterized by empirical formulas. We define a collective enhancement factor by $K = \rho_{\rm SMMC}/\rho_{\rm HFB}$, where $\rho_{\rm SMMC}$ is the total SMMC state density and $\rho_{\rm HFB}$ is the level density calculated in the finite-temperature Hartree-Fock-Bogoliubov (HFB) approximation. The latter describes the density of intrinsic states and thus the above ratio is a measure of the enhancement in the number of states due to collective states that are built on top of the intrinsic states.

Fig. 5 shows the calculated enhancement factor K as a function of excitation energy E_x for the even neodymium isotopes $^{144-152}$Nd. We also show by arrows the energies of the various thermal phase transitions that occur in the HFB approximation. The thin arrows correspond to the pairing transitions (for both protons and neutrons), while the thick arrows describe the shape transitions. The spherical nuclei $^{144-146}$Nd do not support rotational bands, so the enhancement factor K is due to vibrational states alone. We observe that this vibrational enhancement factor decays to ~ 1 at excitation energies above the proton and neutron pairing transitions. The heavier neodymium isotopes $^{148-152}$Nd are deformed in their ground state. We observe that K has a local minimum above the pairing transitions, which we attribute to the interplay between

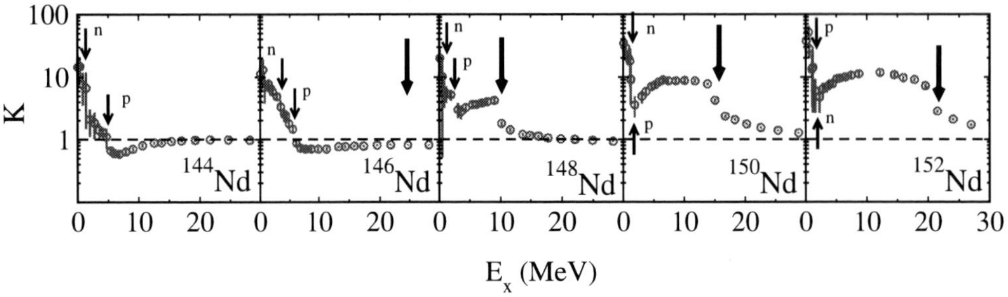

Figure 5. Collective enhancement factor K (open circles) of the even neodymium isotopes $^{144-152}$Nd versus excitation energy E_x (see text for the definition of K). The thin arrows correspond to the proton and neutron pairing transitions, while the thick arrows describe the shape transitions. Adapted from Ref. [14].

the decay of vibrational collectivity and the enhancement due to rotational collectivity. The plateau that follows the local minimum describes the enhancement from rotational states. This rotational enhancement decays in the vicinity of the shape phase transition.

7. Conclusion

The SMMC is a powerful method for calculating statistical and collective properties of nuclei in the framework of the CI shell model in very large model spaces. We reviewed a recent method to circumvent the odd-particle sign problem and its application to the calculation of pairing gaps and level densities of odd-mass nuclei. We discussed recent applications of SMMC to heavy rare-earth nuclei and in particular the microscopic emergence of collectivity in such nuclei.

Acknowledgments

It is a pleasure to dedicate this article to Aldo Covello in recognition of his important contributions to nuclear structure physics. This work was supported in part by the DOE grant No. DE-FG-0291-ER-40608, by the JSPS Grant-in-Aid for Scientific Research (C) No. 25400245, and by the Turkish Science and Research Council (TÜBİTAK) grant No. ARDEB-1001-112T973. Computational cycles were provided by the NERSC high performance computing facility and by the High Performance Computing Center at Yale University.

References

[1] Lang G H, Johnson C W, Koonin S E and Ormand W E 1993 *Phys. Rev.* C **48** 1518
[2] Alhassid Y, Dean D J, Koonin S E, Lang G H, and Ormand W E 1994 *Phys. Rev. Lett.* **72** 613
[3] Nakada H and Alhassid Y 1997 *Phys. Rev. Lett.* **79** 2939
[4] Alhassid Y, Liu S and Nakada H 1999 *Phys. Rev. Lett.* **83** 4265
[5] Alhassid Y, Liu S and Nakada H 2007 *Phys. Rev. Lett.* **99** 162504
[6] Alhassid Y, Fang L and Nakada H 2008 *Phys. Rev. Lett.* **101** 082501
[7] Özen C, Alhassid Y and Nakada H 2013 *Phys. Rev. Lett.* **110** 042502
[8] Hubbard J 1959 *Phys. Rev. Lett.* **3** 77; Stratonovich R L 1957 *Dokl. Akad. Nauk. S.S.S.R.* **115** 1097
[9] Dufour M and Zuker A P 1996 *Phys. Rev.* C **54** 1641
[10] Mukherjee A and Alhassid Y 2012 *Phys. Rev. Lett.* **109** 032503
[11] Alhassid Y, Bonett-Matiz M, Liu S and Nakada H arXiv:1304.7258
[12] Voinov A V, Grimes S M, Brune C R, Massey T and Schiller A 2012 *EPJ Web of Conferences* **21** 05002; Voinov A V (private communication)
[13] Bonett-Matiz M, Mukherjee A and Alhassid Y 2013 *Phys. Rev.* C Rapid Comm. **88** 011302
[14] Alhassid Y, Özen C and Nakada H 2014 *Nuclear Data Sheets* **118** 233
[15] Capote R *et al.* 2009 *Nuclear Data Sheets* **110** 3107, and references therein

Octupole transitions in the ^{208}Pb region

Zs Podolyák[1], C M Shand[1], E Wilson[1], B A Brown[2], H Grawe[3], C J Chiara[5,4], S Zhu[4], B Fornal[6], R V F. Janssens[4], M Bowry[1], M Bunce[1], M P Carpenter[4], N Cieplicka[6], A Y Deo[7,8], G D Dracoulis[9], C R Hoffman[4], R S Kempley[1], F G Kondev[10], G J Lane[9], T Lauritsen[4], G Lotay[1], M W Reed[9], P H Regan[1,11], C Rodríguez Triguero[12], D Seweryniak[4], B Szpak[6], P M Walker[1]

[1] Department of Physics, University of Surrey, Guildford, GU2 7XH, UK
[2] Michigan State University, East Lansing, Michigan 48824-1321, USA
[3] GSI Helmholtzzentrum für Schwerionenforschung GmbH, D-64291 Darmstadt, Germany
[4] Physics Division, Argonne National Laboratory, Argonne, IL 60439, USA
[5] Department of Chemistry and Biochemistry, University of Maryland, College Park, MD 20742, USA
[6] H. Niewodniczański Institute of Nuclear Physics, PL-31342 Kraków, Poland
[7] Department of Physics, University of Massachusetts Lowell, Lowell, MA 01854, USA
[8] Department of Physics, Indian Institute of Technology Roorkee, Roorkee - 247667, India
[9] Department of Nuclear Physics, Research School of Physics and Engineering, Australian National University, Canberra, ACT 0200, Australia
[10] Nuclear Engineering Division, Argonne National Laboratory, Argonne, IL 60439, USA
[11] National Physical Laboratory, Hampton Road, Teddington, Middlesex, TW11 0LW, UK
[12] School of Computing, Engineering and Mathematics, University of Brighton, Brighton, BN2 4GL, UK

E-mail: z.podolyak@surrey.ac.uk

Abstract. The ^{208}Pb region is characterised by the existence of collective octupole states. Here we populated such states in ^{208}Pb + ^{208}Pb deep-inelastic reactions. γ-ray angular distribution measurements were used to infer the octupole character of several E3 transitions. The octupole character of the 2318 keV $17^- \rightarrow 14^+$ in ^{208}Pb, 2485 keV $19/2^- \rightarrow 13/2^+$ in ^{207}Pb, 2419 keV $15/2^- \rightarrow 9/2^+$ in ^{209}Pb and 2465 keV $17/2^+ \rightarrow 11/2^-$ in ^{207}Tl transitions was demonstrated for the first time. In addition, shell model calculations were performed using two different sets of two-body matrix elements. Their predictions were compared with emphasis on collective octupole states.

1. Introduction

The even-even nucleus ^{208}Pb is unusual as its first excited state, at 2615 keV, has spin-parity 3^-. This state decays to the ground-state by a collective E3 octupole transition with a strength of B(E3)=33.8(6) W.u. [1]. The 3^- collective state arises due to the large number of $\Delta l = \Delta j$=3 orbital combinations around Z=82 and N=126. Collective octupole transitions can be also found in the nearby nuclei, often built on single/multi particle/hole states. The excitations formed in this way are often yrast, therefore accessible experimentally [2]. For some transitions, their octupole character was determined directly or indirectly (from knowing the spin-parities of initial and final states). The character of other octupole transitions were proposed based purely on theoretical considerations. The aim of this contribution is to demonstrate the octupole character of several such transitions, using angular distribution analysis. In terms of the shell model, octupole states have a very mixed wave function. We discuss the theoretical description of octupole states within the shell model.

2. Experiment

A ^{208}Pb beam of 1446 MeV energy bombarded a 75 mg/cm^2 thick ^{208}Pb target. Nuclei around ^{208}Pb were populated by transferring few nucleons from target or beam. All reaction products were stopped in the thick target. The beam current was ~0.25 particle-nA on average, and the experiment ran for ~7 days. The γ rays were detected with the Gammasphere array [3], consisting of 101 HPGe detectors and their BGO anti-Compton shields. Tantalum, cadmium and copper absorbers were used in front of the detectors in order to reduce the dominant Pb X-ray yields. The trigger required three coincident γ-rays detected within 2 μs. Further details on the experimental conditions have been given in our previous conference papers [4, 5].

The data were sorted into $\gamma\gamma\gamma$ cubes with different time conditions. In addition, the data were sorted in two-dimensional $\gamma\gamma$ matrices for angular correlation and angular distribution studies.

3. Results

The high energy part of the recorded γ-ray spectrum is shown in figure 1. Several of the transitions visible in the 2.3-2.75 MeV region were previously identified/suggested to be E3 transitions [6]. Those discussed in the present paper are labelled.

In order to prove/establish the octupole character of the transitions angular distribution analysis was performed. As these transitions are strong, usually no gates on particular transitions were needed. The γ-ray intensities were fitted with the $W(\theta) = a_0 + a_2P_2(cos\theta) = a_0[1 + a_2/a_0P_2(cos\theta)]$ function. $P_2(cos\theta) = 0.5[3cos^2(\theta) - 1]$ is the second order Legendre polynomial and θ is the angle between the beam and the emitted γ ray. The a_2/a_0 parameter is negative for stretched dipole transitions and positive for stretched quadrupole and octupole ones. a_2/a_0 is larger for octupole transitions. The exact value of a_2/a_0 depends on the amount of spin alignment generated in the reaction. Therefore we determined it for well known E3 and E2 transitions in ^{208}Pb and other nuclei. Based on this, we expect an a_2/a_0 ~0.36 for E3 and a_2/a_0 ~0.22 for E2 transitions (see below).

The angular distribution results are shown in figure 2. Here we discuss the nuclei individually:

^{208}*Pb:* The E3 character of the 2615 keV transition with $B(E3; 3^- \to 0^+)$=33.8(6) W.u. is well known [1]. Similarly, the 4611 keV state with $I^\pi = 8^+$ is depopulated by a 1413 keV transition with a strength of $B(E3; 8^+ \to 5^-)$=12.6(20) W.u. We use these cases, together with the 583 keV $5^- \to 3^-$ E2 transition, to calibrate our a_2/a_0 coefficients. We note that our angular correlations analysis, which can be directly compared with theory agree with the well known multipolarities of these transitions, as we have shown in [4]. In addition a (17^+) state at 9062 keV is known to decay into a (14^-) state by a 2318 keV transition [1, 7]. The spin-parity assignment was based purely on theoretical considerations. Our angular distribution proves the octupole character of this transition.

^{206}*Pb:* The 15^- state at 6431 keV is depopulated by a stretched E3 2403 keV transition [8]. Therefore this transition is also used as a test case.

^{207}*Pb:* The $13/2^+$ state at 1633 keV is based on the neutron-hole $\nu i_{13/2}$. According to the latest Nuclear Data Sheets evaluation from 2011 [9], this state is populated by a 2485 keV transition from the 4118 keV $(15/2^-)$ $\nu j_{15/2}$ state. In contrast, this state has been suggested in 1992 as $(19/2^-)$ with $\nu i^{-1}_{13/2} \times 3^-$ character by Schramm et al. [10]. The spin-parity assignment, and indeed the assignment of the transition to the ^{207}Pb nucleus, was based on comparison with particle-octupole vibration coupling calculations. Here, our measurement settles the controversy. The 2485 keV transition has an a_2/a_0=0.36(2) coefficent, therefore it has octupole character. The E1 character, suggestesd by the compilation, can be clearly disregarded.

^{209}*Pb:* The 1423 keV $15/2^-$ state has a configuration $\nu j_{15/2}$ [11] and decays by an E3 transition to the ground-state. This case is also a calibration point for the a_2/a_0 coefficient. Rejmund et al. [6] proposed a 2419 keV transition populating the $15/2^-$ state. Based on theoretical expectations, they assigned $(21/2^+)$ to this new 3842 keV state. The angular distribution of the 2419 keV γ ray, as measured in the present experiment, confirms its octupole character.

^{207}*Tl:* The $11/2^-$ $\pi h^{-1}_{11/2}$ state at 1348 keV is isomeric. According to the 2011 Nuclear Data Sheets evaluations [9] it is populated by a 2465 keV transition from a state at 3813 keV. The compilation does not suggest any spin-parity assigment for this state. In contrast, the authors who observed this state in 2000 interpreted it as $17/2^+$ [6] based on the particle-vibration coupling model. Our angular distribution measurement of the 2465 keV transition establishes its octupole character.

^{209}*Bi:* The ground state is $9/2^-$ based on the $\pi h_{9/2}$ orbital. The $15/2^+$ state at 2741 keV has $\pi h_{9/2} \times 3^-$ character, and it decays by an E3 transition with strength of 25(4) W.u. [11]. Therefore this case is also a calibration point for the a_2/a_0 coefficient.

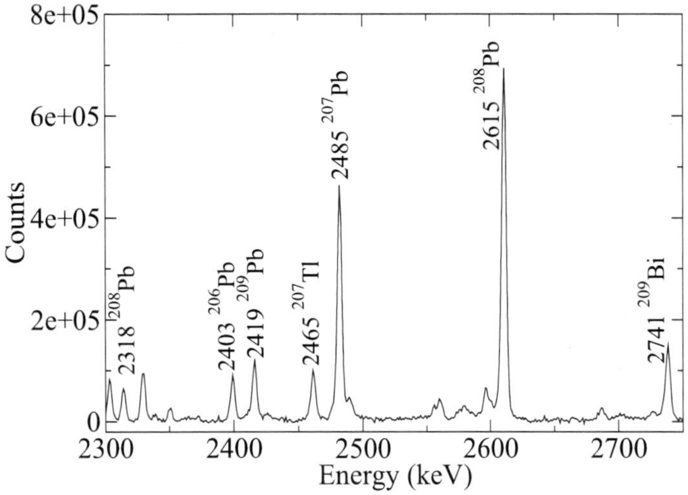

Figure 1. The high-energy region of the full projection of the $\gamma\gamma\gamma$ cube. The octupole transitions discussed in the paper are labelled.

The description of collective states is difficult in the shell model. Consequently the usual approach is to consider the octupole as a vibrational phonon. Particle octupole-phonon coupling calculations proved to be very powerful in predicting/explaining the energies of such excitations [6]. Here we performed shell model calculations using two different sets of interactions (and model spaces). The OXBASH code [12] was employed. The single particle/hole energies relative to ^{208}Pb were taken from experiments. Calculations were performed for ^{208}Pb and for the four neighbouring single particle/hole nuclei ^{207}Tl, ^{209}Bi, ^{207,209}Pb. The details of the two shell model calculations are given below:

KHH7B: The model space considered consisted of proton orbitals $d_{5/2}$, $h_{11/2}$, $d_{3/2}$, $s_{1/2}$ below Z=82 and $h_{9/2}$, $f_{7/2}$, $i_{13/2}$ above, and neutron orbitals $i_{13/2}$, $p_{3/2}$, $f_{5/2}$, $p_{1/2}$ below N=126 and $g_{9/2}$, $i_{11/2}$, $j_{15/2}$ above. These are in total 14 orbitals, with four Δj = Δl=3 pairs across

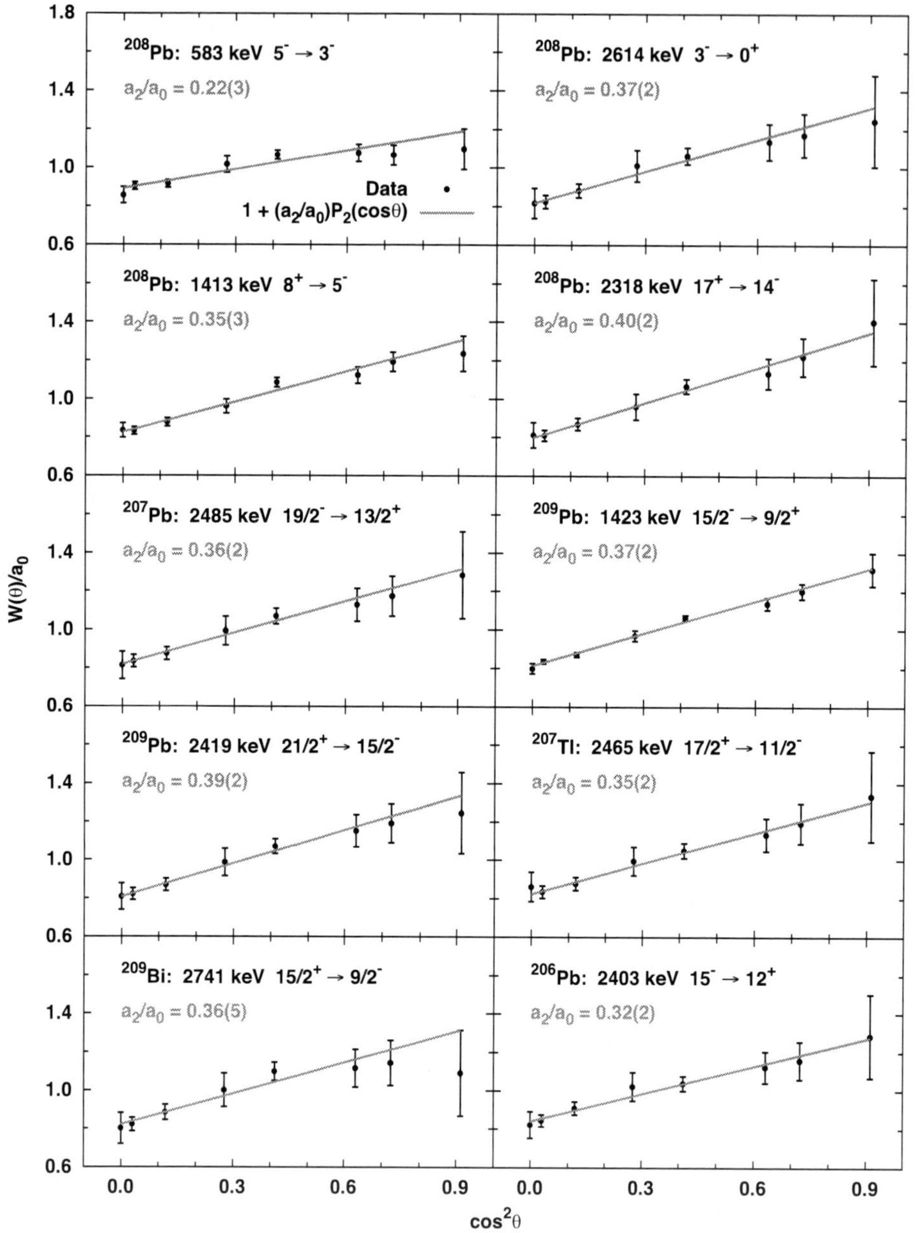

Figure 2. Angular distributions of octupole transitions in the ^{208}Pb region. For comparison the angular distribution of the 583 keV E2 transition is also shown.

Z=82 and N=126. The cross shell two-body interaction matrix elements (TBMEs) are based on the H7B interaction [13], while the neutron-proton ones on the Kuo-Herling interaction [14]. Mixing between proton and neutron particle-hole excitations were not considered. We note that these calculations describe accurately valence particle excitations (when no core-breaking is needed). They were used extensively on nuclei above Z=82 [15], and below Z=82 along the N=126 line [16, 17, 18, 19], as well as on both in the N>126 [20] and N<126 [19] regions.

KHM3Y: The model space consisted of proton orbitals $\mathbf{g_{7/2}}$, $d_{5/2}$, $h_{11/2}$, $d_{3/2}$, $s_{1/2}$ below Z=82 and $h_{9/2}$, $f_{7/2}$, $i_{13/2}$, $\mathbf{f_{5/2}}$, $\mathbf{p_{3/2}}$, $\mathbf{p_{1/2}}$ above, and neutron orbitals $i_{13/2}$, $p_{3/2}$, $f_{5/2}$, $p_{1/2}$, $\mathbf{h_{9/2}}$, $\mathbf{f_{7/2}}$ below N=126 and $g_{9/2}$, $i_{11/2}$, $j_{15/2}$, $\mathbf{g_{7/2}}$, $\mathbf{d_{5/2}}$, $\mathbf{d_{3/2}}$, $\mathbf{s_{1/2}}$ above. The additional orbitals, compared to the KHH7B calculations, are shown in bold. There are in total 24 orbitals, with

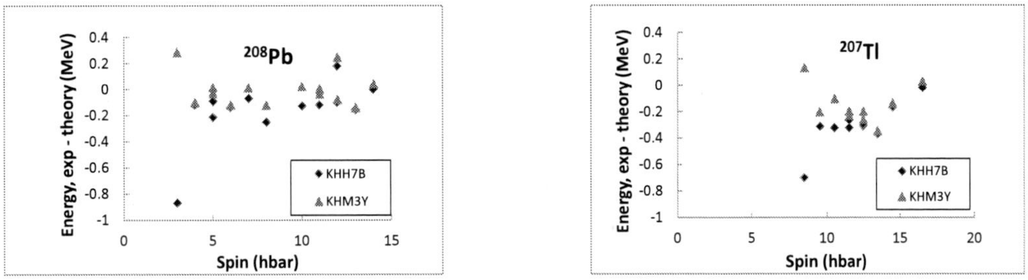

Figure 3. Energy difference between experimental [1, 24] and shell-model excitation energies for core excited states in ^{208}Pb and ^{207}Tl. For details of the two shell model calculations, labelled KHH7B and KHM3Y, see the text.

eight $\Delta j = \Delta l$=3 pairs across Z=82 and N=126. The cross shell two-body matrix elements are based on the M3Y interaction [21], while the neutron-proton interactions are based on the Kuo-Herling interaction as given in [22]. Such calculations gave a good description of single and double octupole states in ^{208}Pb [23]. In the present work the calculations were extended to the neighbouring nuclei.

The main characteristics of the two calculations are: *(i)* both sets of calculations give a good description of the non-collective core-excited states in ^{208}Pb; *(ii)* the energies of the collective octupole states is overestimated by the KHH7B interaction and underestimated by the KHM3Y one; *(iii)* the ordering of the three-particle states in ^{207,209}Pb, ^{207}Tl, ^{209}Bi are similar in the two calculations, with the KHH7B interaction predicting sligthly higher excitation energies.
The two calculations are compared with the experimental level energies for ^{208}Pb and ^{207}Tl [24] in figure 3.

The biggest differences between the two calculations are regarding the octupole states. In order to understand the underlying reasons we looked into the wave functions. For example in ^{207}Tl the KHM3Y interaction for the $17/2^+$ state, interpreted as $\pi h^{-1}_{11/2} \times 3^-$ [6, 24], gives a very fragmented wave function. The highest contributions are from $\pi h^{-1}_{11/2}\nu p^{-1}_{3/2}g_{9/2}$ (15.5%), $\pi h^{-1}_{11/2}\nu d^{-1}_{3/2}g_{9/2}$ (14.1%), $\pi h^{-1}_{11/2}\nu f^{-1}_{5/2}g_{9/2}$ (9.0%), $\pi h^{-1}_{11/2}\pi s^{-1}_{1/2}f_{7/2}$ (8.9%), $\pi h^{-1}_{11/2}\nu i^{-1}_{13/2}j_{15/2}$ (7.5%). The $\Delta j = \Delta l$=3 contribution is above 50%. In contrast the KHH7B interaction gives a predominantly $\pi h^{-1}_{11/2}\nu p^{-1}_{1/2}g_{9/2}$ (80.8%) character. The $\Delta j = \Delta l$=3 contribution is of the order of 5%. The introduction of 2p-2h states in these calculations are expected to increase the mixing considerably.

Table I summarises the results on the discussed octupole states and transitions. The energies of collective octupole states are overpredicted and slightly underpredicted by the KHH7B and KHM3Y interaction, respectively. We note the good description of the 8^+ and 5^- states in ^{208}Pb, suggesting limited collectivity in them. This is in line with the measured $B(E3)$=12.6(20) W.u., which is ~2.5 times smaller than the collective $B(E3; 3^- \rightarrow 0^+)$=33.8(6) W.u. [1] in the same nucleus.

4. Conclusions

The octupole character of several transitions in the ^{208}Pb region was demonstrated using angular distribution analysis. The octupole character of the 2318 keV $17^- \rightarrow 14^+$ in ^{208}Pb, 2485 keV $19/2^- \rightarrow 13/2^+$ in ^{207}Pb, 2419 keV $21/2^+ \rightarrow 15/2^-$ in ^{209}Pb, and 2465 keV $17/2^+ \rightarrow 11/2^-$ in ^{207}Tl transitions was shown for the first time. Two sets of shell model calculations, to describe

Table 1. Summary of the E3 transitions investigated in the present work. The experimental excitation energies are compared with two sets of calculations. All energies are in keV.

Nucleus	$I_i^\pi \rightarrow I_f^\pi$	E_γ	a_2/a_0	E_i	E_f	E_i	E_f	E_i	E_f
				experimental		KHH7B		KHM3Y	
^{208}Pb	$3^- \rightarrow 0^+$	2614	0.37(2)	2614	0	3483	0	2329	0
^{208}Pb	$8^+ \rightarrow 5^-$	1413	0.35(3)	4861	3198	4864	3415	4733	3230
^{208}Pb	$17^+ \rightarrow 14^-$	2318	0.40(2)	9061	6743	-	6746	-	6701
^{207}Pb	$19/2^- \rightarrow 13/2^+$	2485	0.36(2)	4118	1633	5003	1633	3996	1633
^{209}Pb	$15/2^- \rightarrow 9/2^+$	1423	0.37(2)	1423	0	1423	0	1423	0
^{209}Pb	$21/2^+ \rightarrow 15/2^-$	2419	0.39(2)	3842	1423	4783	1423	3748	1423
^{207}Tl	$17/2^+ \rightarrow 11/2^-$	2465	0.35(2)	3813	1348	4514	1348	3681	1349
^{209}Bi	$15/2^+ \rightarrow 9/2^-$	2741	0.36(5)	2741	0	3653	0	2539	0
^{206}Pb	$15^- \rightarrow 12^+$	2403	0.32(2)	6431	4027	-	-	-	-

both collective octupole states and less collective core breaking states, were performed and compared. In the future, the possible use of modern effective interactions, e.g. [25], will shed more light on the role of interactions in describing collective octupole states.

Acknowledgments

Acknowledgments This work is supported by the Science and Technology Facilities Council (STFC), UK, US Department of Energy, Office of Nuclear Physics, under Contract No. DE-AC02-06CH11357 and DE-FG02-94ER40834, NSF grant PHY-1404442. The contributions of the Argonne National Laboratory technical staff are gratefully acknowledged.

References

[1] Martin M J 2007 *Nucl. Data Sheets* **108** 1583
[2] Broda R et al. 2004 *Eur. Phys. J. A* **20** 145
[3] Lee I Y 1990 *Nucl. Phys. A* **520** 641c
[4] Wilson E et al. 2013 *Acta Phys. Pol. B* **44** 381
[5] Wilson E et al. 2014 *EPJ: Web of Conferences* **66** 02110
[6] Rejmund M et al. 2000 *Eur. Phys. J. A* **8** 161
[7] Wrzesinski J et al. 2001 *Eur. Phys. J. A* **10** 259
[8] Kondev F G 2008 *Nucl. Data Sheets* **109** 1527
[9] Kondev F G and Lalkovski S 2011 *Nucl. Data Sheets* **112** 707
[10] Schramm M et al. 1992 *Z. Phys.* **344** 121
[11] Martin M J 1991 *Nucl. Data Sheets* **63** 723
[12] Brown B A et al. 2004 *OXBASH for Windows, MSU-NSCL report* 1289
[13] Hosaka A, Kubo K I, Toki H 1985 *Nucl. Phys. A* **444** 76
[14] Kuo T T S and Herling G H 1971 *US Naval Research Laboratory Report* **2258** unpublished
[15] Cieplicka N et al. 2012 *Phys. Rev. C* **86** 054322
[16] Podolyák Zs et al. 2009 *Eur. Phys. J. A.* **42** 489
[17] Podolyák Zs et al. 2009 *Phys.Lett. B* **672** 116
[18] Steer S J et al. 2008 *Phys.Rev. C* **78** 061302(R)
[19] Steer S J et al 2011 *Phys.Rev. C* **84** 044313
[20] Al-Dahan N et al. 2009 *Phys. Rev. C.* **80** 061302
[21] Bertsch G et al 1977 *Nucl. Phys. A* **284** 399
[22] Warburton E K, Brown B A 1991 *Phys. Rev. C* **43** 602
[23] Brown B A 2000 *Phys. Rev. Lett.* **85** 5300
[24] Wilson E et al. 2014 *to be published*
[25] Coraggio L, Covello A, Gargano A, Itaco N 2009 *Phys. Rev. C* **80** 021395(R)

Pre-equilibrium Emission and α-clustering in Nuclei

F Gramegna[1], D Fabris[2], T Marchi[1], M Degerlier[3], O V Fotina[4;5], V L Kravchuk[6], M D'Agostino[7], L Morelli[7], S Appannababu[1], G Baiocco[7], S Barlini[8], M Bini[8], A Brondi[9], M Bruno[7], G Casini[8], M Cinausero[1], N Gelli[8], R Moro[9], A Olmi[8], G Pasquali[8], S Piantelli[8], G Poggi[8], S Valdrè[8] and E Vardaci[9]

[1] INFN - Laboratori Nazionali di Legnaro, Legnaro (Padova), Italy
[2] INFN sezione di Padova,Padova, Italy
[3] Nevsehir Haci Bektas University, Science and Art Faculty, Physics Dep. Nevsehir, Turkey
[4] Physical Department, Lomonosov Moscow State University, Moscow, Russia
[5] Skobeltsyn Institute of Nuclear Physics, Lomonosov Moscow State University, Moscow,Russia
[6] National Research Center "Kurchatov Institute", Moscow, Russia.
[7] Dipartimento di Fisica, Università di Bologna and INFN sezione di Bologna, Bologna, Italy
[8] Dipartimento di Fisica, Università di Firenze and INFN sezione di Firenze, Firenze, Italy
[9] Dipartimento di Fisica, Università di Napoli and INFN sezione di Napoli, Napoli, Italy

E-mail:gramegna@lnl.infn.it

Abstract. The study of nuclear states built on clusters bound by valence neutrons in their molecular configurations is a field of large interest, which is being renewed by the availability of exotic beams: clustering is, in fact, predicted to become very important at the drip-line, where weakly bound systems will prevail. Although for light nuclei at an excitation energy close to the particle separation value there are experimental evidences of such structure effects, this is still not the case for heavier nuclear systems. Many attempts have been done using preformation alpha clustering models, but there is still a lack of experimental data capable to give a direct feedback. In particular, searching for alpha clustering effects in medium mass systems is still a challenge, which can give new hints in this subject. In the past we have studied the reactions 250, 192 and 130 MeV ^{16}O + ^{116}Sn, observing a significant increase in the fast emitted α-particle yield. This effect was ascribed to the presence of pre-formed α-clusters in the ^{16}O projectile nucleus. In order to investigate these aspects, in a model independent way, a new experimental campaign has been performed with the GARFIELD + RCo set up, to compare results from two different reactions: a double magic α-cluster (^{16}O) and a non-magic α-cluster projectile (^{19}F) at the same beam velocity (16AMeV) have been chosen, impinging respectively on ^{65}Cu and ^{62}Ni targets, thus leading to the same $^{81}Rb^*$ compound nucleus. The angular distributions and the light charged particles emission spectra in coincidence with evaporation residues have been measured and analyzed. The preliminary results of the data analysis and the main features of the theoretical model used for their interpretation are presented.

1. Introduction

The idea that cluster of nucleons might be pre-formed prior to emission from nuclei has been discussed since many years and was originally proposed by Hafstad and Teller in 1938 [1]. More recently a large interest has been re-addressed to the problem of clustering in nuclei, due to the study of weakly bound nuclei at the drip lines, where clustering might be the preferred structural mode, especially in the case of light nuclei [2]. Examining the binding energies of nuclei in their ground state as a function of the mass number, the particular behaviour found shows a systematic trend, well described by the liquid drop model as due to a shell structure effect. In particular the specific property of the nucleon–nucleon force, for which a saturation arises due to the coupling to zero of both the spin and the isospin quantum numbers, produces a very strong binding of α-particles which can therefore be recognized as a unique cluster subsystem in nuclei. The α-particle is the main ingredient in the concept proposed by Ikeda in his diagram [3], where highly clustered states are predicted at excitation energies around the energy threshold for the decomposition into specific cluster channels. In the extended Ikeda diagram, moreover, it is suggested that in neutron-rich systems neutrons may act as valence particles which can be exchanged between the α-particle cores, in a similar way electrons are exchanged in atomic molecules. In the nuclear case the covalent neutrons stabilize the unstable multi-cluster states, giving rise to nuclear structures which may be described by molecular concepts. These concepts are well reproduced in model independent approaches like the Fermionic Molecular Dynamics (FMD) by Feldmeier et al. [4,5] or the Antisymmetrized Molecular Dynamics (AMD) with effective N-N forces by Horiuchi and Kanada-En'Yo [6-7].

Up to now, these structures are mainly described by theory, but still a lack of experimental data is present, mainly due to the low intensity of exotic beams presently available. Therefore, while waiting for the availability of the next generation of radioactive beam facilities like SPES [8], HIE-ISOLDE [9] and SPIRAL2 [10], it is of particular interest to search for α-clustering effects in non-traditional observables, like those deriving from pre-equilibrium process studies, which may bring new information on the cluster formation process.

2. Previous measurements

In a previous campaign the formation of the ^{132}Ce compound nucleus by means of ^{16}O induced reaction on ^{116}Sn at various incident energies has been studied with the aim of identifying the amount of pre-equilibrium emission in asymmetric entrance channel reactions. During this campaign an extra yield was observed for pre-equilibrium α-particle emission, which was not reproduced by a Hybrid Exciton Model based prediction [11]. The model was using a modified version of the PACE2 code, where the main variation was the introduction of a non-equilibrium stage before the complete thermalization and compound nucleus formation. The relaxation process which occurs during the fusion reaction is firstly accounted for by the exciton model, based on the Griffin prescription [12], in which the description of the angular distribution of the fast emitted particles is still an intricate question [13]. The main parameter to be set is the initial number of excitons ($n_0=n_{particles}+n_{holes}$), that can be estimated from the empirical trend described in the work by N. Cindro et al. [14] and it is mainly related to the projectile properties. In the case of the considered ^{16}O induced reactions this number is n_0=17=16p+1h.

Starting from such initial exciton number, a general good description was obtained by model prediction when compared to double differential cross section proton energy spectra at all the incident energies considered. On the contrary, an enhanced fast α-particle production was observed experimentally, especially at the most forward measured angles, which was not accounted for by the calculated distributions. A possible explanation of this enhanced α-particle emission might be the effects induced by the α-cluster structure in the ^{16}O projectile [15].

This effect has therefore been taken into account in the model introducing a pre-formation cluster probability. The combination of different initial configurations has been considered: in particular a second configuration has been chosen in which the ^{16}O projectile is supposed to be divided into a ^{12}C core plus an α particle. The probability of occurrence of this last pattern with respect to the original one (i.e.^{16}O projectile as a whole) is the free parameter to be determined from the comparison to the experiment. The data have been better reproduced with a quite sizeable probability (up to 50%) of α-cluster pre-formation. However, while the shape of the energy spectra seems to be better described, still problems are present in the reproduction of total multiplicities of light charged particle, which needs to be understood. Results are shown in Fig.1 where the experimental multiplicity obtained at 250MeV $^{16}O+^{116}Sn$ are compared with the predicted values both for the evaporative and the fast emission part. As it can be noticed the calculated values of the total (mainly evaporative) multiplicities for alpha and protons are too large and this is due to the PACE2 prediction which seems to underestimate the neutron emission. Similar results have been observed in literature [16].

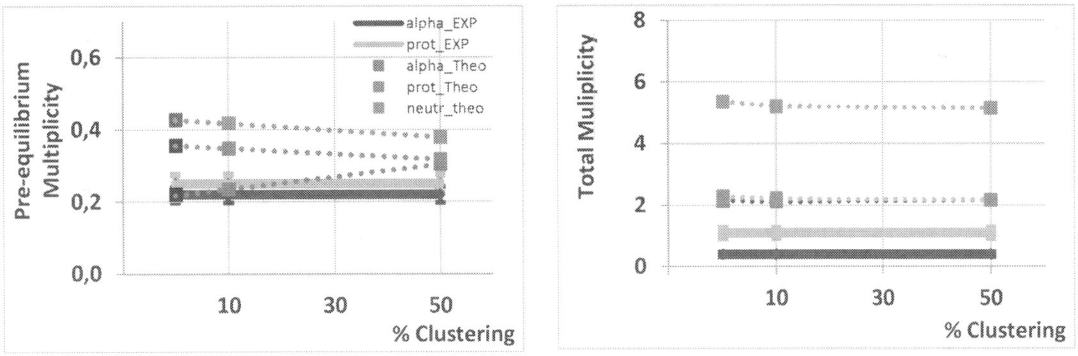

Figure 1. Comparison with calculated and experimental multiplicity for the $^{16}O+^{116}Sn$ reaction at 250 MeV. On the left pre-equilibrium multiplicity, on the right total (pre-equilibrium plus evaporative multiplicities). The calculations have been performed with the Hybrid Exciton Model +PACE2 code (see text) with different pre-formation cluster probability (0%, 10% and 50%).

For what it concerns the fast emission still an overestimation is observed for proton multiplicities, while more reasonable values are obtained with α-particles. However, the percentage of fast α-particles with respect to the total ones emitted in the experimental data is not at all reproduced by the calculation ($[M^{\alpha}_{pre}/M^{\alpha}_{tot}]_{exp}$~0.5;[$M^{\alpha}_{pre}/M^{\alpha}_{tot}]_{theo}$~0.1). Further experimental data and more exclusive observables are therefore needed to compare models and data for a better understanding of the whole process.
In order to obtain, in a model independent way, a confirmation of possible effects of α-cluster structure in the projectile, two different entrance channel reactions have been studied in an energy range where fast emission was predicted: in particular a double magic α-cluster (^{16}O) and a non-magic α-cluster projectile (^{19}F) have been chosen, impinging respectively on ^{65}Cu and ^{62}Ni targets, thus leading to the same $^{81}Rb^*$ compound nucleus.

3. The Experiment

The two fusion reactions $^{16}O + ^{65}Cu$ and $^{19}F+^{62}Ni$ have been studied at 16AMeV incident energy in order to directly compare their light charged particle emission spectra and yield ratios. The same projectile velocity was chosen since the pre-equilibrium emission is expected to mostly depend on this parameter [17]. In this situation, even with an expected small difference in the evaporative part due to the excitation energy (E*=209 MeV and E*=240 MeV respectively for ^{16}O and ^{19}F induced reactions), the fast emission process is predicted to be almost the same for both reactions. Any observed difference and overproduction of fast α-particle between the two cases would suggest, in a model independent way, a possible influence of the projectile α-structure effect.
The experiment has been performed at the Legnaro National Laboratory where the ALPI-TANDEM XTU accelerator complex was used for the ^{16}O and ^{19}F beam production and acceleration. The experimental apparatus was the GARFIELD array, implemented with the Ring Counter (RCo) at the most forward angles, fully equipped with digital electronics [18]. The GARFIELD set-up consists of two large drift chambers employing micro-strip gas chamber (MSGC) as amplified ΔE stage and CsI(Tl)

scintillators as residual energy detectors. It can identify light charged particles and fragments in an angular range from θ = 29° to θ =151°. The RCo is a three–stage annular detector made by Ionization Chamber as first stage, a reverse mounted nTD strip silicon detector as second one, CsI(Tl) crystals as final stage. It covers the angular range between θ = 5° and θ =17°.
The coupled GARFIELD plus RCo apparatus can perform complete high quality particle identification (both Z and A) and energy determination in a nearly 4π coverage (θ = 5°-151°) for light charged particles and, in the most forward direction (θ = 5°-17°) even for fragments with charge up to Z=14. Z identification is provided up to Z=28-30 in the whole subtended angular range.

4. Preliminary Results

Fully identified light charged particles have been measured both in single and in coincidence with Evaporation Residues (ERs), detected in the RCo in the range θ = 8.6°-17° (just beyond the grazing angle). The ERs have been selected setting the proper gates in the IC-Si ΔE–E spectra.

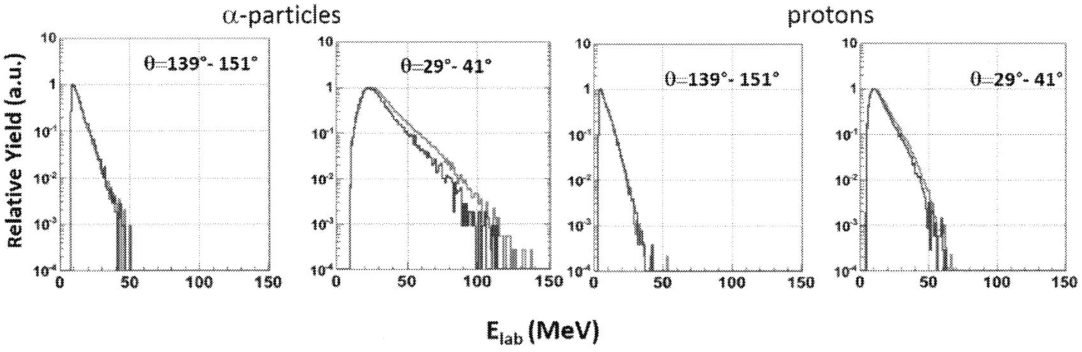

Figure 2. Double differential energy spectra in the laboratory frame (normalized to the maximum) for α-particles (left panel) and for protons (right panel) for the two reactions 256 MeV $^{16}O + ^{65}Cu$ (blue line) and 304 MeV $^{19}F+^{62}Ni$ (red line) at two different detecting angles (θ =29°-41°, θ =139°-151°).

In this preliminary analysis, only the double differential energy spectra obtained in coincidence with ERs for α-particles and protons and detected in the GARFIELD drift chambers have been sorted out. The α-particle and proton spectra, normalized to the maximum, obtained for the two systems have been compared, as shown, as an example, in Fig.2. The two reactions show very similar proton spectra on the whole angular range, except for a small difference at the most forward angle, which can be ascribed to the slightly larger excitation energy of the ^{19}F induced reaction. A much more evident difference is, on the contrary, observed in the case of the α-particle emission spectra, especially at the most forward GARFIELD detection angles. The calculated emission spectra performed with an evaporative code like PACE2 (or PACE4), which mainly takes into account the difference in the excitation energy of the Compound Nucleus, confirms a small difference for the two systems and compatible to that one experimentally observed in the proton decay channel. When compared to the data, they also support the idea that a second fast emission source is needed for both systems, to better describe mainly the α-particle decay channel. In the meanwhile that the data sorting and calibration procedure will be concluded, which will permit an complete experimental comparison between the two systems, a first evaluation of the expected amount of fast emission in the two cases was performed by means of the prediction of the Hybrid Exciton Model. The calculation was done starting from an initial exciton number of n_0=17(16p+1h) in the case of $^{16}O+^{65}Cu$. With this configuration the calculation seems to reasonably describe the shape of the α-particle spectra, except for an underestimation at the most

forward angle in GARFIELD, but, with the same initial parameter, it strongly overestimates the proton pre-equilibrium emission. In Fig.3 the comparison of the calculation is shown for α-particles and protons in the $^{16}O+^{65}Cu$ system at the most forward GARFIELD angular range (θ= 29°-41°).
Performing the same comparison in the case of the $^{19}F+^{62}Ni$ reaction, where an initial exciton configuration of n_0=20(19p+1h) was used, a quite similar result is obtained, as shown in Fig. 4 for the same angular range as in Fig.3. In the ^{19}F induced reaction case, the experimental fast α-particle overproduction is even larger than in the ^{16}O induced reaction, while the fast protons are, again, largely overestimated. A tentative explanation for the observed difference between the α-particle decay in the two systems may be the lower energy needed to break up the ^{19}F nucleus into α+^{15}N (4.01 MeV) with respect to the ^{16}O to be divided into α+^{12}C (7.2 MeV).

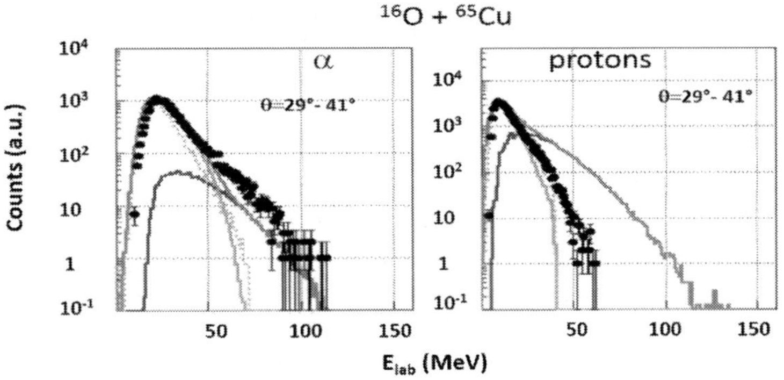

Figure 3. Comparison of experimental laboratory energy distribution (black dots) at θ=29°- 41° of α-particles (left panel) and protons (right panel) for the system $^{16}O+^{65}Cu$ with preliminary calculations from the Hybrid Exciton Model (red line total, green line evaporative (PACE2), orange dots evaporative (PACE4), blue line pre-equilibrium). The normalization has been done at the area (exp. – total).

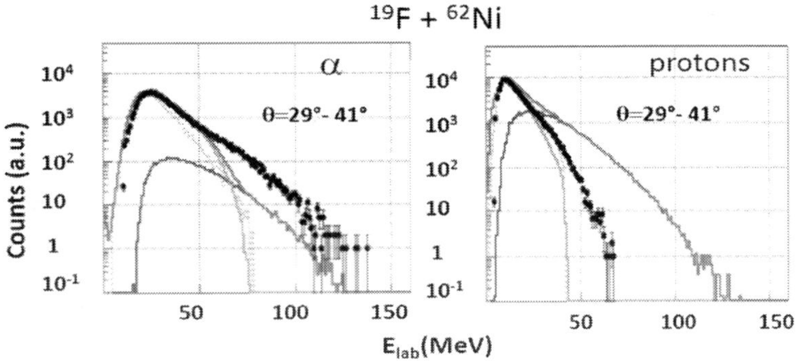

Figure 4. Same as Fig.3 for the $^{19}F+^{62}Ni$ system.

By changing the initial configuration parameter (i.e. diminishing the exciton number) the description of the α-particle slightly improves, but, on the contrary, this worsen the proton description. Once again, like it was observed in the $^{16}O+^{116}Sn$ case, a unique initial parameter in the Hybrid Exciton model is not able to describe both the proton and α-particle channels suggesting that some more attention has to be paid to the clustering structure effects. For example in the used calculation no deuteron emission is considered, which may strongly influence the relative proton decay probability. However part of the problem may be also in the total multiplicity, which is mainly related to the evaporative code used after the fast process steps [16], as observed in our previous data. This is to be further analysed.

5. Conclusions

In order to probe possible α-clustering effects in medium nuclei we are studying the secondary particle emission from the 256MeV $^{16}O+^{65}Cu$ and 304MeV $^{19}F+^{62}Ni$ systems. From the preliminary comparison between the two reactions a difference has been evidenced related to the fast α-decay channel, which can be linked to the difference in the projectile structure. In the meanwhile that the data analysis and calibration for the whole apparatus is completed, the experimental spectra detected in GARF in coincidence with ERs have been compared with the Hybrid Exciton Model proposed by O.V. Fotina [11,15]. At least in the GARFIELD angular range analysed up to now, using an initial exciton number estimated by the empirical trend reported in [14], the α-particle spectra seem to be reasonably reproduced as far as the shape is concerned, except for a small part at the most forward angles. On the contrary, using the same initial parameters, the fast emission of protons is largely overestimated. In a similar way, the model was not able to reproduce both protons and α-particle with the same initial parameters in our previous measurement related to the $^{16}O+^{116}Sn$ reaction study: in that case, however a better description of protons was obtained, while some added pre-formation probability was needed to describe the α-particle spectra. Moreover, looking more into details, an overestimate of the charged particle total multiplicity was also evidenced for that system, together with a lower weighting factor between fast and thermal emission especially for α-particles ($[M^{\alpha}_{pre}/M^{\alpha}_{tot}]_{exp}\sim 0.5$;$[M^{\alpha}_{pre}/M^{\alpha}_{tot}]_{theo}\sim 0.1$).

This demonstrates that a more complete analysis is needed to understand the process: in particular, due to the better performances reached in the present experiment in terms of identification and resolutions, the possibility of studying all different light charged particles decay channels will be possible. Moreover the larger angular range in which the particles have been identified will permit to better disentangle and study the pre-equilibrium emission. It will be needed to extract multiplicities, angular distributions and to study exclusive α–α, α–N correlation channels. From the theoretical point of view the Hybrid Exciton Model calculations has to be upgraded, following all particles and studying the coupling of fast emission with different evaporation codes. Moreover, it is under study the possible application of other theoretical approaches, like for example the dynamical AMD [19] code.

Due to the high resolution of the apparatus, even more peripheral collisions like the decay of the projectile (break-up) can be studied, in order to compare the ^{16}O and ^{19}F cases for what concern their α-structure through a complete reconstruction of the projectile–like decay.

References

[1] Hafstad LR, Teller E 1938 *Phys. Rev.* **54** 681
[2] von Oertzen W, Freer M, Kanada-En'yo Y, 2006 *Phys. Rep.* **432** 43-113.
[3] Ikeda K, Tagikawa N, Horiuchi H, Prog. 1968 *Theor. Phys. Suppl* **464**
[4] Feldmeier H et al., 1995 *Nucl. Phys. A* **586**, 493
[5] Neff T, Feldmeier H and Roth R 2005 *Nucl. Phys. A* **752** 94
[6] Horiuchi H, Kanada- En'Yo Y, 1997 *Nucl. Phys. A* **612**, 394c
[7] Kanada- En'Yo Y, Horiuchi H, 2001 *Prog. Theor. Phys. Suppl.* **142** 206
[8] https://web.infn.it/spes/
[9] https://**hie-isolde**.web.cern.ch/
[10] www.ganil-**spiral2**.eu/**spiral2**
[11] Fotina OV et al. 2010 *Int. Journ. Mod. Physics E* **19** 1134
[12] Griffin JJ , 1966 *Phys. Rev. Lett.* **17** 478
[13] Blann M, Chadwick MB, 2000 *Phys. Rev. C* **62** 034604
[14] Cindro N et al., 1991 *Phys. Rev. Lett.* **66** 868
[15] Fotina O et al. – 2014 *EPJ Web of Conferences* **66**, 03028
[16] Vardaci E et al. 2010 *EPJ A* **43** 127; Di Nitto A et al. 2011 *EPJ A* **47** 83
[17] Cabrera J et al. 2003 *Phys Rev. C* **68** 034613
[18] Bruno M, Gramegna F et al., 2013 *EPJ A* **49** 128 and ref. therein.
[19] Ono A, 2012 *EPJ Web of Conf.* **31** 00023

Single-particle strength in neutron-rich ^{71}Cu from the (d,^{3}He) proton pick-up reaction

S Franchoo[1], P Morfouace[1], M Assié[1], F Azaiez[1], D Beaumel[1], S Boissinot[2], R Borcea[3], G Burgunder[4], L Cáceres[4], N De Séréville[1], Z Dombradi[5], J Elseviers[6], B Fernandez-Dominguez[7], A Gillibert[2], S Giron[1], S Grévy[8], F Hammache[1], O Kamalou[4], V Lapoux[2], L Lefebvre[1], A Lepailleur[4], C Louchart[2], G Marquinez-Duran[9], I Martel-Bravo[9], I Matea[1], D Mengoni[10], L Nalpas[2], D Napoli[11], M Niikura[1], F Recchia[10], A Sanchez-Benitez[9], J-A Scarpaci[1], D Sohler[5], O Sorlin[4], M Stanoiu[3], I Stefan[1], C Stodel[4], J-C Thomas[4] and Z Vajta[5]

[1] IPN, Orsay, France
[2] Irfu, Saclay, France
[3] Ifin-HH, Bucharest, Romania
[4] Ganil, Caen, France
[5] MTA Atomki, Debrecen, Hungary
[6] University of Leuven, Belgium
[7] University of Santiago, Spain
[8] CENBG, Bordeaux, France
[9] University of Huelva, Spain
[10] University of Padova, Italy
[11] LNL, Legnaro, Italy

E-mail: franchoo@ipno.in2p3.fr

Abstract. We have measured the ^{72}Zn(d,^{3}He)^{71}Cu proton pick-up reaction in inverse kinematics at 38 MeV/u. A dedicated set-up was developed to overcome the experimental challenges posed by the low cross section of the reaction and the low energy of the outgoing ^{3}He particles. The excitation-energy spectrum was reconstructed and spectroscopic factors were obtained after analysis of the angular distributions in the finite-range Distorted-Wave Born Approximation (DWBA). The results show that unlike for the $\pi f_{5/2}$ orbital, the $\pi f_{7/2}$ single-particle strength is not appreciably affected by the addition of neutrons beyond $N = 40$.

1. Introduction

With a long and fruitful history to its credit, the shell model still accounts for much of our understanding of nuclear structure at low excitation energy [1]. The bulk of the interaction within the nuclear many-body system is well represented by a mean field, the potential of which gives rise to a shell structure, from which in turn well-known magic numbers appear that define the closed shells. These magic numbers, however, are not absolute and universal but are subject to a slow evolution as one moves away from the valley of stability into the realm of exotic isotopes, where the numbers of protons and neutrons are out of their ordinary balance. The fate of a presumed doubly magic nucleus like ^{78}Ni therefore carries much interest, as its closed-shell character comes under scrutiny. While the ^{78}Ni isotope has been observed for the first time two decades ago [2], only its half-life is known so far [3] and direct spectroscopic information is not yet available.

The neutron-rich copper isotopes are characterised by one proton outside the nickel core and present an ideal opportunity to probe the movement of a single proton in the field of a closed-shell nucleus. In particular, the $Z = 28$ closure is the first shell gap that originates from the spin-orbit interaction and its evolution away from stability gives insight into the isospin dependence of the spin-orbit force. Proceeding along the copper isotopic chain, one is able to explore the influence of the occupation of the neutron orbitals on the nuclear structure. Of particular interest is the transition at $N = 40$. Because the first excited state in ^{68}Ni is located at a rather high energy and moreover shows a spin and parity of 0^+, the existence of a subshell gap at $N = 40$ was rapidly acknowledged [4]. Much experimental work has been done since, revealing a third 0^+ state at 2511 keV [5] and readjusting the energy of the second 0^+ state to 1604 keV [6]. This has in turn spurred new theoretical activity, leading to a richer interpretation that advances the coexistence of spherical and strongly deformed shapes in ^{68}Ni in the frame of the Monte-Carlo Shell Model (MCSM) [7].

In the β decay of ^{71}Cu, it was observed that the first excited state of spin and parity $5/2^-$, which is situated in a stable manner at an energy between 1 and 1.2 MeV in the ^{63,65,67,69}Cu isotopes, suddenly comes down to 534 keV [8]. This distinctive feature coincides with the addition of neutrons beyond $N = 40$. Because of the high spectroscopic factor of $C^2S = 1.5$ in the ^{70}Zn(d,^{3}He)^{69}Cu proton pick-up reaction [9], the level was interpreted as chiefly corresponding to the $\pi f_{5/2}$ single-particle state. For the level at 1190 keV in ^{71}Cu, seen in the $E2$ cascade that desexcites a $19/2^-$ microsecond isomer, a $7/2^-$ assignment was made [10, 11]. Based on lower limit for the $\log ft$ value of 5.9 in the β decay of ^{71}Ni, the ground state of which is conjectured to consist of a $9/2^+$ configuration, and the similarity of the γ branching pattern between ^{69}Cu and ^{71}Cu, a spin and parity of $7/2^-$ was suggested for the level at 981 keV. This was corroborated by calculations within the particle-core coupling model (PCM) that linked the state to the $f^{-1}_{7/2}$ proton hole, onto which a 2p–1h quasiband would dwell in the same manner as for ^{69}Cu [12]. The Coulomb excitation of ^{71}Cu accordingly revealed a value of 10.7(12) Weisskopf units for the 1190-keV level, while the 981-keV state remained unobserved [13].

In the present work, we set off to measure the spectroscopic factors of the $7/2^-$ levels with the aim of obtaining the strength function of the $\pi f_{7/2}$ single-particle state. We chose to do so in the ^{72}Zn(d,^{3}He)^{71}Cu proton pick-up reaction, as it should selectively populate the hole states in copper.

2. Experimental set-up

The experiment took place at the Ganil laboratory in Caen, France. A radioactive beam of ^{72}Zn at 38 MeV/u and a rate of 1.5 10^5 particles per second was obtained from the fragmentation of a primary ^{76}Ge beam on a beryllium target of 733 μm thickness. The reaction products were selected in the Lise spectrometer, resulting in a beam purity of 55 % with ^{74}Ga as main contaminant. After travelling through two Cats beam-tracking detectors [14], the beam arrived at a deuterated polypropylene target of 0.26 mg/cm^2 thickness.

The target was surrounded by four Must-2 telescopes [15], covering forward angles of 8 to 50° in the laboratory. The energy of the ^{3}He particles coming out of the reaction, however, was less than 20 MeV for laboratory angles lower than 46°. Since they would have been stopped already in the first stage of Must-2, a double-sided silicon strip detector (DSSSD) of 300 μm, four silicon strip detectors with nominal thickness of 20 μm were therefore added in front of the Must-2 array. From transmission measurements with an α source it appeared that the actual thickness deviated in places with up to 25%. Only after a pixel-by-pixel mapping on a 1 mm^2

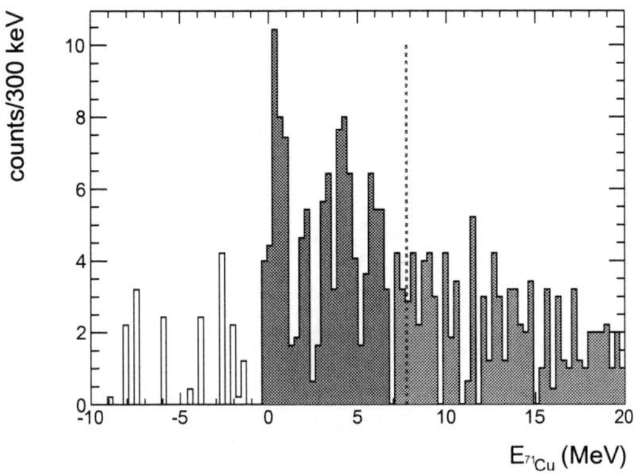

Figure 1. Excitation-energy spectrum of ^{71}Cu. In red we show the part of the spectrum that was assigned to particular levels and for which angular distributions were extracted. The area in pink is statistically not conclusive. The counts below 0 were excluded from the ^{71}Cu spectrum on grounds of energy. The dotted blue line indicates the threshold for neutron emission.

grid was carried out to correct for the variation in energy loss, the necessary resolution for particle identification could be achieved. Two Must-2 telescopes were installed at a forward angle of 85° to measure also elastic scattering.

At 0°, a ionisation chamber filled with CF_4 gas at a pressure of 100 mbar and followed by a NE 104 plastic scintillator of 2 cm of depth allowed for the identification of the outgoing heavy ions. In spite of the high count rate a resolution of $Z/\Delta Z \approx 30$ could be realised, which was sufficient for the present purpose. The signal in the ionisation chamber was digitised at a sampling rate of 40 MHz and was also used for pile-up rejection. The electronic dead time was kept at 15% throughout the experiment.

3. Results

The ^{3}He particles of interest were identified by combination of their time of flight, ΔE signal in the 20-μm strip detectors and E deposit in the Must-2 DSSSD. For each selected event, the excitation energy was reconstructed. After the subtraction of background stemming from reactions on the carbon nuclei in the target, five peaks could be distinguished in the spectrum. The width of the first peak is larger than the experimental resolution of 360 keV, indicating the presence of a doublet. From the known level scheme, the doublet can include two or more states among the $3/2^-$ ground state, the $1/2^-$ state at 454 keV and the $5/2^-$ state at 534 keV.

The angular distributions of the emitted ^{3}He particles were fitted with functions calculated within the finite-range Distorted-Wave Born Approximation (DWBA). For the incoming channel, the relativistic Daehnick optical potential was used [16]. For the outgoing channel, we took the Perey and Perey parametrisation [17].

We found that the ground-state doublet could be fitted with a combination of $L = 1$ and $L = 3$ distributions. The $L = 1$ component shows a spectroscopic factor of $C^2S = 0.8(2)$ and encompasses both the $\pi p_{3/2}$ ground state and the $1/2^-$ excited state of $\pi p_{1/2}$ nature. The $L = 3$ contribution yields $C^2S = 1.4(6)$ and we attribute it to the $5/2^-$ level, for which we infer a dominant $\pi f_{5/2}$ configuration. We note that the high spectroscopic factor exhausts the available $f_{5/2}$ proton strength. The next three peaks were best fitted with $L = 3$ functions. Cumulatively they add up to a spectroscopic strength of $C^2S = 6.9(8)$ and we assume that they correspond to

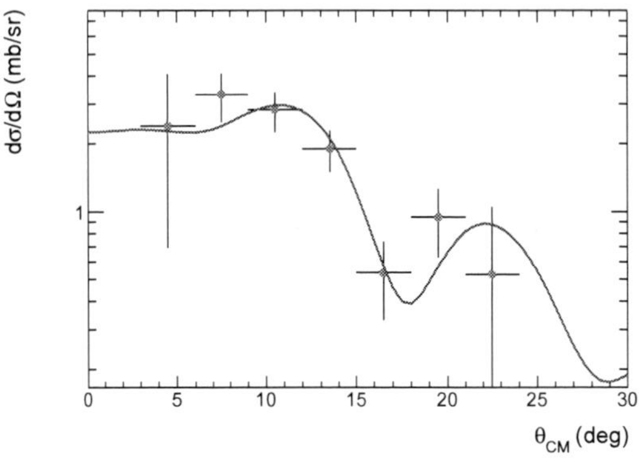

Figure 2. Summed angular distribution for the second, third and fourth peak in the excitation-energy spectrum of ^{71}Cu. The blue line corresponds to a DWBA calculation for $L = 3$.

the $f_{7/2}$ force, of which we thus determined 86%. The experimental centroid of the measured $f_{7/2}$ force is found at 3.8 MeV, which represents a lower limit in as far as part of the strength stays undetected. The fifth peak in the excitation-energy spectrum was fitted with a superposition of $L = 0$ and $L = 2$ distributions and likely originates from hole excitations in the deeper *sd* shell.

In our experiment, no pick-up strength was detected between 0.9 and 1.7 MeV. The low number of counts in this region is compatible with background. Therefore our data do not support the presence of a strong $L = 3$ component in the wave function of the presumptive $7/2^-$ state at 981 keV.

4. Discussion

The experimental data show that the $f_{7/2}$ proton single-particle strength remains at several MeV of excitation energy in ^{71}Cu. It does not come down appreciably and does not follow the sharp duck of the $\pi f_{5/2}$ state, preventing a rapid closure of the $Z = 28$ gap. It should be kept in mind, however, that the $\pi f_{7/2}$ orbital resides farther away from the Fermi surface and its influence on the nuclear structure at low energy is less immediate. Instead it appears more susceptible to fragmentation and therefore any change in its centroid is less visible than it is for its $\pi f_{5/2}$ spin-orbit partner.

We now come back to the nature of the $7/2^-$ state at 981 keV in ^{71}Cu, for which a single-particle character was proposed in earlier work as mentioned above. We interpret the structure of copper in a weak-coupling scheme of a proton times the excited states in the nickel core and refer to the available data [6, 9] and the MCSM calculations [7]. Of the two $7/2^-$ states at 1.74 and 1.87 MeV in ^{69}Cu, the one at 1.87 MeV with small spectroscopic factor of $C^2S = 0.45$ could correspond to the $\pi p_{3/2}$ ground state coupled to the oblate structure in ^{68}Ni the 0_2^+ bandhead of which is proposed at 1604 keV. For the other one, the $\pi f_{7/2}^{-1}$ broken-core configuration at 1.74 MeV with large spectroscopic factor of $C^2S = 2.7$, there is no immediate counterpart in the calculations. None of both matches the prolate shape that is calculated to prevail in ^{68}Ni near 3 MeV. Since the MCSM now puts forward that it is the prolate shape that would come down in energy in ^{70}Ni to take the place of the 0_2^+ state, there is no indication that the corresponding low-lying prolate structure in ^{71}Cu would correspond to a $f_{7/2}$ proton-hole state. Our data confirm this analysis by showing that within the experimental sensitivity indeed there is no $\pi f_{7/2}$ transfer strength at low energy in ^{71}Cu.

Moreover, we observe that two γ bands are known in ^{71}Cu from experiments that made use of deep-inelastic collisions. The first one shows an $E2$ sequence built on the $3/2^-$ ground state [10, 11]. The second one forms an $\Delta L = 1$ pattern on top of the first $5/2^-$ level at 534 keV [18]. The $7/2^-$ levels that make part of these bands are those at 1190 and 981 keV, respectively. If we assume a dominant $M1$ character for the second band, Coulomb excitation to the $7/2^-$ level at 981 keV would involve a structurally different configuration, which would not be as favourable as the excitation to the $7/2^-$ level at 1190 keV. Taking together the preceding arguments, we conclude that there is insufficient evidence for a $\pi f_{7/2}^{-1}$ interpretation of the level at 981 keV, as it was first proposed by the PCM calculations and later inferred from its non-observation in Coulomb excitation.

5. Conclusion

Notwithstanding the low cross section of the order of 10 mb and the low energy of the outgoing light particles that did not exceed 7 MeV/u for the angles of interest, it has been possible to measure the spectroscopic factors for the (d,^{3}He) proton pick-up reaction with a ^{72}Zn radioactive beam in the neutron-rich copper region beyond $N = 40$. Three levels carrying $\pi f_{7/2}$ strength were found in ^{71}Cu. Their spectroscopic factors were determined from the comparison of the experimental angular distributions with finite-range DWBA calculations. We found 86% of the $f_{7/2}$ proton strength and a centroid of the $\pi f_{7/2}$ force that remains situated at several MeV of excitation energy, precluding an early closure of the $Z = 28$ shell gap in the neutron-rich copper isotopes.

References

[1] Sorlin O and Porquet M-G 2008 *Prog. in Part. and Nucl. Phys.* **61** 602
[2] Engelmann C *et al* 1995 *Z. Phys.* A **352** 351
[3] Hosmer P *et al* 2005 *Phys. Rev. Lett.* **112501** 94
[4] Bernas M, Dessagne P, Langevin M, Payet J, Pougheon F and Roussel P 1982 *Phys. Lett.* B **113** 279
[5] Mueller W *et al* 2000 *Phys. Rev.* C **61** 054308
[6] Recchia F *et al* 2013 *Phys. Rev.* C **88** 041302
[7] Tsunoda Y *et al* 2014 *Phys. Rev.* C **89** 031301
[8] Franchoo S *et al* 1998 *Phys. Rev. Lett.* **81** 3100
[9] Zeidman B and Nolan J A 1978 *Phys. Rev.* C **18** 2122
[10] Grzywacz R *et al* 1998 *Phys. Rev. Lett.* **81** 766
[11] Ishii T, Asai M, Hossain I, Kleinheinz P, Ogawa M, Makishima A, Ichikawa S, Itoh M, Ishii M and Blomqvist J 1998 *Phys. Rev. Lett.* **81** 4100
[12] Oros-Peusquens A and Mantica P 2000 *Nucl. Phys.* **A669** 81
[13] Stefanescu I *et al* 2008 *Phys. Rev. Lett.* **100** 112502
[14] Ottini S *et al* 1999 *Nucl. Instr. and Meth. in Phys. Res.* A **431** 476
[15] Pollacco E *et al* 2005 *Eur. Phys. J.* A **25** 287
[16] Daehnick W, Childs J and Vrcelj Z 1980 *Phys. Rev.* C **21** 2253
[17] Perey C and Perey F 1976 *At. Data and Nucl. Data Tabl.* **17** 1
[18] De Angelis G 2012 *Phys. Scr.* T 150 014010

Exploration of nuclear structure beyond the limits of stability via magnetic moment measurements: ^{82}Sr and ^{90}Sr

N. Benczer-Koller and G. J. Kumbartzki

Department of Physics and Astronomy, Rutgers University, New Brunswick, NJ, US

E-mail: nkoller@physics.rutgers.edu

Abstract. Magnetic moment measurements over the last fifty years have provided considerable information on the microscopic structure of low-lying excited states of nuclei across the periodic table. At the same time there have been many developments in theoretical calculations, particularly in nuclei having either protons or neutrons or both filling orbitals at or near closed shells. Recent experimental challenges were to find means to populate states in nuclei far from stability. The development of radioactive beams has been slow and few nuclei have been accessed by that route. However, novel approaches, in particular the capture of an α particle from a ^{12}C target by stable nuclei in a beam, have opened new pathways. In this work, the region around $N = 50$, which includes the isotopes of Kr, Sr and Zr, has been investigated. The g factors of 2_1^+ and 4_1^+ states in the unstable isotopes of ^{82}Sr and ^{90}Sr were measured by the transient field technique and compared to the results of large scale shell-model calculations. In addition, the g factor of the 4_1^+ state in ^{86}Kr was measured.

1. Introduction

Over the last 28 years and 10 "International Spring Seminars on Nuclear Physics" our group has focused on the measurements of magnetic moments of short-lived states in many nuclei, from light to heavy. These nuclei exhibit behaviour ranging from single-particle excitations to collective modes. The magnetic moments have also been examined in isotopic and isotonic series of nuclei as a function of the energy and spin of their excited states. Often the results have been expected and predicted by specific models; occasionally, however, surprising results have opened new windows into the understanding of the nucleon arrangements within a given nucleus. On one hand, nuclei with neutron and/or proton numbers far from magic numbers tend to have g factors $\sim Z/A$ which are predicted for collective excitations or in the IBA model. On the other hand, the g factors of states in nuclei near closed shells are very sensitive to the details of the shell-model configurations chosen for their description. Thus, precise experimental information can guide shell-model calculations. But discrepancies between experimental observations and the theoretical results appear vividly in studies of long chains of isotopes such as, for example, in the Ca, Zr, Sn and Nd isotopes.

The advances in the field have been supported by developments and continuous improvements in both the experiments and the data analysis. The magnetic moment measurements using the transient field technique made a leap forward with the use of beam projectiles as probe ions. Nuclei in isotopic pure beams were Coulomb excited in inverse kinematics using a standard C target. That approach leads to higher detection yields of the reaction products and, ultimately, to improved statistics. It also opens the access to rare isotopes available as beams. But very few such beams of sufficient intensity have been produced so far: ^{76}Kr, $^{38,40}S$, ^{132}Te, ^{134}Te, ^{126}Sn and ^{72}Zn [1–7].

In view of the difficulties encountered in the preparation of beams of radioactive nuclei, a roundabout way was developed to access nuclei beyond stability. This opportunity ensued from the use of α-capture reactions in which specific beam ions capture an α particle from a ^{12}C target, a process which often leads to nuclei off stability. This reaction, which can be applied in the same kinematic conditions as Coulomb excitation, has been exploited in light nuclei in the Ti, Zn ,Ge region up to ^{100}Pd [8–10].

In the present work, Sr isotopes at the limits of high and low Z/N ratio, $^{82}_{38}$Sr and $^{90}_{38}$Sr, have been studied. The resulting magnetic moments obtained for these nuclei have been compared to the moments in the neighboring $_{36}$Kr and $_{40}$Zr isotopes in the same mass region which encompasses the $f_{5/2}$, $p_{3/2}$, $p_{1/2}$ and $g_{9/2}$ orbitals for protons and $g_{9/2}$, $d_{5/2}$, $g_{7/2}$, $d_{3/2}$, $s_{1/2}$, and $h_{11/2}$ for neutrons.

A compilation of the g factors of 2^+_1 states that were measured, before this work was started, in even-even nuclei from Zn to Cd is shown in Fig. 1. The results obtained in the current experiment and previous data on Kr isotopes are juxtaposed in Fig 2. At the same time important developments have taken place in the shell-model calculations describing these nuclei.

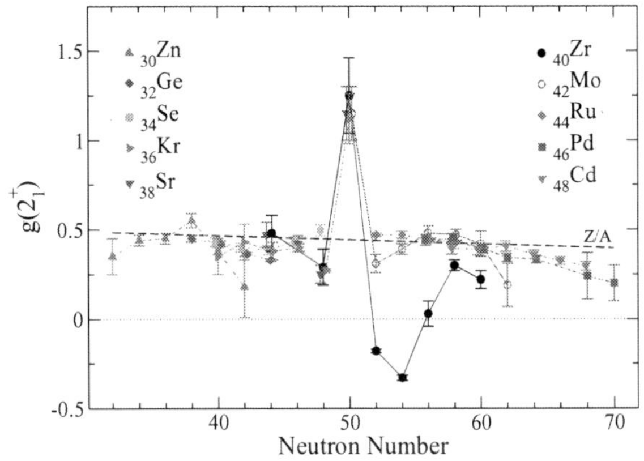

Figure 1. Compilation of all g factors of 2^+_1 states in the even Zn to Cd isotopes [11].

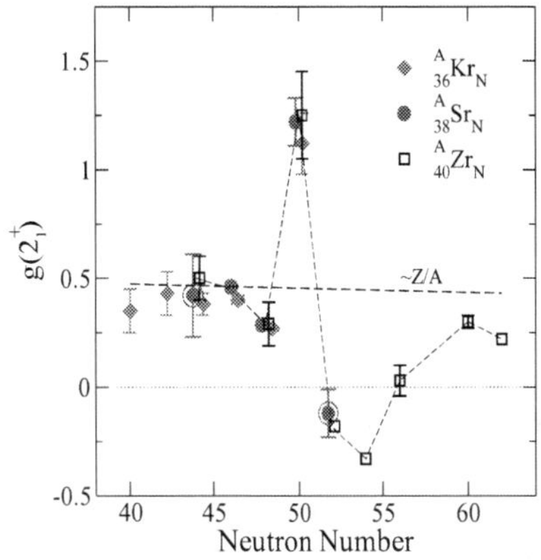

Figure 2. g factors of 2^+_1 states in even Kr, Sr and Zr isotopes. The data for $^{82}_{38}$Sr$_{44}$ and $^{90}_{38}$Sr$_{52}$ (large circles) were obtained in this work.

2. Experimental details

Beams of ^{78,86}Kr were accelerated in the K500 Texas A&M University cyclotron. Experiments were carried out with $\sim$ 1 pnA beams at three different beam energies of 3.0, 3.1, and 3.2 MeV/u. These energies were chosen because they straddle the Coulomb barrier for the beam - target combination of ^{A}Kr - ^{12}C.

Under these conditions, both the Coulomb excitation of Kr isotopes and α-transfer reactions leading to ^{82}Sr and ^{90}Sr can occur. The latter reaction proceeds via capture by a beam ion

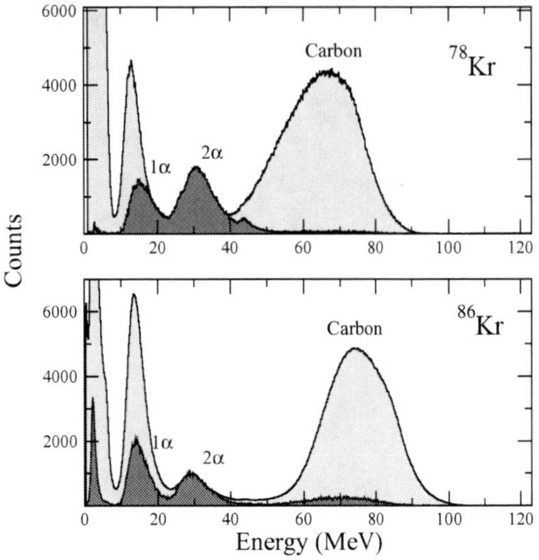

Figure 3. (light grey) Particle spectra obtained with ^{78}Kr and ^{86}Kr beams at 3.2 MeV/u. (dark grey) Particle spectra gated on γ rays belonging to ^{82}Sr or ^{90}Sr ions [11] .

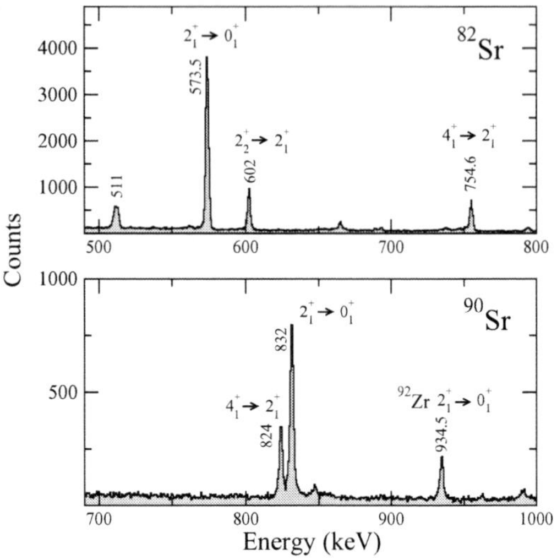

Figure 4. Partial γ spectra observed in coincidence with α particles from the α-capture reactions leading to ^{82}Sr and ^{90}Sr [11].

of an α particle from the C target, with the residual ^{8}Be breaking up immediately into two α particles, ^{12}C(^{78,86}Kr,^{8}Be)^{82,90}Sr.

The C recoils that follow Coulomb excitation reactions and the two α particles from the ^{8}Be breakup were detected in a PIPS silicon detector placed at 0° with respect to the beam. The γ rays emitted by the recoiling excited beam nuclei were detected in four germanium clover detectors located at appropriate angles where the slope of the particle-γ angular correlation is favorable.

A typical particle spectrum identifying the various particle components, is shown in light grey in Fig. 3. The dark color spectra correspond to particle spectra, mostly α particles and protons, obtained by gating on γ rays from ^{82}Sr and ^{90}Sr [11] . Spectra of γ rays from ^{82}Sr and ^{90}Sr in coincidence with the two-α particle-peak are shown in Fig. 4.

Angular correlations were measured for both Coulomb-excitation and α-transfer reactions. These latter reactions lead to nuclei with a lower spin alignment than that observed in Coulomb excitation. However, this reduction in sensitivity is compensated for by the increased α-transfer reaction cross-section.

The details of the experimental setup and of the data acquisition and analysis are extensively described in [11].

3. Results

The g factors of Coulomb-excited 2_1^+, 2_2^+ and 4_1^+ states in ^{78}Kr and the g factor of the 2_1^+ state in ^{86}Kr, measured in the present experiment, reproduce well the values obtained previously [12]. In addition, the magnetic moment of the 4_1^+ state in ^{86}Kr was measured for the first time. The g factors of the 2_1^+ and 4_1^+ states in ^{82}Sr and ^{90}Sr were obtained. The new results are listed in Table 1. The g factors of 2_1^+ states in even-even Kr, Sr and Zr isotopes are shown in Fig. 2.

Table 1. Summary of the new g factors measured in this experiment.

I^{π}	${}^{86}_{36}\mathrm{Kr}_{50}$	${}^{82}_{38}\mathrm{Sr}_{44}$	${}^{90}_{38}\mathrm{Sr}_{52}$
2_1^+		$+0.44(19)$	$-0.12(11)$
4_1^+	$+1.03(14)$	$+0.53(39)$	$-0.02(17)$

4. Theoretical calculations

The nucleus ${}^{82}_{38}\mathrm{Sr}_{44}$ is located in the middle of a major shell, $28 \leq N, Z \leq 50$. Collective models predict $g \sim Z/A$, a result which was indeed observed in this work for both the 2_1^+ and 4_1^+ states, and had been previously observed as well for ${}^{84}_{40}\mathrm{Zr}_{44}$.

The ${}^{86}\mathrm{Kr}$ nucleus, where the neutrons fill the magic N=50 shell and the protons occupy the $p_{3/2}$ and $f_{5/2}$ shells, is amenable to relatively simple shell-model calculations (SM) described below in Section 4.1. For ${}^{90}\mathrm{Sr}$, a full large-space shell-model calculation was carried out and is described in Section 4.2.

4.1. ${}^{86}_{36}\mathrm{Kr}_{50}$

A simple SM calculation for the 2_1^+ and 4_1^+ states in ${}^{86}\mathrm{Kr}$ was carried out [13]. In that calculation the neutrons are inert and there are two proton holes in the $p_{3/2}, f_{5/2}$ orbital space. The Surface Delta Interaction (SDI) was used and it was assumed that the energies of the $p_{3/2}$ and $f_{5/2}$ orbitals are identical. The free-nucleon spin g factor for the proton $(g_s)_{\pi} = +5.586$ was used. The magnetic moments μ of the 2_1^+ and the 4_1^+ states are then given by

$$\mu(2_1^+) = \mathrm{a}^2\ \mu(f_{5/2})^{-2} + \mathrm{b}^2\mu\ [(f_{5/2})^{-1}(p_{3/2})^{-1}] + \mathrm{c}^2\mu(p_{3/2})^{-2} \text{ and}$$
$$\mu(4_1^+) = \mathrm{d}^2\mu[(p_{3/2})^{-1})(f_{5/2})^{-1}] + \mathrm{e}^2\mu(f_{5/2})^{-2}.$$

The values of the coefficients a, b, c, d, and e were obtained by matrix diagonalization resulting in

$\mu(2_1^+) = +2.0$ and $g(2_1^+) = +1.0$, and $\mu(4_1^+) = +4.0$ and $g(4_1^+) = +1.0$.

Both calculations are in agreement with the measurements.

4.2. Large-scale shell-model calculation

Calculations were carried out using the model space outside the ${}^{78}\mathrm{Ni}$ core with the protons in the $f_{5/2}$, $p_{3/2}$, $p_{1/2}$, and $g_{9/2}$ orbitals and the neutrons in the $d_{5/2}$, $g_{7/2}$, $d_{3/2}$, $s_{1/2}$, and $h_{11/2}$ orbitals. The effective interaction for this valence space was constructed by monopole corrections of the realistic G-matrices based on the CD-Bonn potential. A standard M1 operator and the j-coupled code NATHAN were used with $g_{s\,eff} = 0.75\ g_{s\,free}$ [11, 14].

The results of the calculations are shown in Table 2. The calculations are in good agreement with the experimental values for the 2_1^+ and 4_1^+ states in ${}^{86}\mathrm{Kr}$ and ${}^{92}\mathrm{Zr}$. However, the calculations over predict the magnitude of the g factor of the 4_1^+ state of ${}^{90}\mathrm{Sr}$. These observations suggest that proton excitations may be underestimated in the latter calculation.

5. Conclusions

New magnetic-moment measurements of 2_1^+ states in the ${}^{82,90}\mathrm{Sr}$ nuclei, beyond the limits of stability, have been performed as well as a measurement of the magnetic moment of the 4_1^+ state in semi-magic ${}^{86}\mathrm{Kr}$. The new results fit well into the systematics for nuclei in this region in so

Table 2. The results of large-scale shell-model calculations for ^{86}Kr, ^{90}Sr and ^{92}Zr, using the model space outside the ^{78}Ni core with the protons in the $f_{5/2}$, $p_{3/2}$, $p_{1/2}$, and $g_{9/2}$ orbitals and the neutrons in the $d_{5/2}$, $g_{7/2}$, $d_{3/2}$, $s_{1/2}$, and $h_{11/2}$ orbitals [11, 14].

	$^{86}_{36}$Kr$_{50}$		$^{90}_{38}$Sr$_{52}$		$^{92}_{40}$Zr$_{52}$	
	Exp't	SM	Exp't	SM	Exp't	SM
$g(2_1^+)$	+1.10(5)	+1.03	−0.12(11)	−0.09	−0.18(1)	−0.24
$g(4_1^+)$	+1.03(14)	+0.99	−0.02(17)	−0.34	−0.50(11)	−0.43

far as ^{82}Sr can be considered a collective nucleus lying in the middle of a major shell, and ^{90}Sr has the characteristics of a nucleus where single-particle excitations dominate.

Large-scale shell-model calculations for the $N = 52$ isotones ^{90}Sr and ^{92}Zr suggest that the $Z = 40$ core might be more robust than the $Z = 38$ core because proton excitations play a more important role in reducing the magnitude of the magnetic moments in ^{90}Sr than in ^{92}Zr.

This work also suggests that, while waiting for the advent of radioactive beams of sufficient intensity to carry out magnetic moment measurements, the α-transfer reaction mechanism should be studied in more depth in order to establish optimum running conditions and increase the sensitivity of the experiments.

We thank Aldo Covello for his many contributions to nuclear physics over the last three decades and for providing a unique environment where the community could share ideas and thrive.

Acknowledgments

The work was supported in part by the U.S. National Science Foundation. The authors thank their collaborators from many institutions who participated in the experiment.
The support of the Texas A&M Cyclotron Institute staff is very much appreciated.

References

[1] Kumbartzki G, Cooper J R, Benczer-Koller N, Hiles K, Mertzimekis T J, Taylor M J, Speidel K H, Maier-Komor P, Bernstein L, McMahan M A, Phair L, Powell J and Wutte D 2004 *Phys. Lett. B* **591** 213
[2] Davies A D, Stuchbery A, Mantica P F, Davidson P M, Wilson A N, Becerril A, Brown B A, Campbell C M, MCook J, Dinca D C, Gade A, Liddick S N, Mertzimekis T J, Müller W F, Jerry J R, Tomlin B E, Yoneda K and Zwahlen H 2006 *Phys. Rev. Lett.* **96** 112503
[3] Benczer-Koller N, Kumbartzki G J, Gürdal G, Gross C J, Stuchbery A E, Krieger B, Hatarik R, O'Malley P, Pain S, Segen L, Baktash C, Beene J, Radford D C, Yu C H, Stone N J, Bingham C R, Danchev M, Grzywacz R and Mazzocchi C 2008 *Phys. Lett. B* **664** 241
[4] Stuchbery A E, Allmond J M, Galindo-Uribarri A, Padilla-Rodal E, Radford D C, Stone N J, Batchelder J C, Beene J R, Benczer-Koller N, Bingham C R, Howard M E, Kumbartzki G J, Liang J F, Manning B, Stracener D W and Yu C H 2013 *Phys. Rev. C* **88**(11) 051304
[5] Kumbartzki G J, Benczer-Koller N, Torres D A, Manning B, O'Malley P D, Sharon Y Y, Zamick L, Gross C J, Radford D C, Robinson S J Q, Allmond J M, Stuchbery A E, Speidel K H, Stone N J and Bingham C R 2012 *Phys. Rev. C* **86**(3) 034319
[6] Fiori E, Georgiev G, Stuchbery A E, Jungclaus A, Balabanski D L, Blazhev A, Cabaret S, Clément E, Danchev M, Daugas J M, Grevy S, Hass M, Kumar V, Leske J, Lozeva R, Lukyanov S, Mertzimekis T J, Modamio V, Mouginot B, Nowacki F, Penionzhkevich Y E, Perrot L, Pietralla N, Sieja K, Speidel K H, Stefan I, Stodel C, Thomas J C, Walker J and Zell K O 2012 *Phys. Rev. C* **85**(3) 034334

[7] Illana A, Jungclaus A, Orlandi R, Perea A, Bauer C, Briz J A, Egido J L, Gernhäuser R, Leske J, Mücher D, Pakarinen J, Pietralla N, Rajabali M, Rodríguez T R, Seiler D, Stahl C, Voulot D, Wenander F, Blazhev A, De Witte H, Reiter P, Seidlitz M, Siebeck B, Vermeulen M J and Warr N 2014 *Phys. Rev. C* **89** 054316
[8] Speidel K H, Leske J, Schielke S, Bedi S C, Zell O, Maier-Komor P, Robinson S J Q, Sharon Y Y and Zamick L 2006 *Phys. Lett. B* **633** 219
[9] Leske J, Speidel K H, Schielke S, Kenn O, Gerber J, Maier-Komor P, Robinson S J Q, Escuderos A, Sharon Y Y and Zamick L 2005 *Phys. Rev. C* **71** 044316
[10] Torres D A, Kumbartzki G J, Sharon Y Y, Zamick L, Manning B, Benczer-Koller N, Gürdal G, Speidel K H, Hjorth-Jensen M, Maier-Komor P, Robinson S J Q, Ahn T, Anagnostatou V, Elvers M, Goddard P, Heinz A, Ilie G, Radeck D, Savran D and Werner V 2011 *Phys. Rev. C* **84**(4) 044327
[11] Kumbartzki G J, Benczer-Koller N, Burcher S, Ratkiewicz A, Rice S L, Sharon Y Y, Zamick L, Speidel K H, Torres D A, Sieja K, McCleskey M, Cudd A, Henry M, Saastamoinen A, Slater M, Spiridon A, Torilov S Y, Zherebchevsky V I, Gürdal G, Robinson S J Q, Pain S D and Burke J T 2014 *Phys. Rev. C* **89** 064305
[12] Mertzimekis T J, Benczer-Koller N, Holden J, Jakob G, Kumbartzki G, Speidel K H, Ernst R, Macchiavelli A, McMahan M, Phair L, Maier-Komor P, Pakou A, Vincent S and Korten W 2001 *Phys. Rev. C* **64** 024314
[13] Zamick L 2014 private communication
[14] Sieja K, Nowacki F, Langanke K and Martínez-Pinedo G 2009 *Phys. Rev. C* **79** 064310

Ischia14 IOP Publishing
Journal of Physics: Conference Series **580** (2015) 012014 doi:10.1088/1742-6596/580/1/012014

The SPES Radioactive Ion Beam facility of INFN

G. de Angelis[1] for the SPES collaboration,

G. Prete[1], A. Andrighetto[1], M.Manzolaro[1], S.Corradetti[1], D.Scarpa[1], M.Rossignoli[1], A.Monetti[1], M.Lollo[1], M.Calderolla[1], J.Vasquez[1], D.Zafiropoulos[1], L.Sarchiapone[1], D.Benini[1], P.Favaron[1], M.Rigato[1], R.Pegoraro[1], D.Maniero[1], L.Calabretta[2], M.Comunian[1], M.Maggiore[1], A.Lombardi[1], L.Piazza[1], A.M.Porcellato[1], C.Roncolato[1], G.Bisoffi[1], A.Pisent[1], A.Galatà[1], M.Giacchini[1], G.Bassato[1], S.Canella[1], F.Gramegna[1], J.Valiente[1], J.Bermudez[1], P.F. Mastinu[1], J.Esposito[1], J.Wyss[1], A. Russo[2], S.Zanella[1]

1. INFN Laboratori Nazionali di Legnaro, I-35020, Legnaro (Pd), Italy
2, INFN Laboratori Nazionali del Sud, I-95123, Catania, Italy

E-mail: giacomo.deangelis@lnl.infn.it

Abstract. A new Radioactive Ion Beam (RIB) facility (SPES) is presently under construction at the Legnaro National Laboratories of INFN. The SPES facility is based on the ISOL method using an UCx Direct Target able to sustain a power of 10 kW. The primary proton beam is provided by a high current Cyclotron accelerator with energy of 40 MeV and a beam current of 0.2-0.5 mA. Neutron-rich radioactive ions are produced by proton induced fission at an expected fission rate of the order of 10^{13} fissions per second. After ionization and selection the exotic isotopes are re-accelerated by the ALPI superconducting LINAC at energies of 10A MeV for masses in the region A=130 amu. The expected secondary beam rates are of the order of 10^{7} - 10^{9} pps. Aim of the SPES facility is to deliver high intensity radioactive ion beams of neutron rich nuclei for nuclear physics research as well as to be an interdisciplinary research center for radio-isotopes production for medicine and for neutron beams.

1. Introduction

SPES is an acronym for Selective Production of Exotic Species. The aim of the SPES project is to provide high intensity and high-quality beams of neutron-rich nuclei to perform forefront research in nuclear structure, reaction dynamics and interdisciplinary fields like medical, biological and material sciences. SPES is a second generation ISOL radioactive ion beam facility. It represents an intermediate step toward the future generation European ISOL facility EURISOL [1]. The SPES project is part of the INFN Road Map for Nuclear Physics; it is supported by the Italian national laboratories LNL (Legnaro) and LNS (Catania). It is based on the ISOL method with an UCx Direct Target able to sustain a power of 10 kW. The primary proton beam is delivered by a Cyclotron accelerator with an energy of 40 MeV and a beam current of 200 μA. Neutron-rich radioactive ions will be produced by Uranium fission at an expected fission rate in the target of the order of 10^{13} fissions per second. The exotic isotopes will be re-accelerated by the ALPI superconducting LINAC at energies of 10A MeV and higher, for masses in the region of A=130 amu, with an expected rate on the secondary target of 10^{7}-10^{9} pps.

Journal of Physics: Conference Series **580** (2015) 012014 doi:10.1088/1742-6596/580/1/012014

2. Challenges of the SPES project

To understand the properties of a nucleus, apart from establishing the interaction between its components, it is necessary to determine the arrangement of the nucleons, i.e. the structure. Presently our knowledge about the structure of nuclei is mostly limited to nuclei close to the valley of stability or nuclei with a deficiency of neutrons. Only recently the availability of beams of unstable ions has given access to unexplored regions of the nuclear chart, especially on the neutron-rich side. Starting from a nucleus on the stability line and adding successively neutrons, one observes that the binding energy of the last neutron decreases steadily until it vanishes and the nucleus decays by neutron emission. The position in the nuclear chart where this happens defines the neutron drip line. It lies much farther away from the valley of stability than the corresponding drip line associated with protons, owing the absence of electrical repulsion between neutrons. The location of the neutron drip line is largely unknown as experimental data are available only for nuclei with mass up to around 30. The interest in the study of nuclei with large neutron excess is not only focused on the location of the drip line but also on the investigation of the density dependence of the effective interaction between the nucleons for exotic N/Z ratios. In fact, changes of the nuclear density and size in nuclei with increasing N/Z ratios are expected to lead to different nuclear symmetries and new excitation modes. While in the case of some very light nuclei a halo structure has been identified, for heavier nuclei the formation of a neutron skin has been predicted. The evolution of nuclear properties towards the neutron drip line depends on how the shell structure changes as a function of neutron excess. This evolution has consequences on the ground state properties of the nuclei and on the single-particle and collective excitations. In particular, studies of neutron-rich nuclei beyond the doubly magic ^{132}Sn are of key importance to investigate the single-particle structure above the N=82 shell closure and find out how the effective interaction between valence nucleons behaves far from stability. New modes of collective motion with large B(E1) strength are also expected in connection with the formation of a neutron skin, namely oscillation of the skin against the core as recently proposed for the neutron rich Ni and Sn isotopes. Presently, neither the thickness nor the detailed properties of the neutron skin of exotic nuclei are known. This information is needed to enable a quantitative description of compact systems like neutron stars, where exotic nuclei forming a Coulomb lattice are immersed in a sea of free neutrons, a system which is expected to display the properties of both finite and infinite (nuclear matter) objects. At the energy of SPES, it will be possible to address important questions related to the study of neutron-rich matter such as nuclear forces, level density, viscosity, barrier, neutron pairing and collective modes. With the high intensity beams delivered by SPES a challenging and broader range of studies in nuclear spectroscopy and reaction mechanism will be performed. Interesting areas where new data will be collected are those in the very neutron rich regions, where shell evolution is an issue. Effects of how the pairing interaction is modified in the nuclear medium will receive significant inputs by measurements of multi-nucleon transfer reactions to specific nuclear states. Effects of rotational damping in the decay of high energy levels, for instance the dynamical dipole emission, will be studied by changing the N/Z of projectile and target. Neutron rich nuclei will be used to investigate the tunneling process in sub-barrier fusion reactions in presence of very positive Q-values, an issue interesting also for astrophysics.

3. The Spes Radioactive Ion Beam Facility

SPES is an INFN project based on the ISOL (Isotope Separation On Line) technique for radioactive beam production and it is presently under construction at the Legnaro National Laboratories. A primary beam provided by a driver accelerator induces nuclear reactions inside a thick target. The reaction products are extracted from the target by thermal process due to the high temperature of the target-ion source system (exceeding 2000° C). Once reached the 1^+ source, the products are ionized and extracted. After an isotopic selection and a further ionization to n^+ , the exotic beam is injected into a post-accelerator.

The SPES facility is based on a proton beam to induce fission on a UCx target (Direct Target). The driver is a 70 MeV proton cyclotron with two exit ports providing a total current of 750 μA. Two

ISOL production targets are planned to be installed and operated alternatively. As the cyclotron can supply two beams at the same time, a second independent facility can be operated in parallel. In particular the high intensity proton beam will be used to produce radioisotopes for nuclear medicine as well as a neutron source. Concerning the latter possibility, the expected neutron production at SPES will exceed 10^{14} n/s with energy up to 40 MeV and a peak between 0.1 and 10 MeV according to target and moderator. The proton beam of the SPES cyclotron is suited for the production of new radioisotopes for medical applications. Taking advantage of the high energy and current, it opens up the possibility to produce ^{82}Sr, ^{67}Cu, and ^{47}Sc isotopes to be used in therapy and diagnostics.
A pictorial sketch of the facility is shown in figure 1.

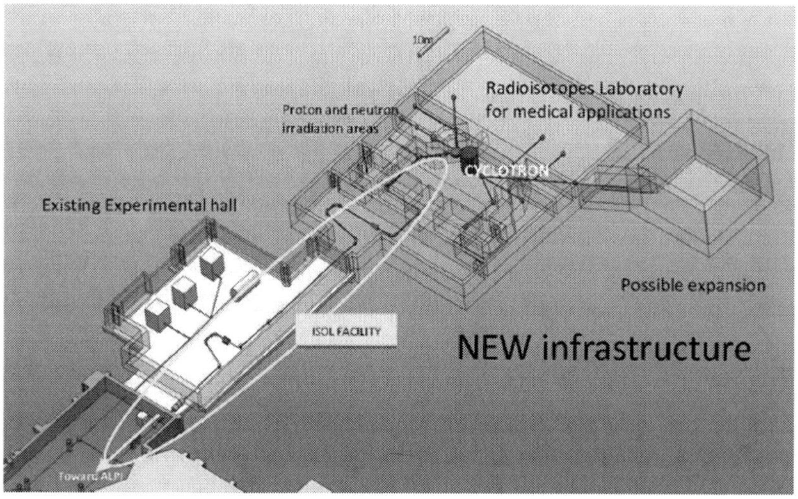

Figure 1. Pictorial sketch of the SPES facility. The ALPI superconducting LINAC (not shown) is on the left side.

3.1. The Isol front-end
The ISOL front-end is the system which couples the proton beam with the UCx target, the ion source and the first part of the transport line of the exotic beam. The whole system has been completed and is now in operation (with stable beam) at LNL for off-line tests. The most critical element of the SPES ISOL facility is the Direct Target. It follows the basic design of the CERN ISOLDE one but represents an innovation in term of capability to sustain the primary beam power. The design is carefully oriented to optimize the radiative cooling, taking advantage of the target system's high operating temperature. The SPES target is designed in order to maximize the release efficiency and the heat dissipation. It consists of multiple thin disks housed in a cylindrical graphite box. It is composed of 7 UCx disks (diameter and thickness of 40 and 1.3 mm, respectively), appropriately spaced in the axial direction in order to dissipate, by thermal radiation, the average power of 8 kW due to the energy release of the proton beam which, passing through them, induces nuclear reactions. The graphite box is housed inside a tubular hollow tungsten ohmic heater. In fact, due to the intense heat exchange by radiation, the energy released by the proton beam is not sufficient to make the system reach the operating temperature. This allows for a better thermal control of the target operation. An extensive simulation study of the target behavior for thermal and release properties is at the basis of the target-ion-source design. Experimental work to benchmark the simulations was carried out in collaboration with HRIBF [2], the Oak Ridge National Laboratory ISOL facility (USA). Partial results on the measured production rates (and on the expected rates of the reaccelerated SPES beam) are shown in figure 2.

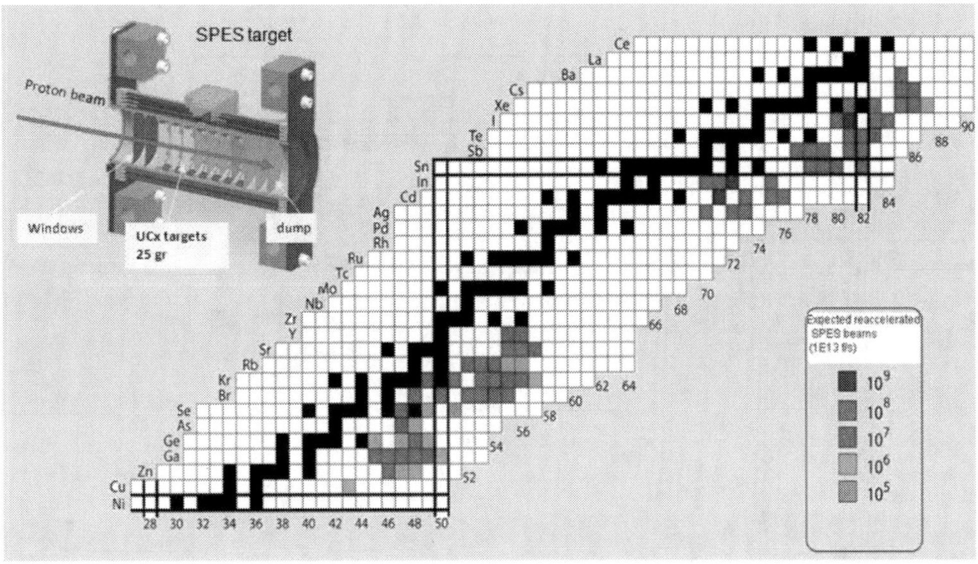

Figure 2. Expected beam on target intensities. Experimental data scaled to: 0.2 mA proton current, 25% transport efficiency. ISOL target operated at 2000^{0}C, ionization and extraction with 1^+ plasma source.

3.2 The ion-source system

The interaction of the proton beam with the UCx target will produce neutron-rich fission fragments subsequently extracted by thermal motion and ionized at 1^+ charge state by the Isol source directly connected to the production target. The hot-cavity ion source chosen for the SPES project was designed at CERN (ISOLDE) [3]. The source has the basic structure of the standard high temperature RIB ion sources employed for on-line operation. The ionizer cavity is a tungsten tube (34 mm length, 3 mm inner diameter and 1 mm wall thickness) resistively heated to near 2000^{0}C. The isotopes produced in the target diffuse in the target material and then effuse through the transfer tube into the ionizer cavity where they undergo surface or laser ionization. The Surface ionization process can occur when an atom comes into contact with a hot metal surface. In the positive surface ionization, the transfer of a valence electron from the atom to the metal surface is energetically favorable for elements with an ionization potential lower than the work function of the metal. For alkalis and some rare earth elements, high ionization efficiencies can be achieved using the surface ionization technique. Such 1^+ source has good efficiency and selectivity for the elements as Rb, Cs, Ba. For most of the other elements, the laser resonant photo-ionization, using the same hot cavity cell, is a powerful method to achieve sufficiently selected ionization. This technique, now under development, has been implemented with the aim to produce beams as pure as possible (chemical selectivity) also for metal isotopes. To ionize elements with high ionization potential, as rare gasses, the plasma source is needed. This source ionizes all the elements without any selectivity. At SPES both surface and plasma sources have been developed and are in operation at the off line test stand. The laser source is presently under development.

3.3. Beam selection and transport

The first mass selection is performed by a Wien Filter with 1/100 mass resolution, installed just after the first electrostatic quadrupole triplet inside the production bunker with the aim to confine the larger part of radioactivity inside the high shielded area. The transfer line toward ALPI is equipped with several beam handling systems to purify the beam. A Beam Cooler and a High Resolution Mass Separator (HRMS) with 1/20000 mass resolution will be installed inside the new building.

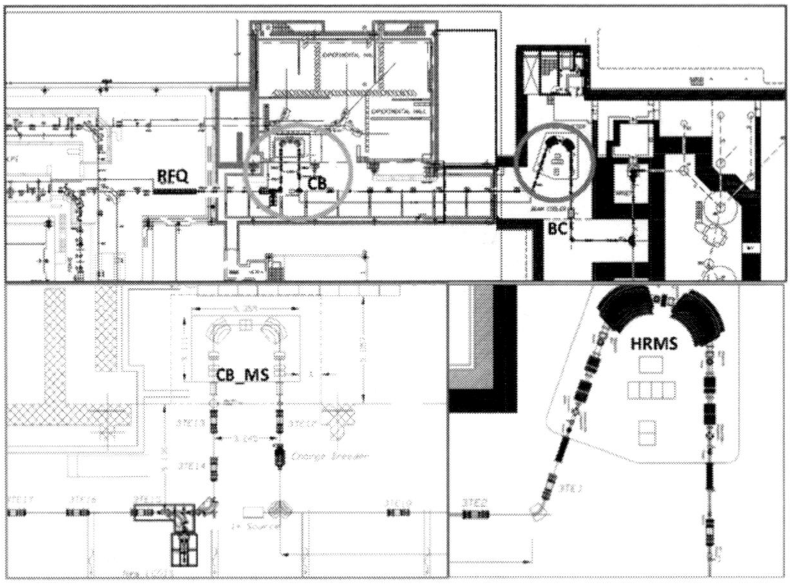

Figure 3. Layout of radioactive beam handling and transfer line. High resolution mass separator and the Charge Breeder mass separator are shown in the lower panels.

Special efforts have been dedicated to design a mass spectrometer with an effective mass resolution of at least 1/20000. Such design takes advantage of a Beam Cooler, to improve the beam emittance, and of a 260 keV HV platform on which the mass separator is mounted. The mass separator is a scaled-up version of the separator designed for the CARIBU facility [4] in ANL (USA). In the SPES configuration the calculated resolution is of the order of 1/20000 constrained by an emittance of 3 mm mrad and an energy spread of 1.3 eV. Before the injection in the ALPI superconducting LINAC it is necessary an increase of the charge state from 1^+ to n^+. This is performed by means of a Charge Breeder. The SPES Charge Breeder is based on ECR method [5] and aims to produce ions with A/q less than 7 at A~130. The design and construction of the Charge-Breeder will be performed in the framework of the SPES-SPIRAL2 collaboration at LPSC (Grenoble, France) following an up-graded version of the Phoenix booster in operation at LPSC. After the Charge Breeder a second mass separator with 1/1000 mass resolution, will be installed to clean the beam from the contaminants introduced by the Charge Breeder itself. The same basic configuration used for the HRMS is adopted. Finally an RFQ pre-accelerator will increase the beam energy to match the ALPI acceptance. The layout of the radioactive beam line is showed in figure 3.

3.4. Exotic beams reacceleration

The secondary beams are injected into the ALPI superconducting LINAC by a new RFQ that will operate in a CW mode (100% duty factor) at a resonant frequency of 80 MHz. This frequency is the same as that of the superconducting structures of the lowest energy section of ALPI. The injection energy of the ions is set to 5.7 keV/u. This choice is a compromise between the desire to reduce the ion energy to simplify the low energy injection and the RFQ bunching section design and the need to increase the beam rigidity in the 1^+ transport line to reduce space charge effects. The extraction energy of the RFQ is set to 727 keV/u to optimize the beam dynamics of the superconducting LINAC ALPI.
ALPI is the superconducting heavy ion LINAC in operation at LNL since the early nineties. It is composed of three sections of quarter wave resonators (QWR) for a total of 80 cavities installed and operates routinely at an equivalent acceleration voltage of 50 MV. The LINAC is constructed in a

bended configuration: it is composed of two branches connected by an achromatic and isochronous U-bend. It uses three different kinds of cavities: Low Beta, Medium Beta and High Beta cavities, according to the different velocity along the acceleration path.
The ALPI LINAC has gone through a continuous upgrade in the number and performance of its accelerating cavities, and consequently in the maximum achievable beam energy. In the last years the cavities of the medium energy QWR section have been upgraded using a new Nb sputtered coating in substitution to the original Pb sputtered layer. A further upgrade program is on the way, to improve the accelerating fields of the present QWRs and adding more cavities in the Low Beta section. The final equivalent voltage will exceed 70 MV. In figure 4 the expected beam energies are reported.

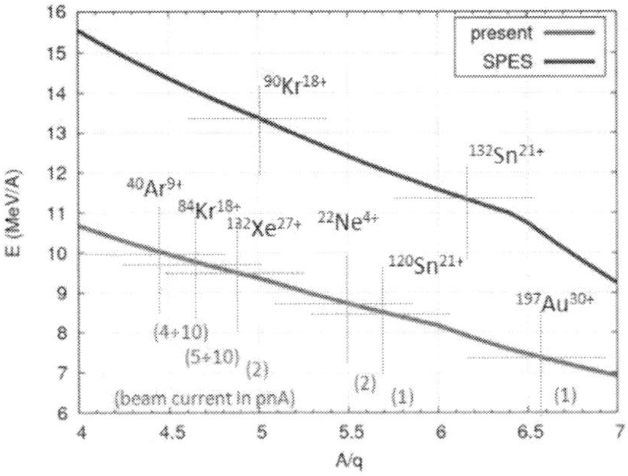

Figure 4. Beam energy performance of ALPI superconductive LINAC.

3.5 Status of the SPES project
The SPES facility is presently under construction at LNL. The building is at an advanced phase and will be completed at end of 2014. The cyclotron, provided by the BEST company in Canada, is under construction (the magnet is completed and the magnetic field was positively mapped). It is expected to be transferred to LNL, after the factory commissioning, in the end of 2014. The authorizations to operate the cyclotron have been already obtained allowing the acceleration of the proton beam on non-fissile target at high power and testing the UCx target up to a current of 5 μA. A further authorization request to run the UCx target at design power will be presented to the Italian authorities within this year. The physics design of the radioactive ion beam transfer line (1^+ and n^+) is completed and the reshaping of the ALPI beam line in the experimental hall is initiated to allow the installation of the Charge Breeder and of the mass separator. The ISOL front-end is completed and under operation in the off-line laboratory to test ion sources, assembly and controls. The first reaccelerated exotic beam is expected around 2018 [6].

References
[1] www.eurisol.org
[2] Scarpa D, Biasetto L, Corradetti S, Manzolaro M, Andrighetto A, Carturan S, Prete G, Zanonato P and Stracener D W 2011, *Eur. Phys. J.* **47** 32
[3] Lindroos M 2004, CERN-AB-2004-086
[4] Savard G, Pardo R, Baker S, Davids C, Levand A, Peterson D, Phillips D, Sun, T, Vondrasek R. Zabransky B and Zinkann G 2011, *Hyperfine Interactions* **199** 301
[5] Lamy T, Bouly J L, Curdy J C, Sole P, Sortais P, Thuillier T, Vieux-Rochaz J L and Voulot D 2004, *Rev. Sci. Instrum.* **75** 1624
[6] https://web.infn.it/spes/

Progress on $M1$ Research

N Pietralla[1], G Rainovski[2], M Reese[1], C Stahl[1], T Beck[1], J Beller[1], C Romig[1], and V Werner[1]

[1] Institut für Kernphysik, Technische Universität Darmstadt, 64289 Darmstadt, Germany
[2] St. Kliment Ohridski University of Sofia, Sofia, Bulgaria

E-mail: pietralla@ikp.tu-darmstadt.de

Abstract. Magnetic dipole transitions in nuclei can offer crucial information on nuclear structure, *e.g.*, on spin-flip processes or on nuclear charge currents. Driven by the need for more comprehensive or more precise data on $M1$ transitions in nuclei, we have recently tried to advance pre-existing experimental methods to meet the present needs. This contribution reports on the current status of our efforts on three aspects: (i) access of spin-flip $M1$ strength in radioactive nuclei, (ii) quadrupole-collectivity of the $M1$ scissors mode, and (iii) highest-precision measurements for strong ground-state $M1$ excitations.

1. Introduction

Magnetic dipole ($M1$) radiation provides crucial information on nuclear physics and insight into nuclear structure phenomena. An important example is the $M1$ excitation of the nucleon itself to its $J^\pi = 3/2^+$ Δ-resonance. It represents an isovector transition from the $T = 1/2$ isospin doublet of the nucleon to the Δ-resonance with isospin $T = 3/2$. Its matrix element amounts to about 3 nuclear magnetons ($\mu_N = e\hbar/2m_N c$). This transition corresponds to a quark spin-flip excitation in the schematic static quark model.

Also the simplest nucleus, the deuteron, features an isovector $M1$ excitation from its $J^\pi_T = 1^+_0$ ground state to its first excited $J^\pi_T = 0^+_1$ resonance. In a simplistic way, this transition can also be considered as a quark spin-flip transition, now flipping the spin of one nucleon versus that of the other. Consequently its $M1$ strength is comparable to that of the nucleon. This isovector $M1$ strength manifests itself most notably in the isovector $M1$ transition between the quasi-deuteron [1] partner states of ^{6}Li, its $J^\pi_T = 1^+_0$ ground state and its $J^\pi_T = 0^+_1$ state at 3.56 MeV excitation energy, which again has an $M1$ transition matrix element of about 4 μ_N in size.

Besides spin-flip dynamics between spin-orbit partner orbitals, $M1$ transitions are sensitive to changes in charge currents inside the nucleus, too. The electromagnetic charge distinguishes protons from neutrons in the nuclear environment. Hence, the investigation of $M1$ transitions between nuclear states [2] can provide us with information on the proton/neutron structure of nuclear states and on the local spin-orbit splitting of effective single-particle orbitals. Together these effects yield access to the dominant degrees of freedom in the formation of nuclear structures in the nuclear valence shell. The paramount importance of the evolution of effective single-particle energies as a function of nucleon numbers on the emerging nuclear structure has become apparent through the advent of spectroscopic information on exotic nuclei over the last two decades. The tensor force between nucleons [3] and effective three-body forces [4] have been shown to be essential ingredients for a global understanding of nuclear structure across the nuclear chart.

In this contribution we have addressed our recent progress on the investigation of nuclear structures that dominantly decay by $M1$ radiation: the new approach of relative γ-ray self-absorption for a precision measurement of the $M1$ excitation strength of ^{6}Li, the first attempt to access the electromagnetic quadrupole radiation strength of the proton/neutron mixed-symmetric orbital 1^+ scissors mode in a quadrupole-deformed nucleus, and a new approach for directly measuring the local spin-orbit splitting in neutron-rich radioactive nuclei by the two-target relativistic projectile-Coulomb excitation method. We have performed corresponding experiments at the Superconducting DArmstadt LINear electron ACcelerator (S-DALINAC) at TU Darmstadt, at the High Intensity γ-ray Source (HIγS) at Duke University, and at the FRagment Separator (FRS) at the GSI Helmholtz Center for Heavy Ion Research at Darmstadt in the scope of the PreSPEC/AGATA campaign.

2. Relativistic $M1$ Coulomb excitation experiment on ^{85}Br with PreSPEC/AGATA

Exotic nuclei with ratios of neutron to proton numbers differing considerably from what is encountered near the line of stability have been found to often feature hitherto unexpected structure, *e.g.*, strong collectivity at the conventional "magic numbers" or aspects of closed shell nuclei even when the proton and neutron numbers do not coincide with them. These phenomena are attributed to an evolution of shell structure which itself arises from a systematic dependence of the effective single-particle energies on the nucleon numbers by weakening or strengthening the local spin-orbit splitting as a function of valence nucleons.

It is apparently desirable to experimentally measure the energy splitting of spin-orbit partner orbitals in exotic nuclei. Spin-flip transitions between spin-orbit partners correspond to $M1$ transitions with a size on the order of one Weisskopf unit, *i.e.*, $M1$ transitions that are comparatively strong. Therefore, the occurence of strong $M1$ transitions in the range of one nuclear magneton can serve as a unique signature for spin-orbit partner orbitals. The direct measurement of such large $M1$ transition rates for γ-ray energies of about one MeV or more is a difficult task, in particular, in heavy exotic nuclei. That is because such transition rates lead to half-lives below one hundred femtoseconds where techniques for direct lifetime measurements start to fail. Classical techniques for unique identication of spin-flip excitations, such as direct measurements of $M1$ radiative widths in photon scattering or electron scattering reactions, or high-resolution charge-exchange reactions, are not yet applicable to the investigation of radioactive isotopes. The measurement of $E2/M1$ mixing ratios δ from the analysis of angular distributions is significantly hindered at relativistic beam energies due to large Lorentz boost and by low statistics in experiments with radioactive ion beams. We have recently proposed that the method of relativistic $M1$ projectile-Coulomb excitation may offer a viable alternative.

The ratio of cross sections for magnetic and electric Coulomb excitation is in first order proportional to $(v/c)^2$ times the ratio of the corresponding matrix elements squared. For sufficiently large $M1$ matrix elements the magnetic contribution to the cross section can be detectable at relativistic velocities, while traditional sub-Coulomb barrier Coulomb excitation is dominated by far by electric excitations. Because of the different energy dependence for electric and magnetic multipoles in the Coulomb excitation process, the ratio of Coulomb-excitation cross-sections at different beam energies allows to extract the ratio of the excitation characters contributing to the respective excitation and, hence, the multipole mixing ratio of the transition and the total electromagnetic transition matrix elements. Exploiting these properties of the Coulomb-excitation process, the straightforward way to extract $E2$ and $M1$ matrix elements is to perform two Coulomb-excitation experiments at different beam energies and determine the ratio of measured cross-sections. This approach demands sufficient beam-time and two cross-sections measurements. In γ-ray spectroscopy experiments with detectors that have a sufficiently good

Journal of Physics: Conference Series **580** (2015) 012015 doi:10.1088/1742-6596/580/1/012015

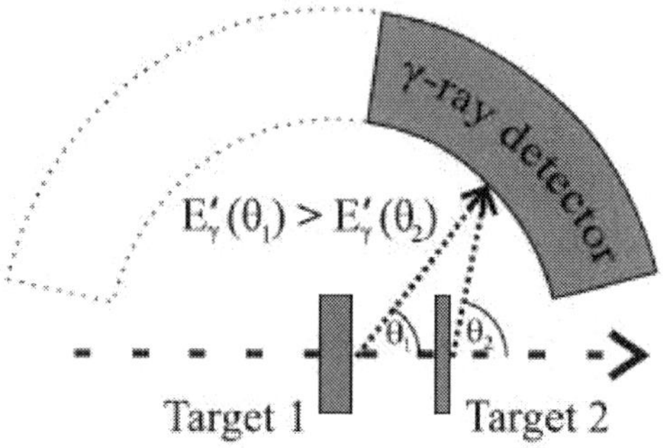

Figure 1. The relativistic ion beam impinges from the left on two targets that are spatially separated by about 10 cm. The first target is thick enough to slow down the beam such that $E2$ and $M1$ contributions to the Coulex cross sections are sufficiently different. Gamma-radiation from the two targets can be distinguished by the different Doppler-shifts due to the considerably different observation angles seen by the granular γ-ray detector array.

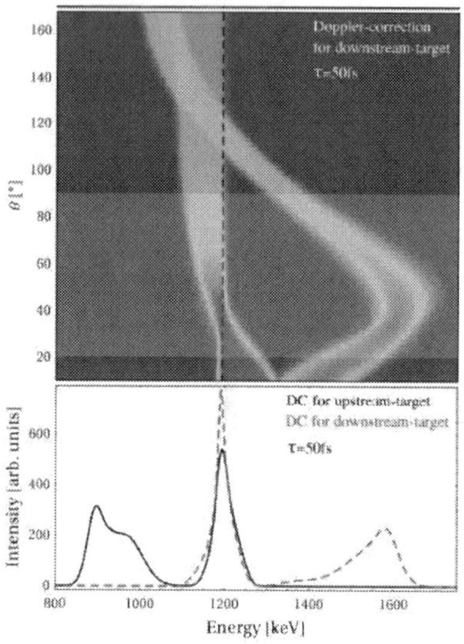

Figure 2. Simulation for the separation of γ-rays from relativistic ions that were Coulomb excited either in the first or in the second target. The ratio of the peak areas is sensitive to the $M1$ matrix element to be measured.

energy resolution, like HPGe detectors, the two cross-section measurements can be combined in one single experiment with two spatially separated targets and a sufficiently large energy-loss in the first target [5]. The principle of the two-target relativistic Coulex multipolarity is exemplified in Figs. 1 and 2.

We report here on the first experiment aiming at the measurement of a proton spin-flip $M1$ transition rate in a relativistic $M1$ projectile-Coulomb excitation reaction. It was carried out on a secondary ion beam of radioactive neutron-rich $^{85}Br_{50}$ nuclei at the neutron shell closure $N = 50$ in the sequence of isotones from stable nuclei towards the neutron-rich doubly magic nucleus ^{78}Ni. Due to the crossing of effective proton single-particle energies of the $\pi(2p_{3/2})$ orbital

Figure 3. Photograph of the PreSPEC/AGATA set-up at the FRS at GSI. The radioactive-ion beam enters from the right. The secondary double target (inset) is surrounded by scintillators and HPGe-detectors of the Hector and AGATA arrays. The secondary reaction products are mass and charge analyzed event-by-event by the LYCCA device [10] to the left.

with the $\pi(1f_{5/2})$ orbital along the chain of heavy nickel isotopes it is interesting to study the evolution of the spin-orbit splitting between the $\pi(2p_{3/2})$ and the higher-lying $\pi(2p_{1/2})$ orbital in this region of the nuclear chart.

Relativistic beams of ^{85}Br with velocities of up to 70% of the speed of light have been abundantly provided by the FRS at GSI [6] following the primary one-proton knock-out reaction of a beam of fully stripped ^{86}Kr ions in a beryllium production target in March 2014. High-resolution in-flight γ-ray spectroscopy at relativistic energies requires precise Doppler-shift correction capabilities. Gamma-rays were measured in-beam with the highly-segmented novel PreSPEC/AGATA device [7, 8, 9]. We report on this new experimental technique, on our experiences with it, and on the status of the data analysis. Figure 3 shows a photograph of the set-up.

The two targets for secondary Coulomb excitation consisted of massive gold plates of about 1 mm thickness, each. The data analysis is currently underway. Calibrations for proper Doppler correction have not been completed, yet. Figure 4 shows a γ-ray spectrum from 5% of the data taken without Doppler correction. A γ-ray line at 547.5(3) keV corresponding to the Coulomb excitation of the ^{197}Au target nuclei is clearly visible. Its yield implies that the sought-after γ-ray line from relativistic $E2/M1$-Coulomb excitation of the $J^\pi = 1/2^-$ level of ^{85}Br at 1.191 MeV excitation energy should be detectable after Doppler corrections if that state would sufficiently purely correspond to the spin-flip excitation of the $\pi(2p_{3/2})$, $J^\pi = 3/2^-$ ground state.

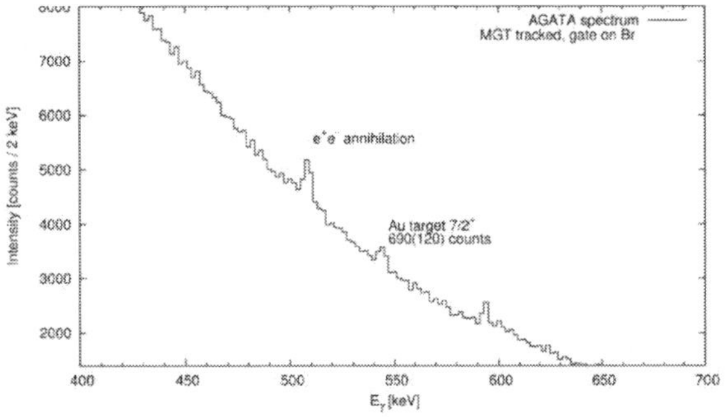

Figure 4. Gamma-ray spectrum without Doppler correction from a part of the data taken spring 2014 with the PreSPEC/AGATA array on the relativistic Coulomb excitation reaction of a beam of radioactive ^{85}Br ions on two gold targets. Coulomb excitation of gold is clearly visible and can serve for the determination of absolute $E2$ and $M1$ relativistic Coulomb-excitation cross sections for ^{85}Br. The data analysis is underway.

3. Quadrupole decay strength of the $M1$ scissors mode of ^{156}Gd

Orbital charge currents can be another source of nuclear $M1$ radiation besides spin-$M1$ modes as discussed above. The most prominent example of orbital $M1$ strength is given by the $J^\pi = 1^+$ nuclear scissors mode discovered in the deformed nucleus ^{156}Gd by Richter and collaborators at Darmstadt in the 1980s [11]. The orbital $M1$ excitation strength is proportional to the square of the quadrupole deformation of the ground state. While the $M1$ strength of the scissors mode has been studied in detail over the last 30 years, the electric quadrupole matrix element between the rotational band of the scissors mode and the ground band is still unknown. It can be obtained from the measurement of the $E2/M1$ multipole mixing ratio of the $1^+_{\rm sc} \rightarrow 2^+_1$ transition. We have measured this multipole mixing ratio for the first time in a deformed nucleus.

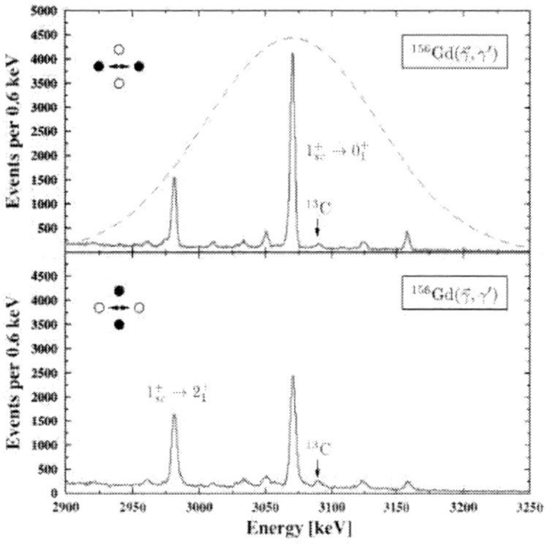

Figure 5. Gamma-ray spectroscopy of the scissors mode of ^{156}Gd excited by the quasi-monoenergetic linearly-polarized photon beam of the HIγS facility at Duke University, Durham, NC. The beam's energy-profile is indicated by the dashed curve. The $1^+_{sc} \to 0^+_1$ $M1$ transition to the ground state at 3070 keV and the $E2/M1$ mixed transition to the 2^+_1 state of the ground state rotational band are dominantly observed. The relative azimuthal distribution about the (horizontal) polarization plane of the incident γ-ray beam is sensitive to the $E2$ contribution of the $1^+_{sc} \to 2^+_1$ transition.

The experiment has been performed on the nuclide ^{156}Gd using the quasi-monoenergetic linearly-polarized photon beam of the High-Intensity γ-ray Source (HIγS) at Duke University, Durham, NC [12]. Figure 5 shows a relevant part of the data. The $E2$ matrix element between the scissors mode and the ground state band of ^{156}Gd is appreciably smaller than 1 W.u. The data and results are currently being prepared for publication.

4. Relative self-absorption experiment on the quasi-deuteron $M1$ excitation of ^{6}Li

The first excited 0^+ state of ^{6}Li with isospin $T = 1$ is the first nuclear state in the chart of nuclides that predominantly decays electromagnetically. Its $M1$ decay width provides a sensitive test for

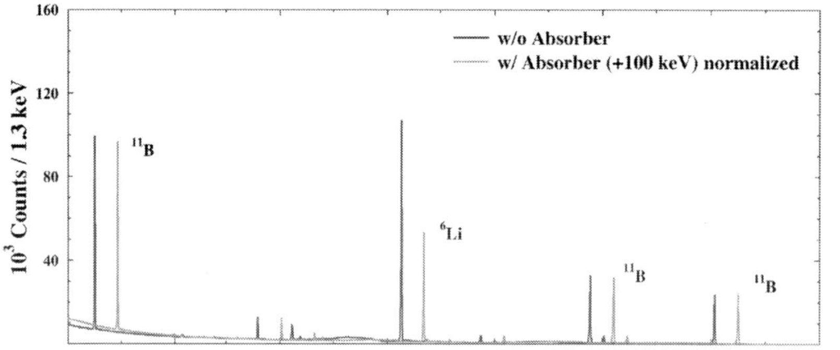

Figure 6. Nuclear resonance fluorescence off ^{6}Li in a nuclear self-absorption experiment at the Darmstadt bremsstrahlung site at the S-DALINAC. The 3563-keV γ-ray line from the $J^\pi_T = 0^+_1 \to 1^+_0$ isovector $M1$ decay transition in ^{6}Li is very pronounced. The lithium target was surrounded by ^{11}B for normalization purposes. The figure shows two γ-ray spectra from one and the same target, one of them taken with a bremsstrahlung beam modified by an upstream lithium absorber (and shifted by +100 keV for better visibility). The absorption of 3563-keV photons by the absorber is proportional to the natural line-width of this state and leads to a precisely measurable reduction of NRF intensity relative to a normalization standard.

modern nuclear theory. It has recently been shown that chiral currents can contribute up to 10% to $B(M1)$ values [13]. A precision measurement of the $M1$ decay strength in ^{6}Li to better then 1% accuracy is desirable. We have recently demonstrated that this goal can be achieved in nuclear self-absorption [14] experiments. Figure 6 shows the γ-ray spectra that were obtained. The self absorption of the $M1$ excitation of ^{6}Li has been determined to sub-% level.

5. Summary

In our contribution we have discussed our recent progress on the experimental investigation of magnetic dipole phenomena in nuclei. While our data are currently under analysis it is obvious that new instruments or the advances of traditional experimental techniques have the potential for new approaches to and deeper insight into nuclear $M1$ phenomena.

Acknowledgments

We have reported progress on three different experiments addressing $M1$ observables in some ways and we thank all those who have contributed to these experiments. In particular, we thank D. Bazzaco, P. Boutachkov, L. Cortes, C. Domingo Pardo, J. Gerl, M. Gorska, I. Kojouharov, W. Korten, N. Lalovic, M. Lettmann, S. Pietri, D. Ralet, and D. Rudolph, for their contributions to the PreSPEC/AGATA experimental campaign at the FRagment Separator at GSI, and the entire PreSPEC collaboration for support. We thank all those who have contributed to the photon-scattering experiments at TU Darmstadt and at Duke Univ., in particular, M. Arnold, J. Conrad, V. Derya, F. Hug, B. Löher, T. Kürzeder, H. Pai, D. Savran, M. Scheck, W. Tornow, H.R. Weller, and M. Zweidinger, for their support. We gratefully acknowledge the financial support by the German Research Council (DFG) under grant No. SFB 634, by the Helmholtz International Center for FAIR (HIC for FAIR) funded in the scope of the LOEWE program by the State of Hesse, by the German Ministry for Education and Research (BMBF) under grant No. 05P12RDFN8, by the EU through the ENSAR grant, and by the Bulgarian NSF under grant No. DDVU0269.

References

[1] Lisetskiy A F, Jolos R V, Pietralla N, and von Brentano P 1999 *Phys. Rev.* C **60** 064310
[2] Heyde K, von Neumann-Cosel P, and Richter A 2010 *Rev. Mod. Phys.* **82** 2365
[3] Otsuka T 2013 *Phys. Scr.* T **152** 014007
[4] Otsuka T, Suzuki T, Holt J D, Schwenk A, and Akaishi Y 2010 *Phys. Rev. Lett.* **105** 032501
[5] Stahl C, Pietralla N, Rainovski G, Reese M 2014 *Nucl. Instrum. Methods Phys. Res.* A, subm. for publ.
[6] Geissel H *et al.* 1992 *Nucl. Instr. and Meth. Phys. Res.* B **70** 286
[7] Akkoyun S *et al.* 2012 *Nucl. Instr. and Meth. Phys. Res.* A **668** 26
[8] Domingo-Pardo C *et al.* 2012 *Nucl. Instr. and Meth. Phys. Res.* A **694** 297
[9] Pietralla N *et al.* 2014 *EPJ Web of Conferences* **66** 02083
[10] Golubev P *et al.* 2013 *Nucl. Instr. and Meth. Phys. Res.* A **723** 55
[11] Bohle D, Richter A, Steffen W, Dieperink A E L, Lo Iudice N, Palumbo F, and Scholten O 1984 *Phys. Lett.* B **137** 27
[12] Weller H R, Ahmed M W, Gao H, Tornow W, Wu Y K, Gai M, and Miskimen R 2009 *Prog. Part. Nucl. Phys.* **62** 257
[13] Pastore S, Pieper S, Schiavilla R, and Wiringa R 2013 *Phys. Rev.* C **87** 035503
[14] Vodhanel R, Brussel M K, Moreh R, Sellyey W C, and Chapuran T E 1984 *Phys. Rev.* C **29** 409

The Nuclear Pairing Gap - How Low Can It Go?

B. Alex Brown

Department of Physics and Astronomy and National Superconducting Cyclotron Laboratory, Michigan state University, East Lansing, Michigan 48824-1321, USA

E-mail: brown@nscl.msu.edu

Abstract. The pairing gap for ^{53}Ca obtained from new experimental data on the masses of $^{52-54}$Ca has the smallest value yet observed. This is explained in the framework of the nuclear shell model with schematic and realistic Hamiltonians as being due to shell gaps around the low-j orbital $1p_{1/2}$. Minima in the pairing gaps for all nuclei are shown and discussed.

1. Introduction

One of the most robust signatures of pairing in nuclei is the odd-even oscillation in the one-neutron separation energies as a function of neutron or proton number. This is illustrated in figure 1 which shows the binding energies and one-neutron separation energies for the calcium isotopes. The figure also shows the results of a shell-model calculation in the $(0f_{7/2}, 0f_{5/2}, 1p_{3/2}, 1p_{1/2})(fp)$ model space with the GX1A Hamiltonian (also referred to as GXPF1A in the literature [1]) compared with experiment. The oscillation in the one-neutron separation energies can be quantified in terms of the energy differences

$$D_n(N) = (-1)^{N+1}[S_n(Z, N+1) - S_n(Z, N)]$$

$$= (-1)^{N+1}[\mathrm{BE}(Z, N-1) + \mathrm{BE}(Z, N+1) - 2\,\mathrm{BE}(Z, N)],$$

where $S_n(N) = \mathrm{BE}(Z, N) - \mathrm{BE}(Z, N-1)$ is the one-neutron separation energy. N is the number of neutrons and Z is the number of protons. This quantity turns out to be always positive and reflects the fact than the even nuclei are always more bound on the average than the neighboring odd nuclei. I will distinguish the results for even and odd N values denoted by, D_{ne} and D_{no}, respectively. In the literature one commonly finds the related quantity known as the odd-even mass parameter or pairing gap $\Delta_n(N) = \frac{D_n(N)}{2}$ (see Figures 2.5 in [2]). I use D rather than Δ because its values are directly connected to simple underlying quantities associated with pairing and shell gaps. Equivalent equations for D_p as a function of proton number are obtained by fixing N and varying Z. This contribution is based on material I have published in [3] and [4].

2. The Calcium Isotopes

Figure 2 shows values of D_n for the calcium isotopes ($N > 20$) obtained from experiment and from two commonly used effective Hamiltonians in the fp model space, KB3G [5] and GX1A [1] and with a new ab-initio Hamiltonian that includes three-nucleon interactions [6]. I also show the excitation energies for the lowest 2^+ states of the even nuclei. The experimental data is from the 2012 mass table [7] together with recent data for $^{51-54}$Ca [8], [9]. The experimental energy

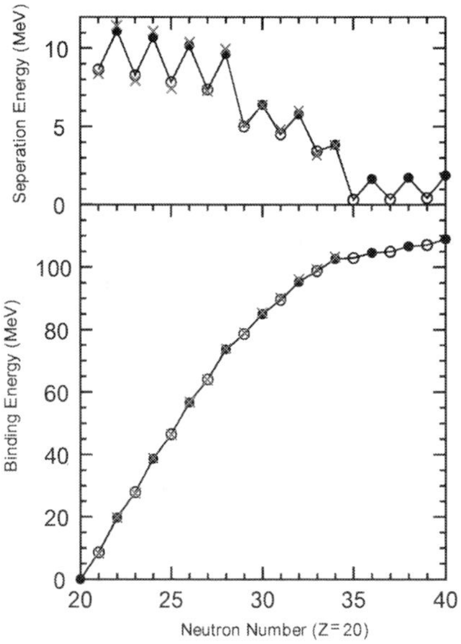

Figure 1. The bottom panel shows the ground-state energies for the calcium isotopes obtained from the pf shell-model calculation with the GX1A Hamiltonian relative to ^{40}Ca with filled circles even N and open circles for odd N, all connected by a line. The crosses are the experimental data. The top panel shows the one-neutron separation energies for GX1A and experiment.

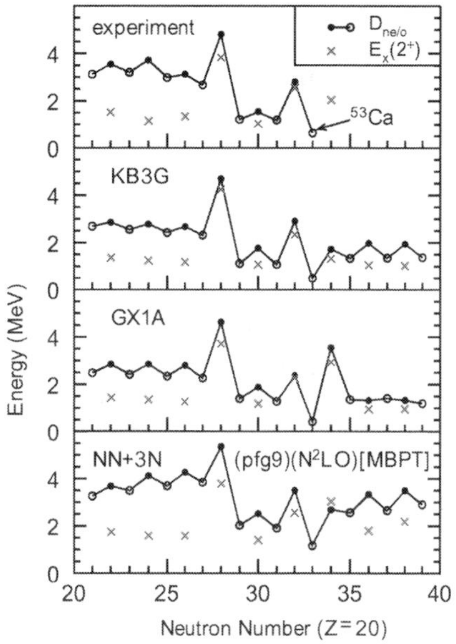

Figure 2. D_n and $E_x(2^+)$ for the calcium isotopes as a function of neutron number. D values are shown by the filled circles for even N and open circles for odd N, all connected by a line. The $E_x(2^+)$ values are shown by the crosses.

shown for the 2^+ state in ^{54}Ca was also recently measured [10]. When the GX1A Hamiltonian was developed, the experimental D_n value for ^{51}Ca was almost half of the value predicted. But the new mass measurements for ^{51,52}Ca [8] showed that the previous experimental mass for ^{52}Ca was incorrect, and gave a D_n value for ^{51}Ca that is now in good agreement with the predictions (as shown in figure 2). The new data point for ^{53}Ca shown in the upper panel of figure 2 obtained from the new ^{53,54}Ca mass measurements [9] turns out to be the lowest value for D_n observed in all nuclei with even Z. The results are in good agreement with the KB3G and GX1A shell model predictions.

I will use the shell model with schematic and realistic Hamiltonians to understand the trends observed for D, and in particular the low value for ^{53}Ca. This will be used to qualitatively understand the trends for minima in the D values for all nuclei.

To obtain insight into the reasons for the patterns observed in figure 2, I start with the simple "surface-delta-function" (SDI) model for the interaction [11]. The SDI differs from the delta interaction by the replacement of the radial integrals by a constant. The results for D obtained with SDI when the single-particle energies are degenerate are shown in the bottom panel of figure 3. The SDI $J = 0$, $T = 1$ two-body matrix elements for orbitals a and b with spins j_a and j_b are $< aa \mid \text{SDI} \mid bb >= C\sqrt{(2j_a + 1)(2j_b + 1)}$, where C is a constant. The interaction strength C is chosen to give a value for D that will turn out to be similar to that obtained with the effective pf shell Hamiltonians for calcium. The excitation energies of the 2^+ states are also constant with the SDI and degenerate single-particle energies. The results for D would be the same if the pf orbitals were replaced by a single orbital with $j = 19/2$, but the constant 2^+ energy would be higher (3 MeV). The D value is determined by the number of m states that participate in the pairing. The interaction energies obtained with SDI are $E(n) = \frac{nV_o}{2}$ for even n, and $E(n) = \frac{(n-1)V_o}{2}$ for odd n, and thus $D = -V_o$. V_o is the paired interaction strength for two particles with $J = 0$. All of the odd N nuclei have four degenerate states with $J^\pi = 1/2^-$, $3/2^-$, $5/2^-$ and $7/2^-$.

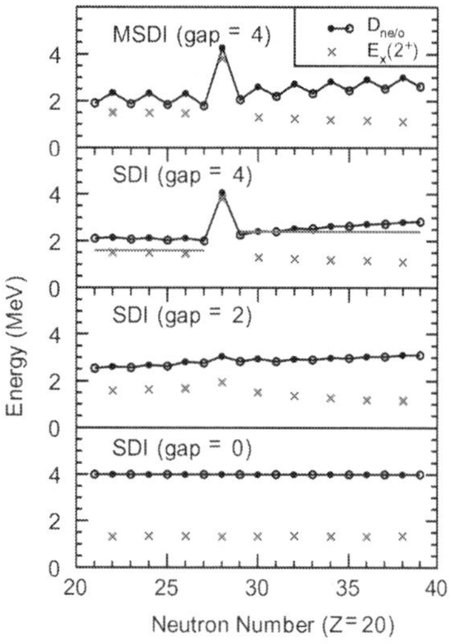

Figure 3. D_n and $E_x(2^+)$ obtained with the SDI interaction as a function of the shell gap between the $0f_{7/2}$ orbital and the degenerate $(0f_{5/2}, 1p_{3/2}, 1p_{1/2})$ set of orbitals. The second panel from the top also shows the results for D_n (thin lines) obtained when the shell gap is infinite. The results for the MSDI interaction are shown in the top panel.

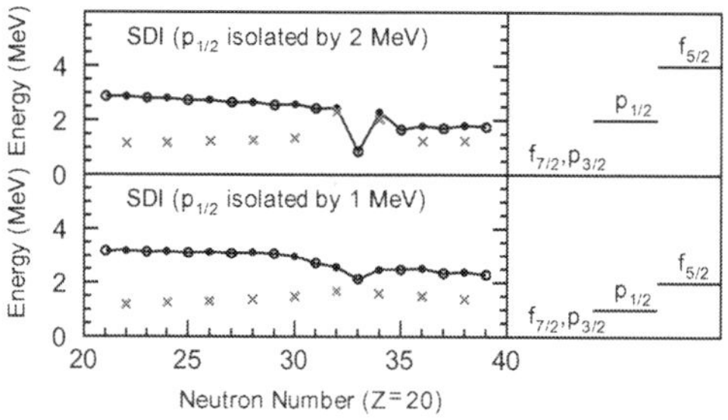

Figure 4. D_n and $E_x(2^+)$ obtained with the SDI interaction as a function of the shell gaps below and above $1p_{1/2}$ orbital. The dip is at $N = 33$.

Figure 3 shows the numerical results obtained as the shell gap between the $0f_{7/2}$ orbital and a degenerate group of $0f_{5/2}, 1p_{3/2}, 1p_{1/2}$ orbitals is increased from zero to four MeV. The value of D_e at $N = 28$ begins to rise when the value of the shell gap becomes greater than D_o. In the infinite gap limit the total original value of 4 MeV is divided between the lower group (1.6 MeV for the $0f_{7/2}$) and the upper group (2.4 MeV for $0f_{5/2}, 1p_{3/2}, 1p_{1/2}$). When the shell gap becomes large at $N = 28$ the value of D is nearly equal to that of the 2^+ excitation energy. Experiment and theory in figure 2 show clear signatures of shell gaps at $N = 28$ and $N = 32$. The GX1A calculation also shows a shell gap at $N = 34$ in contrast to the KB3G and NN+3N models that do not show a gap at $N = 34$. Mass measurements for the more neutron-rich calcium isotopes are required for the experimental value at $N = 34$. The experimental energy of the 2^+ state in

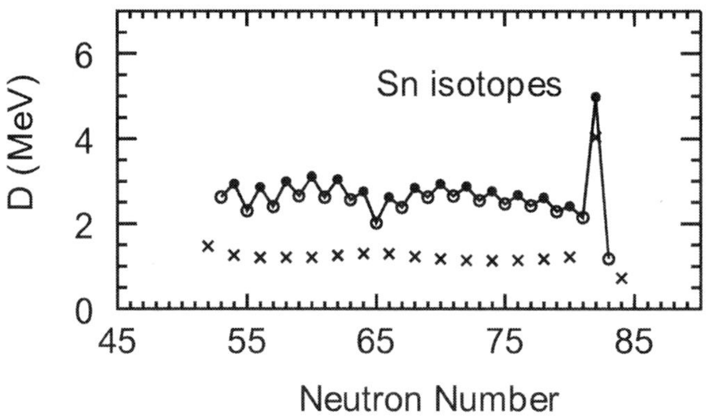

Figure 5. Experimental values of D_n for the Sn isotopes. The crosses show the energies of the 2^+ states.

^{54}Ca [10] and indirect evidence from production cross section data [12] indicate that the shell gap at $N = 34$ is about 0.5 MeV smaller than that given by GX1A.

In order to obtain a dip in the D_o value one needs to make shell gaps below and above a low-j orbital. This is shown in figure 4 where there are three groups of orbitals with $1p_{1/2}$ in the middle split by one and two MeV from the other orbitals. In the limit of a completely isolated $1p_{1/2}$ orbital $D_o = -C\sqrt{(2j_a+1)(2j_b+1)} = -2C = 0.4$ MeV.

The main defect of the SDI model is the lack of oscillations in the D values that are observed in experiment. In the 1960's this was recognized as a basic failure of the delta and SDI Hamiltonians. It was empirically fixed by adding a constant to the interaction to make the so-called modified-delta interaction [13] of the form $V(\mid \vec{r}_1 - \vec{r}_2 \mid) = A\delta(\mid \vec{r}_1 - \vec{r}_2 \mid) + B$. With the SDI form of the radial integral this becomes the so-called modified-surface-delta interaction (MSDI). The modern interpretation of this constant is that it comes from core-polarization corrections and three-body interactions. It is essential to obtain a good saturation property for the binding energies. The D values obtained with $B = 0.2$ MeV and with a shell gap of four MeV are shown at the top of figure 3. This constant simply adds a term $n(n-1)B/2$ to the all energies. Thus, $D_o = -V_o - B$ for odd n and $D_e = -V_o + B$ for even n. Half the sum of neighboring even and odd D gives the pairing contribution: $D_a = \frac{1}{2}[D_e(N) + D_o(N-1)] = -V_o$, and half of the difference gives quadratic dependence: $D_b = \frac{1}{2}[D_e(N) - D_o(N-1)] = B$. Figure 2-5 in Bohr and Mottelson [2] is based upon the D_a combination ($\Delta = D_a/2$).

Thus, the MSDI model with shell gaps gives a semi-quantitative understanding of all trends observed in $D_{e/o}$. The low D_o value for ^{53}Ca observed in experiment and theory in figure 2 is due to occupation of the $1p_{1/2}$ orbital at $N = 33$. The experimental value for ^{53}Ca is $D_o = 0.65(10)$ MeV compared to the calculated values (in MeV) of 1.170 (MBPT), 0.425 (GX1A) and 0.489 (KB3G). The key quantity for the effective Hamiltonians is the $< (1p_{1/2})^2 \mid V \mid (1p_{1/2})^2 >$ effective two-body matrix element. It is 0.151 MeV for KB3G and 0.053 MeV for GX1A compared to -0.20 MeV with the MSDI model. If the $1p_{1/2}$ orbital was completely isolated, its D_o value would be negative for GX1A and KB3G. The small positive D_o values obtained with GX1A and KB3G are due to mixing with the other orbitals.

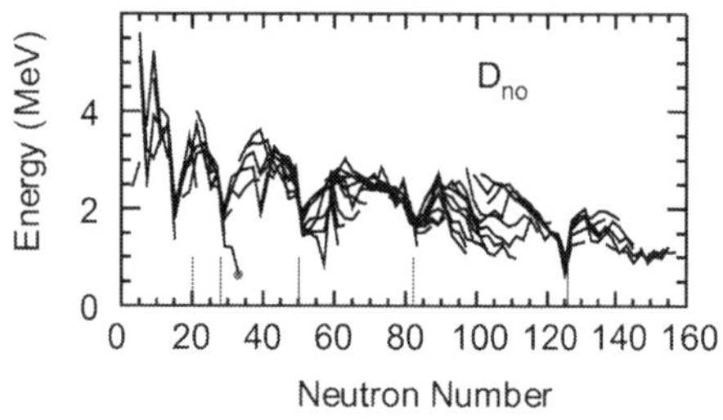

Figure 6. Experimental values of D_{no} for all nuclei with even Z. Values are plotted as a function of the number of neutrons ($N > Z$) and connected by lines for a given Z value. The value for ^{53}Ca is shown by the red dot. The vertical lines show the location of the magic numbers 20, 28, 50, 82 and 126.

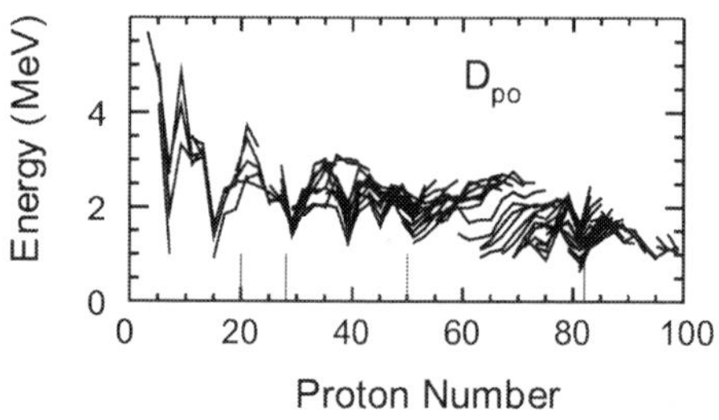

Figure 7. Experimental values of D_{po} for all nuclei with even N. Values are plotted as a function of the number of protons ($N > Z$) and connected by lines for a given N value. The vertical lines show the location of the magic numbers 20, 28, 50 and 82.

3. The Tin Isotopes

The experimental results for the Sn isotopes are shown in figure5. One observes a small dip at $N = 65$ due to small shell gaps around the $2s_{1/2}$ orbital with the $1d_{5/2}, 0g_{7/2}$ orbitals just below and the $1d_{3/2}, 0h_{11/2}$ orbitals just above. The well-known neutron shell gap appears at $N = 82$. The reduction in the pairing from $N = 81$ to $N = 83$ is due to the high number of m states associated with the nearly degenerate $2s_{1/2}, 1d_{3/2}, 0h_{11/2}$ set of orbitals just below $N = 82$ (18 m sates) compared to the relatively isolated $1f_{7/2}$ orbital just above $N = 82$ (with 8 m states). As shown in [14], for $Z = 54$ (Xe) D is about the same at $N = 81$ and $N = 83$. This change of slope between $Z = 50$ and $Z = 54$ is described by the shell-model calculations carried out by the Naples group [15]. Part of the is due to the spreading out of the effective single-particle energies for the $2s_{1/2}, 1d_{3/2}, 0h_{11/2}$ orbitals as Z increases. The energy of the $2s_{1/2}$ orbital moves above those of the other two orbitals, resulting in some isolation of the low-j $2s_{1/2}$ state for $Z = 54$. This change was termed an "anomaly" in [14] because it could not be understood on the basis of energy-density functional (EDF) calculations. What the "anomaly" indicates is that the single-particle energies are incorrect in those EDF models.

4. Global Properties

The experimental D_{no} values for all nuclei obtained from the 2012 mass table [7] and the new calcium experiment [9] are shown in figure6. Plots of the experimental D_n values for each isotope are shown on my website [16]. The value of D_o for ^{53}Ca at $N = 33$ is the lowest value obtained for all nuclei and ties with that of ^{207}Pb that has a value of $D_o = 0.63$ MeV. Due to the change in nuclear size, the SDI pairing interaction strength decreases roughly as $(A)^{-1/2}$. With this scaling the minimum is 50% lower in ^{53}Ca compared to ^{207}Pb.

The minimum at $N = 33$ is only observed for calcium ($Z = 20$). With increasing Z, the $1p_{1/2}$ and $0f_{5/2}$ orbitals cross over, and at $N = 39$ there is a small minimum in D_o coming from gaps around the $1p_{1/2}$ orbital in ^{68}Ni ($Z = 28$) and ^{70}Zn ($Z = 30$). Robust minima are observed for $N = 15$ due to the $1s_{1/2}$ orbital and $N = 127$ due to the $2p_{1/2}$ orbital. All of these nuclei have a $J = 1/2$ spin. The $0d_{5/2}$ - $1s_{1/2}$ gap disappears for carbon ($Z = 6$) [17], and the $N = 15$ dip should be gone for ^{21}C. Mass measurements of ^{21,22}C are required.

Another relatively low value of D_o occurs for $N = 57$ at ^{97}Zr, and is associated with a relatively isolated neutron $2s_{1/2}$ orbital at that point. The dip at $N = 57$ only occurs for $Z = 40$. As Z increases, the energy of the $2s_{1/2}$ orbital crosses above the energy of the $0g_{7/2}$ orbital and creating a small dip at $N = 65$ (57+8) in ^{115}Sn (figure5). As Z decreases, the energy of the $2s_{1/2}$ orbital should cross under the energy of the $1d_{5/2}$ orbital, creating a minimum at $N = 51$ (57-6) for $Z = 28$. Mass measurements of $^{78-80}$Ni are required. It is remarkable that the energy of the neutron $2s_{1/2}$ orbital migrates from being the level just above the Fermi surface at $Z = 28$ to the level just below the Fermi surface at $Z = 54$. This is due to the ℓ-dependence of the single-particle energies in a Woods-Saxon type potential.

The experimental D_{po} values for all nuclei obtained from the 2012 mass table [7] are shown in figure7. There are robust minima at $Z = 7$ (nitrogen) due to the $0p_{1/2}$ orbital, at $Z = 15$ (phosphorus) due to the $1s_{1/2}$ orbital, and at $Z = 81$ (thallium) due to the $2s_{1/2}$ orbital. The robust minimum at $Z = 29$ (copper) is due to the isolated $1p_{3/2}$ orbital. The minimum at $Z = 39$ (yttrium) starts at $N = 38$ and is due to the $2p_{1/2}$ orbital. $^{87-97}$Y all have $1/2^-$ ground state spins. Generally the shell gaps are smoothed out by the energy splitting of the Nilsson orbitals in deformed nuclei. However the $Z = 39$ dip remains in $^{99-101}$Y which are presumably deformed and have uncertain ground-state spins.

5. Summary

In summary, I have discussed how the oscillations in the neutron separation energy depend on shell gaps and effective interactions. The D_{no} value obtained for ^{53}Ca is the smallest value yet measured and is in good agreement with predictions based on previously obtained effective Hamiltonians for the pf shell. Its small value is due to shell gaps on both sides of the low-j $0p_{1/2}$ orbital. Other mass regions where small D_o values are observed due to isolated low-j orbitals were discussed.

Acknowledgements: I acknowledge support from NSF grants PHY-1068217 and PHY-1404442.

References

[1] Honma M, Otsuka T, Brown B A and Mizusaki T 2002 *Phys. Rev. C* **65** 061301(R); 2004 *Phys. Rev. C* **69** 034335; 2005 Euro. Phys. Jour. A **25** Suppl. 1, 499
[2] Bohr A and Mottelson B R 1988 *Nuclear Structure, Vol. I* (World Scientific Publishing Company)
[3] Brown B A 2013 *Phys. Rev. Lett.* **111** 162502
[4] Brown B A, "Pairing in Light and Exotic Nuclei" in "Fifty Years of Nuclear BCS, Pairing in Finite Systems", edited by R. A. Broglia and V. Zelevinsky 2013 World Scientific
[5] Poves A, Sanchez-Solano J, Caurier E and Nowacki F 2001 *Nucl. Phys. A* **694**, 157
[6] Holt J D , Otsuka T, Schwenk A and Suzuki T 2012 *Jour. Phys. G* **39** 085111; Holt J D , Menendez J and Schwenk A 2013 *Jour. Phys. G* **40** 075105; and A. Schwenk private communication
[7] Wang M, Audi G, Wapstra A H, Kondev F G, MacCormick M, Xu X and Pfeiffer B 2012 *Chinese Physics C* **36** 1603
[8] Gallant A T *et al.* 2012 *Phys. Rev. Lett.* **109** 032506
[9] Wienholtz F *et al.* 2013 *Nature* **346** 498
[10] Steppenbeck D *et al.* 2013 *Nature* **502** 207
[11] Plastino A, Arvieu R and Moszkowski S A 1966 *Phys. Rev.* **145** 837
[12] Tarasov O B *et al.* 2013 *Phys. Rev. C* **87** 054612
[13] Glaudemans P W M, Brussaard P J and Wildenthal B H 1967 Nucl. Phys. **A102** 593
[14] Hakala J, *et al.* 2012 *Phys. Rev. Lett.* **109** 032501
[15] Coraggio L, Covello A, Gargano A and Itaco N 2013 *Phys. Rev. C* **88** 041304(R)
[16] http://people.nscl.msu.edu/~brown/resources/resources.html
[17] Stanoiu M, *et al.* 2008 *Phys. Rev. C* **78** 034315

Nuclear Structure Studies of Some Neutron Rich Nuclei Produced in ^{252}Cf Spontaneous Fission

A.V. Ramayya, J. H. Hamilton, J. K. Hwang
Department of Physics and Astronomy, Vanderbilt University, Nashville, Tennessee 37235

E-mail: a.v.ramayya@vanderbilt.edu

Abstract. High spin states of neutron-rich nuclei such as ^{133,134}Te, ^{93}Sr, ^{105}Nb have been studied by measuring $\gamma-\ \gamma-\ \gamma$ coincidences (cube), $\gamma-\ \gamma-\ \gamma-\ \gamma$ coincidences (hypercube) and angular correlations from the spontaneous fission of ^{252}Cf with the LBNL Gammasphere detector array. Four types of particle-hole bands built on the known 334.3 keV isomer in ^{133}Te were identified. The level structure of ^{93}Sr is interpreted, in part, as arising from the weak coupling of the $1d_{5/2}$ neutron hole to the yrast states of the ^{94}Sr core. The g-factor of the 4^+ state in ^{134}Te was measured, for the first time, by using a new technique developed for measuring angular correlations with Gammasphere. A new level scheme of ^{105}Nb was established. Three new collective bands were identified with a total of 14 new levels and 36 new γ transitions. In $^{117-122}$Cd, a shift to more slightly deformed structures was found where the excited levels do not fit the long held picture of one, two and three phonon bands.

This paper is dedicated to Prof. Aldo Covello for his achievements, for the past two decades, in nuclear structure calculations.

1. Experimental technique

In the fission of ^{252}Cf about 200 neutron-rich nuclei are produced. Since each nucleus emits 10-50 γ-rays mostly in the energy region of 10 to 2000 keV, the γ spectrum becomes very complex. The only available technique to study these nuclei is through the study of prompt triple and higher fold $\gamma-\ \gamma$ coincidences with large detector arrays such as Gammasphere and Eurogam.

The data for the present high spin studies of ^{133,134}Te, ^{93}Sr, ^{105}Nb were obtained using the Gammasphere detector array with 101 detectors at Lawrence Berkeley National Laboratory. A ^{252}Cf spontaneous fission source with an α activity of 62 μCi was sandwiched between two 10 mg/cm^2 iron foils, which were used to stop the fission fragments and eliminate the need for a Doppler correction. A total of 5.7×10^{11} triple- and higher-fold γ-ray coincidence events were measured. A $\gamma-\ \gamma-\ \gamma$ coincidence matrix (cube) and a $\gamma-\ \gamma-\ \gamma-\ \gamma$ coincidence matrix (hypercube) were constructed. More details about this experiment and data analysis procedure can be found in [1,2,3]. A newly developed technique for measuring angular correlations with the Gammasphere detector array by sorting our high statistics data into 17 angle bins was used here to measure the g factors and assign spins and parities to excited states in neutron-rich nuclei produced in the spontaneous fission of ^{252}Cf. More details of this technique can be found in [4]. In case of Cd isotopes, the low energy excited states were obtained with our ^{252}Cf data and β-decay work.

2. Particle-hole states in ^{133}Te [1]

The nucleus ^{133}Te is expected to have excited states originating from two protons in the $\pi g_{7/2}$ orbital above the Z =50 shell gap and one neutron hole in the $\nu h_{11/2}$ orbital below the N=82 shell gap. The level structure of ^{133}Te is very interesting and extends our understanding of the proton particle-neutron hole states in this region. From the β^- decay of ^{133}Sb [5] the (11/2$^-$) isomer at 334.3 keV and 14 other states between the (7/2$^+$, 5/2$^-$) and (5/2$^+$) levels at 1096 and 2755 keV, respectively, were known. Below the (11/2$^-$) isomer, the $d_{3/2}$ and $s_{1/2}$ neutron-hole states were observed. In [1], new high spin levels in ^{133}Te from spontaneous fission work of ^{252}Cf were identified including neutron particle-hole states. Shell-model calculations within the 50-82 shells are in good agreement with the experimental data for the excitation energy of levels up to 4.3 MeV.

Four types of particle-hole bands were observed in ^{133}Te. The spins, parities, and dominant configurations of these bands were assigned by the theoretical predictions of the shell-model calculations. In these calculations, ^{132}Sn is assumed to be a closed core and the valence protons and neutrons occupy the five orbits of the 50 – 82 shells. As a two-body interaction between the valence nucleons, a realistic effective interaction derived from the Bonn-A free nucleon-nucleon (NN) potential was employed [6]. This effective interaction already produced quite satisfactory results for the light N=82 isotones as well as for ^{132}Sb [7]. As regards the single-particle proton and single-hole neutron energies, they were taken from the experimental spectra of ^{133}Sb and ^{131}Sn, respectively [7]. The latter, however, have been slightly modified to accurately reproduce the 1/2$^+$ and 11/2$^-$ neutron-hole states in ^{133}Te.

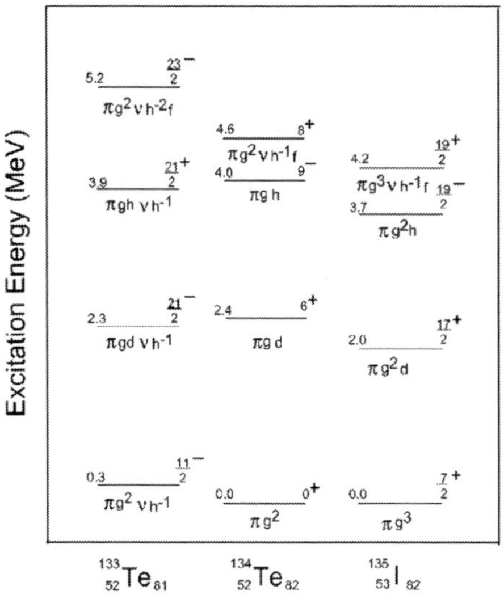

Figure 1. Band head energies of the fourbands in ^{133}Te, ^{134}Te and ^{135}I [1].

The configurations assigned to the bands are confirmed by comparing the ^{133}Te band head energies to those in ^{134}Te and ^{135}I. In figure 1, the band head energies of the four bands in ^{134}Te and ^{135}I are shown for energy comparison. One proton excited states from the 1$g_{7/2}$ orbital to the 2$d_{5/2}$ orbital are identified at 2.3, 2.4, and 2.0 MeV in ^{133}Te, ^{134}Te, and ^{135}I, respectively. The one proton excited states from the $g_{7/2}$ orbital to $h_{11/2}$ orbital are identified at 3.9, 4.0, and 3.7 MeV for ^{133}Te, ^{134}Te, and ^{135}I, respectively. The states corresponding to the $\nu f_{7/2}(h_{11/2})^{-1}$ particle-hole first excited state at 4041 keV of the doubly magic core nucleus, ^{132}Sn, are reported at 4.6 MeV in ^{134}Te and at 4.2 MeV in ^{135}I. The appearance of the neutron particle-hole excited states around 4–5 MeV has been suggested by shell-model calculations employing empirical NN interactions. On this basis, the band beginning at 5.2 MeV in ^{133}Te is interpreted as the particle-hole excitation from the core. It is impressive that the particle-hole excitation energy of 5.2 MeV in ^{133}Te is the highest ever observed. The spins of this band were assigned on the basis that $E1$ transitions dominate between the negative parity bands and the positive parity band, such as those in the cases of ^{134}Te and ^{135}I.

3. High spin states in ^{93}Sr [2]

As a first attempt to interpret the level scheme of ^{93}Sr, we have tried to identify the observed states as arising from the weak coupling of the $1d_{5/2}$ neutron hole to the states of the ^{94}Sr core. While some features of the experimental spectrum may be understood in this simple way, a detailed description can only be obtained in terms of the shell model. In principle, one may think of a large-scale calculation with ^{100}Sn as a closed core, where one has to consider 12 proton holes in the 50 closed shell and five neutrons beyond the 50 shell. A much simpler approach, which greatly reduces the dimensions of the model space, is based on the assumption that Z=38 is, to a large extent, a good magic number. This assumption has indeed been made in most of the existing calculations for the Sr and Zr isotopes. Motivated by the data made available by this experiment, a shell-model calculation was performed for this nucleus as well as for the three adjacent isotopes, ^{90,92,94}Sr, assuming ^{88}Sr as a closed core and letting the 5, 2, 4, and 6 valence neutrons, respectively, occupy the five levels of the 50-82 shell. In these calculations a realistic effective interaction derived from the CD-Bonn *NN* potential [8] was used. As a result, no adjustable parameter appears in the calculations.

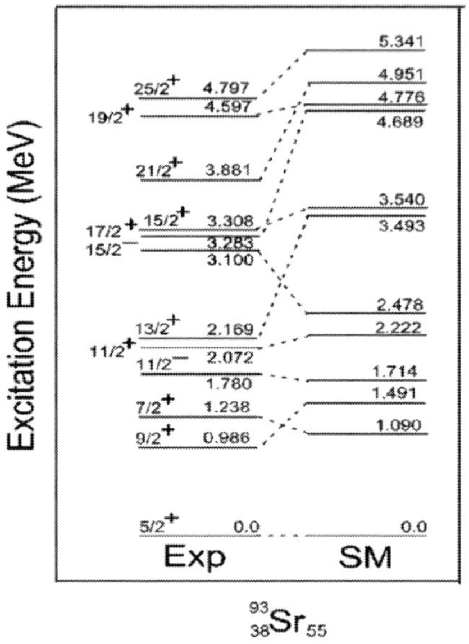

Figure 2. Comparison of experimental levels with the shell-model calculations in ^{93}Sr [2].

Twenty one new γ transitions were identified in ^{93}Sr. It turns out that the calculated results provide only a partial interpretation of the experimental spectra of the above nuclei. In particular, for ^{93}Sr the calculated level energies are not in a good agreement with the experimental ones as shown in Fig. 2. This is likely to be a consequence of our neglecting the proton degrees of freedom. However, only large-scale calculations may shed light on this point.

4. g-factor of the 4^+ state in ^{134}Te [4]

Since the fission fragments were stopped in iron foils, they were therefore subjected to the magnetic hyperfine fields (B_{HF}) in the iron lattice caused by their implantation in substitutional sites. For a nuclear state with lifetime τ , the spin vector of the nucleus will rotate about B_{HF} over the lifetime of the state, with the frequency of the rotation proportional to the strength of the field and the g-factor of the state. For this experiment, the magnetic domains in the iron foils are randomly oriented; the foils were not cooled and there were no applied external fields. The net result of the rotation of the implanted nuclei about the randomly oriented fields is an attenuation of the expected angular correlation. The attenuation factors $G_{2,4}$ are related to the Larmor precession frequency ω_L and the lifetime. Experimentally, the attenuation factors $G_{2,4}$ are defined by fitting the measured angular correlation to the parameters A_2 and A_4. By comparing experimental values with the theoretical values, one can calculate attenuation coefficients. The attenuation of the angular correlation coefficients are used to calculate the g-factors of the 4^+ state in ^{134}Te.

The *g*-factor of the 4^+ state in ^{134}Te was measured here for the first time. The unattenuated coefficients for the $4^+ \to 2^+ \to 0^+$cascade are A_2 = 0.101(3), A_4 =0.006(4). The angular correlation between two gamma rays in a cascade is unchanged by an unobserved intermediate stretched *E*2 transition in the cascade, provided the intermediate transition does not cause attenuation through precession in the 2^+ state. The 2^+ state is so short-lived as to preclude measurable precession. Therefore, the $4^+ \to 2^+$ transition can be skipped in the $6^+ \to 4^+$, $2^+ \to 0^+$ angular correlation without affecting the A_2 and A_4 values. The $6^+ \to 4^+$, $2^+ \to 0^+$ correlation is shown in Fig. 3. This correlation shows attenuation from the theoretical values of A_2 = 0.102, A_4 = 0.009, with the attenuation factor G_2 = 0.28(5). From the two different magnetic hyperfine fields of 34(16) T and 19(6) T [9] previously reported for the 4^+ state, the g-factors of $0.70^{+0.55}_{-0.38}$ and $1.31^{+0.31}_{-0.14}$ were obtained, that consistent with the previously calculated value of 0.83 [10].

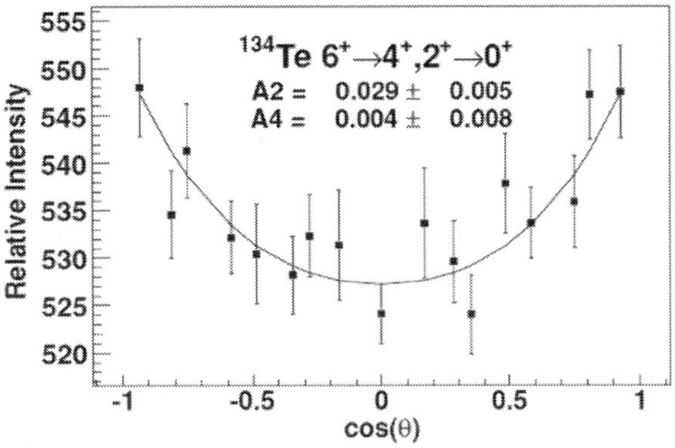

Figure 3. The correlation for the $6^+ \to 4^+$, $2^+ \to 0^+$ cascade in ^{134}Te. This correlation gives an attenuation coefficient of G_2 =0.28(5) [4].

5. Multi-phonon γ -vibrational bands in ^{105}Nb [3]

A new level scheme with three new collective bands was established in ^{105}Nb. The ^{105}Nb with Z = 41 and N = 64 is located in the A = 100 neutron-rich region. Searching for multi-phonon γ -vibrational bands in this nucleus is important for systematically understanding the structural characteristics in this region. In [3], we reinvestigated the level structure of ^{105}Nb. Many new levels and transitions were observed, and the 1γ and 2γ bands proposed. Total Routhian surface and triaxial projected shell model (TPSM) calculations [11] were carried out. These calculations support our 1γ and 2γ assignments. TPSM calculations are performed for ^{105}Nb with the fixed axial deformation ε = 0.350 and the triaxial deformation of 0.130. The calculations indicate that the yrast band in ^{105}Nb originates from the $\pi 5/2^+[422]$ Nilsson orbital with $K^\pi = 5/2^+$.

The obtained energy levels compared with the experimental data are presented in Fig. 4. It can be seen that the calculations are in excellent agreement with the available data. Not only the band-head energies for the yrast and excited bands, but also the intervals between the levels (i.e., the moments of inertia), are well described by the calculation. This strongly supports the interpretation for these bands as 1γ and 2γ collective bands, respectively, built on the yrast K^π = $5/2^+$ structure. Another band based on the 1686.5 keV level may be a three-phonon γ -vibrational band or a three quasiparticle band.

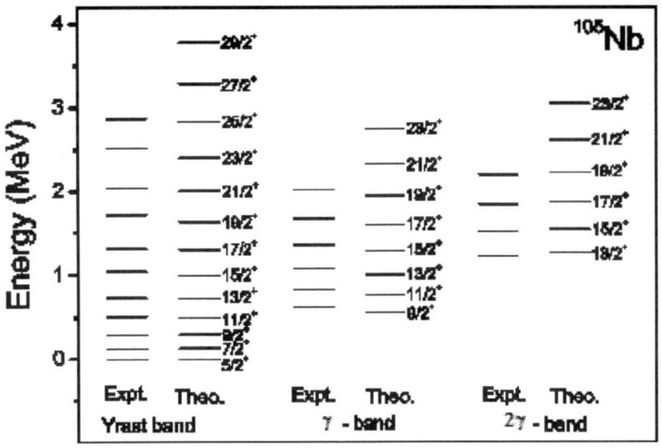

Figure 4. Comparison of the calculated and experimental level energies [3].

6. Low energy collective states in $^{116\text{-}126}$Cd [12]

It has been observed that there are significant deviations from the expected U(5) dynamical symmetry for ^{110,112,114,116}Cd nuclei. However, a significant mixing with intruder states in this region is expected. In [11], we investigated states in the heavier ^{120,124,126}Cd populated via beta decay. The decay patterns of ^{120,124}Cd are similar to lighter Cd isotopes even though the intruder states are much higher in energy. In particular, ^{120}Cd and ^{124}Cd show a similar pattern as lighter Cd isotopes, with no good candidates for 2^+ and 0^+ 3-phonon states. The situation for ^{126}Cd is unclear since 3-phonon states are not observed. Whether this is due to fact that these states are not populated in beta decay or they are simply absent cannot be determined. We have shown that the deviations from U(5) dynamical symmetry for the heavy even-mass Cd isotopes continue to higher masses even though the intruder states are further away in energy. While these states do not appear to be multiphonon states, their exact nature remains unclear. Future studies using other experimental methods (g-factors, etc.) will be needed to better understand the nature of these states. Overall, the multi-phonon vibrational approach fails to explain the low-energy structure of these Cd isotopes. An alternative way to describe the low-lying states in Cd isotopes is to place them as independent band structures rather than as N-phonon states. In all of these cases, there is a ground state band, a quasi-gamma band, and an intruder band as shown in Fig. 5 for ^{116}Cd. This interpretation is more consistent with the experimental data especially for the 0^+_4 and 2^+_4 states, than the three-phonon picture, as the decay pattern from these levels is what would be expected from the decay of independent bands. Also, in $^{116-122}$Cd a shift to more slightly deformed structures was found where the excited levels do not fit the long held picture of one, two and three phonon bands. The low-lying states in Cd may be understood as independent band structures rather than as N-phonon states.

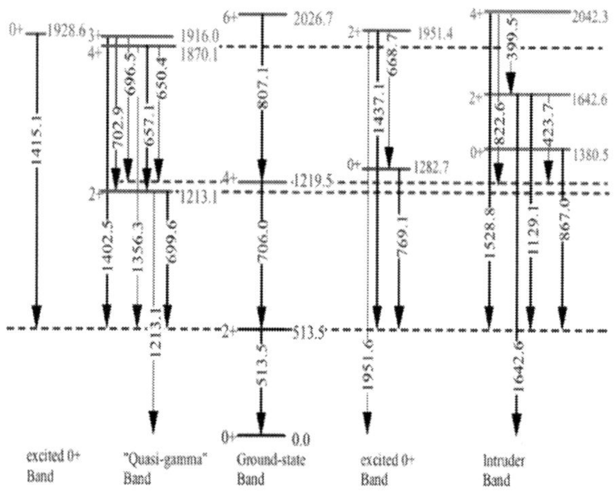

Figure 5. Low energy levels in ^{116}Cd [11].

7. Summary

In summary, level schemes of neutron rich nuclei near the line of β-stability can be studied using spontaneous fission sources such as ^{252}Cf and ^{248}Cm. The γ-spectroscopy techniques can be successfully applied with large detector arrays such as Gammasphere and Eurogam. Since the conference is in honor of Prof. Aldo Covello's contributions to nuclear structure calculations, we have chosen to present the results for a few nuclei wherein he and his group performed several calculations in collaboration with Vanderbilt's group.

Acknowledgments

The results reported in the papers considered here have been obtained in collaboration with groups from the University and INFN of Naples, the Tsinghua University, LBNL, and the Flerov Collaboration. The work at Vanderbilt University is supported by U.S. Department of Energy under Grant and Contract No. DE-FG05-88ER40407.

Journal of Physics: Conference Series **580** (2015) 012017 doi:10.1088/1742-6596/580/1/012017

References

[1] Hwang J K *et al* 2002 *Phys. Rev. C* **65** 034319 and references therein
[2] Hwang J K *et al* 2003 *Phys. Rev. C* **67** 014317 and references therein
[3] Li H J *et al* 2013 *Phys. Rev. C* **88** 054311 and references therein
[4] Goodin C *et al.* 2008 *Phys. Rev. C* **78** 044331 and references therein
[5] *Table of Isotopes*, 8th ed. edited by R.B. Firestone *et al.* (Wiley, New York, 1996)
[6] Machleidt R1989 *Adv. Nucl. Phys.* **19** 189
[7] Andreozzi F *et al.* 1999 *Phys. Rev. C* **59** 746
[8] Machleidt R 2001 *Phys. Rev. C* **63** 024001
[9] Rao G N 1985 *Hyperfine Interact.* **26** 1119
[10] Brown B A *et al.* 2005 *Phys. Rev. C* **71** 044317
[11] Sheikh J A *et al.* 2010 *Phys. Lett. B* **688** 305
[12] Batchelder J C *et al.* 2012 *J. Phys. Conf. Ser.* **387** 012005 and references therein.

Two particle transfer reactions: the search for the Giant Pairing Vibration

Carlos H. Dasso [1], **Hugo M. Sofia** [2] **and Andrea Vitturi** [3,4]

[1] Departamento de Fisica Atomica, Molecular y Nuclear, Facultad de Fisica, Apartado 1065, E-41080 Sevilla, Spain
[2] CNEA, Departamento Fisica Teorica, Tandar, Buenos Aires, Argentina
[3] Dipartimento di Fisica e Astronomia "G.Galilei", Padova, Italy
[4] INFN, Padova, Italy

E-mail: andrea.vitturi@pd.infn.it

Abstract. We discuss different points that are fundamental to determine the actual magnitude of the cross sections measured in two-particle transfer reactions between heavy ions. These are important issues to consider when the aim is the experimental observation of the somewhat elusive Giant Pairing Vibrations.

In this contribution we address the subject of two-particle transfer in reactions involving heavy ions. Among the variety of still open aspects of these reactions, our emphasis is placed here on the identification of the predicted so-called "Giant Pairing Vibrations"(GPV) [1]. The existence of pair correlations is known to provide an enhancement in the magnitude of the ground-state to ground-state transition matrix elements between systems that differ by a number of two nucleons. Interest in the Giant Pairing modes emerged when it was realized that analogous enhancements should also be expected from particle-particle correlations involving transitions to higher single-particle shells, much in the same way as the ones responsible for the existence of the more familiar Giant Surface Vibrations.

The nuclear structure information contained in the size of the transition matrix elements is, however, only a part of the story. To arrive to the actual magnitude of the cross sections, the dynamical characteristics of the waves describing the incoming and outgoing nuclear currents must also be taken into account. How this is implemented in practice depends very much on the characteristics of the formalism employed. We shall here refer, concretely, to the language invoked in the semiclassical treatment, where the matching conditions between the entrance and exit orbits of relative motion are expressed in terms of the bombarding energy, the reaction Q-value and the existence of an optimal match [2]. All of these quantities are determined by the prevalent experimental conditions.

In reactions between ordinary systems, with similar binding energies, the Q-value for particle transfer tends to be small, a situation that favors the particle-transfer between ground states. For precisely the same reason the transfer into the predicted high-energy range of the Giant Pairing Vibrations is normally unfavored and it is therefore not so much of a surprise that these modes have not been readily observed [3].

In reference [4] the authors pointed out the importance of taking into account properly the existence of the optimal Q-values to understand the apparent lack of experimental evidence. Their conclusions, however, were reached in the context of the macroscopic model for pair transfer [5]. The concept of a generalized one-body pair-creation field is not very useful in the investigation of transitions to individual states since the excitation energy of the addition or removal modes is not direcly associated with the reaction Q-value (cf. reference [6], chapt. 9, and [5]). We side-step here the use of the macroscopic pair-transfer model and procure to describe the process in the more conventional language of a transition between two well-defined nuclear states.

For a ground-state to ground-state transition, the Q-value is obtained directly from the masses of the constituent particles of the reacting nuclei, namely $Q = \sum_i m_i c^2 - \sum_f m_f c^2$ where i and f stand to indicate the "initial" and "final" character of the fragments involved. A positive sign of Q reveals if a process liberates energy and may occur spontaneously. Otherwise, the incident beam energy is always available to make even endothermic processes possible. In fact, if the reaction in question is not meant to induce a transition between the ground states but to some net excitation E^* the bombarding energy should be raised accordingly. The relevant parameter χ then changes from that maximum value $\chi = Q$ to $\chi = Q - E^*$. In what follows -

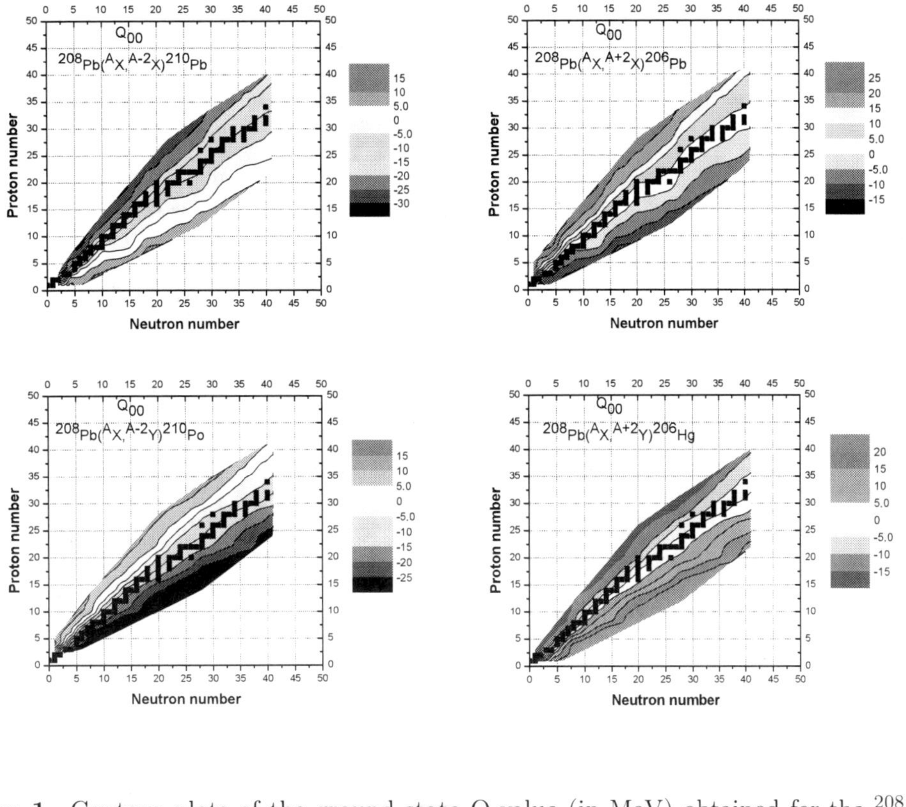

Figure 1. Contour plots of the ground state Q-value (in MeV) obtained for the ^{208}Pb target using as projectiles all known nuclei up to proton and neutron number equal to 50. The four frames correspond to 2-neutron stripping (upper left), 2-neutron pick-up (upper rigth), 2-proton stripping (lower left) and 2-proton pick-up (lower rigth). The black squares correspond to the stable nuclei.

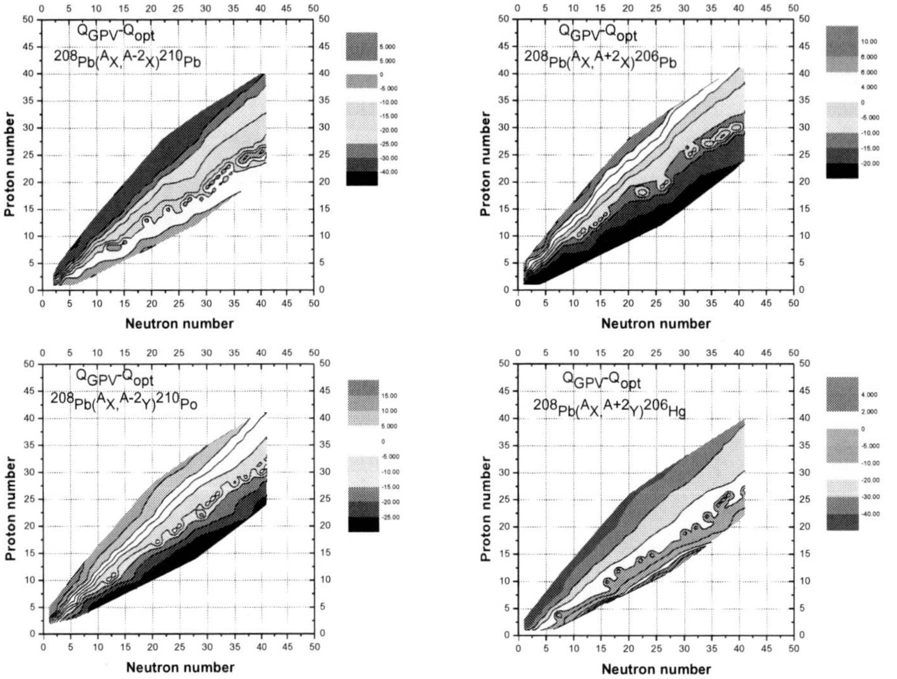

Figure 2. The same than figure 1, but for the quantity Q_{GPV}-Q_{opt} (in MeV). For the calculation of Q_{GPV} the energy of the GPV has been assumed to be 12 MeV. The Q_{opt} has been calculated for a projectile energy (in the c.m. reference frame) corresponding to 20 % above the corresponding Coulomb barrier of the projectile-target interaction.

aiming to the GPV - we shall take a characteristic value $E^* \approx 12$ MeV. We stress that this is only for illustration purposes and that there is no problem adopting any other values if deemed more appropriate.

Since the nuclear masses are available as electronic tables in [7] we can propose a quick survey of the characterisitic numbers involved. Our strategy will be to chose a given target candidate and explore a wide range of possible projectiles. The idea is to put in evidence those reactions that significantly deviate from the aforementioned tendency to favor low-excitation energy transitions and, instead, identify those that would help populate the GPV modes. Let us here note that the existing records automatically end whenever the projectile mass has not been determined.

As an illustration of this procedure we first take, as a target, the nucleus of ^{208}Pb. We then plot the two-particle transfer reaction Q-values as a function of the number of protons and neutrons of the projectile. The results are shown, as contours of the resulting surface, in figure 1. The four frames correspond to two-neutron pick-up, two neutron stripping, two-proton pick-up and two-protong stripping. As it is apparent from the figure positive Q-values are mostly obtained using neutron-rich projectiles in connection with two-neutron stripping and two-proton pick-up, while proton-rich projectiles lead to positive Q-values for two-neutron pick-up and two-proton stripping.

For the optimal Q-values in the absence of angular-momentum transfer, we can use the following expression [2] which is not undisputed but somewhat author-dependent,

$$Q_{opt} = \frac{Z_d(Z_b - Z_A)e^2}{r_0} - \frac{1}{2}\frac{m_d(m_b - m_A)}{m_a + m_A}v_0^2 + m_{aA}\ddot{r}_0\delta r_0 \quad (1)$$

where the indexes A and a refer to the initial particles in the reaction, B and b are the final particles,Z_d and m_d are the total charge and mass of the transfer particles, v_0 and $\ddot{r}_0$ are the velocity and the acceleration at the distance of closest approach r_0 and $\delta r_0 = m_d/2((R_A - R_a + r_0)/m_A - (R_a - R_A + r_0)/m_a)$. in terms of the radii R of the initial reaction partners. As we mentioned earlier, the relevant quantity to investigate the eventual population of a GPV of excitation energy E^* is the difference betwen the parameter $\chi = Q - E^*$ and the optimal Q-value, i.e. $\chi - Q_{opt}$. This quantity is plotted in figure 2 as a function of the number of protons and neutrons of possible projectiles using the same organization of the frames as in figure 1.

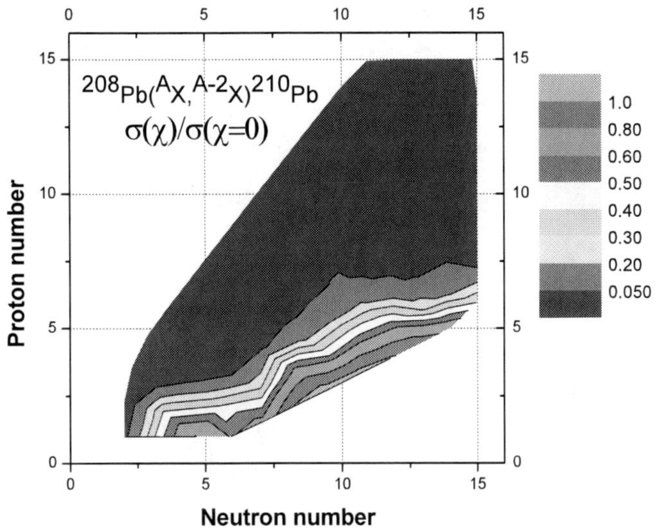

Figure 3. Contour plot of the cut-off exponential factor of equation (2), obtained for the excitation of the GPV in ^{210}Pb in two-neutron stripping reactions, using as projectiles all known nuclei up to proton and neutron number equal to 15. The black squares correspond to the stable nuclei, while the two circles indicate the two Borromean nuclei ^{6}He and ^{11}Li.

One can easily judge the quantitative effect of the numbers displayed in figure 2. This is done taking into account that the actual magnitude of the measured transfer cross sections is modulated by a gaussian factor centered at the optimal Q-value according to the expression [2]

$$\sigma(\chi) = \sigma(\chi = Q_{opt}) \exp\Big[- \frac{(\chi - Q_{opt})^2}{2\hbar^2\kappa\ddot{r}_0}\Big] \quad (2)$$

where κ is the slope of the two-particle transfer effective form factor ($\kappa \approx$ 1fm). It is thus the exponential factor in equation (2) that gives us a practical tool for guidance in the search for the most suitable projectiles. The contours of this quantity are displayed in figure 3 for the case

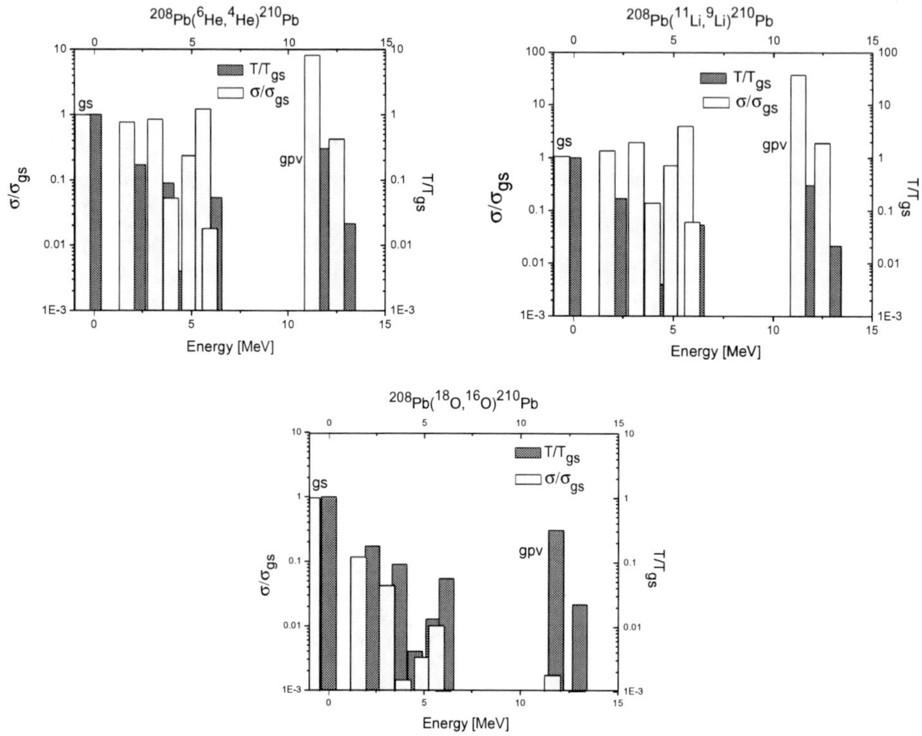

Figure 4. Pair addition strength distribution, normalized to the ground state, to all 0^+ states in ^{210}Pb (red bars) compared to the normalized pair transfer cross section (yellow bars) in the case of ^{6}He (upper left), ^{11}Li (upper right) and ^{18}O (lower frame) .

of two-neutron stripping, for the excitation of the GPV in ^{210}Pb, using as projectiles all known nuclei up to proton and neutron number equal to 15 .

As a further example we show in figure 4 the addition-pair strength distribution, normalized to the ground state, to all 0^+ states in ^{210}Pb compared to the corresponding normalized pair transfer cross sections in the case of ^{6}He, ^{11}Li and ^{18}O. The pair strength has been calculated using discrete particle-particle RPA, starting from single particle levels taken from reference [8], with a pairing constant G adjusted to fit the difference of the ground state energies (^{210}Pb - ^{206}Pb). The calculation shows the prediction of a strong high-lying state at 12 MeV (GPV). As far as the calculation of the relative two particle transfer cross section, in all cases the bombarding energy has been chosen 20 % larger than the corresponding Coulomb barrier. It is apparent from the figures that, due to the Q-value effect, the population of the high-lying in the two-particle transfer process is strongly depressed in the case of the ^{18}O induced reaction, a case in which the Q-value is already negative for the g.s. transition, and hence even more negative for the GPV region. On the converse, in the case of ^{6}He and ^{11}Li projectile the ground state Q-value is positive, leading to a Q-value for the GPV close to the optimal value.

More generally, if we concentrate on the excitation of the Giant Pairing Vibration, we consider again in figure 5 the full variety of projectiles and evidence (in the lighter green color), those that, because of favorable Q-value, lead to a population of GPV larger than that of the ground state.

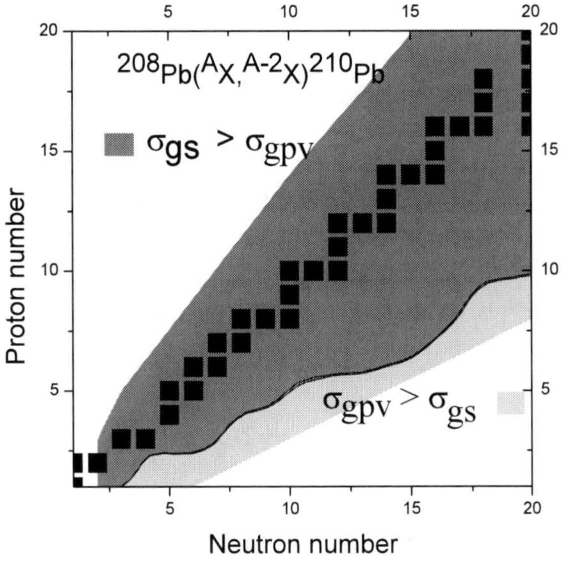

Figure 5. The figure refers to the case of two-neutron stripping reaction on ^{208}Pb with different projectiles. The lighter green area shows the part of the nuclear chart corresponding to the projectiles that, because of favorable Q-value, are expected to lead to a population of GPV larger than that of the ground state.

In this contribution we have tried to provide orientation for a judicious choice of reaction conditions that may shed experimental evidence on the predicted existence of Giant Pairing Vibrations. There is indeed flexibility in some choices that we have made but, in general, our estimates are based on rather well-established features of semiclassical reaction theory. As such we trust that the guidelines we have developed would be useful to optimize the search for these high-lying collective pairing modes.

Acknowledgements

This research has been partially supported by the Spanish Ministerio de Educación y Ciencia and by the European regional development fund (FEDER) under project numbers FIS2011-28738-C02-01, the Consolider Ingenio Programme CSD2007-0042, the Junta de Andalucia P07-FQM-02894 and the INFN, Italy.

References

[1] Broglia R A and Bes D R 1977 *Phys. Lett.* **B69** 129
[2] Broglia R A and Winther A 1991 *Heavy ion reactions. Lecture notes*, Addison-Wesley
[3] Mouginot B etal., *Phys. Rev. C***83** (2011) 037302
[4] Fortunato L, von Oertzen W, Sofia H M and Vitturi A 2002 *Eur. J. Phys. A***14** 37
[5] Dasso C H and Pollarolo G 1985 *Phys. Lett.* **B155** 223; Dasso C H and Vitturi A 1986 *Phys. Lett.* **B179** 337
[6] Dasso C H and Vitturi A, Eds 1987 *Collective aspects in pair-transfer phenomena*, SIF, Conference Proceedings **18**
[7] http://www.nndc.bnl.gov/masses/mass.mas03
[8] Uhrer R A and Sorensen R A 1966 *Nucl. Phys.* **86** 1

Journal of Physics: Conference Series **580** (2015) 012019 doi:10.1088/1742-6596/580/1/012019

The importance of closed shell structures in the synthesis of super heavy elements

J H Hamilton[1], S Hofmann[2] and Y T Oganessian[3]
[1]Department of Physics and Astronomy, Vanderbilt University, Nashville, TN 37235, USA
[2]GSI Helmholtzzentrum fuer Schwerionenforschung, 64291 Darmstadt, Germany
[3]Flerov Laboratory of Nuclear Reactions, Joint Institute for Nuclear Research, 141980 Dubna, Russia

E-mail: j.h.hamilton@vanderbilt.edu

Abstract. The importance of shell closures and gaps in the single-particle energies for protons and neutrons on the stability of elements beyond Z = 100 will be described. Following the development of microscopic models with shell corrections, microscopic-macroscopic models predicted large gaps in the single-particle energy levels for protons and neutrons at Z = 102, 108 and N = 152, 162 for the same deformed shapes. Shell gaps for spherical shapes for N = 184 and Z = 114, 120 or 126 were also predicted to form an "Island of Stability" with very long half lives for fission and alpha decay. Cold fusion reactions involving beams of Ca to Zn and targets of stable ^{208}Pb and ^{209}Bi were pioneered at GSI and used to synthesize new elements for Z = 107 to 112 and in Japan a new isotope of 113. Hot fusion reactions between radioactive actinide targets and neutron-rich ^{48}Ca beams were pioneered in JINR leading to the synthesis of new elements with Z = 113 to 118. Data on two neutron separation energies, spontaneous fission half lives and total half lives of super heavy elements showing the importance of reinforcement of the Z = 102, N = 152 and Z = 108, N = 162 single particle level gaps at the same deformation and Z = 114-126, N = 184 shell gaps in the synthesis of super heavy elements 107 to 118 are presented along with the latest results on their synthesis.

By around 1965, eight new elements beyond U (Z = 92) up to Z = 100 had been discovered. Could one go further? Macroscopic models, like liquid drop models, at that time, predicted that nuclei beyond Z ≈ 100 could not be synthesized because beyond this Z the barrier against fission would go to zero and then nuclei would fission on a time scale $T_{SF} \approx 10^{-19}$ s. Then came the introduction of microscopic models with shell corrections [1]. Macroscopic-microscopic models predicted large gaps in the single-particle energy levels for protons and neutrons for deformed shapes first for Z = 100, N = 152 [2] and then for Z = 102, N = 152 and Z = 108, N = 162 [3]. The reinforcement of the Z = 102, N = 152 and Z = 108, N = 162 level gaps at the same deformations provide the added stability to allow nuclei around these regions to have observable half lives (see Fig. 1). Calculations also predicted for N = 184 and Z = 114, 120 or 126 there were shell gaps for spherical shapes (for example, refs. 4-6). While all the calculations have predicted N = 184 is a spherical closed shell for neutrons, Z = 114, 120

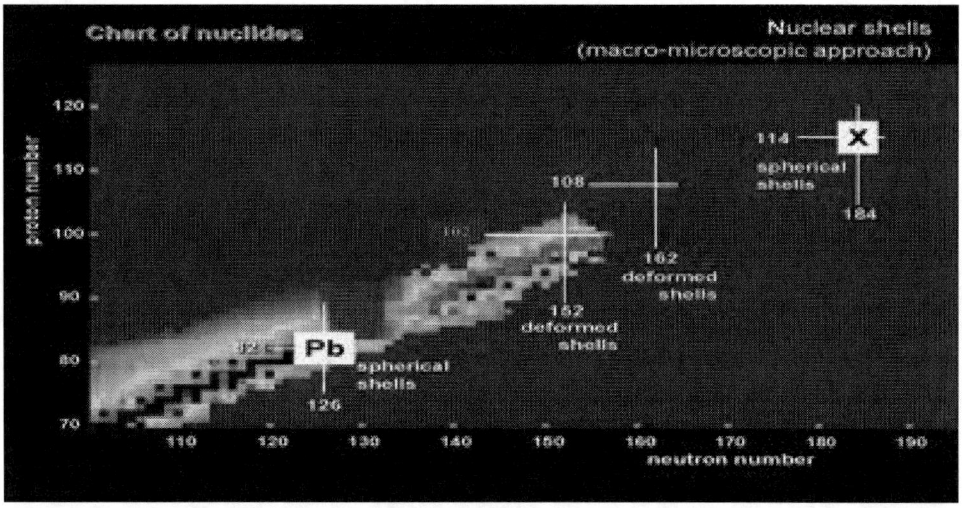

Figure 1. Chart of the nuclides showing the spherical gaps at Z = 82, N = 126 and predicted Z = 114, N = 184 and shell gaps for deformed shapes at Z = 102, N = 152 and Z = 108, N = 162

and 126 have been predicted in different approaches to be the next spherical closed shell for protons (for example, ref. 7-10). Earlier in macroscopic-microscopic calculations an "Island of Stability" was predicted around these spherical shell gaps where nuclei could have very long half lives for fission and alpha decay. In Fig. 2 are shown the single particle levels calculated in a microscopic-macroscopic approach for both neutrons (a) and protons (b) as a

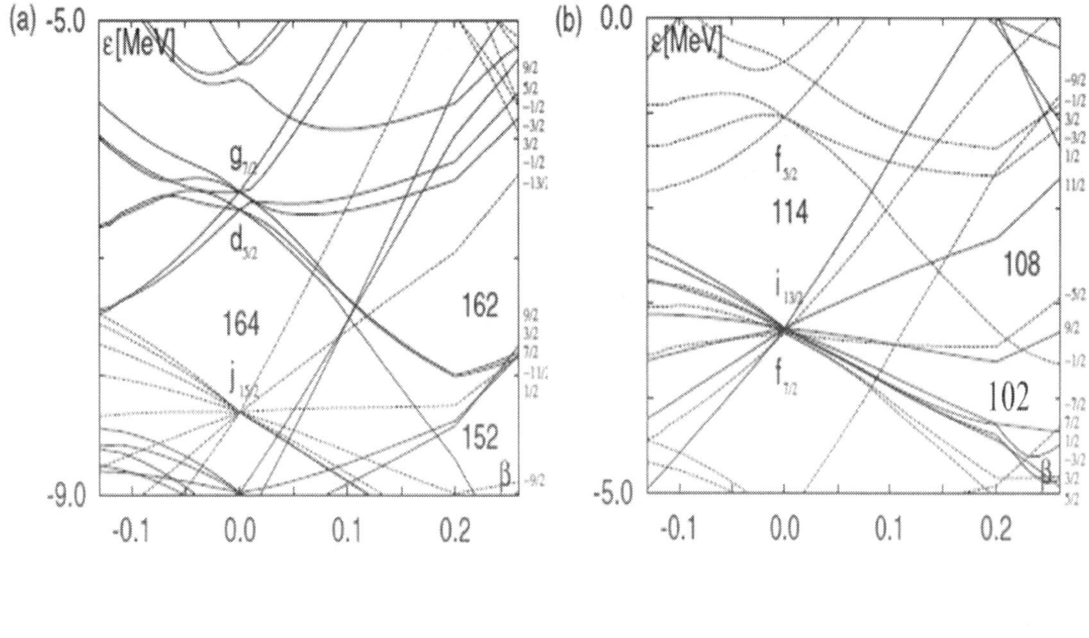

Fig. 2. Neutron (a) and proton (b) single particle energies with shell gaps seen at Z = 102, 108 and N = 152, 162 at β ~ 0.23.

function of deformation β. Note the occurrence of shell gaps for Z = 102, N = 152 and Z = 108, N = 162 all at the same deformation β ~ 0.23. The occurrence of these shell gaps at the same deformation reinforce each other to give added stability for these Z,N combinations. The importance of these gaps and their reinforcement will be seen in the experimental data presented later.

These theoretical predictions led to two approaches to synthesize new elements [11, 12, 13]. Cold fusion reactions involving beams of Ca to Zn and targets of stable ^{208}Pb and ^{209}Bi where the excitation energy was so low only one neutron was evaporated were pioneered at GSI in Germany. The velocity separator SHIP was used at GSI in these studies [11, 12]. These cold fusion reactions were used to synthesize new elements for Z = 107 to 112 [11, 12] and in Japan a new isotope of element 113 [14]. Hot fusion reactions between radioactive actinide targets and neutron-rich ^{48}Ca beams where 3 to 5 neutrons were evaporated were pioneered in Russia [11, 13]. The Dubna gas-filled recoil separator DGFRS was used in the Dubna work [11, 13]. These reactions led to the synthesis of new elements with Z = 113 to 118. The reactions and new elements synthesized in these two approaches and the elements into which they decay are given in Fig. 3. Elements 114 and 116 have recently received names, Florovium and Livermorium, respectively.

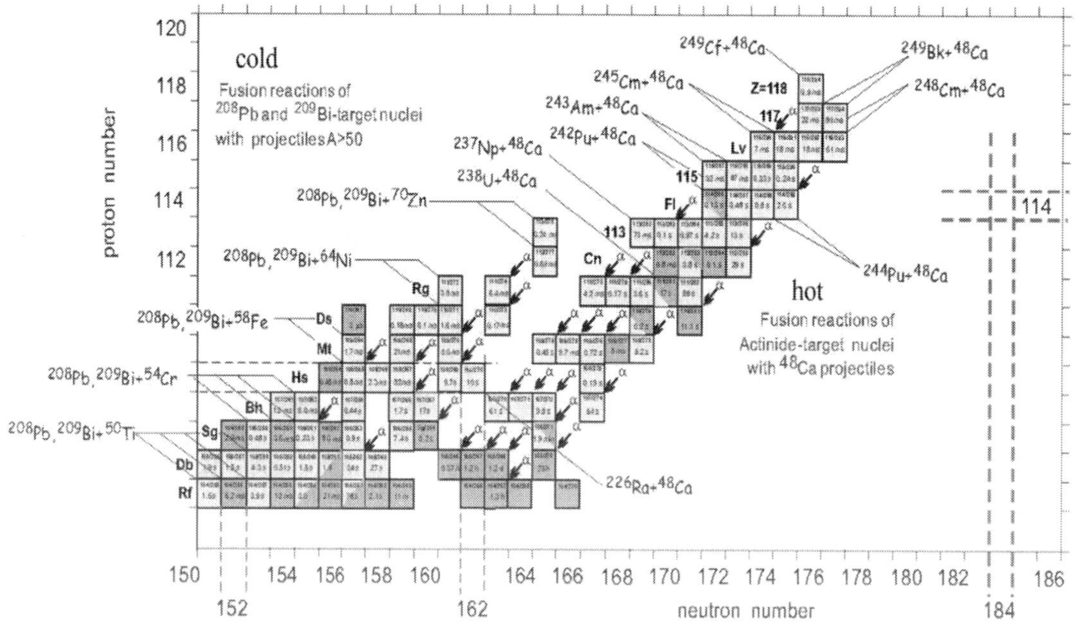

Fig. 3. Most known nuclei (as of 2013) and the reactions required for their production. For each known isotope, the element name, mass number, and half-life are shown. The relatively neutron-deficient isotopes of the elements up to Z = 113 were created in cold fusion reactions. The more neutron-rich isotopes of elements 112-118 were produced in hot fusion reactions. Spherical closed shells for Z = 114 and N = 184 are shown. The bold dashed lines indicate nuclei with increased stability due to large gaps between the single particle energy levels at Z = 108 and N = 152 and 162.

Journal of Physics: Conference Series **580** (2015) 012019 doi:10.1088/1742-6596/580/1/012019

In recent experiments, definitive evidence for the synthesis of the new elements 113, 115 and 117 have been published [15, 16, 17, 18]. The Dubna, Oak Ridge, Vanderbilt, Livermore, Reactor Institute Collaboration carried out two sets of experiments of the ^{243}Am + ^{48}Ca reaction from November 2010 – February 2012 at five beam energies to observe 28 new events of ^{288}115, and five of a new isotope ^{289}115, to firmly establish the earlier report of the synthesis of elements 115 and 113. The ^{289}115 is especially important because this isotope is likewise observed in the ^{249}Bk + ^{48}Ca reaction to synthesizing the isotope ^{293}117 which alpha decays to ^{289}115. The agreement of these two reactions provides cross-bombardment evidence for the discoveries of 117, 115 and 113. Subsequently, a group using the new TASCA separator at GSI also observed 22 decay chains of ^{288}115 in the ^{243}Am + ^{48}Ca reaction [17] with the same decay properties as the 31 chains seen in Dubna to provide independent laboratory confirmation of the new elements of 113 and 115. Studies of the ^{249}Bk + ^{48}Ca reactions were repeated in Dubna to yield 11 new events of ^{293}117 and 3 of ^{294}117 and repeated at TASCA to yield two new similar ^{294}117 decay chains [18] to provide independent confirmation of element 117.

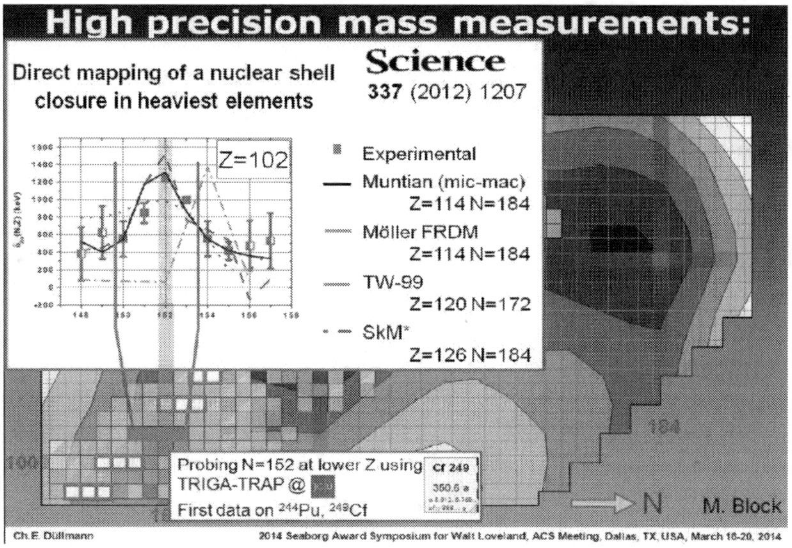

Fig. 4. The two neutron separation energies as a function of N for Z = 102. The strong increase at N = 152 demonstrates the importance of this shell closure as it is reinforced by the Z shell closure

Fig. 4 shows evidence for the double shell gaps at Z = 102 and N = 152 where the two neutron separation energy strongly peaks at N = 152 for Z = 102. In Figure 5 is shown the partial half lives for spontaneous fission vs. N. Note the strong peaks, longer half lives, for all Z for N = 152, 162 and approaching N = 184. The α-decay half lives likewise show a strong increase as N increases toward N = 184 (see Fig. 6). The rapidly increasing half lives for Z = 111, 113, 115 and 117 as a function of N support the predictions of a long-lived "Island of Stability" around N = 184.

We see the synthesis of new elements beyond Z = 100 are strongly dependent on both the neutron and proton shell structures, in particular, the reinforcing shell gaps for deformed shapes for Z = 102, N = 152 and Z = 108, N = 162 and the strong spherical gap at N = 184

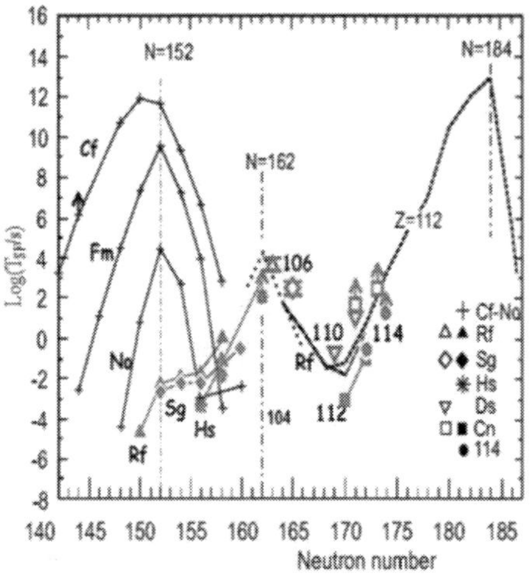

Figure 5. Spontaneous fission half-lives versus N. The solid symbols and crosses denote even-even nuclei; open symbols denote even-odd nuclei. Solid lines are drawn through the experimental points of even-even nuclei. The dashed lines represent theoretical calculations. Abbreviation: SF, spontaneous fission.

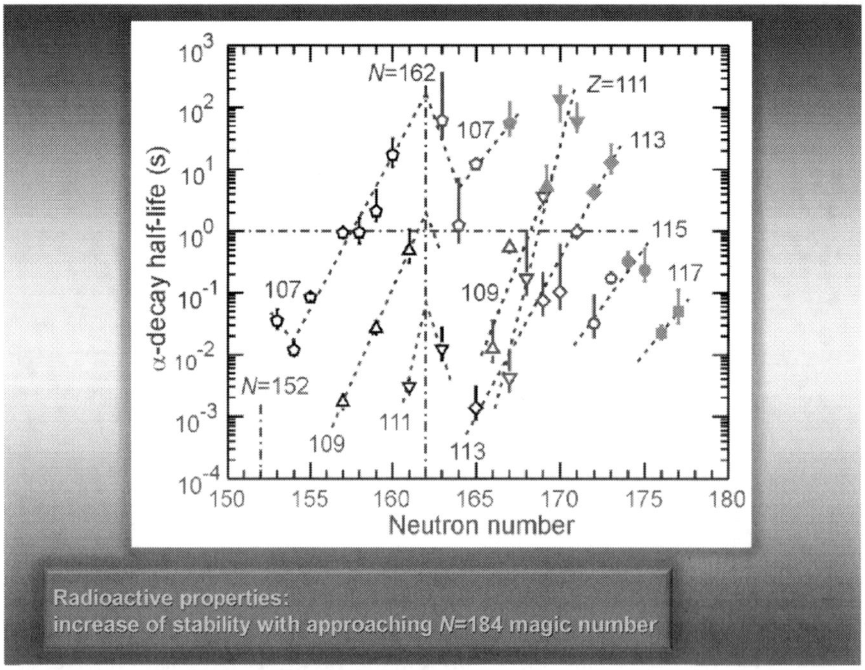

Figure 6. Alpha decay half lives as a function of N.

where the next spherical shell closure for Z is yet to be determined. There is little evidence it is for Z = 114. Recent theoretical studies indicate it will be for Z $\geq$ 120 [9]. Efforts at

TASCA to synthesize elelment 119 in the $^{249}Bk + {}^{50}Ti$ reactions yielded only an upper limit of 70 Femto barn for the production cross-section. The FLNR laboratory is in the process of building a SHE facility that should produce up to 1000 SHE in less than a year. This new facility can provide the opportunity to synthesize new elements 119 and 120 despite their very low production cross-sections.

"The discoveries of the new elements from 107 to 112 in cold fusion and from 113 to 118 in hot fusion are now well established. Cross-bombarded verification, measurements of excitation functions and reproducible results at different laboratories were all important in establishing these discoveries.

From the properties of the heaviest nuclei and their decay products with Z = 104−118 and N = 153−177 covering a mass range A = 257−294 (given in Fig. 3) one sees that for nuclei having Z 40% larger than that of Pb, an impressive extension in nuclear survivability is observed. Although SHE are at the limits of Coulomb stability, shell stabilization lowers the ground-state energy, creates a fission barrier, and thereby enables SHE to exist. The basic theoretical concept of the existence of closed shells in the region of the hypothetical super heavy nuclei and their decisive role in determining the limits of nuclear mass has received clear experimental verification" [11].

Acknowledgement
J.H.H.'s work at Vanderbilt University was supported by the US Department of Energy through grant DE-FG-05-88ER40407. S.H.'s was supported by the Helmholtz Association through a Helmholtz Professorship. We acknowledge the support of the Russian Foundation for Basic Research through grants 11-02-12050 and 11-02-12066.

References

[1] Strutinsky V M 1967 *Nucl. Phys.* A **95** 420
[2] Möller P, Nilsson S G and Nix J R 1974 *Nucl. Phys.* A **229** 292
[3] Patyk Z and Sobiczewski A 1991 *Phys. Lett* B **256** 307
[4] Meldner H *Ark. Fys.* 1967 **36** 593
[5] Nilsson S G 1968 *Nucl. Phys.* A **115** 545
[6] Mosel U and Greiner W 1969 *Z. Phys.* **222** 261
[7] Cwiok, S *et al* 1996 *Nucl. Phys.* A **611 246**
[8] Bender M. *et al* 1999 Phys. Rev C **60** 034304
[9] Kuzmina A N *et al* 2012 *Phys Rev* C **85** 014319
[10] Sobiczewski A 2011 *Radiochim. Acta* **99** 395
[11] Hamilton J H, Hofmann S and Oganessian Y T 2013 *Annu. Rev. Nucl. Part Sci.* **63** 383
[12] Hofmann S 2011 *Radiochim. Acta* **99** 405
[13] Oganessian Y T 2011 *Radiochim. Acta* **99** 429
[14] Morita K *et al* 2012 *F. Phys. Soc. Zpn.* **81** 103201
[15] Oganessian Y T *et al* 2012 *Phys. Rev. Lett.* **108** 022502; Oganessian Y T *et al* 2012 *Phys. Rev. Lett.* **109** 162501
[16] Oganessian Y T *et al* 2010 *Phys. Rev. Lett.* **104** 142502
[17] Rudolph Y T *et al* 2013 *Phys. Rev. Lett.* **111** 112502
[18] Khayagbaatar J. *et al* 2014 *Phys. Rev. Lett.* **112** 172501

Low-energy enhancement of M1 strength

R. Schwengner
Institute of Radiation Physics, Helmholtz-Zentrum Dresden-Rossendorf, 01328 Dresden, Germany

S. Frauendorf
Department of Physics, University of Notre Dame, Notre Dame, Indiana 46556, USA

A.C. Larsen
Department of Physics, University of Oslo, 0316 Oslo, Norway

E-mail: r.schwengner@hzdr.de

Abstract. Magnetic dipole strength functions have been deduced from averages of a large number of $M1$ transition strengths calculated within the shell model for the nuclides ^{90}Zr, ^{94}Mo, ^{95}Mo, and ^{96}Mo. An enhancement of $M1$ strength toward low transition energy has been found for all nuclides considered. Large $M1$ strengths appear for transitions between close-lying states with configurations including proton as well as neutron high-j orbits that re-couple their spins and add up their magnetic moments coherently. The $M1$ strength function deduced from the calculated $M1$ transition strengths is compatible with the low-energy enhancement found in (^{3}He,^{3}He') and (d,p) experiments. The present work presents an explanation of the experimental findings.

Photonuclear reactions and the inverse radiative-capture reactions between nuclear states in the region of high excitation energy and large level density, the so-called quasicontinuum of states, are of considerable interest in many applications. Radiative neutron capture, for example, plays a central role in the synthesis of the elements in various stellar environments [1, 2]. An improved theoretical description of neutron capture reactions is important for next-generation nuclear technologies, such as the transmutation of long-lived nuclear waste [1, 3]. Rates of these reactions are calculated using codes that are based on the statistical reaction theory (e.g. TALYS [4]). A critical input to these calculations is the average electromagnetic transition strengths, described by photon strength functions. For example, modifications of the electric-dipole ($E1$) strength function can cause drastic changes in the abundances of elements produced via neutron capture in the r-process occurring in violent stellar events [5].

In the energy range below about 10 MeV, which is relevant for the applications, the dipole strength function f_1 is dominated by the tail of the isovector electric giant dipole resonance (GDR), which is the collective vibration of the neutron system against the proton system. The damped vibration is described by a Lorentz shape to $f_1(E_\gamma)$ [6, 7, 8], where E_γ is the energy of the photon. Combinations of two or three Lorentz curves are used to describe the double or triple humps of the GDR caused by quadrupole and triaxial deformation of the nuclei [9, 10, 11]. The Generalized Lorentzian (GLO) [12] includes a correction to the Standard Lorentzian (SLO) [6, 7], which accounts for the temperature of the nucleus emitting the photons.

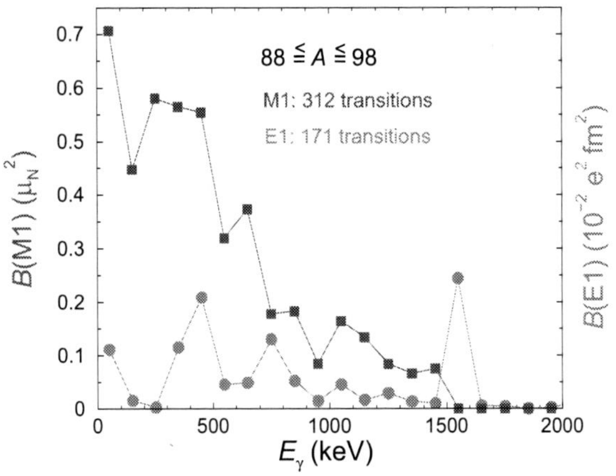

Figure 1. (Color online) Average $B(M1)$ values (blue squares) and $B(E1)$ values (red circles) in 100 keV bins of transition energy deduced from known transitions strengths for nuclides with $88 \leq A \leq 98$ as compiled in Ref. [21].

For the magnetic dipole ($M1$) contribution to f_1, two types of excitations have been considered so far. The scissors mode, which is interpreted as a small-amplitude rotation of the neutron system against the proton system, generates a bump of the $M1$ strength around 3 MeV in deformed nuclei [13]. After it had been well established in the absorption spectra of the ground state, it was recently also identified in the emission from highly excited states (see Ref. [14] and earlier work cited therein). At higher energy, typically around 8 MeV, the $M1$ strength is dominated by the spin-flip resonance [13]. Phenomenological $M1$ strength functions used in statistical-reaction codes are approximated by Lorentz curves with parameters usually derived from systematics [8].

The Lorentz curves used for the $E1$ and $M1$ strength functions decrease when approaching $E_\gamma = 0$. In contrast, an increase of the dipole strength function below 3 MeV toward low γ-ray energy has been found in several nuclides in the mass range from $A \approx 50$ to 100, such as ^{56,57}Fe [15], ^{60}Ni [16], and ^{105,106}Cd [17]. In particular, this low-energy enhancement of the strength function was deduced from experiments using (^{3}He,^{3}He') reactions on various Mo isotopes [18] and was confirmed in an independent experiment using the ^{94}Mo$(d,p)^{95}$Mo reaction [19]. The increase at low γ-ray energies may have a potentially large impact on neutron-capture reaction rates relevant for astrophysical processes [20]. Neither of these measurements were able to distinguish between $E1$ and $M1$ strength. An indication for an $M1$ character of the low-energy enhancement was discussed for the case of ^{60}Ni [16]. The properties of the $E1$ and $M1$ strength functions at very low energy are still a challenging problem.

In a first step we considered known $M1$ and $E1$ transition strengths in nuclei with mass numbers from 88 to 98 [21]. The average $B(M1)$ and $B(E1)$ values of these transitions are shown as a function of the γ-ray energy in Fig. 1. One sees that the average $B(M1)$ values increase toward low energy, whereas the $B(E1)$ values do not show a pronounced trend.

To study the behavior of $M1$ strength in more detail and to get information about the underlying mechanism we have performed shell-model calculations. We have studied the isotopes ^{94}Mo, ^{95}Mo, ^{96}Mo, in which the low-energy enhancement has been observed, and the $N = 50$ nuclide ^{90}Zr. The shell-model calculations were performed by means of the code RITSSCHIL [22] using model spaces composed of the $\pi(0f_{5/2}, 1p_{3/2}, 1p_{1/2}, 0g_{9/2})$ proton orbits and the $\nu(1p_{1/2}, 0g_{9/2}, 1d_{5/2})$ neutron orbits relative to a ^{66}Ni core (SM1), or the $\nu(0g_{9/2}, 1d_{5/2}, 0g_{7/2})$

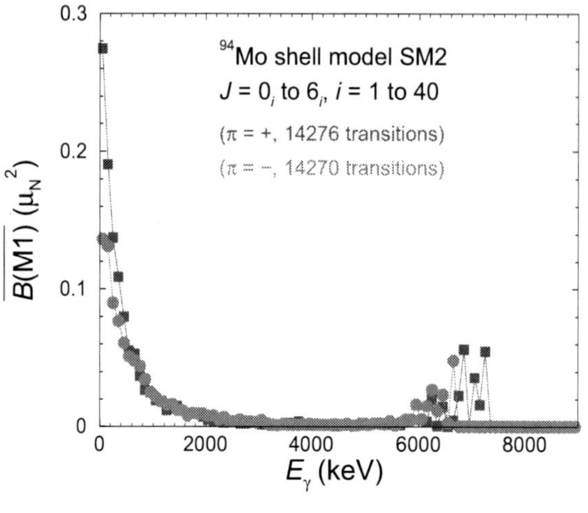

Figure 2. (Color online) Average $B(M1)$ values in 100 keV bins of transition energy calculated for positive-parity (blue squares) and negative-parity (red circles) states in ^{94}Mo.

neutron orbits relative to a ^{68}Ni core (SM2). The configuration spaces were tested in detail in our earlier shell-model studies of nuclei with $N = 46 - 54$ (see Refs. [23, 24, 25, 26] and Refs. therein) and were found appropriate for the description of level energies as well as $M1$ and $E2$ transition strengths in nuclides around $A = 90$. As a further test, we compared the energies of yrast and yrare levels in ^{94,95,96}Mo and ^{90}Zr from the present calculation with the experimental ones, which agree within 300 keV. In the following, we discuss results of SM2.

The calculations included states with spins from $J = 0$ to 6 for ^{90}Zr, ^{94}Mo, ^{96}Mo, and from $J = 1/2$ to $13/2$ for ^{95}Mo. For each spin the lowest 40 states were calculated. The reduced transition probabilities $B(M1)$ were calculated for all transitions from initial to final states with energies $E_f < E_i$ and spins $J_f = J_i, J_i \pm 1$. For the minimum and maximum J_i, the cases $J_f = J_i - 1$ and $J_f = J_i + 1$, respectively, were excluded. This resulted in more than 14000 $M1$ transitions for each parity $\pi = +$ and $\pi = -$, which were sorted into 100 keV bins according to their transition energy $E_\gamma = E_i - E_f$. The average $B(M1)$ value for one energy bin was obtained as the sum of all $B(M1)$ values divided by the number of transitions within this bin. The results for ^{94}Mo are shown in Fig. 2.

For all considered nuclides and each parity a pronounced low-energy enhancement of the average $\overline{B}(M1)$ values is seen. The bump around 7 MeV in ^{90}Zr and ^{94}Mo is caused by $1 \to 0$ and $0 \to 1$ transitions from states dominated by the spin-flip configuration $\nu(0g_{9/2}^{-1}0g_{7/2}^{1})$. The cumulative strength calculated for the $1^+ \to 0_1^+$ transitions in ^{90}Zr is consistent with the value deduced in a recent experiment as shown in Ref. [27]. In ^{95}Mo and ^{96}Mo the bump around 7 MeV does not appear, because the excitation of a $1d_{5/2}$ neutron to the $0g_{7/2}$ orbit is preferred to $\nu(0g_{9/2}^{-1}0g_{7/2}^{1})$.

To find out which states generate strong $M1$ transitions, the average $\bar{B}(M1)$ values for ^{90}Zr and ^{94}Mo are analyzed as a function of the energy of the initial states. A large spike found at 1.5 MeV in the distribution of $\pi = +$ states in ^{94}Mo arises from the $2_2^+ \to 2_1^+$ and $4_2^+ \to 4_1^+$ transitions which link the main configurations $\nu(1d_{5/2}^2)$ in the 2_1^+ and 4_1^+ states with $\pi(0g_{9/2}^2)\nu(1d_{5/2}^2)$ in the 2_2^+ and 4_2^+ states. These findings are consistent with the experimental results given in Refs. [28, 29], which list E_γ = 1196 keV and $B(M1) = 0.56(5)\mu_N^2$ for the corresponding $2_3^+ \to 2_1^+$ transition to be compared with the calculated values of 884 keV and

$0.96\ \mu_N^2$. The bump in the $\pi = +$ distribution between 2 and 3 MeV includes among others the 1_2^+ state with the main configuration $\pi(0g_{9/2}^2)\nu(1d_{5/2}^2)$. It deexcites with $B(M1) = 0.37\mu_N^2$ to the ground state, comparable with the experimental 1_2^+ state described in Refs. [28, 29]. For ^{96}Mo there are analogous similarities of the calculations with the experimental results [30, 31]. The broad enhancements between 2 and 8 MeV (6 MeV) for the $\pi = +$ ($\pi = -$) distributions contain contributions from many states, where all included initial spins contribute approximately the same fraction. The $\overline{B}(M1)$ distributions versus E_i in ^{95}Mo and ^{96}Mo look similar to the ones in ^{94}Mo, but are shifted to somewhat lower excitation energy.

The low-energy enhancement of $M1$ strength is caused by transitions between many close-lying states of all considered spins located well above the yrast line in the transitional region to the quasi-continuum of nuclear states. Inspecting the wave functions, one finds large $B(M1)$ values for transitions between states that contain a large component (up to about 50%) of the same configuration with broken pairs of both protons and neutrons in high-j orbits. The largest $M1$ matrix elements connect configurations with the spins of high-j protons re-coupled with respect to those of high-j neutrons to the total spin $J_f = J_i, J_i \pm 1$. The main configurations are $\pi(0g_{9/2}^2)\nu(1d_{5/2}^2)$, $\pi(0g_{9/2}^2)\nu(1d_{5/2}^1 0g_{7/2}^1)$, and $\pi(0g_{9/2}^2)\nu(1d_{5/2}^2 0g_{9/2}^{-1} 0g_{7/2}^1)$ for positive-parity states in ^{94}Mo. Negative-parity states contain a proton lifted from the $1p_{1/2}$ to the $0g_{9/2}$ orbit in addition. In ^{90}Zr, analogous configurations are generated by exciting protons over the subshell gap at $Z = 40$ and neutrons over the shell gap at $N = 50$, i.e. $\pi(1p_{1/2}^{-2} 0g_{9/2}^2)\nu(0g_{9/2}^{-1} 1d_{5/2}^1)$ and $\pi(1p_{1/2}^{-2} 0g_{9/2}^2)\nu(0g_{9/2}^{-1} 0g_{7/2}^1)$ for positive-parity states and only one $1p_{1/2}$ proton lifted for negative-parity states. The orbits in these configurations have large g factors with opposite signs for protons and neutrons. Combined with specific relative phases of the proton and neutron partitions they cause large total magnetic moments.

The $M1$ strength functions were deduced using the relation $f_{M1}(E_\gamma) = 16\pi/9\ (\hbar c)^{-3}\ \overline{B}(M1, E_\gamma)\ \rho(E_i)$. They were calculated by multiplying the $B(M1)$ value in μ_N^2 of each transition with 11.5473×10^{-9} times the level density at the energy of the initial state $\rho(E_i)$ in MeV^{-1} and deducing averages in energy bins as done for the $\overline{B}(M1)$ values (see above). The level densities $\rho(E_i, \pi)$ were determined by counting the calculated levels within energy intervals of 1 MeV for the two parities separately. For the Mo isotopes, the total level densities $\rho(E_i)$ are well reproduced by the constant-temperature expression $\rho(E_i) = \rho_0 \exp(E_i/T_\rho)$ as long as $E_i <$ 3 MeV. For higher energies the combinatorial level density deviates from this expression and eventually decreases with excitation energy, which is obviously due to missing levels at high energy in the present configuration space. From a fit to the combinatorial values in the range $E_i < 2$ MeV we found for (ρ_0, T_ρ) in (MeV^{-1}, MeV) values of (1.37, 0.67), (1.90, 0.54), and (1.25, 0.58) for ^{94}Mo, ^{95}Mo, and ^{95}Mo, respectively. The level density in the semi-magic ^{90}Zr shows a more complicated energy dependence. The total $M1$ strength function for ^{94}Mo is shown in Fig. 3. As for the $\overline{B}(M1)$ values, there is a pronounced enhancement below 2 MeV, which is well described by the exponential function $f_{M1}(E_\gamma) = f_0 \exp(-E_\gamma/T_f)$. For ^{90}Zr, ^{94}Mo, ^{95}Mo, and ^{96}Mo, the parameters are $f_0 = (34, 37, 39, 55) \times 10^{-9}$ MeV^{-3} and $T_f = (0.50, 0.50, 0.51, 0.48)$ MeV, respectively.

To compare the calculated strength functions with the ones deduced from the (^{3}He,^{3}He') experiments of Ref. [18], the $E1$ contributions have to be added. Because a calculation of the $E1$ strength within the present model space is not possible, we adopted the GLO expression with parameters adjusted to (γ, n) data [32] and the (^{3}He,^{3}He') data above 4 MeV, where our $M1$ contribution is negligible. In the comparison, we focus on the low-energy region observed only via the (^{3}He,^{3}He') reaction, whereas there exist also other experimental data for energies above about 4 MeV [33]. As seen in Fig. 3, the dipole strength functions found in the present calculations resemble the ones deduced from ^{3}He-induced reactions on $^{93-98}$Mo [18, 20] and from a recent ^{94}Mo$(d,p)^{95}$Mo experiment [19], though the experimental data are available for $E_\gamma > 1$

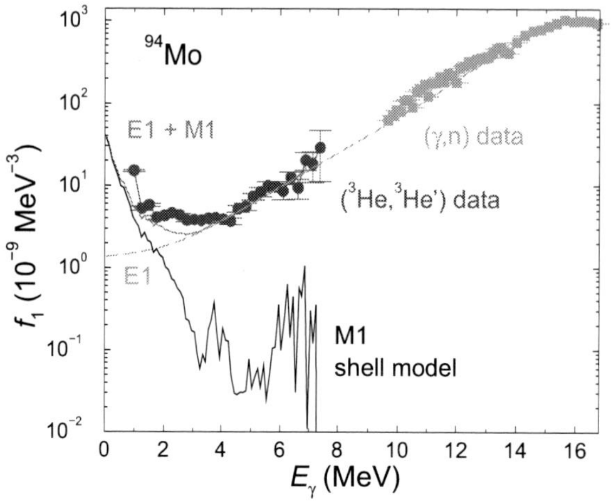

Figure 3. (Color online) Strength functions for ^{94}Mo deduced from (^{3}He,^{3}He') (blue circles) and (γ, n) (green squares) experiments, the $M1$ strength function from the present shell model calculations (black solid line), $E1$ strength according to the GLO expression with parameters $E_0 = 16.36$ MeV, $\sigma_0 = 185$ b, $\Gamma = 5.5$ MeV, $T = 0.35$ MeV (green dashed line), and the total $(E1 + M1)$ dipole strength function (red line).

MeV only. There is a certain freedom in determining the parameters for the GLO, which results in some uncertainty of the magnitude of the GLO in the enhancement region. As the GLO gives only a minor contribution to the total strength below $E_\gamma = 2$ MeV, an acceptable modification of the parameters will not remove the exponential enhancement caused by the $M1$ radiation. It will change the values around 2 MeV, leaving room for other possible enhancement mechanisms. The comparison suggests that at least part of the low-energy enhancement in the experimental dipole strength functions can be explained by $M1$ transitions in the quasi-continuum of states. The analogous low-energy enhancement predicted for ^{90}Zr suggests an experimental study of this nuclide.

The re-coupling of spins leading to large $B(M1)$ values has been discussed in connection with high-spin multiplets (see, e.g. Refs. [23, 24, 25, 26]). An analogous mechanism generates the "shears bands" manifesting "magnetic rotation" [34], which was also observed in the mass-90 region [35, 25]. The "mixed-symmetry" configurations of the interacting boson model arise also from a reorientation of the proton angular momentum with respect to the neutron one. All these phenomena appear in nuclei near closed shells, if there are active high-j proton and neutron orbits near the Fermi surface with magnetic moments adding up coherently. Because these conditions are also prerequisites for the low-energy enhancement, one may expect it to appear in the same nuclei as the phenomena just mentioned. For example, the mixed-symmetry configurations discussed for ^{94}Mo [28, 29] and ^{96}Mo [30, 31] correspond to the dominating configurations $\pi(0g^2_{9/2})\nu(1d^x_{5/2})$ ($x = 2, 3, 4$ for ^{94,95,96}Mo, respectively) that were found causing large $B(M1)$ strengths in the present calculations. The regions in the nuclear chart, where magnetic rotation is expected, are delineated in Fig. 22 of Ref. [34]. In fact, ^{90}Zr and the Mo isotopes discussed in the present work as well as the Fe, Ni, and Cd isotopes, for which the low-energy enhancement was observed [15, 16, 17], belong to these regions. On the other hand, ^{117}Sn [36], ^{158}Gd [37], and the Th, Pa isotopes [14], for which no low-energy enhancement was observed, lie outside these regions. In Ref. [20] it was demonstrated that a low energy-energy enhancement of the dipole strength function comparable with the present one for f_{M1} (GLOup2

in Ref. [20]) increases the astrophysical (n, γ) rate of the r-process by more than a factor of 10. A comparable increase may be expected for nuclei near the neutron drip line located in the mass regions around (Z, N) = (22,48), (26,52), (34,80), (64,118), where magnetic rotation has been predicted and thus, the $M1$ strength should be enhanced at low energy.
The results of the present work were published in Ref. [38].

We thank F. Dönau, H. Grawe, E. Grosse, and E. Litvinova for stimulating discussions. A. C. L. acknowledges support from the Research Council of Norway, project grant no. 205528, and S. F. acknowledges support by HZDR and by the DoE Grant DE-FG02-95ER4093.

References

[1] Arnould M *et al.* 2007 *Phys. Rep.* **450** 97
[2] Käppeler F *et al.* 2011 *Rev. Mod. Phys.* **83** 157
[3] Chadwick M B *et al.* 2011 *Nucl. Data Sheets* **112** 2887
[4] Koning A J, Hilaire S and Duijvestijn M C 2005 AIP Conf. Proc. **769** 1154
[5] Goriely S 1998 *Phys. Lett.* B **436** 10
[6] Brink D M 1955 Ph. D. thesis, Oxford University, unpublished.
[7] Axel P 1962 *Phys. Rev.* **126** 671
[8] Capote R *et al.* 2009 *Nucl. Data Sheets* **110** 3107
[9] Bohr A and Mottelson B R 1975 *Nuclear structure, vol. II,* (W. A. Benjamin, Inc., Reading, Massachusetts)
[10] Eisenberg J M and Greiner W 1975 *Nuclear theory, vol. I,* (North-Holland, Amsterdam)
[11] Junghans A R *et al.* 2008 *Phys. Lett.* B **670** 200
[12] Kopecky J and Uhl M 1990 *Phys. Rev.* C **41** 1941
[13] Heyde K, von Neumann-Cosel P and Richter A 2010 *Rev. Mod. Phys.* **82** 2365
[14] Guttormsen M *et al.* 2012 *Phys. Rev. Lett.* **109** 162503
[15] Voinov A *et al.* 2004 *Phys. Rev. Lett.* **93** 142504
[16] Voinov A *et al.* 2010 *Phys. Rev.* C **81** 024319
[17] Larsen A C *et al.* 2013 *Phys. Rev.* C **87** 014319
[18] Guttormsen M *et al.* 2005 *Phys. Rev.* C **71** 044307
[19] Wiedeking M *et al.* 2012 *Phys. Rev. Lett.* **108** 162503
[20] Larsen A C and Goriely S 2010 *Phys. Rev.* C **82** 014318
[21] National Nuclear Data Center, http://www.nndc.bnl.gov/nudat2/
[22] Zwarts D 1985 *Comput. Phys. Commun.* **38** 365
[23] Schwengner R *et al.* 1995 *Nucl. Phys.* A **584** 159
[24] Schwengner R *et al.* 1998 *Phys. Rev.* C **57** 2892
[25] Schwengner R *et al.* 2002 *Phys. Rev.* C **66** 024310
[26] Schwengner R *et al.* 2009 *Phys. Rev.* C **80** 044305
[27] Rusev G *et al.* 2013 *Phys. Rev. Lett.* **110** 022503
[28] Pietralla N *et al.* 1999 *Phys. Rev. Lett.* **83** 1303
[29] Fransen C *et al.* 2003 *Phys. Rev.* C **67** 024307
[30] Fransen C *et al.* 2004 *Phys. Rev.* C **70** 044317
[31] Lesher S R *et al.* 2007 *Phys. Rev.* C **75** 034318
[32] Beil H *et al.* 1974 *Nucl. Phys.* A **227** 427
[33] Rusev G *et al.* 2009 *Phys. Rev.* C **79** 061302(R)
[34] Frauendorf S 2001 *Rev. Mod. Phys.* **73** 463
[35] Schnare H *et al.* 1999 *Phys. Rev. Lett.* **82** 4408
[36] Agvaanluvsan U *et al.* 2009 *Phys. Rev. Lett.* **102** 162504.
[37] Chyzh A *et al.* 2011 *Phys. Rev.* C **84** 014306
[38] Schwengner R, Frauendorf S and Larsen A C 2013 *Phys. Rev. Lett.* **111** 232504

Observation of bound and unbound 1^- states in ^{208}Pb by particle spectroscopy

Andreas Heusler
Gustav-Kirchhoff-Strasse 7/1, 69120 Heidelberg, Germany
E-mail: A.Heusler@mpi-hd.mpg.de

Abstract. Using the Q3D magnetic spectrograph of the Maier-Leibnitz-Labaratorium at Garching (Germany), experiments on the ^{208}Pb(d,d'), the ^{207}Pb(d,p), the resonant and the non-resonant ^{208}Pb(p,p') reactions were performed. The ^{208}Pb(p,p') reaction was investigated near all seven known isobaric analog resonances in ^{209}Bi. The excitation energies of about 300 states in ^{208}Pb at $E_x < 8$ MeV were determined with an accuracy of about 100 eV. The mean distance between states in ^{208}Pb is about 10 keV up to the neutron threshold ($S(n) = 7368$ keV). The number of 1^- states observed by particle spectroscopy up to the proton threshold ($S(p) = 8.00$ MeV) is close to the number predicted by the schematic shell model without residual interaction. The structure of several 1^- states is deduced from particle spectroscopy. Four out of the seven lowest 1^- states contain about 80% strength of a single configuration.

1. Introduction

In ^{208}Pb, about three hundred particle bound states with spins from 0^- to 14^- and 0^+ to 12^+ and, in addition, unbound states with spins up to 16^- and 17^+ are known [1]. Most states below 7 MeV are described by one-particle one-hole configurations [1, 2, 3, 4, 5]. Yet at rather low energies (4.8-5.3 MeV) multi-particle-hole configurations are firmly identified, namely the neutron pairing vibration state and members of the double-octupole multiplet. The proton pairing vibration state is not yet known. Spin, parity, dominant configuration and several admixing configurations for 72 negative parity states with spins from 0^- to 8^- below 6.4 MeV are known [5] and similarly for about 40 positive parity states with spins from 1^+ to 11^+ below 5.9 MeV [2, 5]. Farther up many states are strongly excited by particle spectroscopy but only few spins are known [1, 6]. I will restrict my discussion to the 1^- states.

The schematic shell model without residual interaction (SSM) [2] predicts sixteen 1^- states up the neutron threshold ($S(n) = 7368$ keV). Nearly the same number of 1^- states is more or less well known [1, 5, 7, 8, 9]. Above the neutron threshold more 1^- states are known from the resonant neutron capture reaction ^{207}Pb(n,γ) where γ-rays up to 0.6 MeV were measured [1, 10]. Shell model calculations yield less bound 1^- states [11, 12]. I will discuss two questions:
(i) How many 1^- states are firmly identified?
(ii) Which particle-hole components are in the 1^- states?

2. Particle spectroscopy with the Q3D magnetic spectrograph

We are using the Q3D magnetic spectrograph of the Maier-Leibnitz-Labaratorium at Garching (Germany) for particle spectroscopy since 2003. The peak shape is highly asymmetric (Fig. 1).

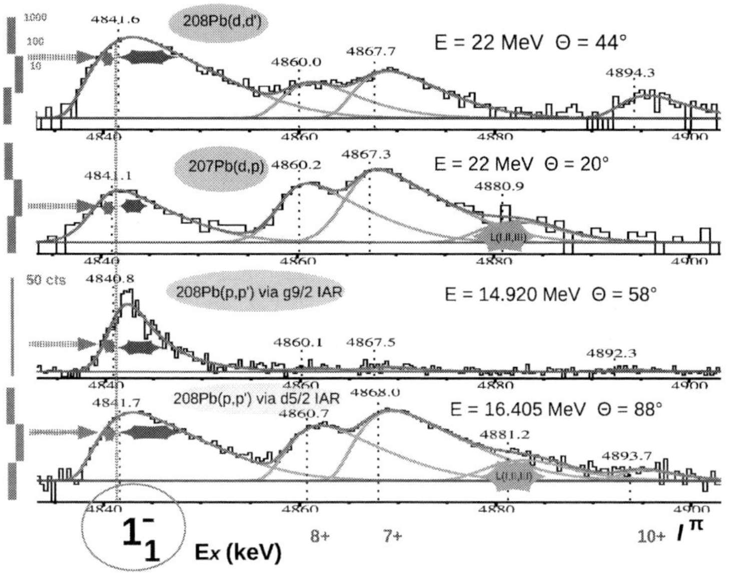

Figure 1. (from top to bottom) Excerpts of spectra for the ^{208}Pb(d,d'), ^{207}Pb(d,p) reactions, and ^{208}Pb(p,p') via the $g_{9/2}$ and $j_{15/2}$ IAR in ^{209}Bi. The energy resolution is 1.5 keV HWHM on the low energy side (magenta marks). The length of the tail depends on the effective target thickness (blue arrows). Each peak is followed by satellites from the electron knockout. L-electrons with 15 keV binding energy produce satellites with relative intensities of about 1% (green marks).

The energy resolution of 1.5 keV half-width at half-maximum (HWHM) on the low energy side is obtained only since ten years. The Q3D magnetic spectrograph itself was build forty years ago. Excitation energies below $E_x = 8$ MeV are determined with a precision of 20-500 eV. Cross sections as low as 0.1 μb/sr are reliably measured within one hour. The peak-to-valley ratio is typically 100:1 and may become 5000:1. From the angular distributions and excitation functions

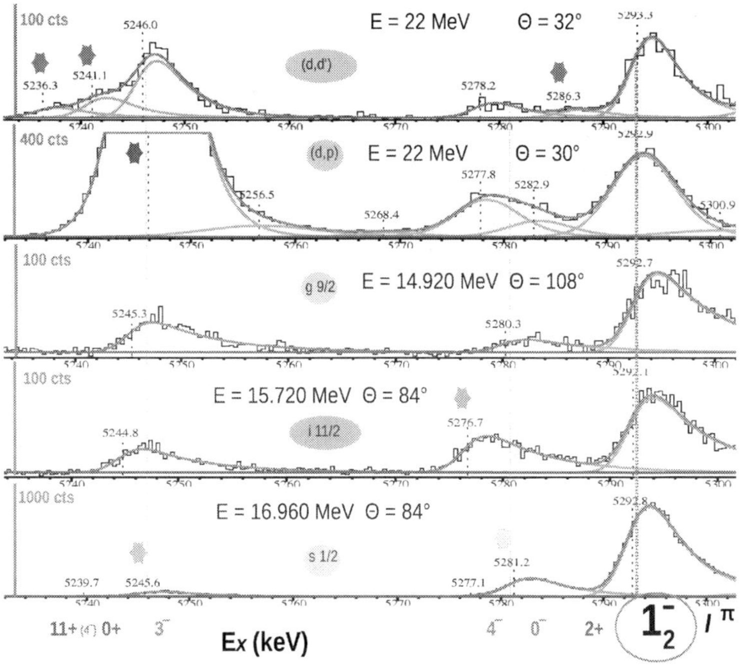

Figure 2. (from top to bottom) Excerpts of spectra near the 1^-_2 state for the ^{208}Pb(d,d'), ^{207}Pb(d,p) reactions, and for ^{208}Pb(p,p') via the $g_{9/2}$, $i_{11/2}$ and $s_{1/2}$ IAR in ^{209}Bi. On the $s_{1/2}$ IAR the 5245 keV 3^- state is twenty times weaker excited, but by the ^{207}Pb(d,p) reaction three times stronger (blue marks). The 5280 keV 0^-_1 state is almost exclusively excited on the $s_{1/2}$ IAR (yellow mark), the 5276 keV 4^-_{10} state on the $i_{11/2}$ IAR (green mark), the 5235 keV 11^+_1 state and two members of the double-octupole multiplet (5241 keV 0^+ and 5286 keV 2^+) by ^{208}Pb(d,d') (magenta marks).

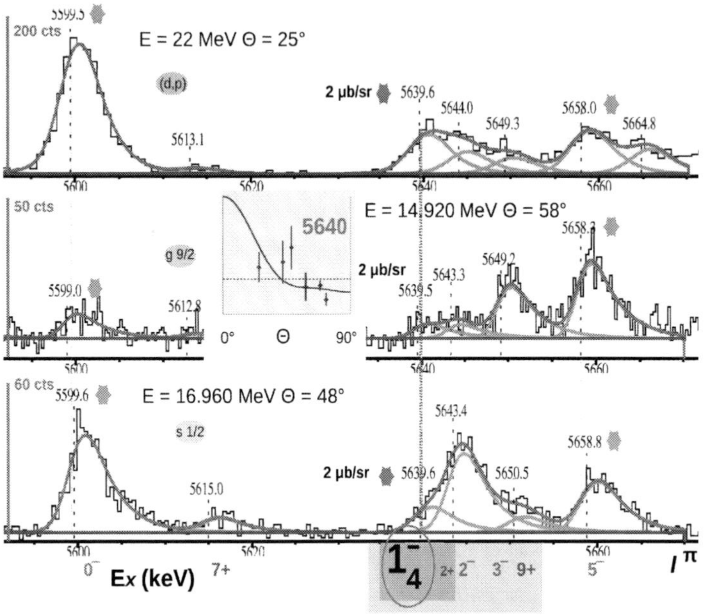

Figure 3. (from top to bottom) Excerpts of spectra near the 1^-_4 state for the ^{207}Pb(d,p) reaction and for ^{208}Pb(p,p') via the $g_{9/2}$ and $s_{1/2}$ IAR in ^{209}Bi. The five states at $5.64 < E_x < 5.65$ MeV build the densest ensemble below 7 MeV while the mean distance between any two states is 10 keV. The inset shows the angular distribution of the 5640 keV 1^-_4 state on the $g_{9/2}$ IAR, the drawn curve the calculated angular distribution for a pure configuration $g_{9/2}f_{7/2}$. The cross section for the 5640 keV 1^-_4 state with the ^{207}Pb(d,p) reaction and on the $s_{1/2}$ IAR is 2 μb/sr (red marks).

up to ten particle-hole components can be determined in each state.

We have performed experiments on the ^{208}Pb(d,d'), ^{207}Pb(d,p), and the resonant and non-resonant ^{208}Pb(p,p') reactions. Inelastic proton scattering via isobaric analog resonances (IAR)

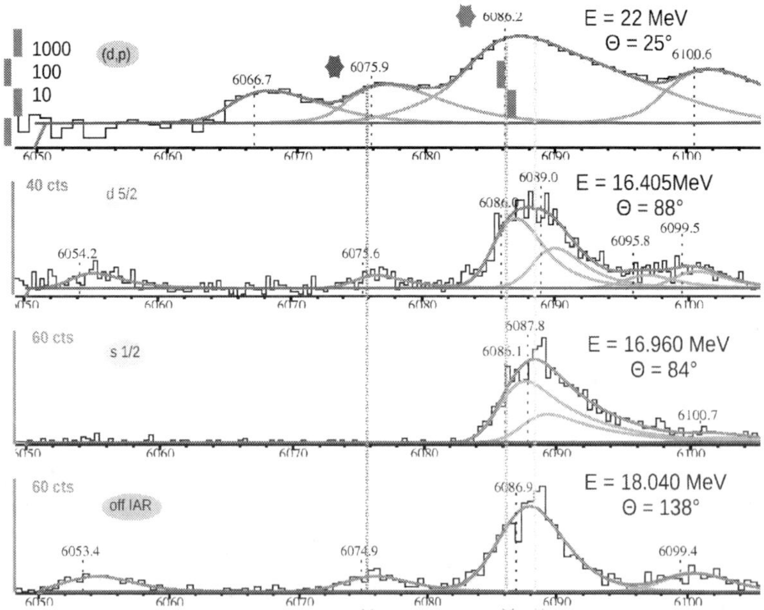

Figure 4. (from top to bottom) Excerpts of spectra for the ^{207}Pb(d,p) reaction (logarithmic scale) and for the ^{208}Pb(p,p') reaction via the $j_{15/2}$ and $s_{1/2}$ IAR in ^{209}Bi, and off resonance. The doublet at $E_x = 6.09$ MeV (green and yellow lines) is resolved on the $j_{15/2}$ and $s_{1/2}$ IARs. The 6076 state (red line) has a vanishing cross section on the $s_{1/2}$ IAR. Peak-to-valley ratios 20:1 and 400:1 are found for the 6076 keV and 6086 keV levels in the ^{207}Pb(d,p) reaction (blue and magenta marks).

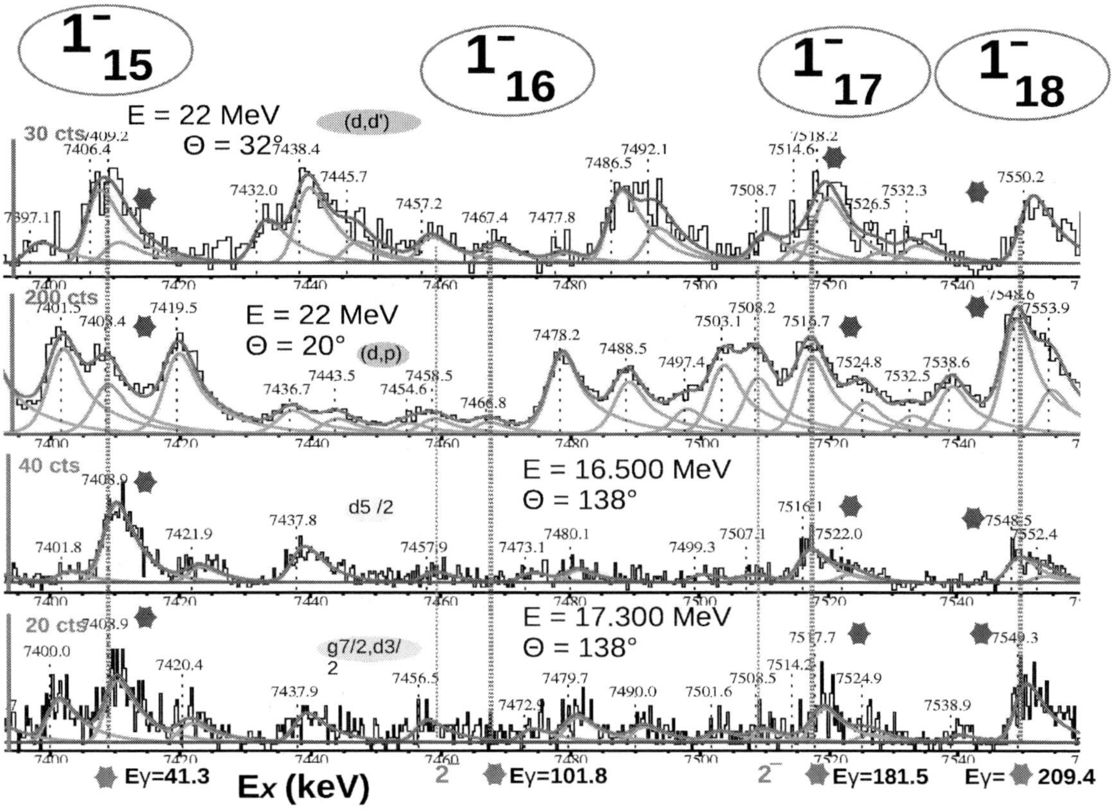

Figure 5. (from top to bottom) Excerpts of spectra for the ^{208}Pb(d, d'), ^{207}Pb(d, p) reactions, and for ^{208}Pb(p, p') via the $d_{5/2}$ and $g_{7/2}$+$d_{3/2}$ IAR in ^{209}Bi taken above the neutron threshold $S(n)$. The γ-energies of the first four 1^- states above the neutron threshold from the resonant neutron capture reaction ^{207}Pb(n, γ) [10] are shown at bottom (magenta marks). Two 2^- states known from ^{207}Pb(n, γ) are identified, too (thin red lines).

is equivalent to a neutron pickup reaction on a target in an excited state. The doublet of the $g_{7/2}$+$d_{3/2}$ IARs is difficult to disentangle. Hence, in effect eight different reactions are studied.

Fig. 1-5 show excerpts from spectra for 3-5 reactions. The full length of a spectrum is about 1 MeV. The mean spacing between the states at $E_x < 7.4$ MeV is 10 keV. The background mostly corresponds to a cross section of about 0.1 μb/sr. The eight reactions excite the states in a highly selective manner. In Fig. 2, the second 1^- state is stronger excited on the $s_{1/2}$ IAR than other states. Several states appear essentially only with one reaction.

In Fig. 3, reliable measurements of cross sections less than 5 μb/sr decide for the identification of the 1^-_4 state. In the ^{208}Pb(n, γ) reaction, two transitions to the ground state are observed in an ensemble of five states at $5639 < E_x < 5650$ keV [1]. The energies are determined with a precision better than 500 eV, the spins as 1^-, 2^+, 2^-, 3^-, and 9^+ [2, 5].

A 1% $s_{1/2}p_{1/2}$ admixture to the 5640 keV 1^-_4 state is derived from the mean cross section of about 2 μb/sr observed both for the ^{207}Pb(d, p) reaction and the ^{208}Pb(p, p') reaction via the $s_{1/2}$ IAR. The angular distribution of the 1^-_4 state on the $g_{9/2}$ IAR is well described by the configuration $g_{9/2}f_{7/2}$ (inset in Fig. 3). The 2^+ state is much weaker excited in particle spectroscopy.

Fig. 4 shows spectra for $6.05 < E_x < 6.10$ MeV. Ref. [1] assigns spin 0^- or 1^- to the 6076 keV state and 1^- to the 6086 keV state, but indeed both states have spin 2^-. The argument for assigning spin 1^- to the 6086 keV state was the strong excitation by the ^{208}Pb(d,p) reaction together with the excitation by ^{208}Pb(α,α') which indicates natural parity. However, the distribution of the $d_{3/2}p_{1/2}$ strength was determined to be nearly 100% in the 5947 keV 1^- state, about 45% $d_{3/2}p_{1/2}$ in the 5924 keV 2^- state leaving another 45% 2^- strength in the 6086 keV state [3]. The spectra on the $d_{5/2}$ and $g_{7/2}$+$d_{3/2}$ IARs (Fig. 4) demonstrate the existence of a close neighbor to the 6086 keV 2^- state within 2 keV. It is weakly excited by the ^{208}Pb(d,p) reaction and has a considerable cross section on the $s_{1/2}$ IAR interpreted as a $s_{1/2}f_{5/2}$ component. Natural parity and an admixing $s_{1/2}f_{5/2}$ configuration assigns spin 3^-.

The 6076 keV state is weakly excited by ^{207}Pb(d,p). The peak to valley ratio is still 20:1 (blue mark in Fig. 4). The assignment of spin 0^- or 1^- [1] is based on the interpretation of the angular distribution for the polarization power in ^{207}Pb(d,p) as $s_{1/2}p_{1/2}$. However, the 6076 keV state is not excited on the $s_{1/2}$ IAR. The vanishing polarisation power is due to an even mixture of $d_{3/2}p_{1/2}$ and $d_{5/2}p_{1/2}$, hence the spin is 2^-.

Fig. 5 shows excerpts of spectra which cover the 1^-_{15}, 1^-_{16}, 1^-_{17}, 1^-_{18} states. They correspond to the four lowest 1^- states above the neutron threshold observed by ^{207}Pb(n,γ) with $E\gamma$ = 41, 102, 181, 209 keV [10]. In particle spectroscopy, at least up to the proton threshold ($S(p)$ =

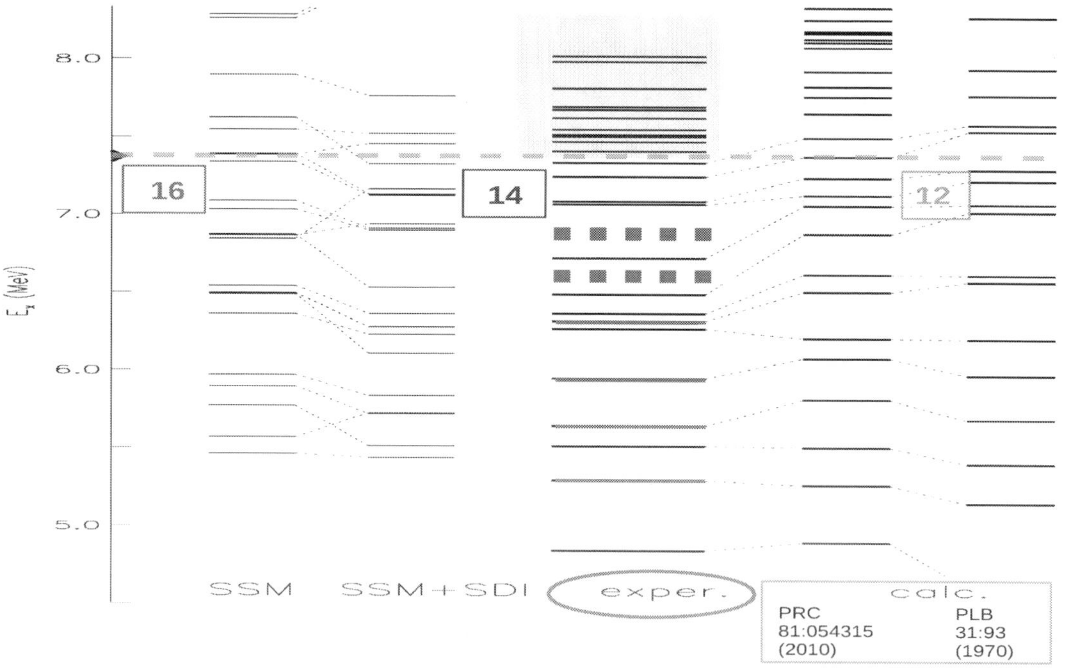

Figure 6. (from left to right) Level scheme of 1^- states in ^{208}Pb calculated by the SSM [2] and the SSM extended by including the surface delta interaction (SDI) [4], from experiment [1, 5, 7, 8, 9], and from shell model calculations with realistic forces [11, 12]. The blue dashed line shows the neutron threshold $S(n)$, the yellow region unbound 1^- states. More 1^- states are expected around $E_x = 6.5 - 7.5$ MeV (red dotted lines). Four states are shown to consist by more than 80% of a single configuration (red lines). The yrast 1^- state calculated by Kuo et al [11] at $E_x = 4279$ keV is not shown.

8.00 MeV), the resolution and the background do not change. Excitation energies are determined with a precision better than 0.5 keV (blue marks).

3. Conclusion

Fourteen bound states are firmly assigned the spin of 1^- [1, 5, 7, 8, 9] and several 1^- states above the neutron threshold are confirmed [1, 10] (Fig. 6). Their excitation energies are determined with an accuracy better than 0.5 keV. The level at $E_x = 6076$ keV assigned spin 0^- or 1^- [1] is shown to have spin 2^-. The level at $E_x = 6.09$ MeV assigned spin 1^- [1] is shown to be a doublet with two states of spin 2^- and 3^- in less than 2 keV distance.

For most of the low-lying 1^- states, up to ten particle-hole components are determined from the analysis of the angular distributions and excitation functions. The angular distributions for ^{207}Pb(d,p) yield the strengths of the configurations where a particle is coupled to the $p_{1/2}$ hole. The comparison of the cross section to calculations yields the orbital angular momenta of the particle(s), the comparison of the analyzing power the spins of the particle(s). For the resonant ^{208}Pb(p,p') reaction, the particle in a particle-hole configuration is determined by the excitation on a certain IAR. The mixing of the hole components is determined from the fit of the angular distribution by a series of Legendre polynomials where the coefficients are products of the well known single-particle widths [6], geometrical coefficients depending on the various spin factors, and the phases between the outgoing proton waves. The strengths of the configurations where a particle is coupled to the $p_{1/2}$ hole agree with values derived from ^{207}Pb(d,p).

Four 1^- states contain more than 80% of a single configuration (Fig. 6). Indeed the transformation matrix from the configurations to these states is close to unity; yet the diagonal is shifted by one: The states with order numbers 2, 3, 4, 7 mainly consist of the configurations 1, 2, 3, 6.

From the comparison of the SSM extended by including the surface delta interaction [4] more bound states are expected. However, only few spins among the two hundred states at $6.4 < E_x < 7.4$ MeV are known [1, 6]. Many of these states are strongly excited by the ^{208}Pb(d,d'), ^{207}Pb(d,p), and ^{208}Pb(p,p') reactions. Especially the states with fractions of the configurations $g_{7/2}f_{5/2}$, $d_{3/2}f_{5/2}$, $g_{7/2}p_{3/2}$, and $d_{3/2}p_{3/2}$ have rather large cross sections [6]. Three of these configurations may couple to spin 1^-.

The very first calculation of the bound 1^- states in ^{208}Pb by T. T. S. Kuo, J. Blomqvist and Gerry Brown [11] does not differ much from the newest calculation by Alex Brown [12]; both indicate only twelve particle bound 1^- states (Fig. 6).

Acknowledgments

I thank P. von Brentano, T. Faestermann, G. Graw, R. Hertenberger, H.-F. Wirth for discussions.

References

[1] Martin M J 2007 *Nucl. Data Sheets* **108** 1583
[2] Heusler A, Graw G, Hertenberger R *et al.* 2010 Phys. Rev. C **82** 014316
[3] Heusler A, Faestermann T, Hertenberger R, Krücken R, Wirth H-F, and Brentano P von 2010 *Eur. Phys. J.* A **46** 17
[4] Heusler A, Jolos R V, and Brentano P von 2013 *Phys. Atomic Nuclei* **76** 807 [2013 *Yad. Fiz.* **76** 860]
[5] Heusler A, Faestermann T, Hertenberger R, Wirth H-F, and Brentano P von 2014 *Phys. Rev.* C **89** 024322
[6] Heusler A, Glöckner H-J, Grosse E, Moore C F, Solf J, and Brentano P von 2014 *Eur. Phys. J.* A **50** 92
[7] Shizuma T, Hayakawa T, Ohgaki H *et al.* 2008 *Phys. Rev. Lett.* **78** 061303
[8] Schwengner R, Massarczyk R, Brown B A *et al.* 2010 *Phys. Rev.* C **81** 054315
[9] Poltoratska I, Neumann-Cosel P von, Tamii A *et al.* 2012 *Phys. Rev.* C **85** 041304
[10] Horen D J, Harvey J A, and Hill N W 1978 *Phys. Rev.* C **18** 722
[11] Kuo T T S, Blomqvist J, and Brown G E 2010 *Phys. Lett.* B **31** 93
[12] Brown B A 2000 *Phys. Rev. Lett.* **85**5300

Study of mixed-symmetry excitations in ^{96}Ru via inelastic proton-scattering

A Hennig[1], M Spieker[1], V Werner[2,3], T Ahn[2,4], V Anagnostatou[2,5], N Cooper[2], V Derya[1], M Elvers[1,2], J Endres[1], P Goddard[2,5], A Heinz[2,6], R O Hughes[2,7], G Ilie[2,8], M N Mineva[9], S G Pickstone[1], P Petkov[1,9], D Radeck[1,2], T Ross[5,7], D Savran[10,11], and A Zilges[1]

[1] Institut für Kernphysik, Universität zu Köln, D-50937 Köln, Germany
[2] Wright Nuclear Structure Laboratory, Yale University, New Haven, Connecticut 06520, USA
[3] Institut für Kernphysik, Technische Universität Darmstadt, D-64289 Darmstadt, Germany
[4] National Superconducting Cyclotron Laboratory NSCL, Michigan State University, East Lansing, Michigan 48824, USA
[5] Department of Physics, University of Surrey, Guildford, GU2 7XH, UK
[6] Fundamental Fysik, Chalmers Tekniska Högskola, SE-41296 Göteborg, Sweden
[7] University of Richmond, Richmond, Virginia 23173, USA
[8] National Institute for Physics and Nuclear Engineering, Bucharest-Magurele, RO-77125, Romania
[9] Institute for Nuclear Research and Nuclear Energy, Bulgarian Academy of Sciences, BG-1784 Sofia, Bulgaria
[10] ExtreMe Matter Institute EMMI and Research Division, GSI, D-64291 Darmstadt, Germany
[11] Frankfurt Institute for Advanced Studies FIAS, D-60438 Frankfurt a.M., Germany

E-mail: `hennig@ikp.uni-koeln.de`

Abstract. Mixed-symmetry states of octupole ($L = 3$) and hexadecapole ($L = 4$) character have been recently proposed in the $N = 52$ isotones ^{92}Zr and ^{94}Mo, based on strong $M1$ transitions to the lowest-lying 3^- and 4^+ states, respectively. In order to investigate similar excitations in the heaviest stable $N = 52$ isotone ^{96}Ru, two inelastic proton-scattering experiments have been performed at the Wright Nuclear Structure Laboratory (WNSL), Yale University, USA and the Institute for Nuclear Physics, University of Cologne, Germany. From the combined data of both experiments, absolute $E1$, $M1$, and $E2$ transition strengths were extracted, allowing for the identification of candidates for MS octupole and hexadecapole states. The structure of the low-lying 4^+ states is investigated by means of *sdg*-IBM-2 calculations.

1. Introduction

The atomic nucleus is a two-component system composed of protons and neutrons. Excitations of the atomic nucleus, which are symmetric and antisymmetric under pairwise exchange of protons and neutrons are denoted as fully-symmetric (FSS) and mixed-symmetry states (MSS), respectively [1, 2]. MSSs are predicted in the proton-neutron version of the interacting boson model (IBM-2) [3, 4, 5, 6] and can be distinguished from FSSs by their F-spin quantum number, which is the bosonic analog of isospin for fermions [4, 5]. An experimental signature of MSSs are strong $M1$ transitions with matrix elements in the order of 1 μ_N to their symmetric counterparts [7, 8].

Mixed-symmetry states have been extensively studied in the $N = 52$ isotones, two neutrons away from the $N = 50$ shell closure (see [8] and references therein). Besides the fundamental one-phonon mixed-symmetry quadrupole excitation 2^+_{ms}, members of the two-phonon $(2^+_{\text{ms}} \otimes 2^+_{\text{s}})$ quintuplet have been identified in ^{92}Zr, ^{94}Mo, and ^{96}Ru as well. The existence of mixed-symmetry states of higher-order multipolarity, namely of octupole ($L = 3$) and hexadecapole ($L = 4$) character, has been recently proposed for the nuclei ^{92}Zr and ^{94}Mo, based on the observation of strong $M1$ transitions to the lowest-lying 3^- and 4^+ states, respectively [9, 10, 11].

Mixed-symmetry octupole excitations have been predicted in the $U_{\pi\nu}(1) \otimes U_{\pi\nu}(5) \otimes U_{\pi\nu}(7)$ dynamical-symmetry limit of the *sdf*-IBM-2 [9]. Along with the $M1$ fingerprint, $E1$ transitions to the 2^+_{s} states are predicted by the model in agreement with the two-body character of the $E1$ operator [12]. In addition, a strong $E1$ transition to the 2^+_{ms} state has been observed in the case of ^{94}Mo. The strong $M1$ transition strength between the lowest-lying 4^+ states in ^{94}Mo has been recently described by introducing g-boson excitations in IBM-2 calculations [11]. This suggests a MS and FS one-phonon hexadecapole admixture in the 4^+_2 and 4^+_1 states, respectively. Additional evidence for this interpretation was provided by shell-model calculations predicting dominant $\sigma = 2$, $j = 4$ configurations for the lowest-lying 4^+ states [13], which are by definition g-boson excitations in the IBM language. Here, σ denotes the seniority.

To study possible mixed-symmetry octupole and hexadecapole states in the heaviest stable $N = 52$ isotone ^{96}Ru, two proton-scattering experiments have been performed. A possible one-phonon hexadecapole admixture in the lowest-lying 4^+ states was investigated in the scope of *sdg*-IBM-2 calculations.

2. Experiments

For the determination of absolute transition strengths, spins and parities of excited states, γ-decay branching ratios, multipole mixing ratios δ, and nuclear level lifetimes τ have to be determined experimentally. For this purpose, the nucleus ^{96}Ru has been studied in two proton-scattering experiments.

One experiment was performed at the WNSL at Yale University. A proton beam with an energy of $E_p = 8.4$ MeV, provided by the ESTU Tandem accelerator, was impinged on an enriched ^{96}Ru target with a thickness of 106 μg/cm^2, mounted on a ^{12}C backing with a thickness of 14 μg/cm^2. To measure the energy of the scattered protons, the target chamber was equipped with five silicon surface-barrier detectors mounted at predominantly backward angles with respect to the beam axis. The de-exciting γ-rays were detected with the YRAST ball array [14], equipped with eight BGO-shielded Clover detectors. $\gamma\gamma$ and pγ coincidence data were acquired in this experiment. Further details on the experimental setup can be found in [15]. From the energy of the scattered protons, the excitation energy of the ^{96}Ru target nuclei was calculated. Gating on the excitation energy in the proton spectra, γ-decay branching ratios were extracted with high sensitivity from the pγ coincidence data. Spin quantum numbers and multipole mixing ratios were determined from the $\gamma\gamma$ coincidence data by means of the $\gamma\gamma$ angular-correlation technique (see e.g., [16, 17]).

The lifetimes of MSSs are expected to be in the sub-picosecond range [8]. Thus, the Doppler-shift attenuation method (DSAM) [18, 19] was applied in a second proton-scattering experiment taking advantage of pγ coincidence data [20]. The same target as used for the prior experiment was bombarded with a proton beam with an energy of $E_p = 7.0$ MeV, provided by the 10 MV Tandem accelerator at the Institute for Nuclear Physics at the University of Cologne. The scattered protons were detected with the new particle detector array SONIC (**S**ilic**ON** **I**dentification **C**hamber), which was equipped with six PIPS silicon particle detectors. SONIC was constructed such that it can be embedded within the existing γ-ray spectrometer HORUS,

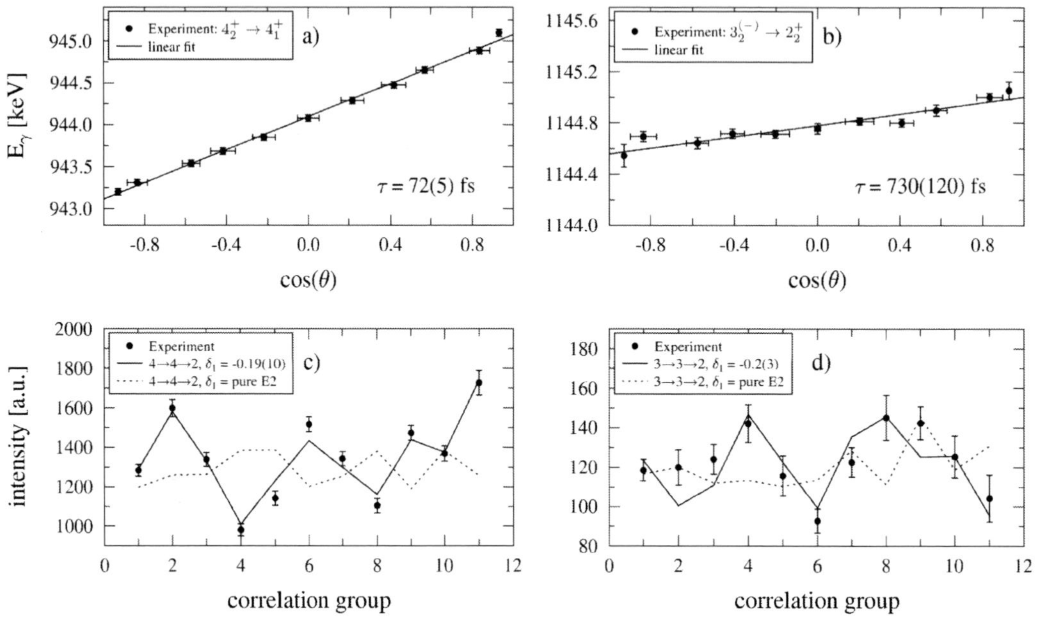

Figure 1. Experimental results for the identification of MS hexadecapole and MS octupole states. The upper panel shows the centroid shifts for γ-rays deexciting the 4_2^+ (a) and $3_2^{(-)}$ states (b) as a function of $\cos(\theta)$. θ is the angle between the direction of the γ-ray emission and the direction of motion of the recoil nucleus. The lifetimes were extracted from the slope of a fit with a first order polynomial. The lower panel shows the extraction of multipole mixing ratios for the de-excitations of the MS hexadecapole (c) and MS octupole (d) candidates to their symmetric counterparts. Both transitions are of predominant $M1$ character (solid lines). Assuming a pure $E2$ character, the experimental data cannot be described at all.

which was equipped with 14 HPGe detectors for the detection of the de-exciting γ-rays.

Applying the DSAM technique to pγ coincidence data yields several advantages. Since the energy and the direction of the scattered particles are known, the velocity and the direction of the ^{96}Ru recoil nucleus can be calculated. Hence, the angle between the direction of the γ-ray emission and the direction of motion of the recoil nucleus can be extracted on an event-by-event basis. Furthermore, the centroid shift of the peaks in the γ-ray spectra can be determined from proton-gated spectra so that feeding from higher-lying states is eliminated. The slowing-down process of the ^{96}Ru recoil nuclei in the target and stopper material was modeled by means of the Monte-Carlo simulation program DSTOP96 [19] which is based on the code DESASTOP [21]. In total, the lifetimes of 30 excited states have been obtained, 24 of them for the first time. In the case of already existing lifetime data, our results are in excellent agreement with the previously measured ones [22, 23, 24].

3. Experimental results for mixed symmetry octupole and hexadecapole states

The experimental results for the identification of mixed-symmetry octupole and hexadecapole candidates are shown in Figure 1. The upper panel shows the centroid shift for de-exciting γ-rays of the 4_2^+ and $3_2^{(-)}$ states. Level lifetimes of $\tau = 72(5)$ fs and $\tau = 730(120)$ fs are obtained, respectively. In the lower panel, the results of the $\gamma\gamma$ angular-correlations are shown, from which the $E2/M1$ multipole mixing ratios are obtained. For the transitions to their symmetric counterparts, a dominant $M1$ character is deduced.

As pointed out in Sec. 1, a strong $M1$ transition to the symmetric octupole state along with an $E1$ transition to the symmetric one-phonon quadrupole state is predicted for the mixed-symmetry octupole state in the $U_{\pi\nu}(1) \otimes U_{\pi\nu}(5) \otimes U_{\pi\nu}(7)$ dynamical-symmetry limit of the *sdf*-IBM-2 [9]. In ^{96}Ru, the $3_2^{(-)}$ state is found at an excitation energy of $E_x = 3077$ keV, which is close to the excitation energies of the MS octupole states in ^{92}Zr (3040 keV) and ^{94}Mo (3011 keV) [10]. A sizeable $M1$ transition strength of $B(M1; 3_2^{(-)} \rightarrow 3_1^-) = 0.14(4)\ \mu_N^2$ is observed for the decay to the symmetric octupole state. Therefore, the $3_2^{(-)}$ state is a likely candidate for the one-phonon MS octupole state in ^{96}Ru. As for ^{94}Mo, a strong $E1$ transition with a strength of $B(E1) = 0.14(3)$ mW.u. is obtained for the decay to the $2^+_{\rm ms}$ state. However, only a weak $E1$ decay with a strength of 0.0017(4) mW.u. is observed for the $3_2^{(-)} \rightarrow 2_1^+$ transition.

The 4_2^+ state is located at an excitation energy of $E_x = 2462$ keV. The newly observed γ-decay to the 2_1^+ state indicates a positive parity for this state. A strong $M1$ transition with a reduced transition strength of $B(M1; 4_2^+ \rightarrow 4_1^+) = 0.90(18)\ \mu_N^2$ is observed along with an $E2$ transition to the 2_1^+ state with a strength of $B(E2; 4_2^+ \rightarrow 2_1^+) = 1.52(19)$ W.u.. The observed $M1$ strength is even stronger than the one obtained for the $2^+_{\rm ms} \rightarrow 2^+_{\rm s}$ transition [22] and comparable to the $4_2^+ \rightarrow 4_1^+$ transition in ^{94}Mo [25]. Thus, the 4_2^+ state is a likely candidate to contain one-phonon hexadecapole MS contributions.

4. *sdg*-IBM-2 calculations

In order to investigate possible one-phonon symmetric and mixed-symmetric hexadecapole contributions in the 4_1^+ and 4_2^+ states, calculations in the framework of the *sdg*-IBM-2 have been performed. Recently, Casperson *et al.* were able to reproduce the strong $M1$ transition between the lowest-lying 4^+ states in ^{94}Mo by introducing g-boson excitations in the IBM-2 without deteriorating the description of the well established quadrupole mixed-symmetry features [11]. Motivated by this approach, we chose the same Hamiltonian for the description of ^{96}Ru:

$$\begin{aligned} \hat{H} &= c\Big((1-\zeta)(\hat{n}_{d_\pi} + \hat{n}_{d_\nu} + \alpha(\hat{n}_{g_\pi} + \hat{n}_{g_\nu})) - \frac{\zeta}{4N}(\hat{Q}_\pi + \hat{Q}_\nu)\cdot(\hat{Q}_\pi + \hat{Q}_\nu) \\ &\quad + \lambda_{sd}\hat{M}_{sd} + \lambda_{sg}\hat{M}_{sg}\Big) \end{aligned} \tag{1}$$

with the quadrupole operator for the proton and neutron bosons:

$$\hat{Q}_\rho = [s_\rho^\dagger \tilde{d}_\rho + d_\rho^\dagger \tilde{s}_\rho]^{(2)} + \beta[d_\rho^\dagger \tilde{g}_\rho + g_\rho^\dagger \tilde{d}_\rho]^{(2)} + \chi_d[d_\rho^\dagger \tilde{d}_\rho]^{(2)} + \chi_g[g_\rho^\dagger \tilde{g}_\rho]^{(2)} \tag{2}$$

with $\rho = \pi, \nu$. For details on the Hamiltonian and the transition operators, see Ref. [11]. The calculations were carried out with the program ArbModel [26]. The number of valence bosons ($N_\pi = 3$ and $N_\nu = 1$) were taken with respect to the doubly-magic nucleus ^{100}Sn. To reduce the number of free parameters, the proton g-factors g_{d_π} and g_{g_π} were chosen to be equal and the neutron effective charges and g-factors as well as the parameters χ_d and χ_g were set equal to zero. The remaining five parameters of the Hamiltonian were fitted to describe the energy of the 2_1^+ state, the $R_{4/2}$ ratio, the $B(E2; 4_1^+ \rightarrow 2_1^+)/B(E2; 2_1^+ \rightarrow 0_1^+)$ ratio, the energy of the 2_3^+ state, which is the one-phonon quadrupole MSS in ^{96}Ru, and the $B(M1; 4_2^+ \rightarrow 4_1^+)/B(M1; 2_3^+ \rightarrow 2_1^+)$ ratio. The proton effective charges and g-factors were fixed to reproduce the $B(E2; 2_1^+ \rightarrow 0_1^+)$ and $B(M1; 2_3^+ \rightarrow 2_1^+)$ strengths, respectively.

The results of the calculation in comparison to the experimental data are compiled in Table 1. The calculated level energies as well as the transition strengths are found in excellent agreement with the data. In particular, the strong $M1$ transition between the lowest-lying 4^+ states is reproduced by the IBM. The predicted value of 1.13 μ_N^2 is close to the experimental value of

Table 1. Results of the *sdg*-IBM-2 calculations in comparison to the data. The experimental and calculated level energies (columns 3 and 4) are given in units of MeV. Columns 5, 6, and 7 show the calculated *s*-, *d*-, and *g*-boson contents in the IBM wave functions. In the last columns, the reduced transition strengths are shown. $M1$ strengths are quoted in units of μ_N^2, $E2$, and $E4$ strengths are given in units of W.u., respectively. The calculated F-spin quantum number is quoted in column 2 as well. An F-spin quantum number of 2 corresponds to maximum F-spin.

		Energies		Boson numbers			Transition strengths			
Level	F	$E_{\rm exp}$	$E_{\rm IBM}$	$\langle n_s\rangle$	$\langle n_d\rangle$	$\langle n_g\rangle$	$J_i^\pi \to J_f^\pi$	$\pi\lambda$	$B(\pi\lambda)_{\rm exp}$	$B(\pi\lambda)_{\rm IBM}$
0_1^+	2	0.000	0.000	3.4	0.3	0.0	-	-	-	-
1_1^+	1	3.154	2.944	1.7	1.8	0.5	$1_1^+ \to 0_1^+$	$M1$	0.17(5)	0.13
2_1^+	2	0.832	0.832	2.6	1.3	0.1	$2_1^+ \to 0_1^+$	$E2$	18.1(5)	18.4
2_2^+	2	1.932	2.165	1.8	2.0	0.2	$2_2^+ \to 2_1^+$	$M1$	0.05(2)	0
							$2_2^+ \to 2_1^+$	$E2$	28(9)	24
2_3^+	1	2.283	2.322	2.5	1.2	0.3	$2_3^+ \to 2_1^+$	$M1$	0.69(14)	0.69
							$2_3^+ \to 0_1^+$	$E2$	1.36(19)	2.53
4_1^+	2	1.518	1.523	2.3	1.2	0.6	$4_1^+ \to 2_1^+$	$E2$	22.6(17)	25.6
							$4_1^+ \to 0_1^+$	$E4$	-	1.09
4_2^+	1	2.462	2.482	2.6	0.5	0.9	$4_2^+ \to 4_1^+$	$M1$	0.90(18)	1.13
							$4_2^+ \to 2_1^+$	$E2$	1.52(19)	1.44
							$4_2^+ \to 0_1^+$	$E4$	-	0.55

0.90(18) μ_N^2. In addition to the strong $M1$ transition, significant g-boson contributions are predicted for the 4_1^+ and 4_2^+ states. They are considerably enhanced compared to, e.g., the g-boson content of the 2_2^+ state, which is known to be the 2^+ member of the $(2_s^+ \otimes 2_s^+)$ two-phonon triplet [27]. The one-phonon hexadecapole admixture is furthermore reflected by the $E4$ transition strengths of the $4_{1,2}^+$ states to the ground-state. The $E4$ transition operator was defined in the same way as in Ref. [11]:

$$\hat{T}(E4) = e_{\pi 1}\left[s_\pi^\dagger \tilde{g}_\pi + g_\pi^\dagger \tilde{s}_\pi\right]^{(4)} . \tag{3}$$

Since no experimental $B(E4)$ values are known for ^{96}Ru, the parameter $e_{\pi 1}$ in Eq. (3) was arbitrarily set to 1 W.u.. Therefore, only relative $E4$ strengths can be compared in Table 1 unless experimental values for the $E4$ strengths become available, which may be accessible, e.g., in electron-scattering experiments. Table 1 also shows the calculated F-spin quantum numbers for excited states in ^{96}Ru. A mixed-symmetry character can be assigned to the 2_3^+ and the 1_1^+ state, which correspond to the one- and two-phonon quadrupole mixed-symmetry states, respectively [22, 23]. In addition, a mixed-symmetry character is also predicted for the 4_2^+ state. Thus, the IBM-2 calculations support the interpretation of the 4_1^+ and 4_2^+ states to show symmetric and mixed-symmetric one-phonon hexadecapole admixtures in their wave functions.

5. Summary

Mixed-symmetry states of octupole and hexadecapole character have been studied in two inelastic proton-scattering experiments. The 3_2^- and 4_2^+ states are likely candidates to show one-phonon mixed-symmetry octupole and hexadecapole contributions, based on strong $M1$ transitions to their symmetric counterparts. Calculations in the framework of the sdg-IBM-2 give further evidence for one-phonon hexadecapole components of symmetric and mixed-symmetric character in the 4_1^+ and 4_2^+ states, respectively.

Acknowledgements

The authors thank R. Casperson and S. Heinze for support with the IBM calculations and N. Pietralla for fruitful discussions. Furthermore, we highly acknowledge the support of the accelerator staff at WNSL, Yale and IKP, Cologne during the beam times. This work is supported by the DFG (ZI 510/4-2), the U.S. Department of Energy Grant No. DE-FG02-01ER40609, and the BMBF Grant No. 05P12RDFN8. S.G.P. and M.S. are supported by the Bonn-Cologne Graduate School of Physics and Astronomy.

References

[1] Heyde K and Sau J 1986 *Phys. Rev. C* **33** 1050–1061
[2] Iachello F 1984 *Phys. Rev. Lett.* **53** 1427–1429
[3] Arima A and Iachello F 1975 *Phys. Rev. Lett.* **35** 1069–1072
[4] Arima A, Otsuka T, Iachello F and Talmi I 1977 *Phys. Lett. B* **66** 205–208
[5] Otsuka T, Arima A and Iachello F 1978 *Nucl. Phys. A* **309** 1–33
[6] van Isacker P, Heyde K, Jolie J and Sevrin A 1986 *Ann. Phys.* **171** 253–296
[7] Heyde K, von Neumann-Cosel P and Richter A 2010 *Rev. Mod. Phys.* **82** 2365–2419
[8] Pietralla N, von Brentano P and Lisetskiy A F 2008 *Prog. Part. Nucl. Phys.* **60** 225–282
[9] Smirnova N A, Pietralla N, Mizusaki T and van Isacker P 2000 *Nucl. Phys. A* **678** 235–257
[10] Scheck M, Butler P A, Fransen C, Werner V and Yates S W 2010 *Phys. Rev. C* **81** 064305
[11] Casperson R, Werner V and Heinze S 2013 *Phys. Lett. B* **721** 51–55
[12] Pietralla N, Fransen C, Gade A, Smirnova N A, von Brentano P, Werner V and Yates S W 2003 *Phys. Rev. C* **68** 031305
[13] Lisetskiy A F, Pietralla N, Fransen C, Jolos R V and von Brentano P 2000 *Nucl. Phys. A* **677** 100–114
[14] Beausang C, Barton C, Caprio M, Casten R, Cooper J, Krücken R, Liu B, Novak J, Wang Z, Wilhelm M, Wilson A, Zamfir N and Zilges A 2000 *Nucl. Instr. and Meth. A* **452** 431–439
[15] Elvers M, Pascu S, Ahmed T, Ahn T, Anagnostatou V, Cooper N, Deng C, Endres J, Goddard P, Heinz A, Ilie G, Jiang E, Küppersbusch C, Radeck D, Savran D, Shenkov N, Werner V and Zilges A 2011 *Phys. Rev. C* **84** 054323
[16] Radeck D, Albers M, Bernards C, Bettermann L, Blazhev A, Fransen C, Heinze S, Jolie J and Mücher D 2009 *Nucl. Phys. A* **821** 1–22
[17] Albers M, Mücher D, Bernards C, Blazhev A, Fransen C, Heinze S, Jolie J, Lisetskiy A, Petkov P, Radeck D and Zell K O 2010 *Nucl. Phys. A* **847** 180 – 206
[18] Alexander T K and Forster J S 1978 *Adv. Nucl. Phys.* **10** 197–331
[19] Petkov P, Gableske J, Vogel O, Dewald A, von Brentano P, Krücken R, Peusquens R, Nicolay N, Gizon A, Gizon J, Bazzacco D, Rossi-Alvarez C, Lunardi S, Pavan P, Napoli D R, Andrejtscheff W and Jolos R V 1998 *Nucl. Phys. A* **640** 293–321
[20] Seaman G G, Benczer-Koller N, Bertin M C and MacDonald J R 1969 *Phys. Rev.* **188** 1706–1710
[21] Winter G 1983 *Nucl. Instr. and Meth.* **214** 537–539
[22] Pietralla N, Barton C J, Krücken R, Beausang C W, Caprio M A, Casten R F, Cooper J R, Hecht A A, Newman H, Novak J R and Zamfir N V 2001 *Phys. Rev. C* **64** 031301(R)
[23] Linnemann A, Fransen C, Gorska M, Jolie J, Kneissl U, Knoch P, Mücher D, Pitz H H, Scheck M, Scholl C and von Brentano P 2005 *Phys. Rev. C* **72** 064323
[24] Adamides E, Sinatkas J, Skouras L D, Xenoulis A C, Gazis E N, Papadopoulos C T and Vlastou R 1986 *Phys. Rev. C* **34** 791–809
[25] Fransen C, Pietralla N, Ammar Z, Bandyopadhyay D, Boukharouba N, von Brentano P, Dewald A, Gableske J, Gade A, Jolie J, Kneissl U, Lesher S R, Lisetskiy A F, McEllistrem M T, Merrick M, Pitz H H, Warr N, Werner V and Yates S W 2003 *Phys. Rev. C* **67** 024307
[26] Heinze S 2008 *Computer program ArbModel* unpublished, University of Cologne
[27] Klein H, Lisetskiy A F, Pietralla N, Fransen C, Gade A and von Brentano P 2002 *Phys. Rev. C* **65** 044315

Quadrupole collectivity in neutron-rich Cd isotopes

S Bönig[1], Th Kröll[1], S Ilieva[1], M Scheck[1,2,3] for the IS411, IS477, IS524 and the MINIBALL collaborations

[1] Institut für Kernphysik, Technische Universität Darmstadt, Darmstadt, Germany
[2] School of Engineering, University of the West of Scotland, Paisley, United Kingdom
[3] SUPA, Scottish Universities Physics Alliance, Glasgow, United Kingdom

E-mail: `sboenig@ikp.tu-darmstadt.de`

Abstract. The proximity to the closed shells at $Z = 50$ and $N = 82$ makes the neutron-rich Cd isotopes a perfect test case for nuclear theories. The energy of the first excited 2^+-state in the even $^{122-128}$Cd shows an irregular behaviour as the Cd isotopes exhibit only a slight increase for ^{122}Cd to ^{126}Cd and even a decrease from ^{126}Cd to ^{128}Cd. This anomaly can so far not be reproduced by shell model calculations. Only beyond mean field calculations with a resultant prolate deformation are capable to describe this anomalous behaviour. In order to gain more information about the neutron-rich Cd isotopes a Coulomb excitation experiment was performed with MINIBALL at REX-ISOLDE, CERN. The extracted transition strengths $B(E2, 0^+_{gs} \rightarrow 2^+_1)$ for ^{122,124,126,128}Cd agree with beyond mean field calculations. The spectroscopic quadrupole moments $Q_s(2^+_1)$ are compared with measurements on odd neutron-rich Cd isotopes.

1. Motivation

The neutron-rich Cd isotopes provide a perfect test for nuclear theory, as they are only two protons away from the shell closure at $Z = 50$ and close to the $N = 82$ magic number. Already twenty years ago a quenching of the shell gap at $N = 82$ has been proposed by Dobaczewski et al. [1]. However, the astrophysical r-process path is expected to pass this region with the waiting-point nucleus ^{130}Cd. Experimental investigations of this semi-magic nucleus found a high Q_β value of ^{130}Cd which supports the idea of a shell gap reduction [2], whereas information on the isomeric decay found no evidence of a total shell quenching [3]. However, the solar abundance peak at $A \approx 130$ is better described including a quenching of the shell gap at $N = 82$ [4]. Such an effect should be reflected in lowered excitation energies and an enhanced transition strength.

The excitation energy of the first excited 2^+-state in the neutron-rich Cd isotopes show an irregular behaviour. The rise towards the $N = 82$ shell closure is not as steep as expected from comparison to the isotones Ba, Te, Xe and Pd, and even decreases from ^{126}Cd ($E(2^+_1) = 652$ keV [5]) to ^{128}Cd ($E(2^+_1) = 645$ keV [5, 6]). The 2^+_1 energy for ^{128}Cd lies lower than for both the isotones $^{132}_{52}$Te ($E(2^+_1) = 974$ keV [7]) with two protons above $Z = 50$ and $^{126}_{46}$Pd ($E(2^+_1) = 974$ keV [8]) with two protons less than $^{128}_{48}$Cd. This anomaly is so far not explained by shell model calculations. Nevertheless, beyond mean field calculations are able to reproduce these findings with a resultant prolate deformation for ^{122,124,126,128}Cd.

The cross section for a Coulomb excitation depends on the transitional and diagonal matrix elements of the nuclear states involved in the excitation process. These matrix elements can be used to determine the transition strength $B(E2, 0^+_{gs} \rightarrow 2^+_1)$ and the spectroscopic quadrupole moment $Q_s\left(2_1^+\right)$ of a nucleus. In order to shed light into the region around doubly-magic ^{132}Sn a Coulomb excitation experiment was performed at the REX-ISOLDE facility located at CERN.

2. Experimental setup and data analysis

At REX-ISOLDE the radioactive ion beams are produced with the ISOL (Isotope Separation OnLine)-technique. Every $n \times 1.2$ s a pulsed 1.4 GeV proton beam from the PS-BOOSTER hits a primary target (in this case UC$_x$), which undergoes fission. The products diffuse through the target and a transfer line into an ion source, where the beam is purified by a selective multistep laser ionisation with the resonant ionisation laser ion source (RILIS). Subsequently the beam is extracted into the beam line and a mass separation is applied. After cooling, bunching, charge breeding and a second mass separation (A/q $\approx$ 4.3) the beam containing the Cd isotopes was further accelerated to 2.8 MeV/u and delivered to the experimental area, where it hits a secondary target. This was chosen to be ^{108}Pd for ^{122}Cd, ^{104}Pd for ^{124}Cd and ^{64}Zn for ^{124,126,128}Cd. The beam energy is considered "safe" for each of the used beam/target combinations, as it is always much below the safe energy. This ensures the nuclear interaction being small and therefore the nuclear potential can be neglected. The average intensity of the beam was between $10^3 - 10^4$ pps. In the scattering process both target and projectile nuclei interact electromagnetically and get excited. The deexcitation γ-rays are detected with the MINIBALL detector array [9]. It consists of eight high purity Germanium triple cluster detectors, each being 6-fold electronically segmented. The array has an efficiency of ~ 7 % at 1.3 MeV. The deexcitation γ-rays are emitted in-flight, which leads to a Doppler shift of their energy. To correct for this effect the angular information of the scattered particles, which is obtained with a double-sided silicon strip detector (DSSSD) [10] with a 16-fold segmentation in θ and 12-fold segmentation in ϕ placed in forward direction, is used. The covered angular range in the laboratory frame was between $\theta_{Lab} \approx 16^\circ - 53^\circ$. Note, that the beam delivered to the secondary target often is not completely pure as it contains impurifications from the production of the beam and from β-decay of the beam particles (*i.e.* indium in the case of a cadmium beam). For the experiments with ^{122,124,128}Cd surface ionised isobaric Cs was present. In the measurement of ^{126}Cd and for parts of ^{124}Cd a quartz transfer line was used, which reduced the Cs contamination to a negligible amount. In the measurement of ^{122,124,126}Cd the deexcitation γ-ray spectrum obtained in the period when the RILIS laser was turned off can be subtracted from the γ-ray spectrum when the laser was turned on. During the laser off period no transitions in Cd are visible in the γ-ray spectra. Therefore the surface ionised Cs as well as the isobaric contaminant In are eliminated [11]. In the ^{128}Cd experiment the Cs contamination was not constant over the time of the measurement and could only be subtracted from the spectra by applying time gates. A gate on late times T (1300 ms $\leq$ T $\leq$ 2300 ms) after proton pulse impact reveals only contributions to the γ-ray spectra from the Cs contamination. Subtracting this from the γ-ray spectrum with a gate set on early times T (200 ms $\leq$ T $\leq$ 1200 ms) after proton pulse impact cancels the Cs contaminant. The isobaric In contamination from the decay of Cd and directly from ISOLDE can be deduced also from the γ-ray spectrum after β-decay. Therefore the beams of ^{124,126,128}Cd were stopped in a thick target and the γ-rays detected. The relative amount of each isotope is extracted by simultaneously fitting it and the efficiency to the γ-ray yields after β-decay. In the case of ^{128}Cd the relative intensities for the γ-ray transitions in ^{128}In and ^{128}Sn were determined. A deviation from literature for transitions in ^{128}Sn was found [12]. In all the experiments a contamination of the 8^- isomeric state of In was present in the beam, which had to be taken into account. The relative amount of Cd for the different experiments are displayed in table 1.

The prompt Doppler-corrected γ-ray spectra show the expected deexcitation of the 2_1^+ state of the used target and the investigated Cd nucleus. Note, that in the case of ^{128}Cd the deexcitation peak at 645 keV proofs the 2_1^+ state of ^{128}Cd being at this energy, which was not clear before.

Table 1. Beam purity and contaminating nuclei for each experiment. ^{122,124,126}Cd see: [11], ^{128}Cd see: [13].

Experiment	$\mathcal{P}$ [%]	Isobaric contaminant
^{122}Cd on ^{108}Pd	58(1)	Cs, In, ^{m}In
^{124}Cd on ^{104}Pd	18(1)	Cs, In, ^{m}In
^{124}Cd on ^{64}Zn	79(4)	In, ^{m}In
(use of quartz transfer line)		
^{126}Cd on ^{64}Zn	45(1)	In, ^{m}In
(use of quartz transfer line)		
^{128}Cd on ^{64}Zn	47	Cs, In, ^{m}In

For the experiments with ^{126}Cd and ^{128}Cd an additional Doppler corrected peak at 243 keV and 324 keV, respectively, is visible. This stems out of the excitation of the isobaric beam contaminant In [11, 13].

The Coulomb excitation cross section of the projectile nucleus σ^{proj} can be determined with the information from the obtained γ-ray spectrum

$$\sigma^{proj} = \frac{1}{\mathcal{P}} \frac{N_{\gamma}^{proj}}{\varepsilon_{\gamma}^{proj}} \left(\frac{N_{\gamma}^{targ}}{\varepsilon_{\gamma}^{targ}} \right)^{-1} \cdot \sigma^{targ}, \tag{1}$$

with N_{γ}^{i} being the γ-ray yield of the projectile/target excitation, respectively, ε_{γ}^{i} is the efficiency of the MINIBALL detectors obtained with a ^{152}Eu source and σ^{targ} is the Coulomb excitation cross section for the target excitation, which can be calculated from the already known target properties. The normalisation to the target excitation is performed in order to reduce systematic uncertainties. The Coulomb excitation cross section is proportional to the matrix elements of the states involved in the excitation. For all the here discussed experiments these states are only the 0^+ ground state and the first excited 2^+ state. This leads to the important matrix elements being the reduced transitional matrix element M_{02} and the diagonal matrix element M_{22}

$$\begin{aligned} M_{02} &= \langle\, 0_{gs}^{+} \parallel \mathcal{M}(E2) \parallel 2_1^{+} \rangle, \\ M_{22} &= \langle\, 2_1^{+} \parallel \mathcal{M}(E2) \parallel 2_1^{+} \rangle, \end{aligned} \tag{2}$$

with $\mathcal{M}(E2)$ being the electric quadrupole operator. Varying these matrix elements in the input for the programs CLX/DCY until the cross section reproduces the experimentally found one gives sets for possible combinations of (M_{02}, M_{22}). Because the cross section exhibits a different sensitivity to the matrix elements at different angles, a division of the scattered particles into distinct center of mass angular ranges restricts the possible combinations of the matrix elements. By performing a maximum likelihood analysis the transition strength

$$B\left(E\lambda, I_i \to I_f\right) = \frac{1}{2I_i + 1} \left|\langle I_i \parallel \mathcal{M}\left(E\lambda\right) \parallel I_f \rangle\right|^2 \tag{3}$$

and the spectroscopic quadrupole moment

$$eQ_s(I) = \sqrt{\frac{16\pi}{5\,(2I+1)}} < II20|II > \langle I \parallel \mathcal{M}(E2) \parallel I \rangle \tag{4}$$

can be extracted.

3. Results, discussion and outlook

The reduced transition strength $B\left(E2, 0^+_{gs} \rightarrow 2^+_1\right)$ has been determined in Coulomb excitation experiments at REX-ISOLDE for the nuclei ^{122,124,126,128}Cd [11, 13]. The values seem to follow the beyond mean field predictions rather than the shell model calculations, where the configuration space is based on a ^{78}Ni core. The proton orbits $\pi\left(1f_{5/2}, 1g_{9/2}, 2p_{1/2}, 2p_{3/2}\right)$ and the neutron orbits $\nu\left(1g_{7/2}, 1h_{11/2}, 2d_{3/2}, 2d_{5/2}, 3s_{1/2}\right)$ were included. In a laser spectroscopy measurement of odd neutron-rich Cd isotopes at COLLAPS [14] small quadrupole moments for the ground state were found. The small spectroscopic quadrupole moments for ^{122,124,126}Cd follow this trend.

Additional information from direct lifetime measurements provide a further constraint on the transitional matrix element. This could already be included in the analysis of ^{122}Cd and ^{126}Cd. However, for ^{124}Cd and ^{128}Cd these information are not available so far. The lifetime of the first excited 2^+ state in these nuclei coud be investigated in the future with a plunger measurement at HIE-ISOLDE [15]. A high source of uncertainty is the lack of statistics in the γ-ray spectrum for such highly exotic beams. The upgrade of ISOLDE to HIE-ISOLDE will in the future provide beams of highly exotic nuclei with a higher intensity and energy to increase the γ-ray yield in the deexcitation spectrum.

To answer the question of which orbitals contribute to the collectivity a campaign to measure the odd Cd isotopes via Coulomb excitation has been started at REX-ISOLDE. In these nuclei the low-lying, high-spin states are accessible via the isomeric $\frac{11}{2}^-$ state, whereas the low-lying, low-spin states can be populated from the $\frac{3}{2}^+$ ground state. ISOLDE provides a beam containing the ground as well as the isomeric state of these nuclei, whose relative amount can be varied with the RILIS. A first successful investigation of ^{123}Cd has been performed already [16].

4. Acknowledgements

This project is supported by BMBF (No. 06 DA 9036I, No. 05 P12 RDCIA, No. 05 P12 RDCIB and No. 05 P12 PKFNE), HIC for FAIR, EU through EURONS (No. 506065) and ENSAR (No. 262010) and the MINIBALL and REX-ISOLDE collaborations.

[1] Dobaczewski J, Hamamoto I, Nazarewicz W and Sheikh J A 1994 *Phys. Rev. Lett.* **72** 7
[2] Dillmann I et al. 2003 *Phys. Rev. Lett.* **91** 16
[3] Jungclaus A et al. 2007 *Phys. Rev. Lett.* **99** 132501
[4] Chen B et al. 1995 *Phys. Lett. B* **355** 37
[5] Kautzsch T et al. 2000 *Eur. Phys. J. A* **9** 201
[6] Cáceres L et al. 2009 *Phys. Rev. C* **79** 011301(R)
[7] Kerek A et al. 1974 *Nucl. Phys. A* **224** 367
[8] Watanabe H et al. 2013 *Phys. Rev. Lett.* **111**
[9] Warr N et al. 2013 *Eur. Phys. J. A* **49** 40
[10] Ostrowski A N et al. 2002 *Nucl. Instr. and Meth. A* **480** 448
[11] Ilieva S et al. 2014 *Phys. Rev. C* **89** 014313
[12] Bönig S 2014 *PhD Thesis* TU Darmstadt
[13] Bönig S *to be published*
[14] Yordanov D T et al. 2013 *Phys. Rev. Lett.* **110** 192501
[15] https://hie-isolde.web.cern.ch/hie-isolde/
[16] Hartig A-L 2014 *Master Thesis* TU Darmstadt

Shell structure and few-nucleon removal in intranuclear cascade

D Mancusi[1], A Boudard[1], J Carbonell[1], J Cugnon[2], J-C David[1] and S Leray[1]

[1] CEA, Centre de Saclay, IRFU/Service de Physique Nucléaire, F-91191 Gif-sur-Yvette, France
[2] University of Liège, AGO Department, allée du 6 Août 17, bât. B5, B-4000 Liège 1, Belgium

E-mail: davide.mancusi@cea.fr

Abstract. It is well known that intranuclear-cascade models generally overestimate the cross sections for one-proton removal from heavy, stable nuclei by a high-energy proton beam, but they yield reasonable predictions for one-neutron removal from the same nuclei and for one-nucleon removal from light targets. We use simple shell-model calculations to investigate the reasons of this deficiency. We find that a correct description of the neutron skin and of the energy density in the nuclear surface is crucial for the aforementioned observables. Neither ingredient is sufficient if taken separately.

1. Introduction

Nuclear reactions between high-energy ($\gtrsim$ 150 MeV) nucleons or hadrons and nuclei are usually described by means of *intranuclear-cascade* (INC) models [1]. In this framework, the projectile is assumed to initiate an avalanche of binary collisions with the nucleons of the target, which can lead to the emission of energetic particles. The nature of INC models is essentially classical. It is typically assumed that nucleons are perfectly localised in phase space and are bound by an average, constant potential; moreover, it is assumed that subsequent elementary collisions are independent.

It was realized some time ago that INC models systematically fail to describe inclusive cross sections for the removals of few nucleons [see e.g. 2, 3]. This is especially surprising in view of the fact that these observables are associated with peripheral reactions and mostly involve collisions between quasi-free nucleons; one would therefore expect intranuclear cascade to provide an accurate description of this particular dynamics. This puzzling result has been known for many years now, but no convincing explanation has ever been put forward.

We will show that the few-nucleon removal process at high energy is sensitive to the the description of the nuclear surface, which we draw from a simple shell-model calculation. We will show that the predictions of an INC model [4] can be substantially improved by casting the shell-model calculation results in a form adaptable to the nuclear model underlying INC.

2. Model description

It is generally assumed that the first stage of high-energy nucleon-nucleus reactions can be described as an avalanche of independent binary collisions. The nuclear model underlying INC is essentially classical, with the addition of a few suitable ingredients that mimic intrinsically

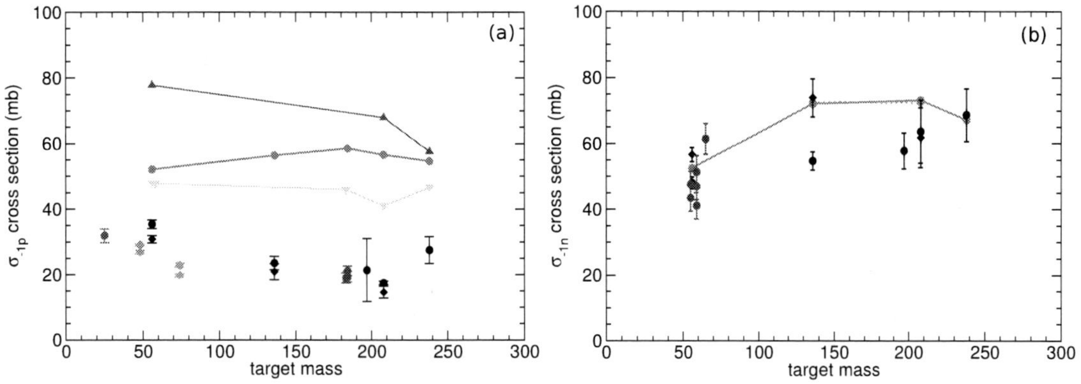

Figure 1. Experimental data for one-proton- (a) and one-neutron-removal cross sections (b) in proton-nucleus reactions above 500 MeV incident energy, as a function of the target mass. Diamonds refer to experimental beam energies around 500 MeV, while circles represent energies between 800 and 1000 MeV. The solid curves represent calculations with INCL (red), Isabel (blue) [9] and Bertini [10] (green) at 1000 MeV. Experimental data taken from Refs. 2, 11–20.

quantum-mechanical features of the initial condition and of the dynamics. At the end of the intranuclear cascade, an excited remnant is left. The de-excitation of this nucleus is typically described by a statistical de-excitation model.

In what follows, we shall make explicit reference to the Liège Intranuclear Cascade model [INCL, 4] and the ABLA07 statistical de-excitation model [5]. The INCL/ABLA07 coupling is in general quite successful at describing a vast number of observables in nucleon-induced reactions at incident energies between ~ 60 and 3000 MeV [6]. The work described hereafter was performed with the latest C++ version of INCL [INCL++, 7].

The INCL model is peculiar in that it explicitly tracks the motion of all the nucleons in the system, which are assumed to move freely in a square potential well. The radius of the well is not the same for all nucleons, but it is rather a function $R(p)$ of the absolute value of the particle momentum (which is a conserved quantity in absence of collisions). The initial particle momenta are uniformly distributed inside a sphere of radius $p_F = 270$ MeV/c. The relation between momentum and radius of the potential well is such that the space density distribution is given by a suitable Saxon-Woods parametrisation; moreover, the nuclear surface is predominantly populated by nucleons whose energy is close to the Fermi energy [8].

3. One-nucleon-removal cross sections

Figure 1 shows the experimental data for one-nucleon removal in proton-induced reactions at energies $\gtrsim 500$ MeV, as a function of the target mass (all targets are β-stable). Calculations with INCL/ABLA07 are shown for comparison. It is clear that the model predictions are in the right ballpark for neutron removal, but they overestimate the proton-removal data by a factor that can be as large as 3–4 for heavy nuclei. Note also that other cascade models similarly overestimate the proton-removal cross sections. Figure 1 suggests that INC models might be affected by a fundamental defect. It is however rather surprising that the deficiency clearly manifests itself in proton removal, but neutron removal seems unaffected.

The analysis of the model calculations indicates that one-proton removal is dominated (about 90% of the cross section) by events with only one proton-proton collision. The two protons leave the nucleus, which however retains some excitation energy. If only one collision took place, the excitation energy is given by the energy of the proton hole, i.e. the difference between the Fermi energy and the energy of the proton that was ejected. This excitation energy is evacuated during the de-excitation stage by neutron evaporation. If the excitation energy is lower than the neutron separation energy, no particle will be evaporated and the energy will be evacuated as gamma rays; in this case the final (observed) residue will therefore be the target nucleus minus one proton. If the excitation energy allows for neutron evaporation, the final residue will be lighter (target minus one proton minus x neutrons). The one-proton-removal cross sections are therefore extremely sensitive to the excitation energy left in the nucleus after the cascade. Note that there is a subtle difference between one-proton and one-neutron removal. One-neutron removal can be realized in two ways: either as a neutron ejection during INC followed by no evaporation (this is analogous to the proton-removal mechanism), or as no neutron ejection during INC followed by evaporation of one neutron.

Our results are essentially independent of the choice of the de-excitation model, since all of them employ very similar separation energies for stable nuclei. Comparison with the experimental data (Fig. 1) seems to suggest that INC underestimates the excitation energy associated with the ejection of a proton; larger excitation energies would lead to increased neutron evaporation and would therefore reduce the one-proton-removal cross section.

4. Refinement of the INC nuclear model

We mentioned at the end of Section 2 that the nuclear surface is predominantly populated by nucleons whose energy is close to the Fermi energy. The ejection of one such nucleon during INC results in little excitation energy for the cascade remnant. However, even deeply-bound nucleons have a non-vanishing probability to be found in the nuclear surface; this aspect is usually neglected by INC models. Another detail that is usually neglected in the INC picture is the presence of neutron (or proton) skins in certain nuclei, such as ^{208}Pb. For surface reactions, this means that the local neutron density is several times larger than the proton density, leading to an enhanced probability for proton-neutron collisions.

4.1. Shell-model calculations

We have estimated the magnitude of both the effects above with a simple shell-model calculation. We assumed a central Saxon-Woods nuclear potential with a spin-orbit term and a Coulomb term for the protons [21]. We numerically solved the radial part of the Schrödinger equation and determined the eigenfunctions and the eigenvalues of the bound states. The single-particle energies correctly reproduce the energies of the particle-hole states in ^{207,209}Pb and ^{207}Tl,Bi.

We would like to use the shell-model proton and neutron densities as inputs for our INC calculation; however, the particle densities in INCL cannot be given by an arbitrary function, so we must somehow adapt the shell-model densities. We chose to fit them with Saxon-Woods distributions (shown in Fig. 2 as dashed lines). The best-fit parameters show that the shell-model densities exhibit a neutron skin in ^{208}Pb. We have thus decoupled the INCL parameters describing the neutron space density from those describing the proton space density. The proton densities have not been modified (because they are already given by fits to the experimental charge radii), but the neutron parameters have been adjusted by the skin thicknesses resulting from the fit shown in Fig. 2.

We have explained in the previous section that the outcome of single-collision cascades is sensitive to the energy of the ejected nucleon. We assume that the probability that a collision ejects a nucleon from a given shell is proportional to the local density of the shell orbital. Furthermore, we neglect rearrangement of the other nucleons in the Fermi sea after the collision; this amounts to assuming that the excitation energy of the hole is simply given by the depth of the hole, measured from the Fermi energy.

With these assumptions, we can estimate the mean and RMS excitation energies that are left in the nucleus if a hole is punched in the Fermi sea at a certain distance from the center. These

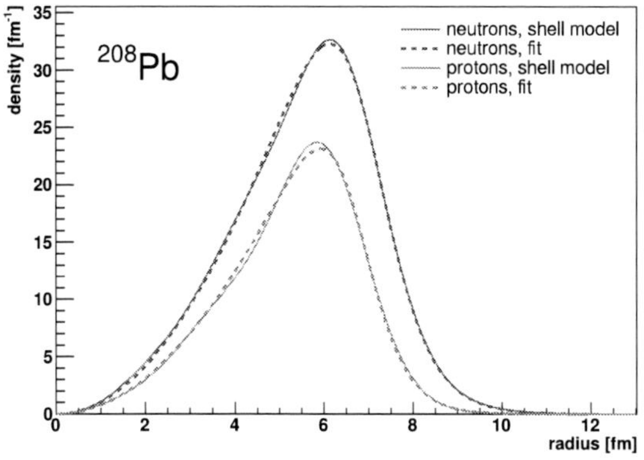

Figure 2. Proton (red) and neutron (blue) densities for ^{208}Pb. The thin solid lines represent the result of the shell-model calculation, while the thick dashed lines are Saxon-Woods fits.

Table 1. Cross sections for one-nucleon removal in 1-GeV p-nucleus reactions, with the following model variants: (a) standard, (b) standard plus neutron skin, (c) standard plus surface fuzziness, (d) standard plus neutron skin and surface fuzziness. Experimental data are taken from Refs. 16, 22.

	^{40}Ca		^{208}Pb	
	$-1p$	$-1n$	$-1p$	$-1n$
(a)	59.8	46.4	59.5	82.1
(b)	58.8	41.4	50.9	112.0
(c)	51.6	38.3	42.1	63.4
(d)	51.9	35.3	33.6	83.8
exp[1]	54.7 ± 7.9	29.8 ± 6.4	17.6 ± 0.5	63.7 ± 9.6

[1] The experimental values for ^{40}Ca refer to an incident energy of 763 MeV.

quantities are plotted in Fig. 3 for the shell-model calculation and for the standard INCL nuclear model ($f = 0$).

It is clear from the results displayed in this picture that the standard INCL nuclear model yields mean and RMS excitation energies that are quite different from those resulting from the shell model. In the surface region, the proton mean and RMS values from INCL are sensibly lower than their shell-model counterparts, which seems to confirm that the excitation energy associated with the ejection of a proton is underestimated by INCL.

4.2. Surface fuzziness

We mentioned in Section 2 that an INCL nucleon moves in a square-well potential whose radius $R(p)$ depends on the nucleon momentum. The function $R(p)$ is uniquely determined by the choice of the space density $\rho(r)$ and by the assumption that nucleon momenta are uniformly distributed in a sharp-surface Fermi sphere. We have shown above that this construction results in excitation energies for one-collision reactions that are much smaller than those resulting from the shell model and, arguably, than those suggested by the available experimental data.

We refine the INCL nuclear model by making $R(p)$ into a random variable. We introduce a *fuzziness parameter* f $(0 \leq f \leq 1)$ and a *fuzzy* square-well radius $R(p, f)$. The precise definition of $R(p, f)$ is outside the scope of this short paper, but it has the following properties: first, for $f = 0$ we recover the standard sharp correlation $(R(p, 0) = R(p))$. Second, for given values of p and f, $R(p, f)$ is a random variable that describes the radius of the square well. The fluctuations in $R(p, f)$ are small if f is close to zero and they are large if f is close to one. Moreover, the fluctuations are constructed in such a way that the space density is still given by $\rho(r)$ and the momentum density is still given by a sharp-surface Fermi sphere.

The construction of the refined INCL nucleus is analogous to the standard preparation algorithm [8]. The only difference is that the radius of the square-well potential is no longer in one-to-one correspondence with the nucleon momentum.

The refined nuclear model introduces fluctuations in the space distribution of nucleons with a

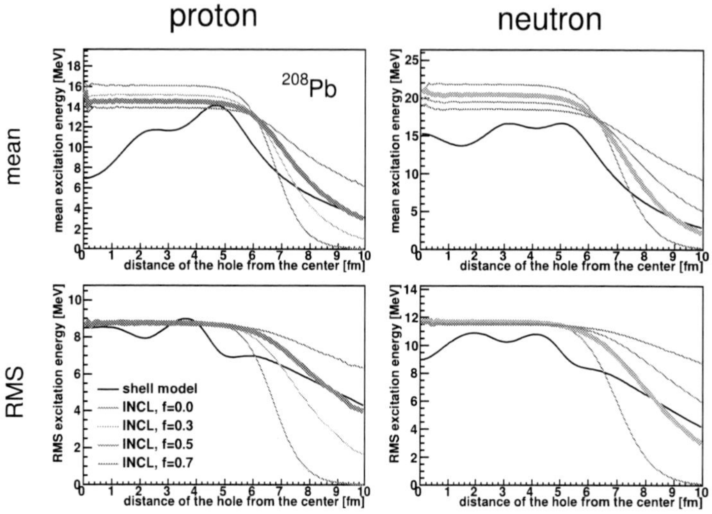

Figure 3. Mean (top) and root-mean-square (bottom) excitation energy induced by the presence of a proton (left) or neutron hole (right) in ^{208}Pb, as functions of the hole position. The black lines represent the shell-model result, while the colored lines are the results generated by the INCL nuclear model for different values of the fuzziness parameter f. The thick lines represent the selected parameter values.

given energy; equivalently, it introduces additional energy fluctuations for the nucleons found at a given position. Figure 3 indeed demonstrates that the average and RMS excitation energies for surface holes *increase* for increasing surface fuzziness, i.e. for increasing fluctuations. No value of the fuzziness parameter yields a good fit to the shell-model result, even if one limits oneself to the surface region. There is some degree of subjectivity in the choice of the best-fit values, which are taken to be $f = 0.5$ for protons and $f = 0.3$ for neutrons. For ^{40}Ca (not shown), the best-fit value was taken to be $f = 0.3$ for both protons and neutrons.

Summarizing, we have refined the INCL nuclear model in two respects. First, we have introduced a neutron skin, as described in Section 4.1. Second, we have introduced surface fuzziness, which increases the energy content of the nuclear surface and the probability for deep-nucleon removal in surface collisions. In the framework of the shell model, this effect is genuinely quantum-mechanical and is due to the penetration of the wavefunction in the classically forbidden region.

5. Results and conclusions

We turn now to the analysis of the results of the refined INC model. Table 1 shows how the neutron skin and the surface fuzziness affect the one-nucleon-removal cross sections in 1-GeV $p+^{40}$Ca and $p+^{208}$Pb. Unfortunately, no experimental data are available for $p+^{40}$Ca at 1 GeV, but since we do not expect a strong dependence on the projectile energy, we can compare to Chen *et al* 's data at 763 MeV [22].

Several observations are due. First, the introduction of the neutron skin in ^{208}Pb boosts the neutron-removal cross section, as expected. This is however undesired, since the cross section calculated by standard INCL is already in moderate excess of the experimental value. Second, surface fuzziness suppresses the cross sections for both one-nucleon-removal channels. This is true both for ^{40}Ca and ^{208}Pb. Third, neither effect is sufficient to compensate for the overestimation of the proton-removal cross section in ^{208}Pb *if considered alone.*

When the two refinements are simultaneously applied to ^{208}Pb, the effect of surface fuzziness for neutron removal almost exactly compensates the effect of the neutron skin, and the final result (83.8 mb) is very close to the value calculated with standard INCL (82.1 mb), which is within two standard deviations (about 30%) of the experimental value. The proton-removal cross section, on the other hand, is reduced by almost a factor of two, which brings it much closer to the experimental data, but not quite in agreement with it. The agreement for the $p+^{40}$Ca cross sections is also improved: the change in the proton-removal cross section is minor ($\sim 10\%$) and stays within the experimental error bar, but the neutron-removal cross section is reduced by $\sim 50\%$, in fair agreement with the experimental value.

In conclusion, we have shown that INCL fails to describe the cross sections for one-nucleon removal at high energy. We have used simple shell-model calculations to show that the key to this deficiency lies in the presence of neutron skins in heavy, stable nuclei and in the energy content of the nuclear surface. In the future we will need to generalize our approach to non-magic nuclei and devise a systematic approach to the description of the properties of the nuclear surface.

References

[1] Serber R 1947 *Phys. Rev.* **72** 1114–1115

[2] Jacob N P and Markowitz S S 1975 *Phys. Rev. C* **11**(2) 541–545

[3] Audirac L, Obertelli A, Doornenbal P *et al.* 2013 *Phys. Rev. C* **88**(4) 041602

[4] Boudard A, Cugnon J, David J C, Leray S and Mancusi D 2013 *Phys. Rev. C* **87** 014606 (*Preprint* 1210.3498)

[5] Kelić A, Ricciardi M V and Schmidt K H 2008 ABLA07 — towards a complete description of the decay channels of a nuclear system from spontaneous fission to multifragmentation *Joint ICTP-IAEA Advanced Workshop on Model Codes for Spallation Reactions* (Trieste, Italy: IAEA) p 181 report INDC(NDC)-0530

[6] Leray S, David J C, Khandaker M, Mank G, Mengoni A, Otsuka N, Filges D, Gallmeier F, Konobeyev A and Michel R 2011 *J. Korean Phys. Soc.* **59** 791–796
IAEA benchmark of spallation models http://www-nds.iaea.org/spallations

[7] Mancusi D, Boudard A, Cugnon J, David J C, Kaitaniemi P and Leray S 2014 *Phys. Rev. C* Submitted

[8] Boudard A, Cugnon J, Leray S and Volant C 2002 *Phys. Rev. C* **66** 044615

[9] Yariv Y and Fraenkel Z 1979 *Phys. Rev. C* **20** 2227–2243

[10] Bertini H W 1963 *Phys. Rev.* **131**(4) 1801–1821
Bertini H W 1969 *Phys. Rev.* **188**(4) 1711–1730

[11] Villagrasa-Canton C *et al.* 2007 *Phys. Rev. C* **75** 044603

[12] Rejmund F, Mustapha B, Armbruster P *et al.* 2001 *Nucl. Phys. A* **683** 540–565

[13] Giot L, Alcántara-Núñez J A, Benlliure J *et al.* 2013 *Nucl. Phys. A* **899** 116–132

[14] Napolitani P, Schmidt K H, Tassan-Got L *et al.* 2007 *Phys. Rev. C* **76** 064609

[15] Audouin L, Tassan-Got L, Armbruster P *et al.* 2006 *Nucl. Phys. A* **768** 1–21

[16] Enqvist T, Wlazło W, Armbruster P *et al.* 2001 *Nucl. Phys. A* **686** 481–524

[17] Taïeb J, Schmidt K H, Tassan-Got L *et al.* 2003 *Nucl. Phys. A* **724** 413–430

[18] Titarenko Y E *et al.* 2002 Experimental and theoretical study of the yields of residual product nuclei produced in thin targets irradiated by 100–2600 MeV protons INDC report INDC(CCP)-434 IAEA, Nuclear Data Section, International Nuclear Data Committee URL https://www-nds.iaea.org/publications/indc/indc-ccp-0434.pdf

[19] Michel R, Gloris M, Lange H J *et al.* 1995 *Nucl. Instrum. Meth. B* **103** 183–222

[20] Reeder P L 1969 *Phys. Rev.* **178**(4) 1795–1801

[21] Blomqvist J and Wahlborn S 1960 *Ark. Fys.* **16**

[22] Chen C X, Albergo S, Caccia Z *et al.* 1997 *Phys. Rev. C* **56**(3) 1536–1543

Shell model description of 0 and 1 ħω states in neutron-rich N = 18 and 19 nuclei with Z = 14 to 19

M Bouhelal[1], F Haas[2], E Caurier[2], F Nowacki[2]

[1]*Laboratoire de Physique Appliquée et Théorique, Université de Tébessa, Tébessa, Algérie*

[2]*IPHC, CNRS/IN2P3, Université de Strasbourg, F–67037 Strasbourg Cedex 2, France*

E–mail: m_bouhelal@yahoo.fr

Abstract. The low excitation energy spectra of N = 18 and 19 sd-shell isotones are investigated. A detailed comparison experiment versus theory using the PSDPF interaction of the energy spectra for ^{32}Si and ^{33}Si, ^{33}P and ^{34}P, ^{34}S and ^{35}S, ^{35}Cl and ^{36}Cl, ^{36}Ar and ^{37}Ar, ^{37}K and ^{38}K will be presented. Based on this comparison, important predictions will be proposed and the general evolution of the main structure properties in these isotones will be discussed.

1. Introduction

The neutron–rich sd-shell nuclei around N = 20 are the subject of active experimental and theoretical studies and exhibit a rich variety of phenomena and physics such as the *island of inversion region*. Currently, much more data essentially obtained from Radioactive Ion Beam (RIB) facilities are available for nuclei near the N = 20 gap. The properties of these nuclei pose a challenge to conventional models of theoretical nuclear structure and a deeper understanding of the underlying physics. Study of such nuclei provides information about the evolution of their structure. For this purpose, we are interested to study this area of nuclei in the framework of the shell model. We remind that the spectra of sd-shell nuclei are characterized by the presence, at low excitation energies, of both normal positive and intruder negative parity states. Intruder 1ħω negative parity states are known throughout the sd–shell. These intruders result essentially from a one-nucleon jump from p to sd or from sd to pf shells. Our PSDPF interaction [1], mainly developed to describe such states, reproduces remarkably well the spectroscopic properties of 0 and 1 ħω states in nuclei throughout the sd–shell and in particular in the upper shell closure region.

We performed shell model calculations using the PSDPF interaction in order to describe the spectroscopic properties of some neutron–rich nuclei such as the N = 20 isotones with Z = 14–16 [2], ^{36}S [3] and ^{34,35}P [4]. We also studied the Phosphorus and Sulfur isotopic chains; the results will be published later. In general, the calculations agree quite well with experiment and some predictions could also be proposed.

Recently, many experiments were implemented to investigate the properties of neutron–rich nuclei through fusion and multi-nucleon transfer reactions. Most of the obtained results could be interpreted in the framework of the shell model using the PSDPF interaction, particularly in the case of ^{33}Si [5],

^{36}Cl [6], ^{37}Cl [7], ^{35}S [8] and ^{34,35,36}P [9] for which all the low-lying spectroscopic properties are well described by PSDPF.

In this paper, we will present a comparison of our results using PSDPF to available experimental data of the spectroscopic properties for N = 18 and 19 isotones with Z = 14 to 19.

Based on this comparison, predictions for unknown spin-parity of several states will be proposed and the general evolution of the main structure properties in these isotones will be discussed.

2. Results and discussion

The performed shell model calculations using the shell-code Nathan [10-12] of the energy spectra with the PSDPF interaction for N = 18 and 19 isotones with Z = 14 to 19 will be presented. We will discuss the results for each nucleus.

The ^{32}Si spectrum is known up to 7100 keV [13]. It contains nine well-identified positive parity states and seven negative parity states as shown on Table 1. The low levels are well described using PSDPF. This is reflected by the small energy ΔE differences. Above ~6000 keV, the calculated excitation energies become higher than the experimental ones. This is due to the fact that their configurations contain more than 0p-0h and 1p-1h components, which is expected in this high-energy range. The $J^\pi = 0^-_1$, 3^+_1, 2^+_4 and 1^-_2 assignments are proposed for the uncertain states at 5791, 5893, 6170 and 6195 keV, respectively. We also propose $J^\pi = 1^+_1$, 2^-_1 and 0^+_3 values for the undetermined states at 5220, 5773 and 5786 keV, respectively. PSDPF predicts a negative parity for the J = 2 state at 5956 keV.

Table 1. Comparison between experimental (Eexp) [13] and calculated (Eth) excitation energies (in keV) for ^{32}Si, with ΔE = Eth - Eexp.

J^π	Eexp	J^π_i	Eth	ΔE	J^π	Eexp	J^π_i	Eth	ΔE
0^+	0	0^+_1	0	0	(0^-)	5791	0^-_1	5416	-375
2^+	1941	2^+_1	2043	102	(3^+)	5893	3^+_1	5602	-291
2^+	4231	2^+_2	4306	75	2	5956	2^-_2	6227	271
0^+	4984	0^+_2	5185	201	3^-	5967	3^-_2	6291	324
(1-4)	5220	1^+_1	5611	391	(2^+)	6170	2^+_4	6985	815
3^-	5289	3^-_1	5495	206	$1^{(-)}$	6195	1^-_2	6717	522
1^-	5413	1^-_1	5241	-172	0^+	6240	0^+_4	7943	1703
2^+	5427	2^+_3	5620	193	2^+	6388	2^+_5	7452	1064
4^+	5502	4^+_1	5589	87	3^-	6477	3^-_3	6395	-82
5^-	5581	5^-_1	5725	144	1^-	6705	1^-_3	6950	245
(1,2,3)	5773	2^-_1	5445	-328	3^-	6860	3^-_4	7435	575
$(0,1,2)^+$	5786	0^+_3	5531	-255	2^+	7083	2^+_6	7887	804

Table 2. Comparison between experimental [13] and calculated spectra (in keV) for ^{33}Si.

J^π	Eexp	J^π_i	Eth	ΔE
$3/2^+$	0	$3/2^+_1$	0	0
$1/2^+$	1010	$1/2^+_1$	798	-212
$7/2^-$	1435	$7/2^-$	1381	-54
$(3/2^-)$	1981	$3/2^-$	1703	-278
$(9/2^-)$	3159	$9/2^-$	3157	-2
$(11/2^-)$	4090	$11/2^-$	4129	39
$3/2^+,5/2^+$	4341	$3/2^+_2$	4192	-149
$3/2^+,5/2^+$	4341	$5/2^+_1$	4222	-119
$(11/2^-)$	4931	$11/2^-$	49 5	44

The ^{33}Si spectrum contains nine levels up to ~5000 keV [13]. Among them, only the three first states are well identified and quite nicely reproduced by PSDPF. Concerning the other states (see Table 2), all the proposed J^{π} agree with the PSDPF predictions. According to this comparison, we suggest that there is a doublet of states at 4341 keV with $J^{\pi} = 3/2^+$ and $5/2^+$.

The ^{33}P spectrum contains 25 levels up to ~5750 keV. Four states at 3990, 5207, 5221 and 5235 keV, were recently observed in the fusion-evaporation $^{18}O(^{18}O,2np)$ reaction [13]. Note that in this type of reactions, states with high spins are selectively populated. These states are not predicted by the PSDPF interaction (not in the valence space). Thus we do not found theoretical counterparts for these levels. However, all other observed states are quite well reproduced by PSDPF. The only exception is for the $5/2^+_4$ and $5/2^+_5$ at 4194 and 5049 keV, which are predicted to be higher than experiment by more than 800 and 1400 keV, respectively, and thus contain more than a $0\hbar\omega$ configuration. The energy differences of the remaining levels vary from 9 keV for the $3/2^+_1$ to 244 keV in the cases of $7/2^-_1$ and $9/2^-_1$. For this N = 18 isotone, we propose $J^{\pi} = 5/2^-_1, 5/2^-_2, 9/2^+_1$ and $7/2^-_2$ for the unidentified states at 5191, 5411, 5498 and 5549 keV, respectively. Recently the experimental ^{34}P spectrum was extended and a detailed comparison with our PSDPF results has been discussed in Ref. [4].

We then examine the ^{34}S spectrum which is well known up to ~6500 keV and comprises 26 levels. The PSDPF results are in quite good agreement with experiment and all the reported states have their theoretical counterparts. Starting from ~6000 keV, the contribution of more complex than 0p-0h and 1p-1h configurations is dominant for the six states with $J^{\pi} = 0^+_4, 2^+_{6,7}, 4^+_2$ and $1^-_{2,3}$. These states are predicted to be higher than their experimental counterparts. Recent results, in particular for the high spin levels, are reported for ^{35}S. A comparison with shell model calculations using the PSDPF interaction and other Hamiltonians is discussed in Ref. [8].

Starting from ^{35}Cl, we approach the neutron and proton ~20 gaps with occupation of the $d_{3/2}$ sub-shell. The ^{35}Cl spectrum is well known up to ~6100 keV and contains a large number of states (~40). More than eleven states have negative parity. PSDPF describes reasonably well the observed excitation energies and important predictions for the uncertain states are proposed. We predict $J^{\pi} = 5/2^+_3$ and $3/2^+_4$ for the doublet of states at 4624 keV as well as $J^{\pi} = 1/2^+_3, 1/2^-_1, 1/2^-_2, 5/2^+_4, 3/2^-_3, 3/2^+_5, 7/2^-_4, 3/2^-_4, 3/2^-_5, 5/2^-_3$ and $5/2^-_4$ for states observed at 4839, 4854, 5010, 5216, 5404, 5600, 5645, 5683, 5758, 5806, 5823 keV, respectively. The parities of the J= 7/2 levels at 4769 and 4881 keV are proposed to be negative and positive, respectively. States at 3979, 5157, 5520 and 5633 keV do not have their theoretical counterparts. We think that these levels have more than 0 and 1 $\hbar\omega$ configurations and thus are predicted by our model at higher excitation energies. This phenomenon is expected, especially, in nuclei near the gaps. New excited states, mainly those of high spins, have been populated in ^{36}Cl through fusion-evaporation reactions. The complete observed spectrum of ^{36}Cl has been compared to the PSDPF prediction and the results are discussed in Ref. [6].

We now discuss the main results of the Z, N= 18 and 19 nuclei, i.e., nuclei close to the Z, N = 20 gaps. The self-conjugate nucleus ^{36}Ar, is well studied, its spectrum comprises, up to ~8200 keV, 40 levels, 17 of them have negative parity. Good agreement is found between PSDPF results and experiment for both positive and negative parity states, except for the 4^-_2 at 6835 keV, which is predicted at 6430 keV. As it was mentioned previously, when approaching the proton and neutron gaps, excitations of few particles across the gaps appear rapidly in the spectrum. Levels observed at 6137, 6611, 6837, 7139, 7432, 7672 and 7710 keV have more complex than 0p-0h and 1p-1h components. However, states at 4951, 6356, 7750, 7767 and 8132 keV are known to be purely collective. It was difficult to find candidates for states at 5194, 6646, 6867 and 7179 keV due to the same reasoning as in ^{35}Cl.

Regarding ^{37}Ar and ^{37}K mirror nuclei, the results are compared to the PSDPF predictions up to ~4000 keV on Table 3. In this excitation energy range, there is a one to one correspondence between mirror levels as well as with the PSDPF ones, except for the $11/2^-$ state not observed in ^{37}K. PSDPF reproduces quite well the observed excitation energies. According to this comparison, the energy of the ^{37}K 7/2^{+} state is proposed to be 2285 keV. The third 3/2^{+}, observed at 3937 keV in ^{37}Ar and known to be at 3839 keV in ^{37}K, is predicted at 4986 keV and is thus probably a more complex than 0p-0h state. We note that the ^{37}Ar nucleus is known up to 5440 keV.

Table 3. Comparison between experimental [13] and calculated excitation energies (in keV) of the ^{37}Ar and ^{37}K mirrors

^{37}Ar		PSDPF			^{37}K		
J^{π}	Eexp	J^{π}_{i}	Eth	ΔE	J^{π}	Eexp	ΔE
3/2^{+}	0	$3/2^{+}_{1}$	0	0	3/2^{+}	0	0
1/2^{+}	1410	$1/2^{+}_{1}$	1489	79	1/2^{+}	1371	118
7/2^{-}	1611	$7/2^{-}_{1}$	1433	-178	7/2^{-}	1380	53
7/2^{+}	2217	$7/2^{+}_{1}$	2416	199	(5/2^{+},7/2^{+})	2285	131
3/2^{-}	2490	$3/2^{-}_{1}$	2788	298	3/2^{-}	2170	618
5/2^{+}	2796	$5/2^{+}_{1}$	2940	144	5/2^{+}	2750	190
5/2^{+}	3170	$5/2^{+}_{2}$	3179	9	5/2^{+}	3240	-61
9/2^{-}	3185	$9/2^{-}_{1}$	2982	-203		2967	15
5/2^{-}	3274	$5/2^{-}_{1}$	3109	-165	5/2^{-}	3082	27
3/2^{-}	3518	$3/2^{-}_{2}$	3383	-135	3/2^{-}	3315	68
7/2^{-}	3527	$7/2^{-}_{2}$	3638	111		3272	366
3/2^{+}	3602	$3/2^{+}_{2}$	3874	272	3/2^{+}	3622	252
11/2^{-}	3706	$11/2^{-}_{1}$	3813	107			
3/2^{+}	3937	$3/2^{+}_{3}$	4986	1049	1/2+,3/2^{+},5/2^{+}	3839	1147

The last isotone studied here is ^{38}K with N= 19. PSDPF describes its energy spectrum between 0 and ~5300 keV. The np-nh excitations appear rapidly at low energy, this is the case of states at 3341, 3430, 3856 and 3931 keV. In this situation, it is almost impossible to find theoretical equivalents for states observed at 3458, 3668, 3704, 3739, 3977, 4175, 4317, 4394 and 5254 keV.

3. Evolution of the 3^{-} and 7/2^{-} states

In this section the evolution of the excitation energies of specific states in the N= 18 and 19 isotones are presented. Let us start with the first excited positive parity state in these isotones. Figure 1 shows the evolution of the experimental versus calculated first positive parity states for even and odd A nuclei.

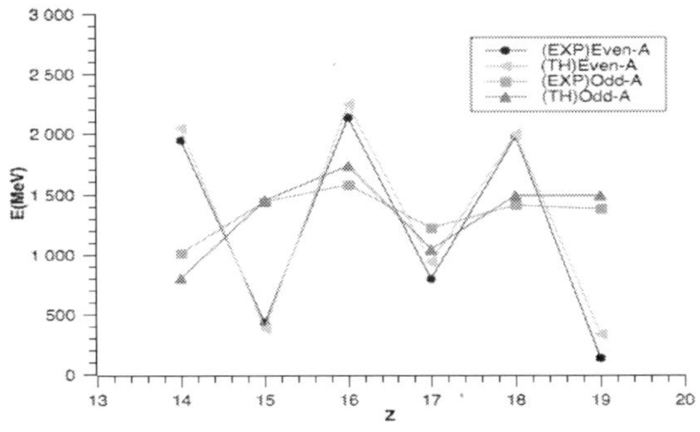

Figure 1. Excitation energy evolution of the first positive parity states for even and odd A in N= 18 and 19 isotones (EXP versus TH).

As it is seen on this figure, the PSDPF interaction reproduces very well all the observed states and their energy variation throughout the sd-shell. In the case of the odd-A nuclei, the excitation energies are practically constant and vary between 1000 and 1500 keV for all the levels. As far as the even-A nuclei are concerned, the excitation energies, as expected for even-even nuclei with N= 18, are higher than those for the odd-odd ones with N= 19.

The evolution of the first 3^- and $7/2^-$ states in N= 18 and 19 isotones is shown on Fig. 2. Note that the $7/2^-$ is the first excited negative parity state in odd-A nuclei. The 3^- is the first excited negative parity state in the even-A isotones except for the odd-odd ^{34}P and ^{36}Cl. Here also PSDPF describes very well the observed excitation energies of these states as well as their variation throughout the sd-shell. As can be seen in Fig. 2, the $7/2^-$ levels appear at low excitation energies close to the first excited positive parity states. This is not the case for the N= 18 isotones ^{33}P and ^{35}Cl, were these states have high excitation energies. This fact is not observed in ^{37}K with Z= 19.

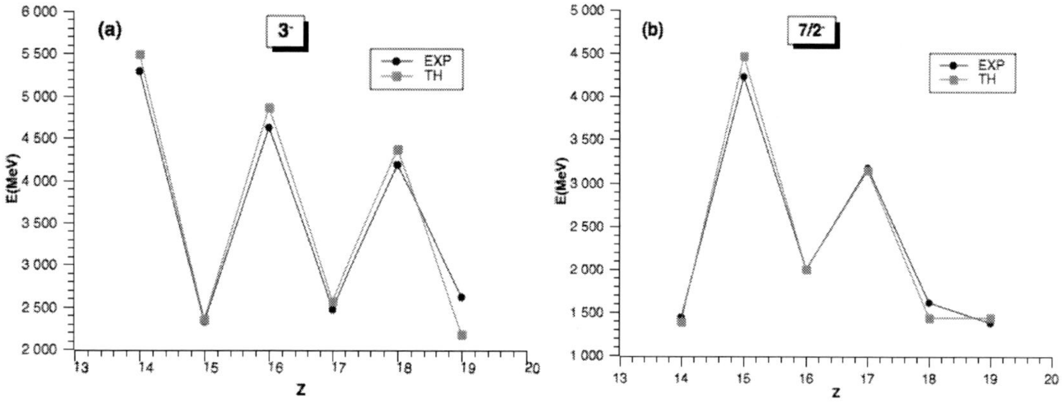

Figure 2. Experimental and calculated excitation energies of (a) 3^- states in N= 18 and 19 isotones and (b) $7/2^-$ states in N= 18 and 19 isotones.

4. Conclusion

The PSDPF interaction describing the 0 and 1 ħω states throughout the sd-shell has been used. The results discussed here are for N=18 and 19 isotones, for which PSDPF reproduces quite well both positive and negative parity states. In the middle of the sd-shell, the 1ħω states have very fragmented (*competition of p-sd and sd-pf jumps)* wave functions, which complicate their description. We would like to retune our interaction in the near future using a new fitting code, developed by E. Caurier, where we can use much more data from nuclei in the middle of the sd-shell. It is also important that future studies cover cases approaching the N and Z ~20 gaps, where the np-nh states become rapidly important even at low energies. To describe them, a new interaction called SDPF-U-Mix, allowing such higher order jumps has to be developed.

References

[1] Bouhelal M, Haas F, Caurier E, Nowacki F and Bouldjedri A 2011 *Nucl. Phys.* A **864** 113
[2] Bouhelal M, Haas F, Caurier E, Nowacki F and Bouldjedri A 2009 *Eur. Phys. J.* A **42** 529
[3] Bouhelal M, Haas F, Caurier E and Nowacki F 2009 *AIP Conf. Proc* (New York) Vol 1165 (American Institute of Physics) p 61
[4] Bouhelal M, Haas F, Caurier E and Nowacki F 2012 *AIP Conf. Proc* (New York) Vol 1491 (American Institute of Physics) p 38
[5] Wang Z M et al. 2010 *Phys. Rev.* C **81** 064301
[6] Aydin S et al. 2012 *Phys. Rev.* C **86** 024320
[7] Szilner S et al. 2013 *Phys. Rev.* C **87** 054322
[8] Aydin S et al. 2014 *Phys. Rev.* C **89** 014310

[9] Chapman R, et al., to be published
[10] Caurier E and Nowacki F 1999 *Acta Phys. Pol.* B **30** 705
[11] Caurier E, Martinez–Pinedo G, Nowacki F, Poves A and Zuker A P 1999 *Phys. Rev.* C **59** 2033
[12] Caurier E, Martinez–Pinedo G, Nowacki F, Poves A and Zuker A P 2005 *Rev. Mod. Phys* **77** 427
[13] http://www.nndc.bnl.gov/nudat2

Level densities and shell corrections of superheavy nuclei

A.N. Bezbakh[1], T.M. Shneidman[1,2,3], G.G. Adamian[1], N.V. Antonenko[1]

[1]Joint Institute for Nuclear Research, Dubna 141980, Russia
[2] Institute of Theoretical Physics, CAS, Beijing 100190, China
[3] Kazan Federal University, Kazan 420008, Russia

E-mail: `bezbakh@theor.jinr.ru`

Abstract. The intrinsic level densities of superheavy nuclei in the α-decay chains of $^{296;298;300}120$ nuclei are calculated using the single-particle spectra obtained with the modifed two-center shell model. The level density parameters are extracted and compared with their phenomenological values used in the calculations of the survival of excited heavy nuclei. The dependences of the level density parameters on the mass and charge numbers as well as on the ground-state shell corrections are studied.

1. Introduction

An extension of the nuclear chart to the elements with $Z > 100$ and the study of their properties provide an important insight into the determination of an island of stability. The experimental trend of nuclear properties (Q_α – values and half-lives) as well as the production cross sections of superheavy elements (SHE) reveal an increasing stability of nuclei approaching $N = 184$, and indicate quite a large shell effects behind $Z = 114$ [1-6]. This means that the predictions of relativistic and nonrelativistic mean-field models [7-11] seem to be valid. In accordance with these self-consistent models a center of the island of stability is expected at $Z = 120 - 126$ and $N = 172$ or 184. In the $(N; Z)$ plane, the line, along which all new SHE were discovered in the actinide-based reactions with ^{48}Ca beam, just approaches this region [1].

In Refs. [12, 13] we modified the two-center shell model (TCSM) [14] to describe the quasi-particle structure of well-studied heavy nuclei in the best way. Based on the TCSM single-particle potential and shape parametrization we calculated the ground-state binding energies and transition energies Q_α for α decays of nuclei with $105 \leq Z \leq 126$. The calculated Q_α are found to be in a good agreement with available experimental data. The dependence of Q_α on the neutron number at given Z indicates the strong shell effects at $N = 184$. Also our calculations result in the stronger shell effects at $Z \geq 120$ than at $Z = 114$.

In this work we use the single-particle spectra calculated with the modified TCSM to investigate the nuclear level densities (NLD) of superheavy nuclei. The NLD is an important characteristic both for the nuclear reaction calculations and for the investigation of the stability of the new supearheavy elements. At low excitations the single-particle level densities are closely related to the shell structure and could indicate the position of the island of stability of heaviest nuclei. The level density, as a function of excitation energy, is required to calculate the survival probability and, correspondingly, the production cross section of heavy nucleus. For nuclei lighter than Cf, the systematic study of NLD was performed in Refs. [15, 16].

In this work we performed the study of the intrinsic level densities of nuclei with $100 \leq Z \leq 130$ whose α-decay chains contain $^{296,298,300}120$ which are of interest for future experiments.

2. A few words about performed calculations.

In our work we use the expression for the nuclear single-particle level density obtained in the superfluid model [17, 18]. Employing the saddle-point method, the intrinsic level density $\rho(U)$ of nucleus with Z protons, N neutrons, and excitation energy U is expressed as

$$\rho = \frac{\exp[S(\beta, \lambda_Z, \lambda_N)]}{(2\pi)^{3/2}\sqrt{D}}. \tag{1}$$

Here, S is the entropy, $\beta = T^{-1}$ is the inverse temperature, λ_Z and λ_N are the chemical potentials for protons and neutrons, respectively, and D is the determinant of the matrix comprised of the second derivatives of the entropy. In the superfluid model adopted here, the entropy S is expressed as

$$S = 2\sum_{k=Z,N}\sum_{\nu}\left\{\ln[1+\exp(-\beta E_{k\nu})] + \frac{\beta E_{k\nu}}{1+\exp(\beta E_{k\nu})}\right\}, \tag{2}$$

where proton ($k = Z$) and neutron ($k = N$) quasi-particle energies $E_{k\nu} = \sqrt{(\varepsilon_{k\nu} - \lambda_k)^2 + \Delta_k^2}$ are calculated using the single-particle levels $\varepsilon_{k\nu}$ obtained with the TCSM. The details of the calculations with the TCSM as well as the parameters used are described in Refs. [12, 13]. The Fermi energies λ_k and correlation functions Δ_k ($k = N, Z$), are calculated at the thermodynamic equilibrium of nucleus determined by solving the system of equations

$$\begin{aligned} Z &= \sum_{\nu}\left(1 - \frac{\varepsilon_{Z\nu} - \lambda_Z}{E_{Z\nu}}\tanh[\frac{1}{2}\beta E_{Z\nu}]\right), \quad N = \sum_{\nu}\left(1 - \frac{\varepsilon_{N\nu} - \lambda_N}{E_{N\nu}}\tanh[\frac{1}{2}\beta E_{N\nu}]\right), \\ \frac{2}{G_Z} &= \sum_{\nu}\frac{\tanh[\beta E_{Z\nu}/2]}{E_{Z\nu}}, \quad \frac{2}{G_N} = \sum_{\nu}\frac{\tanh[\beta E_{N\nu}/2]}{E_{N\nu}}, \end{aligned} \tag{3}$$

where G_Z and G_N are the constants of pairing interaction. In Eqs. (3), the sums run over all the single-particle levels considered.

In Fig. 1, the calculated level densities $\rho_m(U)$ are presented for well-deformed nuclei ^{162}Dy, ^{166}Er together with the experimental data [19]. The agreement with experimental data is rather good, which provides us some confidence in the predictions of $\rho(U)$ in the region of SHE.

3. Dependence of level density parameter on shell corrections

Intrinsic level density is strongly influenced by nuclear shell structure. For magic or nearly magic nuclei, the level density is smaller than for mid-shell nuclei at the same excitation energy. This effect is related to the large single-particle spacings in the nuclei with the closed shell in the ground state. So, irregularities of the single-particle spectra are responsible for the shell corrections and peculiarities of the intrinsic level density. The calculated intrinsic level densities for the superheavy nuclei are presented in Fig. 2 for two nuclei from each of the considered alpha-decay chains of the elements $^{296;298;300}120$. The solid lines correspond to the isotopes $^{296;298;300}120$ and exemplify the behavior of the level density in nuclei with quite a strong shell correction, while the dotted lines are for the isotopes $^{264;266;268}104$ with smaller shell corrections. As seen, at low excitation energies from 10 to 30 MeV the level densities in $Z = 120$ isotopes are smaller within one order of magnitude than those in $Z = 104$ isotopes due to the smaller densities of the single-particle states near the Fermi surfaces. At smaller excitation energies

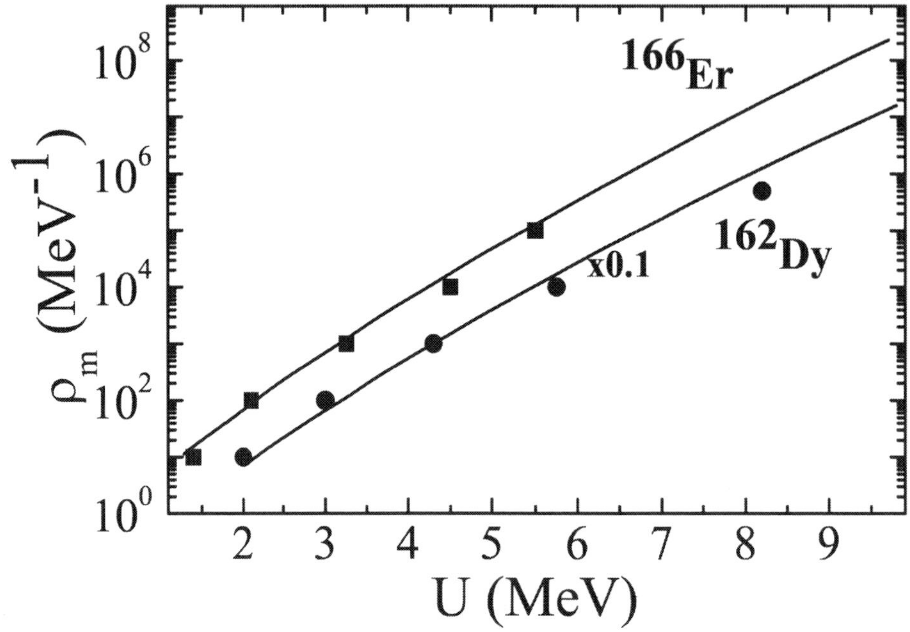

Figure 1. Calculated level densities ρ_m in ^{162}Dy and ^{166}Er as functions of excitation energy (lines) are compared with the experimental data (symbols) from Ref. [19]. For the ^{162}Dy nucleus, the experimental and theoretical results are multiplied by factor 0.1.

$U < 10$ MeV, the effects of pairing are stronger in the nuclei with weaker shell effects and the level densities of nuclei with strong and weak shell effects become comparable. At higher excitation energies, $U > 40 - 50$ MeV, the shell effects are faded and the level densities of all heavy nuclei become similar.

The similar effects as for $Z = 120$ isotopes are found for the isotones with $N = 184$ (Fig. 3). So our calculations specify the shell closure for neutrons at $N = 184$ which is consistent with all previous predictions.

We found that the best fit of the calculated intrinsic level density with the Fermi-gas expression is achieved if one uses the level density parameter calculated as

$$a = S^2/(4U), \tag{4}$$

where S and U are entropy and excitation energy of the nucleus, respectively. Since the thermodynamical equilibrium is assumed for each excitation energy, the level density parameter can be written as

$$a = S_N^2/4U_N + S_Z^2/4U_Z = a_N + a_Z, \tag{5}$$

where $S_N(S_Z)$ and $E_N(E_Z)$ are the entropies and excitation energies of only neutron (proton) subsystems.

The Fig. 4 presents the dependences of shell corrections δE_{sh} and level density parameters a at $U = 10$ MeV and $U = 60$ MeV excitation energies on the atomic number A for three α-decay chains containing the SHE $^{296;298;300}120$ which could be synthesized in near future with available projectiles and targets. The data of Ref. [22] are used for δE_{sh}. One can see the strong correlations between the shell corrections and level density parameter a at excitation energy

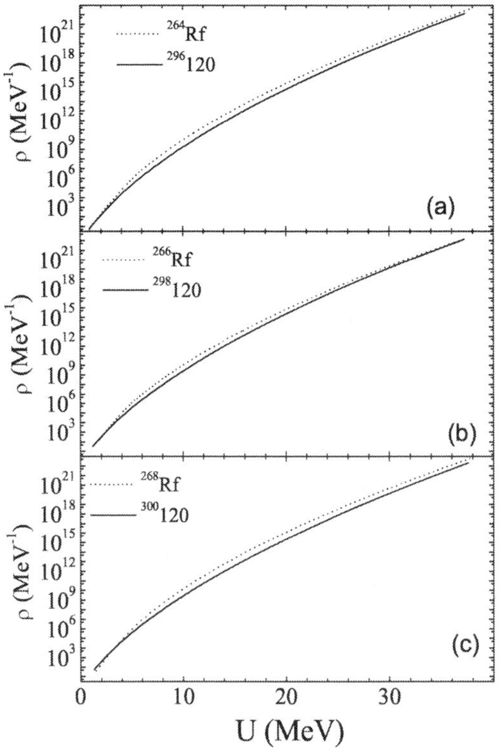

Figure 2. The calculated intrinsic level densities as functions of excitation energy for ^{264}Rf and ^{296}120 (a), ^{266}Rf and ^{298}120 (b), ^{266}Rf and ^{300}120 (c). The results for the isotopes of Rf and Z = 120 nucleus are presented by dashed and solid lines, respectively.

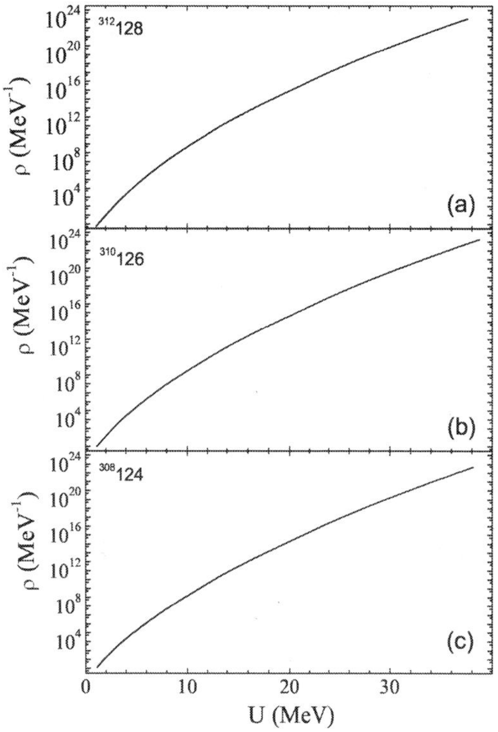

Figure 3. The same as in Fig. 2, but for the N = 184 isotones: ^{312}128 (a), ^{310}126 (b), ^{308}124 (c).

U = 10 MeV. The larger negative shell corrections result in the decrease of the value of a with respect to the neighborhood nuclei. At higher excitation energy (U = 60 MeV), the correlation between δE_{sh} and a is destroyed and (as demonstrated in upper part of Fig. 4) level density parameter becomes rather smooth function of A. Still, the shell structure slightly modifies the general trend of the level density parameter. For example, $a(^{266}\mathrm{Rf})$ is smaller slightly than $a(^{264}\mathrm{Rf})$ at U = 60 MeV because of the closed neutron shell N = 162 in ^{266}Rf nucleus.

Based on the study of the dependencies of a on δE_{sh}, U, and A one can use the following parametrization [20] of the level density parameter:

$$a(A,U) = \tilde{a}(A)\left[1 + \frac{1-\exp\{-U/E'_D\}}{U}\delta E_{sh}\right], \tag{6}$$

where $\tilde{a}(A)$ is the parameter smoothly depending on A. It defines a at large excitations when the shell effects are washed out. By analyzing the level density parameters with the Eq. (4), the value of the damping parameter E'_D = 27 MeV is found. The corresponding asymptotic level density parameter $\tilde{a}(A)$ can be fitted with the following functions [20]:

$$\tilde{a}(A) = \alpha A + \beta A^2, \tag{7}$$

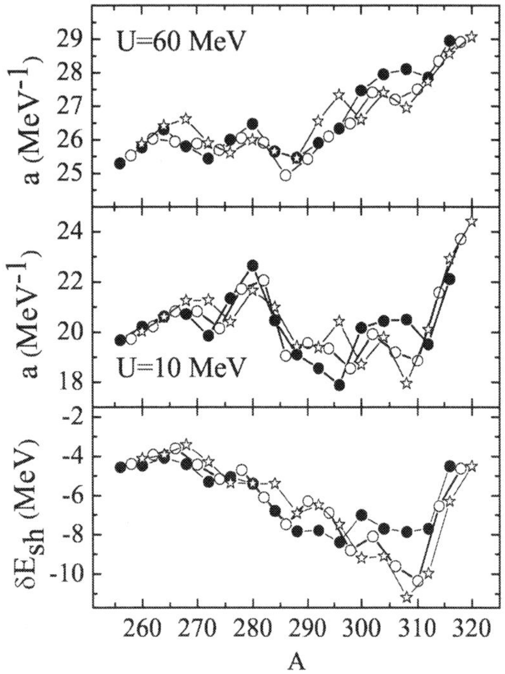

Figure 4. Calculated ground-state shell corrections δE_{sh} (lower part), the level density parameter a at U=10 (middle part) and 60 MeV (upper part), obtained with Eq. (4), as functions of mass number A. The nuclei from alpha-decay chains containing $^{296}120$, $^{298}120$, $^{300}120$ are marked by closed circles, open circles, and stars, respectively.

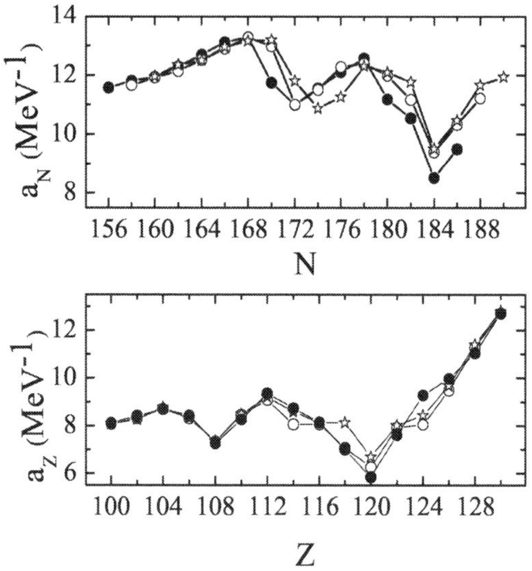

Figure 5. Calculated level density parameters a (U=10 MeV) as a function of neutron number N (upper part) and proton number Z (lower part) for the superheavy nuclei considered. The nuclei from alpha-decay chains containing $^{296}120$, $^{298}120$, $^{300}120$ are marked by closed circles, open circles, and stars, respectively.

where the constants α=0.118 MeV^{-1} and β=-0.53×10^{-4} MeV^{-1} are found with the least square method. These values are close to those proposed in Ref. [20].

The level density parameters for the nuclei with $Z \geq 100$ are of especial interest to look for the position of the next proton and neutron shell closures beyond $Z = 82$. To separate the neutron and proton shell effects one should investigate the proton (a_Z) and neutron (a_N) level density parameters defined in Eq. (5) (see Fig. 5). At $Z = 108$ and 120 there are minima of a_Z in all chains. This reflects quite strong proton shell effects at $Z = 108$ and 120. At $Z = 120$, the minima of a are the deepest and well pronounced. The similar behavior of a occurs near $Z = 82$. The sub-shell at $Z = 114$ exists but provides weaker shell effect than at $Z = 120$. For nuclei with $Z = 124 - 128$, the minima of a are due to the neutron shell at $N = 184$.

4. Summary

The level density parameters were calculated for the nuclei of alpha-decay chains containing $^{296}120$, $^{298}120$, and $^{300}120$. The strong shell effects at $Z = 120$ and $N = 184$ were demonstrated to be in accordance with our previous calculations and shell-model predictions. The dependencies of the level density parameter on the shell correction and excitation energy were studied. The damping factor in the well-known expression (6) [20] was found to be $E'_D = 27$ MeV. For superheavy nuclei considered, the level density parameter is approximately $A/(11-13)$ MeV at excitation energies corresponding to (3-5) neutron evaporation channels.

Acknowledgments
We acknowledge fruitful discussions with E.-G.Zhao and S.- G.Zhou. T.M.S. acknowledges the support from the CAS Fellowship for the Young Scientists and Russian Government Subsidy Program of the Competitive Growth of Kazan Federal University. This work was partly supported by RFBR (Moscow), NSFC (Beijing), and Dynasty Foundation (Moscow).

References

[1] Oganessian Yu Ts 2007 *J Phys.* G **34** R165, 2010, Phys. Rev. Lett. **104** 142502, 2013, Phys. Rev. C **87** 014302.
[2] Hofmann S *et al.* 2007 *Eur. Phys. J.* A **32** 251, 2009, *Lec. Notes Phys.* **764** 203
[3] Armbruster P 2008 *Eur. Phys. J.* A **37** 159
[4] Adamian G G, Antonenko N V and Sargsyan V V 2009 *Phys. Rev.* C **79** 054608
[5] Adamian G G, Antonenko N V, Sargsyan V V and Scheid W 2010 *Nucl. Phys.* **A834** 345c
[6] Dong J, Zuo W and Scheid W 2011 *Phys. Rev. Lett.* **107** 012501; Li Z, Sun B, Shen C H and Zuo W 2013 *Phys. Rev.* C **88** 057303
[7] Reinhard P G 1989 *Rep. Prog. Phys.* **52** 439
[8] Ring P 1996 *Prog. Part. Nucl. Phys.* **37** 193
[9] Bender M, Heenen P H and Reinhard P G 2003 *Rev. Mod. Phys.* **75** 121
[10] Meng J, Toki H, Zhou S G, Zhang S Q, Long W H and Geng L S 2006 *Prog. Part. Nucl. Phys.* **57** 470
[11] Li J J, Long W H, Margueron J and Van Giai N 2013 *arXiv: nucl-th* **1303.2765v1**
[12] Adamian G G, Antonenko N V and Scheid W 2010 *Phys. Rev.* C **81** 024320
[13] Adamian G G, Antonenko N V, Kuklin S N and Scheid W 2010 *Phys. Rev.* C **82** 054304
[14] Maruhn J and Greiner W 1972 *Z. Phys.* A **251** 431
[15] Iljinov A S *et al.* 1992 *Nucl. Phys.* **A543** 517
[16] von Egidy T and Bucurescu D 2005 *Phys. Rev.* C **72** 044311
[17] Decowski P, Grochulski W, Marcinkowski A, Siwek K and Wilhelmi Z 1968 *Nucl. Phys.* A **110** 129
[18] Adeev G D and Cherdantsev P A 1975 *Yadernaya Fizika* **21** 491
[19] Melby E, Guttormsen M, Rekstad J, Schiller A, Siem S and Voinov A 2001 *Phys. Rev.* C **63** 044309
[20] Ignatyuk A B, Smirenkin G N and Tishin A S 1975 *Yadernaya Fizika* **21** 485
[21] Shlomo S and Natowitz J B 1991 *Phys. Rev.* C **44** 2878
[22] Kuzmina A N, Adamian G G, Antonenko N V and Scheid W 2012 *Phys. Rev.* C **85** 014319

Density dependent effective interactions and recollections of the Rutgers-Princeton years

T. T. S. Kuo

Dept. of Physics and Astronomy, Stony Brook Univ., Stony Brook,NY 11794-3800, USA

E-mail: thomas.kuo@stonybrook.edu

Abstract. The density-dependent effective interactions given by the new Brown-Rho (new-BR) scalings and chiral three-nucleon force V_{3N} are compared with the empirical density-dependent force of the Skyrme interactions. The new-BR scaling is based on a Skyrmion-half-Skyrmion model where nuclear matter is treated as a Skyrmion matter for density smaller than a transition density $n_{1/2} \simeq 0.32 fm^{-3}$, while a half-Skyrmion matter for density greater. In this model, the meson mass, nucleon mass and meson-nucleon coupling are all scaled with density, making the resulting two-nucleon interaction density dependent. By integrating out a participating nucleon over the Fermi sea, Holt, Kaiser and Weise have obtained an effective three-nucleon force $\bar{V}_{3N}$ which is also a density-dependent two-nucleon interaction. The equation of state for symmetric nuclear matter given by the new-BR-scaled V_{2N}, $\bar{V}_{3N}$, and (unscaled-V_{2N} + a Skyrme-type density-dependent force) are all found to be closely similar to each other.

1. Introduction

I have known Aldo for a very long time, starting from 1964 when he was a postdoc at the Rutgers University and I was one at the Princeton University. Rutgers and Princeton are located at nearby townships, separated by merely about 20 miles. Their nuclear physics groups were both very active and worked closely together. We had two joint seminars every week: Monday afternoon seminar at Rutgers and Thursday night Bull Session (which lasted typically more than 3 hours) at Princeton. The faculty members in these groups were A. Arima, Ben Bayman, Gerry Brown, J. de Boer, Ruby Sherr, Igal Talmi, George Temmer,... and among the young postdocs were Joe Ginocchio, Tony Green, G. Sartoris, Chun Wa Wong, Larry Zamick,... in addition to Aldo and me. We were having very good times together, including the well-remembered Rutgers-Princeton nuclear-physics soccer games.

At one of the joint seminars, the then newly developed Skyrme effective interaction was discussed. This interaction is of the form (see e.g. [1])

$$
\begin{aligned}
V_{sky} &= \sum_{i<j} V(i,j) + \sum_{i<j<k} V_{3b}(i,j,k), \\
V_{3b} &= t_3 \delta(\vec{r}_i - \vec{r}_j)\delta(\vec{r}_j - \vec{r}_k), \\
&\rightarrow \frac{t_3}{6}(1 + x_3 P_\sigma)\delta(\vec{r}_1 - \vec{r}_2)\rho(\vec{r}_{av}) \equiv D_{sky} \qquad (1)
\end{aligned}
$$

where the last term D_{sky} is a 'density-dependent' zero-range nucleon-nucleon interaction. (Note that the strength of D_{sky} is typically rather strong such as $t_3 \simeq 14000$ MeV fm^6.[1]) Where does

D_{sky} come from ? There were many discussions about this question at the seminar. At that time, I think nobody had an answer, except that it was empirically needed for nuclear saturation. (In other words, without this term the nuclear matter calculated with V_{sky} can not reproduce empirical nuclear matter saturation energy of $E_0/A \simeq -16$ MeV and density $n_0 \simeq 0.16$ fm^{-3}.)

After so many years, I think we are making progress toward answering the above question. As to be described later, density-dependent effective interactions are generated by both the Brown-Rho scalings [2–7] and/or chiral three-nucleon force V_{3N}[8, 9]. It is of interest to study if the effects of such interactions can be reproduced, to certain extents, by a Skyrme-type empirical density dependent force. We shall do so in this report. The organization of the present report is as follows. In Section 2 we shall describe the new-BR scaling [7] and briefly discuss its density dependent effects to nuclear matter and neutron stars. In Section 3 a similar description will be presented for the chiral three-nucleon force. We shall compare the density-dependent effects from new-BR scaling, chiral three-nucleon force, and an empirical Skyrme-type density dependent force. A summary will also be included in this section.

2. New Brown-Rho scaling

In the early version of the Brown-Rho scaling [2, 6] (to be referred to as the old-BR scaling), the scaling function $\Phi(n)$ of the form

$$\Phi(n) = \frac{m^*}{m} = 1 - C\frac{n}{n_0}, \quad C \simeq 0.15 \tag{2}$$

is employed. Here m^* denotes the meson mass in nuclear medium of density n, and m in free space. n_0 is the nuclear-matter saturation density ($0.16 fm^{-3}$). This scaling has had important and desirable effects in density region near n_0. With its inclusion, nuclear matter calculations have given satisfactory saturation properties [10, 11]. It has also played a key role in shell-model calculations for the extra-long life time ($\sim$ 5000yrs) of the $^{14}C -^{14} N$ β-decay [12].

The above scaling is clearly meant for low densities only. (For instance, the above $\Phi(n)$ is undefined at high n.) In fact the nuclear equation of state (EOS) at densities considerably higher than n_0 is still, by far, largely uncertain. The new-BR scaling is an attempt of studying the high-density nuclear EOS, which is needed for describing neutron-star properties. It is based on a Skyrmion-half-Skyrmion lattice model [13, 14] where nuclear matter is found as composed of Skyrmions at densities below the transition density $n_{1/2}$, while as half-Skyrmions for densities above. It has been estimated that $n_{1/2}$ lies typically between 1.3 and $2n_0$ [15].

A first application of this Skyrmion-half-Skyrmion model to nuclear matter and neutron stars was carried out in [7]. In this model we have Skyrmions for $n \leq n_{1/2}$ (Region I) and half-Skyrmions for $n < n_{1/2} < n_\chi$ (Region II), where the chiral restoration density n_χ is set to be $10n_0$. The scaling functions for these two regions are different. We have for region-I

$$\frac{m_M^*}{m_M} = \frac{m_N^*}{m_N} = \Phi_I(n), \quad \Phi_I(n) = \frac{1}{1 + c_I\frac{n}{n_0}} \; (mesons,\; nucleons),$$
$$\frac{g^*}{g} = 1 \; (coupling\; constants\; unscaled), \tag{3}$$

and for region-II

$$\frac{m_M^*}{m_M} = \Phi_{II}(n), \quad \Phi_{II}(n) = \frac{1}{1 + c_{II}\frac{n}{n_0}} \; (mesons),$$
$$\frac{m_N^*}{m_N} = y(n) \simeq 0.8 \; (nucleons), \quad \frac{g^*}{g} = \Phi_{II}(n) \; (coupling\; constant\; g_{N\rho}). \tag{4}$$

Some differences between the above scaling and that of the old-BR scaling [10–12] may be mentioned. In the old-BR scaling, only the mass and cut-off parameter of the ρ, σ and ω mesons are scaled. In addition to these scalings, the nucleon mass and the nucleon-ρ coupling constant $g_{N\rho}$ are also scaled in new-BR as indicated above. Furthermore, in new-BR the scalings in regions I and II, are different, having parameters c_I and c_{II} for the two regions.

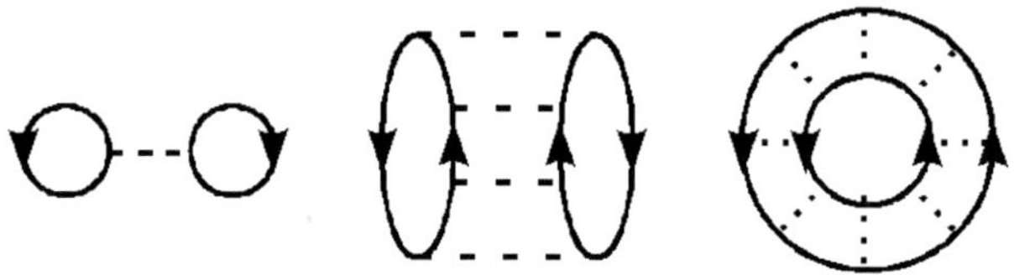

Figure 1. Diagrams included in the all-order *pphh* ring-diagram summation.

We have used a V_{low-k} ring-diagram method [7, 10, 11] to calculate the nuclear EOS with the new-BR scaling. In this method the *pphh* ring diagrams as shown in figure 1 are summed to all orders. There diagram (a), (b) and (c) are respectively a 1st-, 4-th- and 8-th-order *pphh* ring diagram. Our method reduces to a V_{lowk} Hartree-Fock (HF) method if we include only diagram (a). Each vertex of the diagrams is a V_{lowk} interaction [16–19] derived from the Bonn potential [20] with its parameters scaled with density according to the new-BR scaling. (We have employed the Bonn potential as its nucleon-mass parameter can be conveniently scaled.)

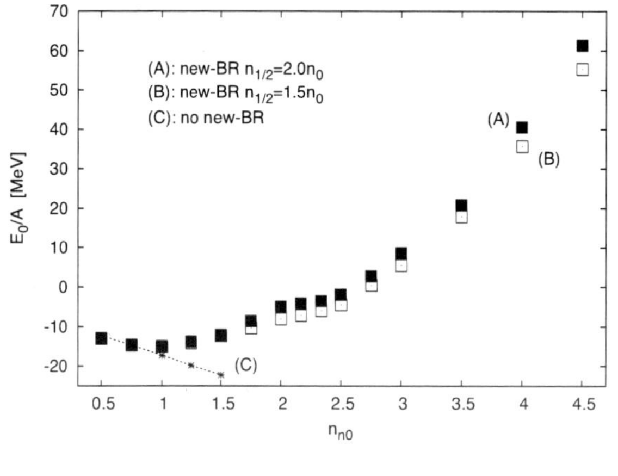

Figure 2. The EOS for symmetric nuclear matter calculated with new-BR scaling.

In figure 2 we present two EOSs so calculated for symmetric nuclear matter. Two choices for the transition density are used, $n_{1/2} = 2n_0$ (A) and $= 1.5n_0$ (B). As seen, the calculation without new-BR scaling (C) is unable to describe the nuclear matter saturation properties (it saturates at at density $\sim 2.5n_0$ and $E_0/A \simeq -22MeV$). The region-I scaling parameters of $c_I(\rho) = c_I(N) = 0.130$, $c_I(\sigma) = 0.121$ and $c_I(\omega) = 0.139$ were used; these values are determined by fitting the empirical nuclear matter properties. The coupling constant $g_{N\rho}$ is not scaled in region I, while it is scaled in region II. The EOS of figure 2 gives saturation properties E_0/A=-15.0 MeV, saturation density $n_{sat} = 0.93n_0$ and compression modulus $K = 206$MeV.

In the $n > n_{1/2}$ region, the meson scaling parameters are taken to be the same as in region I, namely $c_{II} = c_I$. But a different scaling is used for nucleons: in region II we use $m_N^*/m_N = y(n) = 0.77$ (0.78) respectively for $n_{1/2} = 2n_0$ $(1.5n_0)$. These values are used to make the values of E_0/A just below and above $n_{1/2}$ approximately equivalent. But their slopes are clearly discontinuous at $n_{1/2}$, implying a Skyrmion half-Skyrmion phase transition there. In figure 3 we display the pressure-density EOS calculated with new-BR. ($p(n) = n[d\epsilon(n)/dn] - \epsilon(n)$, ϵ being the energy density.) As seen, there is discontinuity at $n_{1/2}$. In fact from the results of figure 3, the Skyrmion half-Skyrmion coexistance region can be determined. For instance, this coexistence region is $\sim 1.7n_0 < n <\sim 2.4n_0$ for the $n_{1/2} = 2n_0$ case.

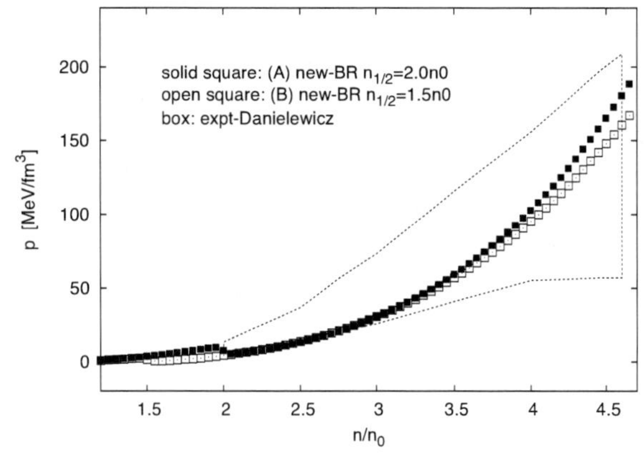

Figure 3. Pressure in symmetric nuclear matter calcualted with new-BR scaling.

In figure 3 our calculated pressure is compared with the empirical constraint of Danielewicz et al. [21]. It is encouraging that our calculated pressure is generally within the range allowed by the constraint. It would be very useful to check experimentally the Skyrmion half-Skyrmion phase transition. This may require much effort, as to locate the pressure 'kink' near $n_{1/2}$ would need a highly precise, and difficult, experimental determination of the nuclear matter pressure there . As another check of our EOS, we have also performed neutron star calcualtions using the new-BR EOS [7]. The calculated mass and radius of the maximum-mass neutron star are ($M = 2.39M_\odot$, $R = 10.9km$) for $n_{1/2} = 2.0n_0$, and ($M = 2.38M_\odot$, $R = 10.9km$) for $n_{1/2} = 1.5n_0$. These masses are considerably larger than the observed masses of $1.97 \pm 0.04M_\odot$ [22] and $2.01 \pm 0.04M_\odot$ [23], indicating that our new-BR neutron EOS is probably too stiff.

3. Chiral three-nucleon force and summary

As we see from the preceeding section, nuclear matter calculations with the new-BR scaled V_{2N} can satisfactorily describe nuclear matter saturation properties but not so with the unscaled V_{2N}. Can the calculations using V_{2N}(unscaled) plus V_{3N} also give satisfactory nuclear saturation properties? In this section we shall address this question. We employ the lowest-order (NNLO) chiral V_{3N} of the form $V_{3N} = V_{3N}^{2\pi} + V_{3N}^{1\pi} + V_{3N}^{ct}$ where

$$V_{3N}^{(2\pi)} = \sum_{i\neq j\neq k} \frac{g_A^2}{8f_\pi^4} \frac{\vec{\sigma}_i \cdot \vec{q}_i \, \vec{\sigma}_j \cdot \vec{q}_j}{(\vec{q_i}^2 + m_\pi^2)(\vec{q_j}^2 + m_\pi^2)} F_{ijk}^{\alpha\beta} \tau_i^\alpha \tau_j^\beta, \tag{5}$$

$$V_{3N}^{(1\pi)} = - \sum_{i\neq j\neq k} \frac{g_A c_D}{8f_\pi^4 \Lambda_\chi} \frac{\vec{\sigma}_j \cdot \vec{q}_j}{\vec{q_j}^2 + m_\pi^2} \vec{\sigma}_i \cdot \vec{q}_j \, \vec{\tau}_i \cdot \vec{\tau}_j, \tag{6}$$

$$V_{3N}^{(\mathrm{ct})} = \sum_{i\neq j\neq k} \frac{c_E}{2f_\pi^4 \Lambda_\chi} \vec{\tau}_i \cdot \vec{\tau}_j, \tag{7}$$

where $g_A = 1.29$, $f_\pi = 92.4$ MeV, $\Lambda_\chi = 700$ MeV, $m_\pi = 138.04$ MeV$/c^2$, $\vec{q_i} = \vec{p_i}' - \vec{p_i}$ is the difference between the final and initial momentum of nucleon i and

$$\begin{aligned} F_{ijk}^{\alpha\beta} &= \delta^{\alpha\beta}\left(-4c_1 m_\pi^2 + 2c_3 \vec{q_i} \cdot \vec{q_j}\right) \\ &+ c_4 \epsilon^{\alpha\beta\gamma} \tau_k^\gamma \vec{\sigma}_k \cdot (\vec{q_i} \times \vec{q_j}) \,. \end{aligned} \tag{8}$$

The parameters $c_1 = -0.76\,\mathrm{GeV}^{-1}$, $c_3 = -4.78\,\mathrm{GeV}^{-1}$, $c_4 = 3.96\,\mathrm{GeV}^{-1}$ are well known, constrained by low-energy NN phase shifts.[24] But the parameters c_D and c_E are not well determined. A range of c_D and c_E values on the Navratil $c_D - c_E$ curve [25] can all fit well the binding energies of ^3H and ^3He.

By integrating out one participating nucleon over the Fermi sea, Holt, Kaiser and Weise [8, 9] have reduced V_{3N} to a density-dependent 2-body force $\bar{V}_{3N}$. Comparing with V_{3N}, $\bar{V}_{3N}$ is much more convenient for nuclear matter calculations. Briefly speaking, they are related by

$$\begin{aligned} V_{3N} &= \frac{1}{36}\Sigma\langle 123|V_{3N}|456\rangle a_3^+ a_2^+ a_1^+ a_4 a_5 a_6, \\ \bar{V}_{3N} &= \frac{1}{4}\Sigma\langle 12|D_{2N}|45\rangle \; a_2^+ a_1^+ a_4 a_5, \\ \langle ab|D_{2N}|cd\rangle &= \sum_{h\leq k_F} \langle abh|V_{3N}|cdh\rangle. \end{aligned} \tag{9}$$

It is seen that $\bar{V}_{3N}$ is a density (k_F) dependent 2-body interaction. Note well there is a n-body counting factor $C(nb)$ to be included in calculations: We have $C(nb) = (1,\ 1/2,$ and $1/3)$ respectively for (2-, 1-, 0-)body vertices. Thus the vertex in (a) of figure 1 is $(V_{2N} + \bar{V}_{3N}/3)$, and each vertex in (b) and (c) is $(V_{2N} + \bar{V}_{3N})$.

We have performed ring-diagram nuclear matter calculations with $(V_{2N}+\bar{V}_{3N})$. The parameters c_D and c_E of V_{3N} (see Eqs.(6-7)) are not well known; they can have a range of values as allowed by the Navratil curve [25]. We have used several sets of allowed c_D and c_E parameters in our ring-diagram nuclear matter calculations and found that the allowed parameters c_D=3.0 and c_E=0.2 are a 'best-fit' choice (among the several sets calculations we have made) in reproducing nuclear saturation properties. We have used these parameters in the present calculation, and with them the saturation propertes given by the $(V_{2N}+\bar{V}_{3N})$ EOS are ($E_0/A = -15.6$ MeV, $n_{sat} = 0.98n_0$, $K = 132$ MeV). Neutron star calculations using the $(V_{2N}+\bar{V}_{3N})$ EOS have also been carried out. The mass and radius of the maximum-mass neutron star so obtained are ($1.99M_\odot$, 9.85 km); that these values are both smaller than the new-BR results mentioned earlier (near figure 3) indicates the new-BR neutron EOS is stiffer than the corresponding $(V_{2N}+\bar{V}_{3N})$ EOS especially at high densities.

As a summary, let us compare the three EOSs in figure 4. The results from new-BR($n_{1/2} = 2n_0$) and $(V_{2N} + V_{3N})$ are remarkably similar to each other, especially for densities

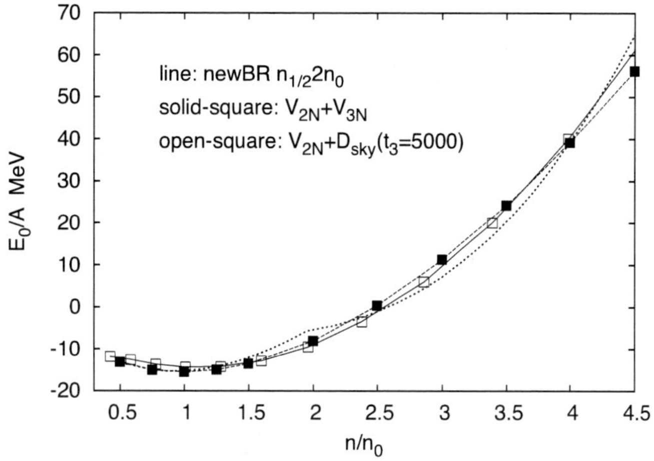

Figure 4. Comparision of ring-diagram nuclear matter calculations.

near n_0. Nuclear matter calculations with V_{2N} alone are unable to give satisfactory saturation properties. and this shortcoming can be amended by including either new-BR or V_{3N}. There may be some underlying equivalence between new-BR and V_{3N}, and its further study will be interesting as well as useful. In figure4, we also compare the above EOSs with the EOS given by (unscaled-V_{2N} + D_{sky}) with t_3=5000 MeV (see equation(1)). It is a 'surprise' that all three are in good qualitative agreement. That the density dependent effects on symmetric nuclear matter from new-BR and/or V_{3N} may be well reproduced by an empirical density-dependent force of the Skyrme type is an encouraging result, indicating that new-BR and/or V_{3N} may provide a microscopic foundation for the empirical Skyrme force.

4. Acknowledgements

I wish to express my sincere gratitude to Aldo for inviting me to this beautiful '11th International Spring Seminar on Nuclear physics'. I also wish to applaud him for having worked so hard in organizing all these 11 seminars (I have attended 10 of them) as well as for his many important contributions to nuclear physics. Aldo has indeed done a lot for all of us. I thank H. Dong, Jeremy Holt, H.K. Lee, R. Machleidt and Mannque Rho for years of close and pleasant collaboration. This work is supported in part by the U.S. Department of Energy under Grant No. DF-FG02-88ER40388.

References

[1] Ring P and Schuck P, *The Nuclear Many-Body Problem* (Springer-Verlag, New York, 1980)
[2] Brown G E, Rho M 1991 *Phys. Rev. Lett* **66** 2720
[3] Hatsuda T and Lee S H 1992 *Phys. Rev. C* **46** R34
[4] Brown G E, Rho M 1996 *Phys. Rep.* **269** 333
[5] Brown G E, Rho M 2002 *Phys. Rep.* **363** 85
[6] Brown G E, Rho M,2004 *Phys. Rep.* **396** 1
[7] Dong H, Kuo T T S, Lee H K, Machleidt R, Rho M 2013 *Phys. Rev. C* **87**, 054332
[8] Holt J W, Kaiser N and Weise W 2009 *Phys. Rev. C* **79**, 054331

[9] Holt J W, Kaiser N and Weise W 2010 *Phys. Rev. C* **81**, 0240021
[10] Siu L W, Holt J W, Kuo T T S and Brown G E 2009 *Phys. Rev. C* **79**, 0540004
[11] Dong H, Kuo T T S and Machleidt R 2009 *Phys. Rev. C* **80**, 065803
[12] Holt J W, Brown G E, Kuo T T S, Holt J D and Machleidt R 2008 *Phys. Rev. Lett* **100**, 062501
[13] Golhaber A S and Manton N S 1987 *Phys. Lett.* **B198**, 231
[14] Lee H K, Park B Y and Rho M 2011 *Phys. Rev. C* **83**, 025206
[15] Park B Y, Min D P, Rho M and Vento V 2002 *Nucl. Phys. A* **707**, 381; Lee H J*et al.* 2003 *Nucl. Phys. A* **723**, 427; Rho M, arXiv:0711.3895 [nucl-th]
[16] Bogner S K, Kuo T T S and Coraggio L 2001 *Nucl. Phys. A* **684** 432
[17] Bogner S K, Kuo T T S, Coraggio L, Covello A and Itaco N 2002 *Phys. Rev. C* **65** 051301R
[18] Bogner S K, Kuo T T S and Schwenk A 2003 *Phys. Rep.* **386** 1
[19] Holt J D, Kuo T T S and Brown G E 2004 *Phys. Rev. C* **69** 034329
[20] Machleidt R *Computational Nuclear Physics 2–Nuclear Reactions* (Langanke, Maruhn, Koonin, eds., Springer NY 1993), Chap.1, p.1
[21] Danielewicz P, Lacey R, and Lynch W A 2002 *Science* **298**, 1592
[22] Demorest P B *et. al* 2010 *Nature* **Vol. 467** 1081
[23] Antoniadis J *et. al* 2013 *Science* **340** 1223232
[24] Rentmeester M C M, Timmermans R G E and de Swart J J 2003 *Phys. Rev. C* **67** 044001
[25] Navratil P, Gueorguiev V G, Vary J P, Ormand W E and Nogga A 2007 *Phys. Rev. Lett* **99**, 042501

A systematic study of mirror and triplet energy differences

S M Lenzi and R Lau

Department of Physics and Astronomy, University of Padua and INFN, Padua, Italy

E-mail: lenzi@pd.infn.it

Abstract. Differences of excitation energy among analogue states in a broad mass range are calculated in the framework of the shell model by introducing isospin breaking interactions. Mirror and triplet energy differences along the N=Z line from mass $A \sim 20$ to $A \sim 60$ have been systematically studied using the same method. It is shown that in all cases an additional term not directly associated to the electromagnetic interaction is needed to reproduce the experimental data.

1. Introduction

The nuclear interaction may be considered with good approximation to be charge symmetric and charge independent. In the nucleus, these symmetries are of course broken by the Coulomb interaction. However, the experimental data show that the isospin symmetry is slightly broken even when Coulomb effects are taken into account. The degree of symmetry breaking by the nuclear interaction can be studied by comparing the binding energies or masses of isobaric nuclei, the so called Coulomb Displacement Energies (CDE) [1]. More recently, with the advent of high-resolution, high-efficient gamma-ray arrays, it has become possible to study differences in *excitation* energy between isobaric analogue states as a function of the angular momentum. These differences are of the order of tens of keV and constitute a very sensitive tool to probe some nuclear structure properties, such as the evolution of the nuclear structure, such as changes of the nuclear radius, the origin of the backbending in rotating nuclei, and other interesting properties [2].

These studies have been performed within the shell model framework, and in particular in nuclei in the $f_{7/2}$ shell, where energy differences between mirror nuclei and T=1 isobaric triplets can be well reproduced by large-scale calculations [2, 3]. From these studies, and after taking into account electromagnetic corrections, it was suggested that an additional isospin breaking term has to be considered to reproduce the data. Two schematic terms, consisting of only one matrix element, have been proposed by Zuker and collaborators [2]. In the case of the differences between excited states in mirror nuclei (MED), the isovector matrix element corresponds to two protons in the $f_{7/2}$ shell coupled to $J = 2$, while for the energy differences of an isobaric triplet (TED), the matrix element corresponds to two protons in the $f_{7/2}$ coupled to $J = 0$. In both cases the strength of the matrix elements is +100 keV and is deduced from the experimental data in mass $A = 42$.

In recent years the experimental studies have been extended from the nuclei in the $f_{7/2}$ shell to other mass regions due to the progress in the experimental techniques and the use of radioactive beams. From the theoretical point of view, there have been studies by Ekman and collaborators [4] in the mass region $A = 57 - 61$ and more recently by Kaneko *et al.* on the excitation energy differences in $A = 67$ [5]. This latter study has been partly based on the work by Zuker *et al.* [2], but have not converged into a unified method for the description of the excitation energy differences in this mass region. In particular, the need of introducing an isospin breaking force does not emerge.

One would think that if a simple isospin breaking force arise so clear in the $f_{7/2}$ shell, it should be also present in the rest of the nuclear chart. Of course the $f_{7/2}$ shell is very particular as it is quite isolated from the rest of the shells and therefore some intrinsic properties can be easily found. In this paper we present some new results on the sd and upper pf shell where we have extended the same method applied to the energy differences in $f_{7/2}$ shell nuclei. This study puts in evidence the general character of the isospin symmetry breaking terms that have to be added to the nuclear interaction to reproduce the excitation energy differences measured so far.

2. Shell model description of MED and TED

The mirror (isovector) energy differences, which account for the charge symmetry breaking as a function of the nuclear spin J are defined as:

$$MED(J) = E_x(J,T,T_z) - E_x(J,T,-T_z). \tag{1}$$

The triplet (isotensor)energy differences, which refer to three isobaric nuclei with $T = 1$ are defined as:

$$TED(J) = E_x(J,T=1,T_z=-1) + E_x(J,T=1,T_z=+1) - 2E_x(J,T=1,T_z=0), \tag{2}$$

where E_x are the excitation energies of the states referred to the ground state or to the lowest state of the same T in each nucleus and $T_z = (N - Z)/2$.

To calculate these differences in the framework of the shell model, we follow the method described in refs. [2, 3]. The terms that contribute to the energy differences can be divided in a monopole Coulomb (V_{Cm}), a multipole Coulomb (V_{CM}), and an additional isospin breaking term V_B. MED and TED are therefore calculated as,

$$MED(J) = \Delta_M < V_{Cm}(J) > +\Delta_M < V_{CM}(J) > +\Delta_M < V_B(J) > \tag{3}$$

$$TED(J) = \Delta_T < V_{CM}(J) > +\Delta_T < V_B(J) > \tag{4}$$

where Δ_M and Δ_T indicate that the differences are obtained as in equation (1) and (2), respectively.

Monopole Coulomb contributions are responsible for single-particle shifts and bulk effects, while the multipole term takes into account the interaction between valence nucleons. The single-particle energy corrections have two different origins.

The energy of the proton orbits is affected proportionally to the square of the orbital momentum l in the harmonic oscillator representation. In particular, the single-particle energy of protons with principal quantum number N, above the closed shell Z_{CS} is modified by the following term [6]:

$$E_{ll} = \frac{-4.5Z_{CS}^{13/12}[2l(l+1) - N(N+3)]}{A^{1/3}(N+3/2)} \tag{5}$$

The relativistic electromagnetic spin-orbit interaction, that affects both the proton and neutron single-particle energies is [1, 7]:

$$V_{ls} = (g_s - g_l)\frac{1}{2m_N^2c^2}(\frac{1}{r}\frac{dV_c}{dr})\vec{l}.\vec{s}, \tag{6}$$

where m_N is the nucleon mass, and g_s and g_l are the gyromagnetic factors of protons and neutrons.

There is an additional monopole effect due to the Coulomb interaction. This is due to changes of the nuclear radius as a function of the angular momentum. The radius of the nucleus depends on the orbitals that are occupied and the occupation number may change with J. The radial term is calculated in the shell model framework as discussed in Ref.[2] from the equation:

$$V_{Cr} = n\alpha_i\left(\frac{m_{i\pi}(g.s.) + m_{i\nu}(g.s.)}{2} - \frac{m_{i\pi}(J) + m_{i\nu}(J)}{2}\right) \quad (7)$$

where $n/2 = |T_z|, m_{i\pi}, m_{i\nu}$ are the proton and neutron occupation numbers of orbit i, respectively, and α_i is a parameter that depends on the particular orbit. It is important to note that due to the way TED are obtained, monopole effects cancel out.

In Ref. [2] it was shown that the effect of the isospin breaking interaction V_B in both MED and TED can be as large at the Coulomb terms. This was deduced from extracting the Coulomb contribution to the MED and TED in mass A=42. For the MED, after subtracting to the experimental value the Coulomb contribution, a large difference (100 keV) was observed for the J=2 state, while for the TED, the maximum difference was found for the J=0 state. The extrapolation to other shells is not straightforward. In mass $A = 42$, the wavefunctions can be approximately considered of pure $f_{7/2}$ nature while this is not the case for other mass regions since all the orbits within the model space can contribute to the V_B term. We generalize the prescription of the $f_{7/2}$, adding the relevant matrix elements of all the orbits within the model space with the same strength that allows a good description of all the available data. For more details see Ref. [2, 3].

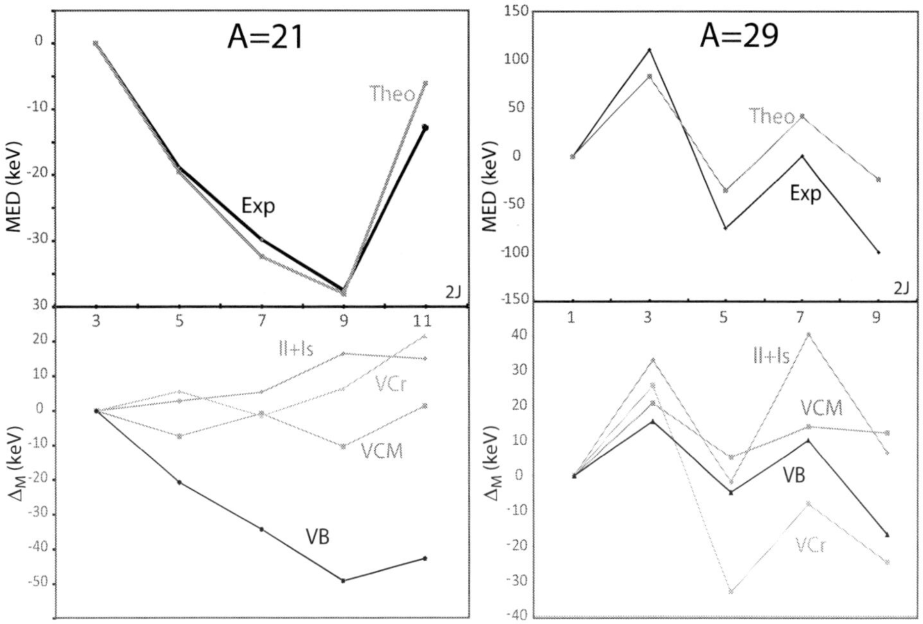

Figure 1. MED (left panels) for the $T = 1/2$ yrast states in the mirror pair ^{21}Na-^{21}Ne. The lower panels show the contribution of the different components to the MED.

3. Results

We calculate the mirror energy differences and the triple energy differences of nuclei in the mass range $A \sim 20 - 60$ where experimental data are available. The calculations were done by using the shell-model code ANTOINE [8, 9] with different effective nuclear interactions. For the *sd* shell, the USD interaction [10, 11] has been used while the KB3G [12] interaction was used for nuclei in the $f_{7/2}$ shell. For $A = 57 - 61$, the GXPF1A [13] interaction has been adopted.

In this section, we show some few illustrative examples of the results obtained for MED and TED of nuclei in the *sd* shell and upper *fp* shell. The result of MED and TED of the $f_{7/2}$ shell can be found in [3].

3.1. MED and TED in the sd shell

The calculations of the MED and TED in the *sd* shell are in very good overall agreement with the data. We report in figure 1 the MED results for mass $A = 21$ and $A = 29$, together with the single contributions. The radial term in this case is due to the changes in occupation number of the $s_{1/2}$ orbit with strength $\alpha = 100$ keV. For the V_B term we consider two matrix elements with two protons in the $d_{3/2}$ and two protons in the $d_{5/2}$ coupled to $J = 2$ with a strength of 100 keV, as that used in the $f_{7/2}$ shell. As can be deduced from figure 1, the contribution of the V_B term is essential to allow a good description of the data, in particular, in $A = 21$.

Following the prescription of Ref. [2, 3], for the TED, the V_B term corresponds to matrix elements with two protons in the $d_{3/2}$, $d_{5/2}$, $s_{1/2}$ coupled to $J = 0$ with a strength of 100 keV for all the orbits. We report in figure 2 the TED for the $T = 1$ triplets in the *sd* shell. The calculated curves are in very good agreement with the experimental data. Interestingly, the contributions from the multipole Coulomb term and the isospin breaking V_B term are very similar.

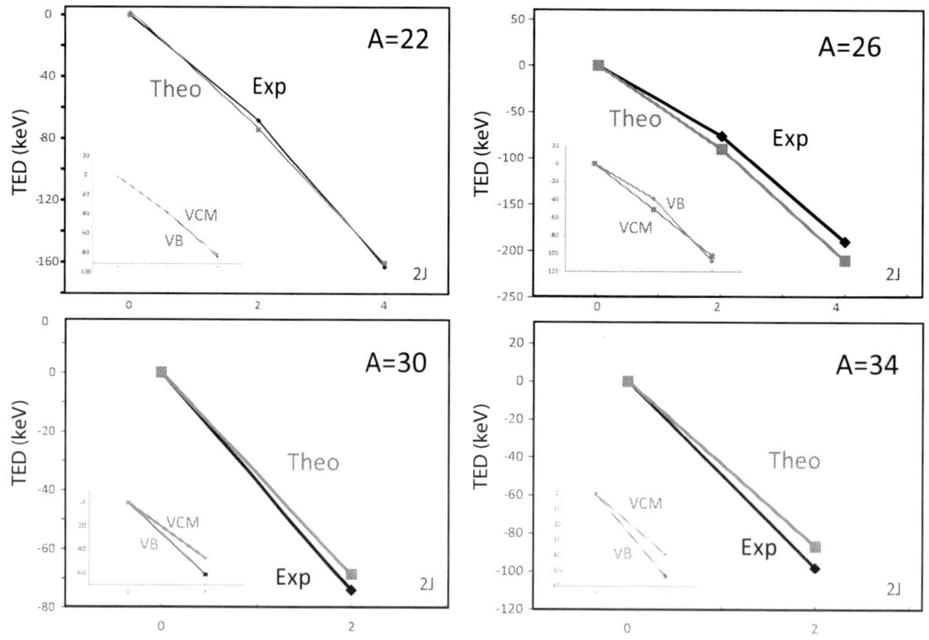

Figure 2. TED for the $T = 1$ yrast states in the isobaric multiplets of mass A= 22, 26, 30 and 34. In the left bottom of each figure the two contributions to the TED in equation (4) are displayed.

3.2. MED and TED of masses in the mass range A=57 to A=61

The calculations of the MED and TED in the upper fp shell are performed following an extension of the prescription of Ref. [2, 3], using the GXPF1A interaction. For the MED, the radial term is due to changes in the occupation number of the $p_{3/2}$ and the $p_{1/2}$ orbits with equal strength α = 200 keV. A different prescription is used for the INC term V_B: instead of increasing the contribution of the J = 2 matrix elements by 100 keV, we decrease by the same amount those coupled to J =0, i.e. we take into account matrix elements with two protons in the $f_{7/2}, p_{3/2}, f_{5/2}$ and $p_{1/2}$ coupled to J = 0 with a strength of -100 keV. This choice is justified by a parallel theoretical work by Bentley and collaborators [14] where a best fit of the available MED data in the $f_{7/2}$, allowing all the matrix elements for two protons in the $f_{7/2}$ to vary, show that it is the relative difference between the J =0 and the J =2 matrix elements which counts. So, an increase of the J =2 matrix element by 100 keV, may be equivalent to a decrease of the J =0 matrix element by 100 keV.

The results are in overall agreement with the available data, with the exception of A = 59. We report in figure 3 the results for mass A=58 together with the single contributions. For the TED, we follow the same prescription as for the other mass regions: the V_B term corresponds to matrix elements with two protons in all the orbits coupled to J = 0 with a strength of 100 keV. The TED for A = 58 are reported in figure 3. As for the sd shell, the contribution of the V_B term is clearly relevant and very similar to that of the Coulomb contribution.

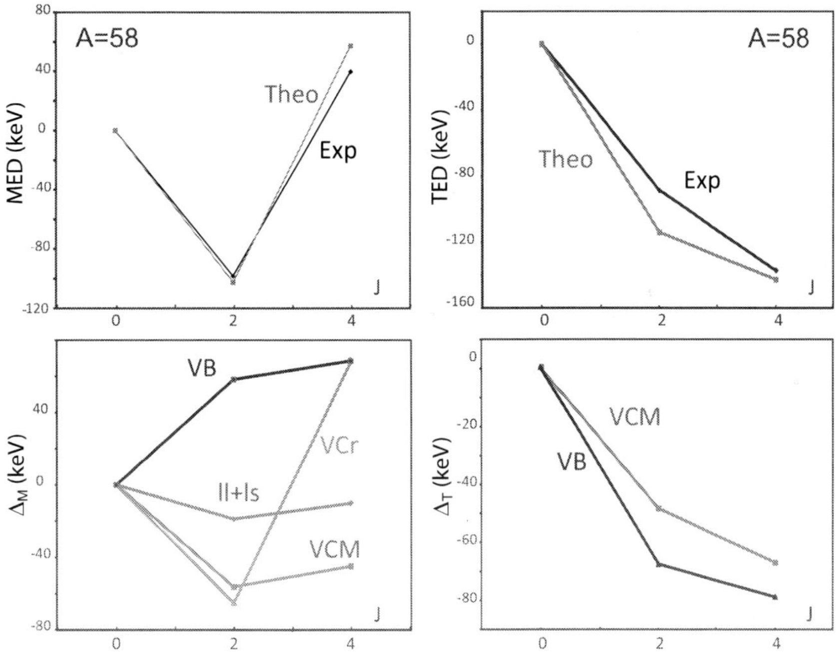

Figure 3. MED and TED for the T = 1 yrast states in the isobaric multiplets of mass A = 58. The lower panels show the contribution of the different components to the MED and TED.

4. Conclusions

We have performed a systematic study of the MED and the TED in different mass regions that cover the sd shell up to the upper fp shell. The theoretical calculations reproduce with the same prescription the TED in the different mass regions. In all cases so far investigated the theoretical calculations are in excellent agreement with the experiment. It is shown that an additional isospin breaking term $V_B = 100$ keV for two protons in the different shells coupled to $J = 0$ is needed to reproduce the data. This term results very similar to that due to the Multipole Coulomb contribution. Here we present some significant examples but it is important to note that the same parameterization reproduces all cases in which data are available. Recently, Kaneko *et al.* have used a similar V_B term for the description of the TED in $A = 66, 70$ within the fp shell [15].

For the MED, the extension and generalization to the *sd* shell of the method introduced in refs. [2, 3], describes very well the data up to mass $A \sim 30$. The need of a generalized isospin symmetry breaking term V_B corresponding to matrix elements with two protons in the same shell coupled to $J = 2$ and strength +100 keV emerges clearly. At higher masses, excitations to the upper fp shell become important. On the other hand, in the upper fp shell, a better agreement with the data is obtained using a V_B term where the protons are coupled to $J = 0$ and with negative strength of -100 keV. This alternative prescription gives also rather good results in the *sd* and the $f_{7/2}$ shells.

The results show a consistent description of the MED and TED in the shell model framework. Moreover, they demonstrate that the evidence of an isospin symmetry breaking term is not constrained to the $f_{7/2}$ shell, but has to be systematically included to account for the experimental results. The microscopic origin of this term is still not understood.

It is important to note that in this analysis, the strengths of the different terms contributing to the MED and TED have not been fitted in order to have the best reproduction of the data. Very schematic values have been used aiming at putting in evidence the general character of the method. A more detailed presentation of the results, that includes new data is in preparation.

References

[1] Nolen J A and Schiffer J P 1969 *Annu. Rev. Nucl. Sci.* **19** 471
[2] Zuker A P, Lenzi S M, Martinez-Pinedo G and Poves A 2002 *Phys. Rev. Lett.* **89** 142502
[3] Bentley M and Lenzi S M 2007 *Prog in Part. and Nucl. Phys.* **59** 497
[4] Ekman J, Fahlander C and Rudolpf D 2005 *Mod. Phys. Lett. A* **20** 2977
[5] Kaneko K *et al.* 2010 *Phys. Rev. C* **82** 061301(R)
[6] Duflo J and Zuker A P 2002 *Phys. Rev. C.* **66** 051394(R)
[7] Ekman J *et al.* 2004 *Phys. Rev. Lett.* **92** 1322502
[8] Caurier E *et al.* 1989-2002, Shell model code ANTOINE, IRES, Strasbourg,
[9] Caurier E and Nowacki F 1999 *Acta. Phys. Pol. Vol.* **30** 705
[10] Wildenthal B H 1984 *Prog in Part. and Nucl. Phys.* **11** 5
[11] Brown B A and Wildenthal B H 1988 *Annu. Rev. Nucl. Part. Sci.* **38** 29
[12] Poves A, Sanchez-Solano J S, Caurier E and Nowacki F 2001 *Nucl. Phys. A* **694** 157
[13] Honma M, Otsuka T, Brown B A and Mizusaki T 2002 *Phys. Rev. C* **65** 061301
[14] Bentley M A *et al.*, to be published.
[15] Kaneko K *et al.* 2014 *Phys. Rev. C.* **89** 031302(R)

Journal of Physics: Conference Series **580** (2015) 012029 doi:10.1088/1742-6596/580/1/012029

Can Realistic Shell-Model Calculations be Predictive?

N Itaco[1,2], L Coraggio[2], A Covello[1] and A Gargano[2]

[1] Dipartimento di Fisica, Università di Napoli Federico II,
Complesso Universitario di Monte S. Angelo, Via Cintia - I-80126 Napoli, Italy
[2] Istituto Nazionale di Fisica Nucleare, Sezione di Napoli
Complesso Universitario di Monte S. Angelo, Via Cintia - I-80126 Napoli, Italy

E-mail: nunzio.itaco@unina.it

Abstract. A major requirement for a nuclear model is, not only to reproduce and describe accurately the available experimental data, but also to provide reliable predictions for physical quantities not yet measured. Over the past two decades, we have performed various realistic shell-model calculations for nuclei in different mass regions, which have all yielded results in good agreement with the available experimental data. In this paper we present some selected results illustrating their predictive power.

1. Introduction

The advances in experimental techniques to produce exotic beams make nowadays possible to plan new experiments to explore the structure of nuclei at and close to the driplines. This is a major challenge of today nuclear physics, since these studies may reveal new features that go beyond our present understanding of nuclear structure. In such a context, a key point is to ascertain the capability of the available nuclear models of providing reliable predictions for physical quantities not yet measured, whose knowledge may be of stimulus and guidance for upcoming experiments.

The shell model is an invaluable theoretical tool to understand the structure of nuclei, giving a framework for a microscopic description of nuclear properties based essentially on the use of effective Hamiltonians. In particular, during the last two decades the realistic shell model, that makes use of effective Hamiltonians derived from the bare nucleon-nucleon (NN) potential in the framework of the many-body theory, has proved to lead to an accurate description of the structure of nuclei in various mass regions both close to and far from the valley of stability (see [1–4] for some recent examples). A major merit of this microscopic approach is that no adjustable parameter is introduced, which makes more interesting to wonder about its predictivity through the examination of the results so far obtained.

Here we shall present some selected cases which clearly evidence the good predictive power of realistic shell-model calculations. These cases will be discussed in section 3, after a brief description of the theoretical framework (section 2). A brief summary is given in the last section.

2. Theoretical framework

As mentioned in the Introduction, a basic ingredient for realistic shell-model calculations is the effective Hamiltonian $H_{\rm eff}$, which is constructed starting from a NN potential V^{NN} that gives a very accurate description of the experimental data of the two nucleon problem. This is the case of all modern NN potentials, as for instance CD-Bonn [5], Argonne V18 [6], Nijmegen [7]. All these potentials, however, exhibit a strong short-range repulsive behavior thus preventing their direct use in the derivation of $H_{\rm eff}$, which is based on a perturbative approach. In the past, this problem has been faced renormalizing the potential by way of the Brueckner reaction matrix, while a new approach is provided today by the availability of low-momentum NN potentials. As a matter of fact, one can start from a realistic model for V^{NN} and integrate out its high-momentum components thus obtaining a smooth potential $V_{\rm low-k}$ that preserves exactly the onshell properties of the original V^{NN} [8, 9]. Alternatively, one can recur directly to a chiral potential based upon the effective field theory (EFT) [10, 11]. In both cases, these low-momentum potentials may be used as input in the derivation of $H_{\rm eff}$ without the need of any further renormalization.

The starting point is the A-nucleon system Hamiltonian. By introducing an auxiliary one-body potential U, this may be written as the sum of a one-body term H_0, describing the independent motion of the nucleons, and a residual interaction H_1:

$$H = \sum_{i=1}^{A} \frac{p_i^2}{2m} + \sum_{i<j=1}^{A} V_{ij}^{NN} = T + V = (T+U) + (V-U) = H_0 + H_1 \quad . \tag{1}$$

In the shell model, the nucleus is schematized as an inert core plus n interacting valence nucleons, and a reduced space, the so-called model space, is defined in terms of a finite subset of d eigenvectors of H_0

$$|\Phi_i\rangle = [a_1^\dagger a_2^\dagger \ldots a_n^\dagger]_i |c\rangle \quad , \tag{2}$$

where $|c\rangle$ represents the inert core, the subscripts 1, 2, ..., n denote the single-particle (SP) valence states and i stands for all the quantum numbers needed to specify the state. In this model space, an effective Hamiltonian may be constructed so as to reduce the diagonalization of the Hamiltonian (1) in a infinite Hilbert space to the solution of the eigenvalue problem:

$$H_{\rm eff} P|\Psi_\alpha\rangle = E_\alpha P|\Psi_\alpha\rangle \quad , \tag{3}$$

where $\alpha = 1, .., d$, the E_α and the corresponding Ψ_α being a subset of the eigenvalues and eigenvectors of the original Hamiltonian.

$H_{\rm eff}$ may be derived by way of a similarity transformation constructing a new Hamiltonian $\mathcal{H} = X^{-1}HX$ that satisfies the decoupling equation between the model space P and its complement Q:

$$Q\mathcal{H}P = 0 \quad , \tag{4}$$

so that the desired effective Hamiltonian is $H_{\rm eff} = P\mathcal{H}P$. A detailed description of the approach for deriving $H_{\rm eff}$ by way of perturbation theory can be found in [12, 13].

Before concluding this section, it is worth spending few words about the issue regarding the choice of the model-space SP energies. The effective Hamiltonian constructed within the above outlined approach contains one- and two-body contributions. However, a subtraction procedure is commonly used in realistic shell-model calculations, so as to retain only the two-body terms of $H_{\rm eff}$, while the SP energies are taken from experiment. This approach, although less fundamental, enables to take into account implicitly the effects of three-body forces on the SP energies.

3. Results

We present some selected results of realistic shell-model calculations, performed in previous works, for nuclei belonging to different mass regions, namely heavy calcium isotopes, neutron deficient tin isotopes, and Sn isotopes beyond $N = 82$.

3.1. Heavy calcium isotopes

In 2009 we published a paper reporting the results of realistic shell-model calculations for neutron-rich Ca isotopes [14]. We performed calculations for calcium isotopes with mass ranging from $A = 49$ to $A = 56$ by employing the fp shell as model space for the valence neutrons outside the doubly magic nucleus ^{40}Ca and the effective interaction derived from the CD-Bonn NN potential. Some predictions for the $^{53-56}$Ca isotopes, that were completely unknown at that time, were shown in [14].

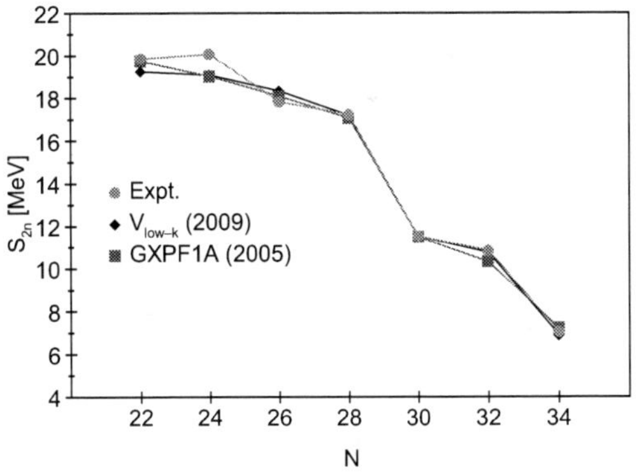

Figure 1. (Color online) Experimental [15, 16] and calculated S_{2n} for calcium isotopes with $N = 22 - 34$.

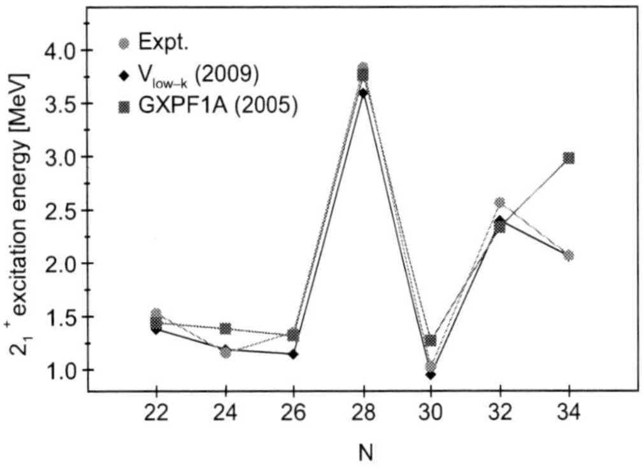

Figure 2. Experimental [17, 18] and calculated excitation energies of the yrast $J^\pi = 2^+$ states for calcium isotopes with $N = 22 - 34$.

During the past few years, there has been a great interest in the study of heavy calcium isotopes from both the experimental and theoretical point of view. This interest is testified, for instance, by two letters that have recently appeared in *Nature*. One of them, published in June 2013, reports the results of an experiment performed at the ISOLDE/CERN facility with ISOLTRAP leading to first mass measurement for ^{53}Ca and ^{54}Ca [16]. These new mass-data reveal a pronounced decrease in the two-neutron separation energy (S_{2n}), quite similar to that observed beyond ^{40}Ca, which was interpreted as a clear signature of the $N = 32$ shell closure.

Few months later, in October 2013, another very interesting letter on the results of a experimental spectroscopic study of ^{54}Ca was published [18]. In this experiment, performed by proton knockout reactions with fast radioactive projectiles, the yrast 2^+ state was identified at an excitation energy of 2.043 MeV. This large value, comparable with that of ^{52}Ca evidences the onset of a sizable subshell closure at $N = 34$ in calcium isotopes.

In figures 1 and 2 we report our calculated S_{2n} for even Ca isotopes and the 2^+-states excitation energies, respectively, as shown in reference [14]. The theoretical values are compared with the presently available data and with the results obtained using the phenomenological shell-model Hamiltonian GXPF1A [19]. As it can be seen, in both cases our predictions turn out to be in quite good agreement with the new data. This is especially noteworthy for the 2^+ excitation energy in ^{54}Ca, for which the two theoretical predictions were remarkably different. This point was discussed in detail in reference [14], where it was shown that the difference between the results of the two shell-model calculations is related to the different $T = 1$ monopole properties of the two effective interactions.

3.2. Neutron-deficient Sn isotopes

The nucleus ^{100}Sn is the heaviest particle-bound doubly magic nucleus with an equal number of neutrons and protons. Therefore its neighborings provide the opportunity to investigate the effective interaction for nuclei in the vicinity of the proton drip line with protons and neutrons filling the same orbitals. In particular, the study of Sn isotopes has attracted a special interest, since it may give direct information on neutron pairing correlations.

The work of reference [20], which dates back to almost 20 years ago, may be placed in this context. The ^{102,103,104,105}Sn nuclei were studied within the framework of the shell model using two effective interactions derived from the Bonn A [21] and the Paris [22] NN potentials. The aim was essentially to assess the role of realistic effective interactions in the mass region around ^{100}Sn. At that time, in fact, the practical value of these interactions was still not completely assessed. The calculations were performed using as model space the five neutron orbitals of the 50-82 shell with SP energies determined through an analysis of the low-energy spectra of the tin isotopes with $105 \leq \mathrm{A} \leq 111$. No information on the spectrum of the single-neutron nucleus ^{101}Sn was in fact available. Now we know the position of the $\frac{7}{2}^+$ state at 0.171 MeV excitation energy with respect to the $\frac{5}{2}^+$ ground state. This value is not far from the adopted one in [20] for the $\epsilon_{g_{7/2}} - \epsilon_{d_{5/2}}$ single-particle spacing, namely 0.2 MeV.

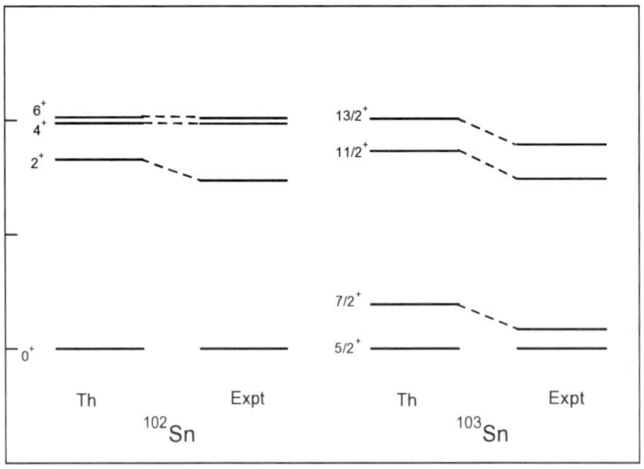

Figure 3. Experimental [17] and calculated low-energy spectra of ^{102,103}Sn.

A satisfactory agreement between theory and experiment was obtained for ^{104,105}Sn by using the Bonn A potential. In particular, the calculated spectrum of ^{104}Sn was very close to the experimental one, the only exception being the position of the 2^+ state which was overestimated by about 250 keV. Analogously, in ^{105}Sn the largest discrepancies were found for the energies of states arising from the coupling of single-particle states to the 2^+ state of ^{104}Sn.

In [20], the predictions for the two lighter nuclei ^{102}Sn and ^{103}Sn, for which only recently some spectroscopic information has become available, were also reported. In figure 3, the experimental spectra of ^{102,103}Sn [17] are compared with the calculated ones. Note that we have included only theoretical states with the experimental counterpart.

It is worth noting that the quality of agreement between theory and experiment is similar to that found for ^{104,105}Sn. Clearly we overestimate the energies of the 2^+ state in ^{102}Sn, which reflects on ^{103}Sn. This may be traced back to $Z = 50$ cross-shell excitations which are not explicitly taken into account in the adopted model space [23].

3.3. Tin isotopes beyond $N = 82$

During the last two decades, there has been substantial progress in gaining experimental information on nucleon-rich nuclei far from the stability line, which has evidenced changes in the shell structure when approaching the neutron drip line. In this context, nuclei in the regions of the shell closures with a large N/Z ratio, as for instance tin isotopes beyond $N = 82$, are a subject of great interest, opening the opportunities to better understand the forces that bind the nucleons together in very extreme conditions. The experimental information on heavy tin isotopes are, however, still very limited owing to the present experimental difficulties to access such short-lived nuclei. Further information, which may help us to gain insight into the structure these nuclei, will be provided in not a distant future thanks to the continuous advances in the experimental techniques. It is, therefore, challenging to make predictions which may stimulate and provide guidance to future experiments.

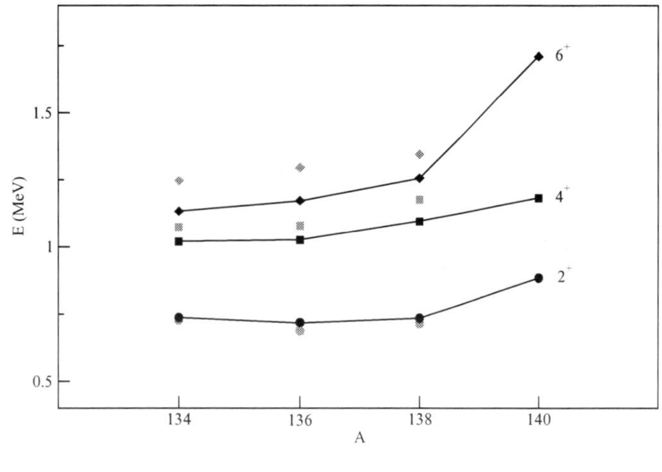

Figure 4. (Color online) Experimental (red symbols) [17, 26] and calculated (black symbols) excitation energies of the yrast $J^\pi = 2^+, 4^+$and 6^+ states for tin isotopes with $A = 134 - 140$.

In 2011, we reported on a shell-model study of Sn isotopes with $N \geq 82$ employing a realistic effective interaction derived from the CD-Bonn NN potential renormalized via the $V_{\rm low-k}$

approach [24]. In these calculations, ^{132}Sn was taken as closed core and the valence neutrons were let to move in the six orbitals of the 82-126 shell. Based on the very good agreement of our results with the observed energies for ^{134}Sn, which was the only known nucleus, we investigated the evolution of the yrast 2^+ state and of the other J $\neq 0$ members of the lowest-lying multiplet when adding neutron pairs up to $N = 90$. As a main outcome of our calculations, we predicted an almost flat behavior for the excitation energies of the yrast $2^+, 4^+$, and 6^+ states (see figure 4). We found only a slight increase at $N = 90$, due to the sub-shell closure of the $f_{7/2}$ orbital. As a matter of fact, our prediction for ^{140}Sn excludes the possible existence of a $N = 90$ shell closure as suggested in [25].

Very recently, studying the projectile fission of a ^{238}U beam at the RIKEN Radioactive Isotope Beam Factory, the delayed γ-ray casacdes originating from the decay of 6^+ isomeric states have been observed in ^{136,138}Sn [26]. The observed excitation energies, reported in figure 4, are very well reproduced by calculated values of reference [24]. This supports our results for the hitherto unknown spectrum of ^{140}Sn.

4. Summary

The predictive power of a nuclear model is a parameter of the utmost importance in the evaluation of its merits. It would be, in fact, highly desirable to know the capability of a nuclear model in providing reliable predictions as well as its limits. This is especially true nowadays, when, thanks to the availability of radioactive ion beams and the advances in detection techniques, experiments are being planned to explore the unknown structure of nuclei when moving towards the drip lines. In this paper we have studied the predictive power of realistic shell-model calculations, by reporting some selected results for heavy calcium isotopes, neutron deficient Sn isotopes and tin isotopes beyond $N = 82$. The results shown here were already the subject of previous works, where the comparison between theory and experiment was, obviously, limited to the data available at that time. In this contribution, we have shown that our predictions give an accurate description of quantities recently measured. This makes us confident in the good predictive power of our approach.

References

[1] Coraggio L, Covello A, Gargano A, Itaco N and Kuo T T S 2009 *Phys. Rev. C* **80** 044320
[2] Coraggio L, Covello A, Gargano A and Itaco N 2010 *Phys. Rev. C* **81** 064303
[3] Coraggio L, Covello A, Gargano A and Itaco N 2013 *Phys. Rev. C* **87** 034309
[4] Coraggio L, Covello A, Gargano A and Itaco N 2014 *Phys. Rev. C* **89** 024319
[5] Machleidt R 2001 *Phys. Rev. C* **63** 024001
[6] Wiringa R B, Stoks V G J and Schiavilla R 1995 *Phys. Rev. C* **51** 38
[7] Stoks V G J, Klomp R A M, Terheggen C P F and Swart J J D 1994 *Phys. Rev. C* **49** 2950
[8] Bogner S, Kuo T T S, Coraggio L, Covello A and Itaco N 2002 *Phys. Rev. C* **65** 051301(R)
[9] Bogner S, Furnstahl R and Schwenk A 2010 *Prog. Part. Nucl. Phys.* **65** 94
[10] Coraggio L, Covello A, Gargano A, Itaco N, Kuo T T S, Entem D R and Machleidt R 2007 *Phys. Rev. C* **75** 024311
[11] Machleidt R and Entem D R 2011 *Phys. Rep.* **503** 1
[12] Coraggio L, Covello A, Gargano A, Itaco N and Kuo T T S 2009 *Prog. Part. Nucl. Phys.* **62** 135
[13] Coraggio L, Covello A, Gargano A, Itaco N and Kuo T T S 2012 *Ann. Phys.* **327**(9) 2125
[14] Coraggio L, Covello A, Gargano A and Itaco N 2009 *Phys. Rev. C* **80** 044311
[15] Audi G, Wapstra A H and Thibault C 2003 *Nucl. Phys. A* **729** 337
[16] Wienholtz F, Beck D, Blaum K, Borgmann C, Breitenfeldt M, Cakirli R B, George S, Herfurth F, Holt J, Kowalska M, Kreim S, Lunney D, Manea V, Menendez J, Neidherr D, Rosenbusch M, Schweikhard L, Schwenk A, Simonis J, Stanja J, Wolf R N and Zuber K 2013 *Nature (London)* **498** 346
[17] Data extracted using the NNDC On-line Data Service from the ENSDF database, file revised as of August 29, 2014.
[18] Steppenbeck D, Takeuchi S, Aoi N, Doornenbal P, Matsushita M, Wang H, Baba H, Fukuda N, Go S, Honma M, Lee J, Matsui K, Michimasa S, Motobayashi T, Nishimura D, Otsuka T, Sakurai H, Shiga Y,

Journal of Physics: Conference Series **580** (2015) 012029 doi:10.1088/1742-6596/580/1/012029

Söderström P A, Sumikama T, Suzuki H, Taniuchi R, Utsuno Y, Valiente-Dobon J J and Yoneda K 2013 *Nature* **502** 207

[19] Honma M, Otsuka T, Brown B A and Mizusaki T 2005 *Eur. Phys. J. A* **25, s01** 499

[20] Andreozzi F, Coraggio L, Covello A, Gargano A, Kuo T T S, Li Z B and Porrino A 1996 *Phys. Rev. C* **54** 1636

[21] Machleidt R 1989 *Adv. Nucl. Phys.* **19** 189

[22] Lacombe M, Loiseau B, Richard J M, Mau R V, Côtè J, Pirés P and Tourreil R D 1980 *Phys. Rev C* **21** 861

[23] Bader V M, Gade A, Weisshaar D, Brown B A, Baugher T, Bazin D, Berryman J S, Ekström A, Hjorth-Jensen M, Stroberg S R, Walters W B, Wimmer K and Winkler R 2013 *Phys. Rev. C* **88** 051301

[24] Simpson G S, Gey G, Jungclaus A, Taprogge J, Nishimura S, Sieja K, Doornenbal P, Lorusso G, Söderström P A, Sumikama T, Xu Z Y, Baba H, Browne F, Fukuda N, Inabe N, Isobe T, Jung H S, Kameda D, Kim G D, Kim Y K, Kojouharov I, Kubo T, Kurz N, Kwon Y K, Li Z, Sakurai H, Schaffner H, Shimizu Y, Suzuki H, Takeda H, Vajta Z, Watanabe H, Wu J, Yagi A, Yoshinaga K, Bönig S, Daugas J M, Drouet F, Gernhäuser R, Ilieva S, Kröll T, Montaner-Pizá A, Moschner K, Mücher D, Naïdja H, Nishibata H, Nowacki F, Odahara A, Orlandi R, Steiger K and Wendt A 2014 *Phys. Rev. Lett.* **113**(13) 132502

[25] Covello A, Coraggio L, Gargano A and Itaco N 2011 *J. Phys. Conf. Ser.* **267** 012019

[26] Sarkar S and Sarkar M S 2010 *Phys. Rev. C* **81** 064328

Journal of Physics: Conference Series **580** (2015) 012030 doi:10.1088/1742-6596/580/1/012030

Spectroscopic properties of neutron rich nuclei beyond ^{132}Sn and seniority mixing

H. Naïdja[†,‡,⋆], **F. Nowacki**[†] **and K. Sieja**[†]

† Université de Strasbourg, IPHC, 23 rue du Loess 67037 Strasbourg, France
CNRS, UMR7178, 67037 Strasbourg, France.
‡ GSI Helmholtzzentrum für Schwerionenforschung GmbH, 64291 Darmstadt, Germany
⋆ Université Constantine 1, (LPMS) 1 route Ain El Bey, 25000 Constantine, Algeria

E-mail: houda.naidja@iphc.cnrs.fr

Abstract. Within the shell model, our calculations have been focused on the study of the spectroscopic properties of the even-even tin isotopes around ^{132}Sn. In particular the calculations of the low-lying states $0^+, 2^+, 4^+, 6^+$ energies, and differents electromagnetic transitions $B(E2, J+2 \rightarrow J)$ have been carried out. Two important points are discussed: the effect of the core excitations on the structure and transition rates in this region, and the possible closure of the sub-shell at N=90.

1. Introduction

The heavy mass region of neutron-rich nuclei around ^{132}Sn, represents an interesting area for experimental and theoretical nuclear structure research. Theoretically, lot of studies have been performed in this region, within the mean field using RPA or QRPA [1, 2, 3], or within the shell model [4, 5, 6], using realistic and empirical interactions. Sensitivity studies of the $(r-process)$ show that the nuclear structure data in the mass $A = 140$ region is most impactful on the calculated abundances.

In this exotic region, new experimental data has been obtained recently in RIKEN [7], by the observation of three delayed gamma rays each from ^{136}Sn and ^{138}Sn, originating from 6^+ isomeric state, and are the first gamma ray transtions observed in these very neutron rich nuclei. These results provided us valuable information to study the nuclear structure of this mass region, and for the development of our empirical shell model interaction.

2. Shell Model tools

Our calculations have been performed using shell model codes ANTOINE and NATHAN [8, 9], including the $0h_{11/2}, 1f_{7/2}, 0h_{9/2}, 1f_{5/2}, 2p_{3/2}, 2p_{1/2}, 0i_{13/2}$ orbitals for neutrons and $0g_{9/2}, 0g_{7/2}, 1d_{5/2}, 1d_{3/2}, 2s_{1/2}$ orbitals for protons, which are taken above the closed core of ^{110}Zr. The advantage to take such a model space, is the fact that we have the possibility to do calculations using ^{110}Zr or ^{132}Sn as cores, just by closing or opening $0h_{11/2}$ orbit for neutrons and $0g_{9/2}$ orbit for protons. Taking into account core excitations constitutes a challenge in the diagonalisation of Hamiltonian matrix due to its extremely large size.

Here we have employed an interaction, which is derived from CD-Bonn potential, renormalized following the so-called V_{low-k} approach with a cutoff momentum of $\Lambda = 2.2 fm^{-1}$,

Journal of Physics: Conference Series **580** (2015) 012030 doi:10.1088/1742-6596/580/1/012030

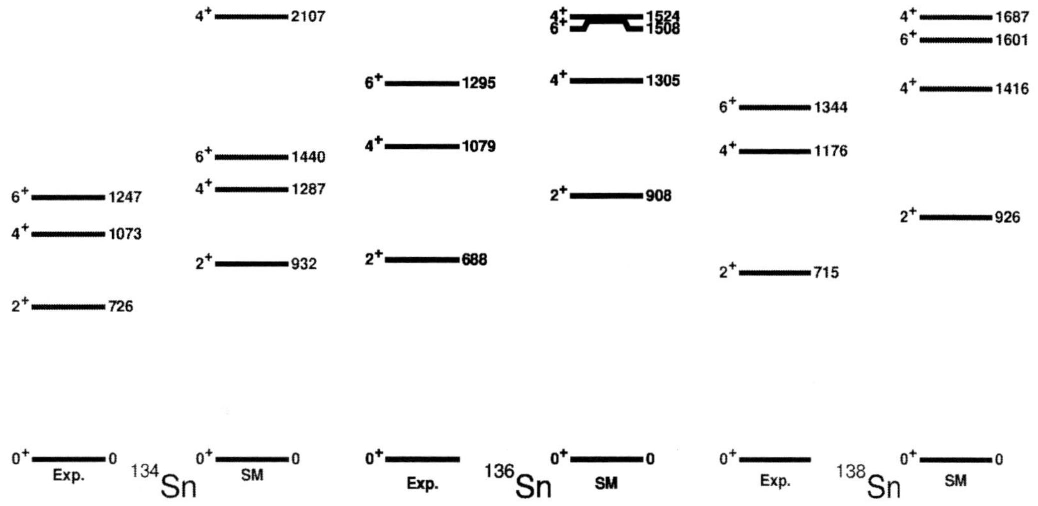

Figure 1. Level schemes of the tin isotopes, calculated using NNS110 interaction.

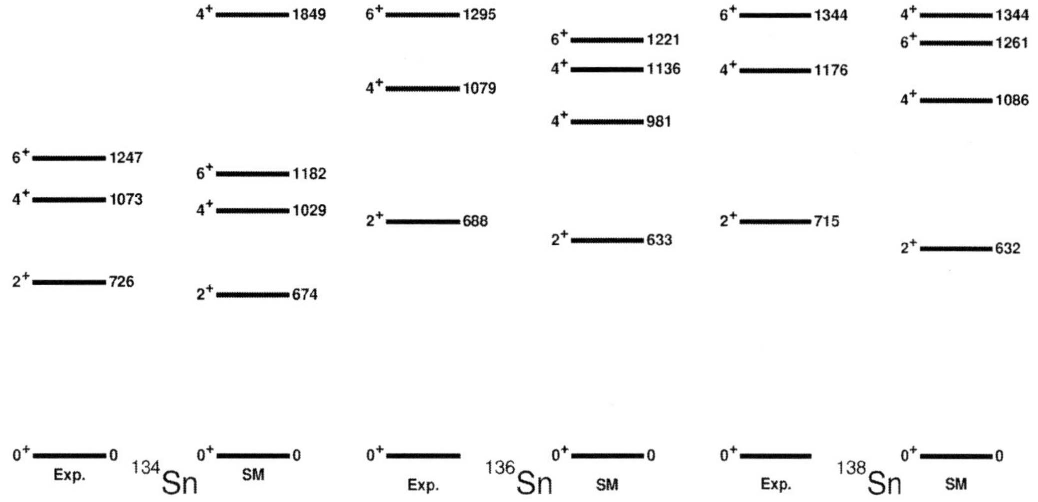

Figure 2. Level schemes of the tin isotopes, calculated using NNS110P interaction, obtained by reducing the pairing force in NNS110 interaction.

and adopted to the model space by many body perturbation theory techniques, including all the diagrams up to the second order [11].

3. Applications and results

3.1. Energy levels and transition rates in $^{134,136,138}Sn$

In the calculations with ^{110}Zr core, we used as input single particle energies from GEMO model [15], which is the J. Duflo and A.P Zuker systematic parametrisation of single particle and single hole energies over the nuclear chart. Further, monopole corrections have been applied to V_{low-k} realistic interaction, to reproduce the experimental structure of ^{133}Sn and ^{133}Sb nuclei [10]. As it is displayed in figure 1, this interaction noted by NNS110 does not reproduce correctly the excitation energies of the tin isotopes compared to experimental data [10, 7].

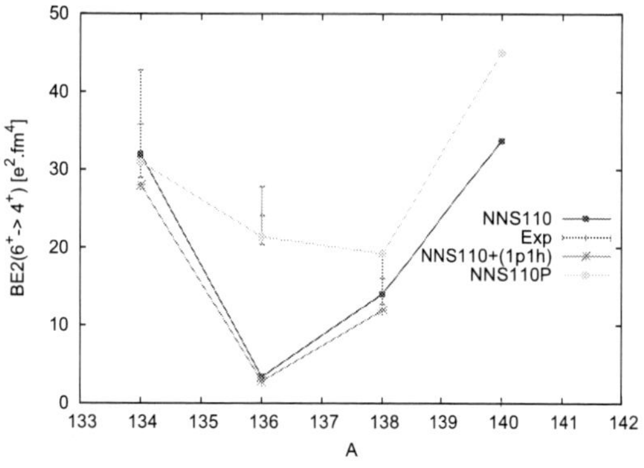

Figure 3. Variation of the electromagnetic transitions as function of the mass number, calculated using: NNS110, opening ^{132}Sn core (1p1h excitations), and NNS110P interactions.

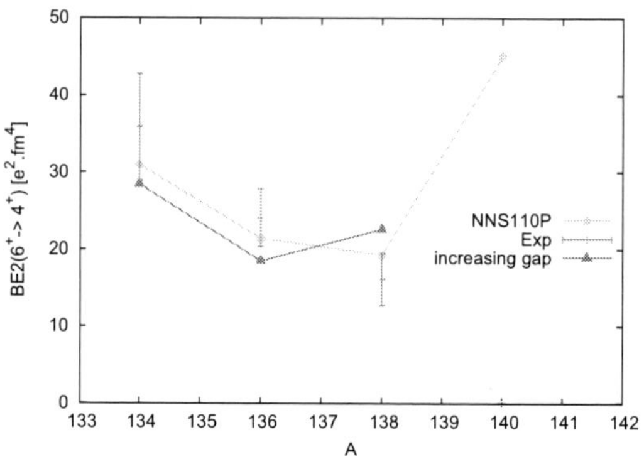

Figure 4. The same as figure 3, but increasing the $f_{7/2} - p_{3/2}$ gap, compared to NNS110P results.

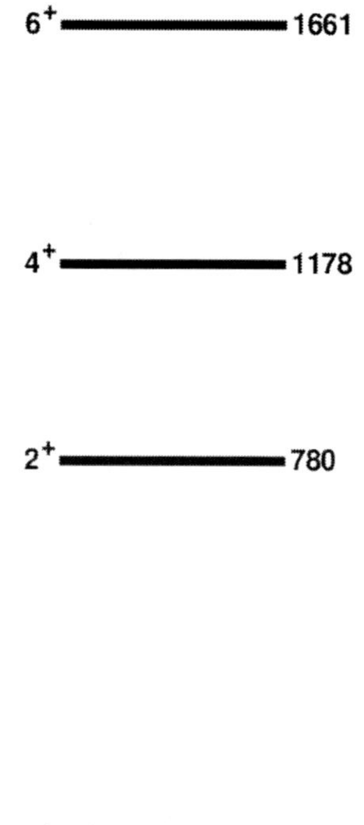

Figure 5. Energy levels of ^{140}Sn calculated using NNS110P interaction.

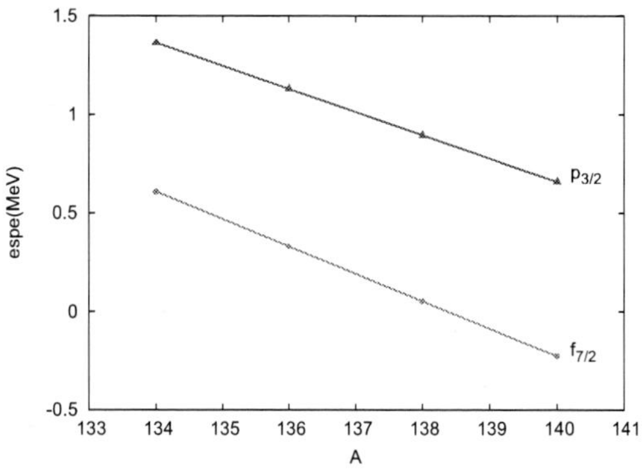

Figure 6. Variation of the gap between $f_{7/2}$ and $p_{3/2}$ as function of the mass number of tin isotopes.

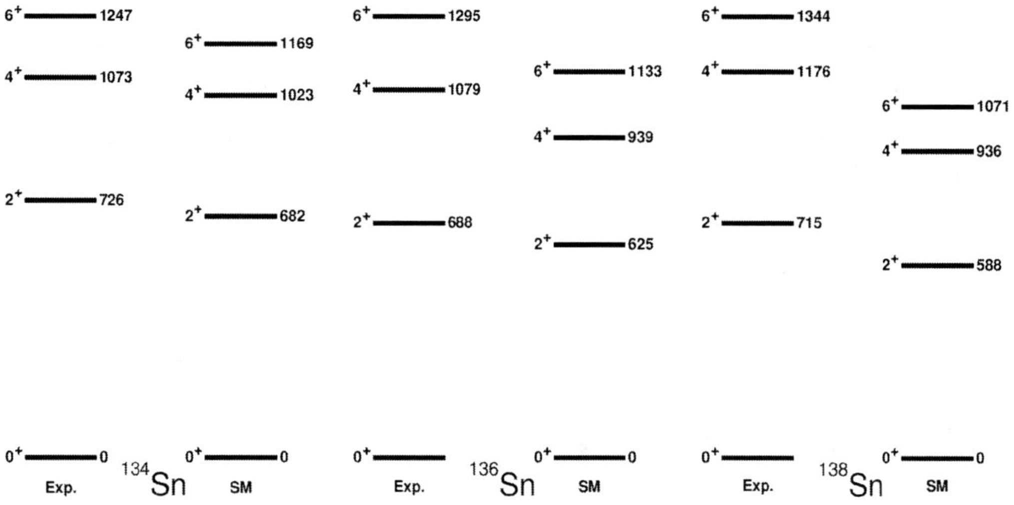

Figure 7. The effect of increasing the $f_{7/2} - p_{3/2}$ gap, to the level schemes of the tin isotopes.

The variation of the electromagnetic transition rates from 6^+ to 4^+ states, as function of the mass number calculated within the shell model, and compared to experimental data [7, 10] is reported in figure 3. As one can notice, with effective charge $e_n = 0.64e$, our calculations predict well the observed transitions from ^{134}Sn and ^{138}Sn, but they underestimate it for the mid-shell nucleus ^{136}Sn. We have performed additional calculations, including $1p1h$ excitations from $h_{11/2}$ neutrons and $g_{9/2}$ protons to the above valence space, with effective charges $e_n = 0.5e$ and $e_p = 1.5e$ for neutrons and protons, respectively. In this case, the transition rates in ^{134,138}Sn are good enough without enhanced effective charges, but still fail to reproduce the exact observed value in ^{136}Sn. It appears that reducing the diagonal and off diagonal pairing strength of the $f_{7/2}$ matrix elements by about 130 keV (denoted hereafter by NNS110P), gives us a good agreement with the measured value of ^{136}Sn transition.

We try to explain this result in the seniority model, where in the two first cases (using NNS110 interaction and opening the core), the $v = 2$ is the dominant component of the first excited states in ^{136}Sn (82% in 4_1^+ state, and 95% in 6^+ state). It is well known that in the seniority selection rule, transitions between 6^+ and 4_1^+ states with the same seniority is forbidden, which leads to a vanishing transition strength between them. Reducing the pairing force in the realistic interaction modifies the seniority structure and leads to a more complex two 4^+ states: Their wave functions contain 4_1^+(53% of $v = 2$ and 46% of $v = 4$), and 4_2^+(43% of $v = 2$ and 57% of $v = 4$), leading to a larger transition rate.

Now, our calculations using the NNS110P interaction are in fair agreement with the observed energy levels of the tin isotopes (figure 2). The $(f_{7/2})^n$ (n is the valence neutron number varying from 2 to 6) is the principal component of the ground state and the first excited states for all the tin isotopes, where its percentage decreases from 90% to about 50%, going from ^{134}Sn to ^{138}Sn. The inversion between the two states 6^+ and 4_2^+ states in ^{136}Sn constitutes another consequence of the seniority mixing discussed in previous paragraph. A very similar effect was observed in [14], where the structure and location of the $6_1^+(v = 2)$ and $6_2^+(v = 4)$ states in ^{72,74}Ni, ^{94}Ru and ^{96}Pb has important implications for the isomeric properties of the $8^+(v = 2)$ state.

Opening ^{132}Sn core and taking 1p1h excitations into account, has a minor effect ($\simeq$ 50 keV) on the excitation energies, compared to NNS110 interaction.

3.2. Sub-shell closure at N=90

The evolution of the shell structure is a hot topic in modern nuclear structure studies. The authors of [5] have proposed the closure of the sub-shell N=90 in ^{140}Sn, by the creation of an important gap between $f_{7/2}$ and $p_{3/2}$ orbitals. The trend is very similar to that observed in the ^{22}O and ^{48}Ca, where neutron shell gaps increase by adding the neutrons in the high j orbit ($1d_{5/2}$ in ^{22}O and $1f_{7/2}$ in ^{48}Ca). It has been shown in Refs.[12, 13] that the 3-body forces play an important role in creation of these gaps.

Figure 5 displays our predictions of the level scheme of ^{140}Sn calculated using NNS110P interaction. Excited states are characterized by excitations from $f_{7/2}$ to $p_{3/2}$ shell, which leads to an increased transition probability $B(E2, 6^+ \rightarrow 4^+)$ in this nucleus (see figure 3).

As one can notice from figures 2 and 5, the spacing between 0^+ and 2^+ remains nearly constant ($\simeq$700 keV) for all the tin isotopes. There is just small increase of the 2^+ and 4^+ states in ^{140}Sn, and it becomes more significant in 6^+ state, which has different configuration $(f_{7/2})^{-2}(p_{3/2})^2$. At the same time, the spacing between $f_{7/2} - p_{3/2}$ shells remains nearly constant, when moving from ^{134}Sn to ^{140}Sn (see figure 6). These results constitute an indication of possible no sub-shell closure in ^{140}Sn.

However, the results in [5] show that using an empirical interaction SMPN, there is a sudden increase of the 2^+ at N=90, and the gap $f_{7/2} - p_{3/2}$ grows by about 2.2MeV, leading to double magicity in ^{140}Sn.

We have imposed this spacing by changing the monopole part. As a consequence, the 2^+ energy increases in ^{140}Sn to 1815 keV, but at the same time, we lose the spectroscopic properties (figure 7) and the transition rates (figure 4) of the remaining tin isotopes.

4. Conclusions

The discovery of new isomers in very exotic magic nuclei [7] is a key for the development of the theory far from stability, the nuclear shell model in particular. Here, we have investigated excitation energies and transition rates in ^{134,136,138}Sn isotopes, and our results using NNS110P interaction seem to be in fair agreement with experiment.

Reducing the pairing force of the realistic interaction was necessary to reproduce the observed transition rates in ^{136}Sn, which leads to a seniority mixing effect. This slight modification of the pairing force has an apparent effect on the excitation energies, and to the position of the second 4^+ state.

For the first time, we have also studied the effect of the opening of ^{132}Sn core on the transition rates, which showed that it plays a minor role in the evolution of the excitation energies and transition rates, confirming the strong magicity of ^{132}Sn.

Finally, we have studied how the modification of the $f_{7/2} - p_{3/2}$ gap influences the results and we conclude that a large $N = 90$ gap is incompatible with the observed excitation energies and $B(E2)$ transitions in the tin isotopes.

Acknowledgments
H.N acknowledges support from the Helmhotz Association through the Nuclear Astrophysics Virtual Institute (VH-VI-417).

References

[1] Terasaki J, Engel J, Nazarewicz W and Stoitsov M 2002 *Phys. Rev* C **66** 054313
[2] Hofmann F, Lenske H 1998 *Phys. Rev* C **57** 3381
[3] Dobaczewski D, Nazararewicz W, Werner T R, Berger J F, Chinn C R and decharge J 1996 *Phys. Rev* C **53** 2809
[4] Kartamyshev M P, Engenland T, Hjorth-Jensen M and Osnes E 2007 *Phys. Rev* C **76** 024313
[5] Sarkar S and Sarkar M Saha 2010 *Phys. Rev* C **81** 064328
[6] Covello A, Coraggio L, Gargano A and Itaco N 2011 *Journal. Phys. Conf. Series* **267** 012019
[7] Simpson G S et al. 2014 *Phys. Rev. Lett* **113** 132502
[8] Caurier E, Martinez-Pinedo G, Nowacki F, Poves A and Zuker A P 2005 *Rev. Mod. Phys* **77** 427
[9] Nowacki F and Caurier E 1999 *Acta. Phys. Polonica* B **30** 705
[10] www.nndc.bnl.gov
[11] Hjorth-Jensen M, Kuo T T S, Osnes E 1995 *Phys. Rep* **261** 125
[12] Zuker A P 2003 *Phys. Rev. Lett* **90** 042502
[13] Otsuka T, Suzuki T, Holt J D, Schwenk A, Akaichi Y 2010 *Phys. Rev. Lett* **105** 032501
[14] Lisetskiy A F, Brown B A, Horoi M and Grawe H 2004 *Phys. Rev* C **70** 044314
[15] Duflo J and Zuker A P 1999 *Phys. Rev.* C **59** R2347

Shell model calculations for neutrinoless double beta decay

Sabin Stoica
Horia Hulubei Foundation, P.O. Box MG12, 077125 Bucharest-Magurele, Romania
E-mail: sabin.stoica@unescochair-hhf.ro

Abstract. The study of the neutrinoless double beta ($0\nu\beta\beta$) decay is of great interest because of its potential to provide us with information about the lepton number conservation and neutrino properties as the neutrinos character (are they Dirac or Majorana particles?) and their absolute mass. Since the $0\nu\beta\beta$ decay has not yet been discovered experimentally, one can only extract limits of the absolute neutrino mass. For that, one needs accurate calculations of both nuclear matrix elements (NMEs) and phase space factors (PSFs) which appear in the theoretical lifetime expressions, corroborated with experimental lifetime limits. In this paper I first present recent shell model (ShM) calculations of the NMEs and PSFs for $0\nu\beta\beta$ decay performed by our group, in the hypothesis that the mechanism of its occurrence is the exchange of light Majorana neutrinos between two nucleons inside the nucleus. Also, the consensus on the use of different nuclear structure ingredients/parameters in the computation of the NMEs is discussed. Then, I present new limits of the Majorana neutrino mass parameter derived from the analysis of $0\nu\beta\beta$ decay of nine isotopes.

1. Introduction

Double beta decay (DBD) is a nuclear natural decay by which an even-even nucleus transforms into another even-even nucleus with the same mass A but with its nuclear charge Z changed by two units. It occurs whatever single β decay can not occur due to energy reasons or if it is highly forbidden by angular momentum selection rules. There are several theoretical possible DBD modes, that can be classified according to the number and type of the leptons released in the decay. We have the so-called two neutrino double beta decays ($2\nu\beta\beta$), where in the final states one finds two electrons/positrons and two anti-neutrinos/neutrinos, and the $0\nu\beta\beta$ decay modes, where besides electrons/positrons one finds no neutrinos in the final states. The positron DBD modes can be accompanied by electron capture EC/β and/or ECEC decay modes. The $2\nu\beta\beta$ decay modes conserve the lepton number (ΔL $= 0$) and are allowed in the initial formulation of the Standard Model (SM). They are already measured for eleven isotopes with values of the half-lives ranging between $10^{18} - 10^{24}$ yr. By contrary, no $0\nu\beta\beta$ decay is confirmed until present by independent measurements. The $0\nu\beta\beta$ decay modes are particularly of great interest because they can provide us with information about the lepton number conservation and neutrino properties as, for example, the neutrinos character (are they Dirac or Majorana particles?) and their absolute mass [1]-[2]. Its discovery would imply that the lepton number may not conserve (ΔL $= 2$) and neutrino is a Majorana particle. They can theoretically occur through several mechanisms, the most investigated being the exchange of light left-handed neutrinos between two nucleons in the nucleus. In this paper we refer to the $0\nu\beta\beta$ decay which occur through this mechanism and where two electrons are released. The lifetimes for this decay mode can be expressed as a product of three factors: a phase space (PSFs) factor, depending on the $Q_{\beta\beta}$ value of the decay and on the nuclear charge Z, a factor representing the NMEs and a Majorana neutrino mass parameter which is related to the elements of the neutrino mixing matrix and to absolute neutrino mass eigenvalues [1]-[2]. To derive accurate neutrino mass parameters and/or predict $0\nu\beta\beta$ decay lifetimes, one needs precise calculations of both NMEs and PSFs. The computation of the NMEs is a challenge in the theoretical study of the DBD since long time. There are several nuclear structure methods for these calculations, the most employed being proton-neutron Quasi Random Phase Approximation (pnQRPA) [3]-[7], Interacting Shell Model(ISM) [8]-[11], Interacting Boson Approximation (IBA) [12]-[14], Projected Hartree Fock Bogoliubov (PHFB) [15] and Energy Density Functional (EDM) method [16]. There are still large differences in the literature between the NME values computed with different methods and by different groups, and these have been largely discussed in the literature (see for example [2], [17]). On the other hand PSFs have been calculated since long time using different approximations [18]-[23], and they were considered until recently to be calculated with enough precision. However, they were recently recalculated within an improved approach by using exact electron Dirac wave functions (w.f.), taking into account the finite nuclear size and electron screening effects [24], and differences were found especially for heavier nuclei. We also recalculated by developing new routines for computing the relativistic (Dirac) electron w.f. with inclusion of the nuclear finite size and screening effects and, in addition, with the use of a Coulomb potential derived from a realistic proton density distribution in the daughter nucleus [25]-[26], and found similar differences as compared with older calculations. So, it worth to take into account these new calculations for extracting the neutrino mass parameter.

In this paper I present recent ShM calculations of NMEs and PSFs performed by our group. The NMEs for three isotopes, i.e. ^{48}Ca, ^{76}Ge and ^{82}Se, have been computed with a Shell Model (ShM) code developed recently by our group (distinct from the Madrid-Strasbourg code) [27]-[28]. The values of the PSFs computed with the code described in Refs. [25]-[26] but with an improved numerical precision. Then, I present new up-date limits of the Majorana neutrino mass parameter derived from the analysis of $0\nu\beta\beta$ decay of nine isotopes.

2. Formalism

The lifetimes for $0\nu\beta\beta$ decay can be expressed as [1]:

$$\left(T_{1/2}^{0\nu}\right)^{-1} = G^{0\nu}(Q_{\beta\beta}, Z) g_A^4 \mid M^{0\nu} \mid^2 \left(\langle m_\nu \rangle / m_e\right)^2 \ , \tag{1}$$

where $G^{0\nu}$, expressed in units of $[yr]^{-1}$, are the PSFs for this decay mode, $Q_{\beta\beta}$ is the energy decay, Z is the nuclear charge, m_e is the electron mass and $\langle m_\nu \rangle$ is the Majorana light neutrino mass parameter, which can be expressed as a linear combination of the light neutrino masses and the elements from the first row of the PMNS (Pontecorvo-Maki-Nakagawa-Sakata) neutrino matrix [29]. $M^{0\nu}$ are the NMEs which depend on the nuclear structure of the parent and daughter nuclei:

$$M_\alpha^{0\nu} = \sum_{m,n} \left\langle 0_f^+ \| \tau_{-m} \tau_{-n} O_{mn}^\alpha \| 0_i^+ \right\rangle \ , \tag{2}$$

where $\alpha = GT, F, T$ are the contributions associated with the Gamow-Teller (GT), Fermi(F) and Tensor(T) two-body transition operators O_{mn}^α, and the summation is performed over all the nucleon states. The computation of the reduced matrix elements of the operators O^α can be decomposed into products of reduced matrix elements within the spin and relative coordinate spaces. Their explicit expressions are [1], [11]:

$$O_{12}^{GT} = \sigma_1 \cdot \sigma_2 H(r) \ , \ O_{12}^F = H(r) \ , \ O_{12}^T = \sqrt{2/3} \left[\sigma_1 \times \sigma_2\right]^2 \cdot (r/R) H(r) C^{(2)}(\hat{r}) \tag{3}$$

The most difficult is the computation of the radial part of these two-body transition operators, which contains the neutrino potentials, defined by integrals of momentum carried by the virtual neutrino exchanged between the two nucleons [6]:

$$H_\alpha(r) = \frac{2R}{\pi}\int_0^\infty j_i(qr)\frac{h_\alpha(q)}{\omega}\frac{1}{\omega+\langle E\rangle}q^2dq \equiv \int_0^\infty j_i(qr)V_\alpha(q)q^2dq \ , \tag{4}$$

where $R = r_0A^{1/3}$ fm ($r_0 = 1.2fm$), $\omega = \sqrt{q^2+m_\nu^2}$ is the neutrino energy and $j_i(qr)$ is the spherical Bessel function (i = 0, 0 and 2 for GT, F, and T, respectively). Usually, in calculations one uses the closure approximation which consists of the replacement of the excitation energies of the states in the intermediate odd-odd nucleus contributing to the decay, by an average expression $\langle E\rangle$. The detailed expressions of $h_\alpha(\alpha = GT, F, T)$ can be found, for example, in refs. [1], [6], [11]. They contain nuclear ingredients such as the finite nucleon size (FNS) and short range correlations effects, as well as inclusion of higher order terms in the nuclear currents (HOC), which are important for an accurate computation of the NMEs. FNS effect is taken into account through nucleon form factors, G_V and G_A, which depend on the momentum:

$$G_A\left(q^2\right) = g_A\left(\Lambda_A^2/(\Lambda_A^2+q^2)\right)^2, \ G_V\left(q^2\right) = g_V\left(\Lambda_V^2/(\Lambda_V^2+q^2)\right)^2 \tag{5}$$

In calculations either the quenched ($g_A = 1$) or unquenched ($g_A = 1.25 - 1.273$), values of the axial-vector constant have been used, while the values of the cut-off parameters are $\Lambda_V = 850MeV$ and $\Lambda_A = 1086MeV$ [1]. The SRC effects are included by correcting the single particle w. f.: $\psi_{nl}(r) \to [1+f(r)]\,\psi_{nl}(r)$. The correlation function $f(r)$ can be parametrized in different ways: the Jastrow prescription with the i) Miller-Spencer (MS) parametrization [30] and the CCM parametrizations, derived with realistic ii) CD-Bonn and iii) AV18 NN potentials [6]:

$$f(r) = -c\cdot e^{-ar^2}\left(1-br^2\right) \tag{6}$$

The three parametrizations mentioned above are associated with different sets of the a, b, c paramaters. We mention that another method, unitary operator method (UCOM) is also successfully used to include SRC effects [4]. The inclusion of HOC brings additional terms in the h_{GT} component and leads to the appearance of the h_T component in the expressions of the neutrino potentials, as it is described in detail in refs. [1], [31]. Besides the nuclear effects mentioned above, a number of parameters as g_A, r_0, (Λ_A, Λ_B) and $< E >$ are involved as well, in the NME calculations . The use of different values for these parameters may result in important differences in the calculated NME values. For example, the use of a quenched or an unquenched value for g_A is still an unsolved matter.

3. Numerical results and discussions

The largest uncertainties in the derivation of the Majorana neutrino mass parameter and/or prediction of $0\nu\beta\beta$ decay half-lives come from the NME computation. Their values depend on the method of calculation and on the different nuclear ingredients/parameters mentioned in the previous section. Fortunately, at present there is a consensus on the use of the nuclear ingredients/parameters [32] which helps to restrict the range of the different NME values from literature. For example, one recommends the inclusion in calculation of the HOC, FNS and the use of softer parametrizations like CCM [6] and UCOM [4] for SRCs. Concerning the input nuclear parameters, one recommends the use of an unquenched value for the axial vector constant (g_A = 1.25-1.273, [41]) and a value of $r_0 = 1.2fm$ for the nuclear radius constant. Also, one can mention that the results are less sensitive to the changes (within a few MeV)

Table 1. The NMEs obtained with different methods and with softer SRC parametrizations, specified in the second column. In calculations an unquenched value of g_A was used.

Method	SRC	^{48}Ca	^{76}Ge	^{82}Se	^{96}Zr	^{100}Mo	^{116}Cd	^{130}Te	^{136}Xe	^{150}Nd
[28]ShM	CD-Bonn	0.81	3.13	2.88					2.21[34]	
[10]ShM	UCOM	0.85	2.81-3.52	2.64				2.65	2.19	
[14]IBM-2	CD-Bonn	2.38	6.16	4.99	3.00	4.50	3.29	4.61	3.79	2.88
[17]QRPA	CD-Bonn		5.93(3.27)	5.30(4.54)	2.19	4.67	3.72	4.80	3.00	3.16[35]
[40]QRPA	UCOM		5.36(4.11)	3.72	3.12	3.93	4.79	4.22	2.80	
[16]GCM	CD-Bonn	2.37	4.60	4.22	5.65	5.08	4.72	5.13	4.20	1.71
[15]PHFB	CD-Bonn				2.98	6.07		3.98		2.68

Table 2. Majorana neutrino mass parameters together with the other components of the $0\nu\beta\beta$ decay halftimes: the $Q_{\beta\beta}$ values, the experimental half-life limits, the PSFs and the nuclear matrix elements.

Nucleus	$Q_{\beta\beta}[MeV]$	$T^{0\nu\beta\beta}_{exp}[yr]$	$G^{0\nu\beta\beta}[yr^{-1}]$	$M^{0\nu\beta\beta}$	$\langle m_\nu \rangle\ [eV]$
^{48}Ca	4.272	$> 5.8\ 10^{22}$[42]	2.46E-14	0.81-0.90	$< [9.6 - 10.69]$
^{76}Ge	2.039	$> 2.1\ 10^{25}$[46]	2.37E-15	2.81-6.16	$< [0.24 - 0.52]$
^{82}Se	2.995	$> 3.6\ 10^{23}$[43]	1.01E-14	2.64-4.99	$< [1.09 - 2.05]$
^{96}Zr	3.350	$> 9.2\ 10^{21}$[44]	2.05E-14	2.19-5.65	$< [4.21 - 10.9]$
^{100}Mo	3.034	$> 1.1\ 10^{24}$[43]	1.57E-14	3.93-6.07	$< [0.41 - 0.63]$
^{116}Cd	2.814	$> 1.7\ 10^{23}$[51]	1.66E-14	3.29-4.79	$< [1.28 - 1.87]$
^{130}Te	2.527	$> 2.8\ 10^{24}$[50]	1.41E-14	2.65-5.13	$< [0.32 - 0.62]$
^{136}Xe	2.458	$> 1.6\ 10^{25}$[47]	1.45E-14	2.19-4.20	$< [0.16 - 0.31]$
^{150}Nd	3.371	$> 1.8\ 10^{22}$[45]	6.19E-14	1.71-3.16	$< [3.10 - 5.73]$

of the $\langle E \rangle$) value, used in the closure approximation, and to (small) variations of the values of the cut-off parameters $\Lambda_{V,A}$. According to this consensus, we display in Table 1 the NME values obtained with different nuclear methods. The first row contains our results for ^{48}Ca, ^{76}Ge and ^{82}Se, performed with our code, described in more detail in [27], [28], while the NME values for the other nuclei are taken from different references, as it is specified. The NME values written in parenthesis for ^{76}Ge and ^{82}Se isotopes represent the values obtained with QRPA methods, when the s.p. energies were adjusted to the occupancy numbers reported in ref. [33]. One remarks, the NME values obtained with QRPA and ShM methods get closer when s.p. occupancies are adjusted to experiment. This is an important step in putting in agreement the ShM and QRPA results and should be verified for other isotopes as well. In Table 2 we present the results for the Majorana neutrino mass parameters ($\langle m_\nu \rangle$), derived from Eq. (3) together with the values of $Q_{\beta\beta}$, the PSFs ($G^{0\nu}$), NMEs ($M^{0\nu}$) and experimental half-lifes ($T^{0\nu}_{1/2}$) for nine isotopes for which data exist. The ranges of the NME values were taken from Table 1. According to the consensus concerning the use of different nuclear ingredients/parameters, we reduced the interval of their spread to about a factor of 2 (with one exception). This results in reducing the uncertainty in deriving the constraints on the Majorana light neutrino mass parameters, while taking into account NME values obtained with all the main nuclear methods on the market. The PSF values were obtained by recalculating them with our code, described in [25], but with an improved numerical precision obtained by enhancing the number of the interpolation points on a case-to-case basis, until the results become stationary. We remark that the obtained values are close to both those reported previously in refs. [24] and [25], for the most of isotopes, but differ from other previous calculations [20], [21], [23] by up to 28%. We mention that our method used for PSF calculations follows the method described in ref. [24], but we built up our own numerical routines. Our results confirm the PSF results reported in [24] obtained with a more rigorous method than those used previously [18]-[23]. Hence, we claim, it is justified the PSF re-calculation with improved methods and the use of the new values in the derivation the neutrino mass parameters. For the experimental half-lives we took the most recent results reported in literature. One observes that the stringent constraints are obtained from the ^{136}Xe isotope, followed by the ^{76}Ge one. This is due to both the experimental sensitivity of the experiments measuring these isotopes and to the reliability of the theoretical calculations of the corresponding PSF and NME quantities. One observes that the experiments measuring these isotopes are already exploring the quasi-degenerate scenarios for the neutrino mass hierarchy (which is around 0.5 eV).

4. Conclusions

We reported new calculations of the NMEs and PSFs for the $0\nu\beta\beta$ decay mode in the hypothesis that it occurs through the exchange of Majorana light neutrinos between the two nucleons inside the nucleus. The NMEs for three isotopes, ^{48}Ca, ^{76}Ge and ^{82}Se, were calculated with the code described in detail in [27], [28]. The PSFs were recalculated with an approach described in ref. [25] but with an improved numerical accuracy, with an increased number the number of interpolation points for each isoptope. We used exact electron w.f. obtained by solving a Dirac equation when finite nuclear size and screening effects are included and, in addition, a Coulomb potential derived from a realistic proton distribution in the daughter nucleus has been employed. Then, we derived new upper limits of the Majorana light neutrino parameters from the analysis of $0\nu\beta\beta$ decay of nine isotopes. For that, other NME values were selected from literature taking advance of the existed consensus in the community on the use of the nuclear ingredients involved in calculations, such as HOC, FHS and SRCs effects, and on the values of several nuclear input parameters. This allows us to restrict the range of spread of the NME values, reported in the literature, and to reduce the uncertainty in deriving constraints on the Majorana neutrino mass parameters. Our results may be useful to have an up-to-date image on the current neutrino mass sensitivities associated with $0\nu\beta\beta$ decay measurements for different isotopes, and to better estimate the range of the neutrino masses that can be explored in the future DBD experiments.

Acknowledgments

This work was done with the support of the MEN and UEFISCDI through the project IDEI-PCE-3-1318, contract Nr. 58/28.10/2011 and Project PN-09-37-01-02/2009.

References

[1] Vergados J, Ejiri H and Simkovic F 2012 *Rep. Prog. Phys.* **75**, 106301
[2] Vogel P 2012 *J. Phys. G: Nucl. Part. Phys.* **39** 124002
[3] Rodin V A, Faessler A, Simkovic F and Vogel P 2003 *Phys. Rev.* C **68** 044302; 2007 *Nucl. Phys.* A **793** 213
[4] Kortelainen M, Civitarese O, Suhonen J and Toivanen J 2007 *Phys. Lett.* B **647** 128;
Kortelainen M and Suhonen J 2007 *Phys. Rev.* C **75** 051303; *Phys. Rev.* C **76** 024315
[5] Simkovic F, Faessler A, Rodin V A, Vogel P and Engel J 2008 *Phys. Rev.* C **77** 045503
[6] Simkovic F, Faessler A, Muther H, Rodin V A and Stauf M 2009 *Phys. Rev.* C **79** 055501
[7] Stoica S and Klapdor-Kleingrothaus H V 2001 *Nucl. Phys.* A **694** 269
[8] Caurier E, Zuker A P, Poves A and Martinez-Pinedo G 1994 *Phys. Rev.* C **50** 225;
Retamosa J, Caurier E and Nowacki F 1995 *Phys. Rev.* C **51** 371
[9] Caurier E, Menendez J, Nowacki F and Poves A 2008 *Phys. Rev. Lett.* **100** 052503
[10] Menendez J, Poves A, Caurier E and Nowacki F. 2009 *Nucl. Phys.* A **818** 139
[11] Horoi M and Stoica S 2010 *Phys. Rev.* C **81** 024321
[12] Barea J and Iachello F 2009 *Phys. Rev.* C **79** 044301
[13] Barea J, Kotila J and Iachello F 2012 *Phys. Rev. Lett.* **109** 042501
[14] Barea J, Kotila J and Iachello F 2013 *Phys. Rev. C* **87** 014315

[15] Rath P K, Chandra R, Chaturvedi K, Lohani P, Raina P K and Hirsch J G 2013 *Phys. Rev.* C **88** 064322
[16] Rodriguez T R and Martinez-Pinedo G 2010 *Phys. Rev. Lett.* **105** 25250
[17] Faessler A, Rodin V and Simkovic F 2012 *J. Phys.* G **39** 124006
[18] Primakov H and Rosen S P 1959 *Rep. Prog. Phys.* **22** 121
[19] Haxton W C and Stephenson Jr G J 1984 *Prog. Theor. Part. Nucl. Phys.* **12** 409
[20] Doi M, Kotani T and Takasugi E 1985 *Prog. Theor. Phys. Suppl.* **83** 1
[21] Doi M and Kotani T 1992 *Prog. Theor. Phys.* **87**, 1207; ibidem 1993 **89** 139
[22] Tomoda T 1991 *Rep. Prog. Phys.* **54** 53
[23] Suhonen J and Civitarese O 1998 *Phys. Rep.* **300** 123
[24] Kotila J and Iachello F. 2012 *Phys. Rev.* C **85** 034316
[25] Stoica S and Mirea M 2013 *Phys. Rev.* C **88** 037303
[26] Pahomi T E, Neacsu A, Mirea M and Stoica S 2014 *Rom. Rep. Phys.* **66** 370
[27] Neacsu A, Stoica S and Horoi M 2012 *Phys. Rev.* C **86** 067304
[28] Neacsu A and Stoica S 2014 *J. Phys. G: Nucl. Part. Phys.* **41** 015201
[29] King S F and Luhn C 2013 *Rep. Prog. Phys.* **76** 056201
[30] Miller G A and Spencer J E 1976 *Ann. Phys.* **100** 562
[31] Simkovic F, Pantis G, Vergados J D and Faessler A 1999 *Phys. Rev.* C **60** 055502
[32] Giuliani A and Poves A 2012 *AHEP* **2012** 857016
[33] Schiffer J P et al. 2008 *Phys. Rev. Lett.* **100** 112501; B P Kay et al. 2009 *Phys. Rev.* C **79** 021301(R)
[34] Horoi M and Brown B A 2013 *Phys. Rev. Lett.* **110** 222502
[35] Fang D L, Faessler A and Rodin V A 2010 *Phys. Rev.* C **82** 051301(R)
[36] Menendez J, Poves A, Caurier E and F Nowacki 2009 *Phys. Rev.* C **80** 048501
[37] Caurier E, Poves A and Zuker A P 1990 *Phys. Lett.* B **252** 13
[38] Suhonen J and Civitarese O 2008 *Phys. Lett.* B **668** 277
[39] Simkovic F, Faessler A and Vogel P 2009 *Phys. Rev.* C **79** 015502
[40] Civitarese O and Suhonen J 2009 *J. Phys.: Conf. Series* **173** 012012
[41] Liu J et al. (UCNA Collaboration) 2010 *Phys. Rev. Lett.* **105** 181803
[42] Umehara S et al. 2008 *Phys. Rev.* C **78** 058501
[43] Barabash A S and Brudanin V B 2011 *Phys. At. Nucl.* **74** 312
[44] Argyriades J et al. (NEMO3 Collaboration) 2010 *Nucl. Phys.* A **847** 168
[45] Argyriades J et al. (NEMO3 Collaboration) 2009 *Phys. Rev.* C **80** 032501
[46] Macolino C et al. (GERDA Collaboration) 2014 *Mod. Phys. Lett.* A **29** 1430001
[47] Gando A et al. 2013 *Phys. Rev. Lett.* **110** 062502
[48] Wilkerson J F et al. (Majorana Collaboration) 2012 *J. Phys.: Conf. Series* **375** 042010
[49] Barabash A S 2010 *Phys. Rev.* C **81** 035501
[50] Andreotti E 2011 *Astropart. Phys.* **34** 822
[51] Danevich F A et al. 2003 *Phys. Rev.* C **68** 035501

Journal of Physics: Conference Series **580** (2015) 012032 doi:10.1088/1742-6596/580/1/012032

Shell-model study of spin modes in nuclei and nuclear forces

Toshio Suzuki[1], Takaharu Otsuka[2], Michio Honma[3] and Naofumi Tsunoda[4]

[1]Department of Physics, College of Humanities and Sciences, Nihon University, Sakurajosui 3-25-40, Setagaya-ku, Tokyo 156-8550, Japan
[2]Department of Physics and Center for Nuclear Study, the University of Tokyo, Hongo, Bunkyo-ku, Tokyo 113-0033, Japan
[3]Center for Mathematical Sciences, University of Aizu, Aizu-Wakamatsu, Fukushima 965-8580, Japan
[4]Department of Physics, the University of Tokyo, Hongo, Bunkyo-ku, Tokyo 113-0033, Japan

E-mail: suzuki@phys.chs.nihon-u.ac.jp

Abstract. Spin-dependent modes in nuclei are studied by shell-model method with the use of new shell-model Hamiltonians which properly take into account important roles of tensor interactions. New Hamiltonians can describe spin degrees of freedom in nuclei remarkably well. Nuclear weak processes at stellar environments are investigated based on these successes. New neutrino-nucleus reaction cross sections on ^{12}C are applied to light-element synthesis in supernova explosions. The production rate for ^{11}B/^{7}Li is pointed out to be useful to determine ν-oscillation parameters, in particular, ν-mass hierarchy. New e-capture rates in Ni isotopes are obtained and implications for element synthesis are discussed. The monopole-based universal interaction is applied to study structure of *p-sd* shell nuclei and ^{40}Ar as well as ν-induced reactions on ^{40}Ar. Repulsive corrections in the isospin T=1 monopoles are shown to be important for proper shell evolutions in neutron-rich carbon isotopes. The repulsive correction is pointed out to be due to three-body forces, in particular, the Fujita-Miyazawa force. Roles of the three-body forces on the shell evolution of neutron-rich calcium isotopes, the closed-shell nature of ^{48}Ca and M1 transition in ^{48}Ca are studied on top of the two-body G-matrix obtained by including core-polarization effects in larger spaces ($\leq 24\hbar\omega$). Effects of the inclusion of $g_{9/2}$-shell are also discussed.

1. Introduction

Spin modes in nuclei are studied by shell-model calculations, and important roles of tensor interaction and three-nucleon forces are discussed. Now, several new shell-model Hamiltonians with proper tensor components are available. They are found to be successful in describing proper shell evolutions and spin responses in nuclei. Gamow-Teller (GT) and magnetic dipole (M1) transitions as well as M1 moments in nuclei are well reproduced by the new Hamiltonians.

New shell-model Hamiltonians are applied to study nuclear weak processes in stars. In Sect. 2, neutrino-nucleus reaction cross sections and e-capture reaction rates at stellar environments are evaluated with the new Hamiltonians, and nucleosynthesis in supernova explosions are discussed.

In Sect. 3, the need for repulsive corrections in isospin T=1 monopoles is pointed out, and this repulsion is shown to be attributed to three-nucleon forces. Roles of three-nucleon forces in neutron-rich isotopes are discussed.

2. Spin responses in nuclei and tensor forces

2.1. New shell-model Hamiltonians and roles of tensor interaction

The Hamiltonian for p-shell, SFO[1], can reproduce GT strengths in ^{12}C and ^{14}C, and M1 moments of p-shell nuclei very well. The GXPF1J[2] for fp-shell can reproduce GT strength in Ni isotopes and M1 strength in ^{48}Ca, ^{50}Ti, ^{52}Cr and ^{54}Fe. Recently, monopole-based universal interaction, VMU[3], with tensor force of $\pi+\rho$-meson exchanges is proposed, and it has been found to be successful in description of structure of p-sd shell[4] and sd-fp shell nuclei[5].

The new interactions have a common feature that they have proper tensor components, that is, the monopole terms of the two-body matrix elements have a characteristic orbit dependence. The monopoles for $j_>$-$j_<$ ($j_> = \ell+1/2$, $j_< = \ell$ -1/2) orbits are more attractive than those for $j_>$-$j_>$ or $j_<$-$j_<$ orbits. This feature comes from the general sign rule for the tensor interaction[6], and is also seen in microscopic G-matrices, whose tensor components are constructed from $\pi+\rho$-meson exchanges. Proper shell evolution and change of magic numbers toward drip-lines are realized by this nature of the tensor interaction[3, 6].

2.2. Nuclear weak processes and nucleosynthesis

Spin dependent transition strengths evaluated with new shell-model Hamiltonians are applied to nuclear weak processes such as ν-nucleus and e-capture reactions, and nucleosynthesis in supernova explosions.

The SFO Hamiltonian, which can reproduce GT strength in ^{12}C, is applied to evaluate new ν-^{12}C reaction cross sections[7]. New cross sections for ^{12}C and also for ^{4}He are used to study light-element synthesis in core-collapse supernova explosions. Production yields of ^{11}B and ^{7}Li are found to be enhanced compared with previous estimations[7, 8]. Effects of MSW matter oscillations in He/C layers on nucleosynthesis are also studied. The abundance ratio of ^{7}Li/^{11}B is shown to be sensitive to the ν-oscillation parameters, the mixing angle θ_{13} and ν-mass hierarchy[8, 9]. The ratio is enhanced for normal mass hierarchy at $\sin^2 2\theta_{13} > 0.02$. The mass hierarchy, therefore, can be determined from the ratio as $\sin^2 2\theta_{13} \sim 0.1$[10]. Inverted mass hierarchy is pointed out to be statistically more favored[11] based on recent measurement on supernova SiC X-grains of the Murchison meteorite[12].

Liquid argon is a powerful target for ν-detection. GT strength in ^{40}Ar and cross section for ^{40}Ar (ν, e^-) ^{40}K are evaluated by shell-model calculations with the use of VMU[3]. GXPF1J and SDPF-M[13] interactions are used for fp- and sd-shells, respectively, and VMU is adopted for sd-fp cross-shells together with the two-body spin-orbit interaction of M3Y[14]. This interaction is referred as SDPF-VMU.

Calculated GT strength in ^{40}Ar is found to be consistent with the experimental data obtained by (p, n) reactions[15] as shown in figure 1(a). The present GT strength and calculated charge-exchange reaction cross section for ^{8}B solar ν[5] are found to be enhanced compared with the previous calculation[16]. Contributions from multipoles other than GT (1^+) and isobaric-analog (0^+) transitions evaluated by RPA become important at ν energies, $E_\nu \geq 50$ MeV[5] (see figure 1(b)).

Electron capture rates of fp-shell nuclei at stellar environments are evaluated with GXPF1J[17]. GT strengths in Ni isotopes obtained by shell-model calculations with GXPF1J are generally more fragmented than those by KB3G[18]. In particular, the GT strength in ^{56}Ni obtained with GXPF1J has two peaks while those for conventional Hamiltonians such as KB3G and KBF have only one peak as shown in figure 2(a). Recent (p, n) reaction experiment confirmed the two-peak structure of the GT strength[19]. E-capture rates at high densities and high temperatures are obtained by using the GT strengths of GXPF1J. The capture rates

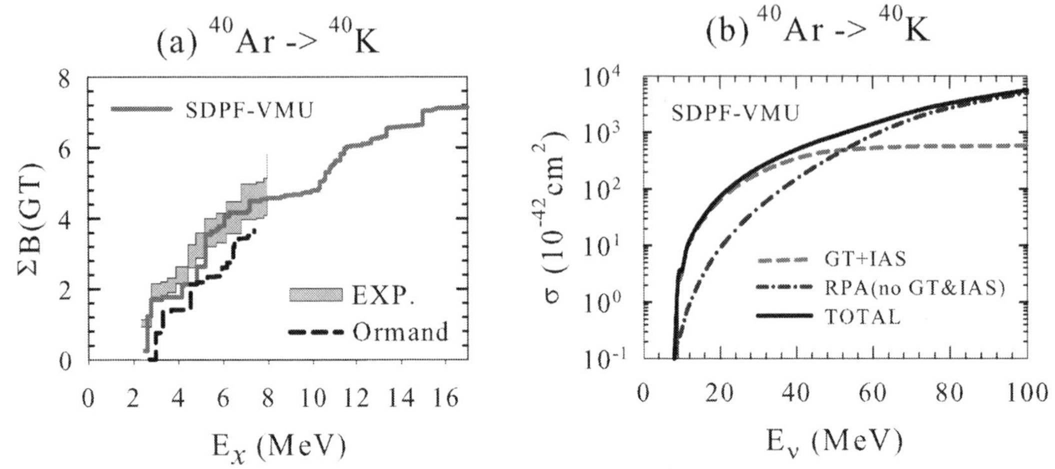

Figure 1. (a) Cumulative sum of $B(GT)$ in ^{40}Ar obtained by the present interaction (SDPF-VMU) and in Ref. [16]. Experimental data are taken from Ref. [15]. (b) Calculated cross sections for ^{40}Ar (ν, e^-) ^{40}K. GT contributions are obtained by shell-model calculations with SDPF-VMU while RPA is used for multipoles other than 0^+ and 1^+. Figures taken from Ref. [5].

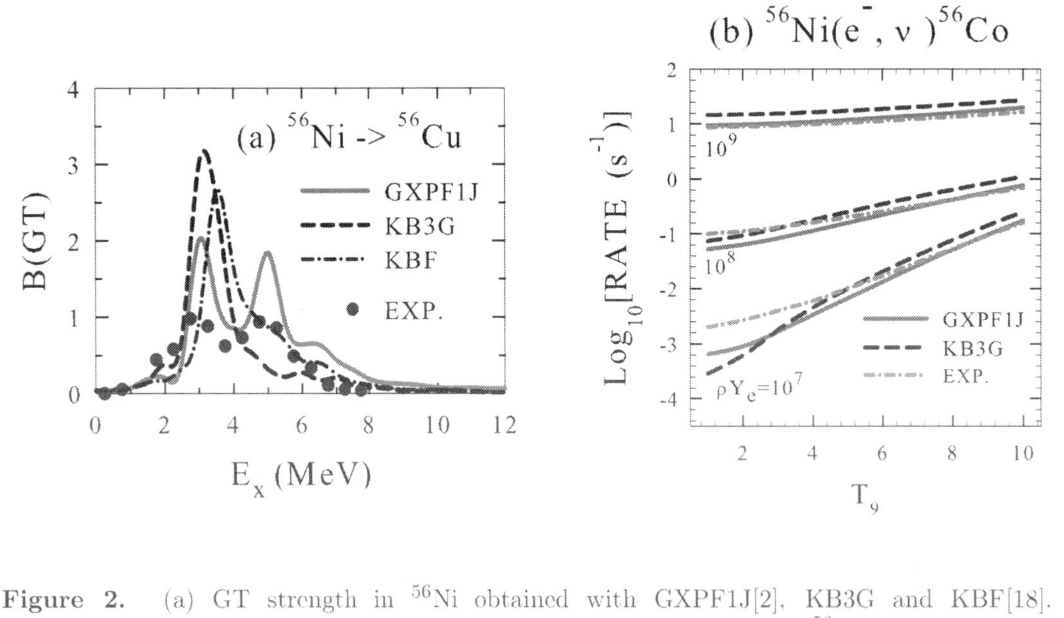

Figure 2. (a) GT strength in ^{56}Ni obtained with GXPF1J[2], KB3G and KBF[18]. Experimental data are taken from Ref. [19]. (b) E-capture rates for ^{56}Ni at densities ρY_e = 10^7, 10^8 and 10^9 g/cm^3 and at temperatures T_9 = 1∼10 (T = $T_9 \times 10^9$ K) obtained with GXPF1J and KB3G as well as with experimental data of Ref. [19] Figure taken from Ref. [17].

become smaller than those with KB3G especially at higher densities as the strengths by GXPF1J are spread over higher excitation energy region (see figure 2(b)).

Accretion of matter to white-dwarfs from their binary stars ignites type-Ia supernova explosions when the white-dwarf mass exceeds the Chandrasekhar limit. An amount of ^{56}Ni is produced in type-Ia supernova explosions. As e-capture process on ^{56}Ni proceeds, neutron-rich nuclei are produced and the lepton-to-baryon ratio (or proton fraction) Y_e gets smaller. If e-capture rates on ^{56}Ni are smaller, production yields of neutron-rich isotopes such as ^{58}Ni decrease and Y_e remains to be a higher value in statistical equilibrium calculations. The problem of over-production of ^{58}Ni, ^{54}Cr and ^{54}Fe compared to the solar abundance[20] could be solved by using smaller e-capture rates of GXPF1J. This problem is now under investigation.

3. Three-nucleon forces and structure of neutron-rich isotopes

3.1. Carbon isotopes

Comparing microscopic G-matrices with good phenomenological shell-model interactions, one notices a general feature: monopole terms of phenomenological interactions are more repulsive (attractive) than those of microscopic ones in isospin T=1 (T=0) channels. The repulsive corrections to the monopoles in T=1 channels are important to reproduce proper shell evolutions in neutron-rich isotopes.

The SFO, whose *sd*-shell part is G-matrix elements, is modified to take into account this repulsive correction in T=1 monopoles in the *sd*-shell matrix elements. Tensor components of *p*-*sd* cross-shell part are also replaced by those of $\pi+\rho$-meson exchanges. The modified version of the interaction, referred as SFO-tls[21], is applied to study structure of neutron-rich carbon isotopes. Due to the repulsive correction in the $0d_{5/2}$-$1s_{1/2}$ (T=1) monopole, effective single-particle energy (ESPE) of $1s_{1/2}$ orbit increases for A= 14∼20 and crosses with the $0d_{5/2}$ orbit at A=16, resulting in lower ESPE for $0d_{5/2}$ than $1s_{1/2}$ at A > 16. The ground state energies for SFO-tls increase after A=22, and ^{22}C turns out to be the drip-line nucleus of carbon isotopes, consisitent with the observation (see figure 3(a)).

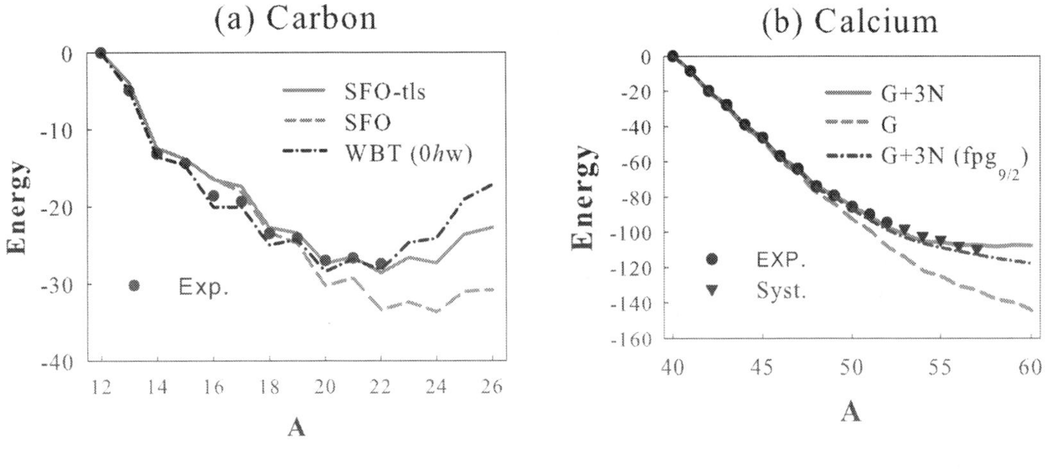

Figure 3. (a) Ground state energies of carbon isotopes obtained by shell-model calculations with SFO-tls, SFO and WBT[24]. (b) Ground state energies of calcium isotopes obtained by shell-model calculations with 2N (G-matrix) and 2N+3N (FM) interactions.

The SFO-tls can explain an anomalous M1 strength in ^{17}C; the B(M1) for $1/2^+ \to 3/2^+_{g.s.}$ is considerably suppressed[22]. Restricting to $d_{5/2}1s_{1/s}$-space, B(M1; $1/2^+ \to 3/2^+$) is shown to vanish exactly[21]. One should note that ESPE of $0d_{5/2}$ and $1s_{1/s}$ orbits are nearly degenerate at A=17. The anomalous suppression of the $B(M1)$ strength in ^{17}C is well explained by SFO-tls[21]. Recently, similar suppression of B(M1) is found to occur in ^{19}C[23].

3.2. Calcium isotopes
The origin of the repulsive correction in T=1 monopoles can be attributed to three-nucleon forces. In particular, the three-nucleon force induced by excitations of Δ_{33}-isobars through two-pion exchanges (Fujita-Miyazawa force[25]) is pointed out to give rise to repulsive corrections to the nucleon-nucleon interaction between valence neutrons[26]. This mechanism is found to be successful to explain the drip-line of oxygen isotopes at A=24[26].

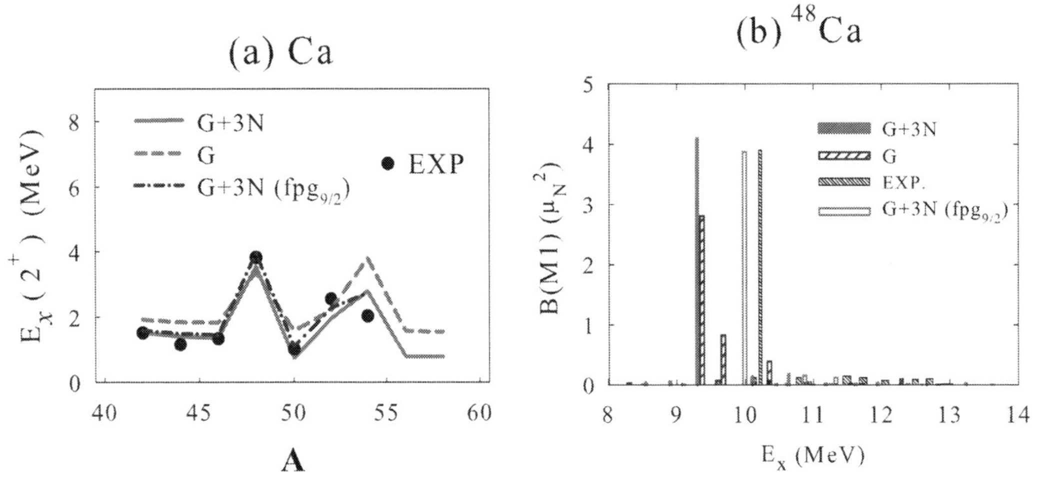

Figure 4. (a) Excitation energies of 2^+_1 states of calcium isotopes obtained with 2N and 2N+3N interactions. (b) $B(M1)$ strength in ^{48}Ca obtained with 2N and 2N+3N interactions. Experimental data are taken from Ref. [30].

Here, we study the effects of the three-nucleon forces in calcium isotopes[28]. G-matrix elements are used for the two-nucleon (2N) interaction while the Fujita-Miyazawa (FM) force is used for the three-nucleon (3N) interaction. Here, core polarization contributions are calculated by including up to third-order Q-box[27], and intermediate states are taken up to $24\hbar\omega$ excitations.

The ground state energies and excitation energies of the 2^+_1 states, $E_x(2^+)$, in calcium isotopes are shown in figure 3(b) and figure 4(a), respectively. Results for fp-shell and also for $fpg_{9/2}$-shell are shown. In case of $fpg_{9/2}$-shell, $g_{9/2}$ orbit is assumed to be degenerate with fp-shell orbits in the G-matrix calculation.

The repulsive contributions from the 3N forces push up the ground state energies, and experimental values are well reproduced with the 3N forces. The $E_x(2^+)$ are also well described with the 3N forces for A =40~50, and a large value of $E_x(2^+)$ in ^{48}Ca close to the experimental value shows the closed-shell nature of ^{48}Ca. Calculated large values of $E_x(2^+)$ for ^{52}Ca and

^{54}Ca show new magic numbers at N=32 and N=34[29]. The deviation between calculated and experimental values of $E_x(2^+)$ in ^{54}Ca obtained with the 2N forces is found to be reduced by inclusion of the 3N forces.

The experimental $B(M1)$ in ^{48}Ca is exhausted by a single state at E_x =10.23 MeV[30]. Calculated $B(M1)$ strength obtained with/without the 3N forces are shown in figure 4(b). The concentration of the $B(M1)$ strength to almost a single peak is reproduced with the 3N forces. Thus, 3N interaction is important to realize the magicity at N=28. In case of $fpg_{9/2}$-shell, the position of the peak comes closer to the experimental value.

Now, non-degenerate treatment of fp and $g_{9/2}$ shells by extended Kuo-Krenciglowa (EKK) method become possible[31]. Treatment of calcium isotopes in $fpg_{9/2}$-shell by the non-degenerate method is now under investigation. Monopole terms obtained with non-degenerate and degenerate methods are found to be significantly different[32], Nevertheless, main features found here, that is, the enhancement of $E_x(2^+)$ in ^{48}Ca, ^{52}Ca and ^{54}Ca and the concentration of $B(M1)$ in ^{48}Ca are common to both the methods.

This work has been supported in part by Grants-in-Aid for Scientific Research (C) 22540290 and (A) 20244022 of the MEXT of Japan, and also by JSPS Core-to-Core Program, International Research Network for Exotic Femto Systems (EFES).

[1] Suzuki T, Fujimoto R and Otsuka T 2003 *Phys. Rev.* C **67** 044302
[2] Honma M, Otsuka T, Mizusaki T, Hjorth-Jensen M and Brown B A 2005 *J. Phys.: Conf. Ser.* **20** 7
Honma M, Otsuka T, Brown B A and Mizusaki T 2002 *Phys. Rev.* C **65** 061301; 2004 ibid. **69** 034335
[3] Otsuka T, Suzuki T, Honma M, Utsuno Y, Tsunoda N, Tsukiyama K and Hjorth-Jensen M 2010 *Phys. Rev. Lett.* **104** 012501
[4] Yuan C, Suzuki T, Otsuka T, Xu F and Tsunoda N 2012 *Phys. Rev.* C **85** 064324
[5] Suzuki T and Honma M 2013 *Phys. Rev.* C **87** 014607
[6] Otsuka T, Suzuki T, Fujimoto R, Grawe H and Akaishi Y 2005 *Phys. Rev. Lett.* **95** 232502
[7] Suzuki T, Chiba S, Yoshida T, Kajino T and Otsuka T 2006 *Phys. Rev.* C **74** 034307
[8] Yoshida T, Suzuki T, Chiba S, Kajino T, Yokomakura H, Kimura K, Takamura A and Hartmann D H 2008 *Astrophys. J.* **686** 448
[9] Suzuki T and Kajino T 2013 *J. Phys.* G **40** 083101
[10] Abe K et al (T2K Collaboration) 2011 *Phys. Rev. Lett.* **107** 041801
An F et al (Daya Bay Collaboration) 2012 *Phys. Rev. Lett.* **108** 171803
[11] Mathews G J, Kajino T, Aoki W, Fujiya W and Pitts J B 2012 *Phys. Rev.* D **85** 105023
[12] Fujiya W, Hoppe P and Ott U 2011 *Astrophys. J.* **730** L7
[13] Utsuno Y, Otsuka T, Mizusaki T and Honma M 1999 *Phys. Rev.* C **60** 054315
[14] Bertsch G, Borysowicz J, McManus H and Love W G 1977 *Nucl. Phys.* **A284** 399
[15] Bhattacharya M, Goodman C D and Garcia A 2009 *Phys. Rev.* C **80** 055501
[16] Ormand W E, Pizzochero P M, Bortignon P F and Broglia R A 1995 *Phys. Lett.* **B345** 343
[17] Suzuki T, Honma M, Mao H, Otsuka T and Kajino T 2011 *Phys. Rev.* C **83** 044619
[18] Poves A, Sanchez-Solano J, Caurier E and Nowacki F 2001 *Nucl. Phys.* **A694** 157
Caurier E, Martinez-Pinedo G, Nowacki G F, Poves A and Zuker A P 2005 *Rev. Mod. Phys.* **77** 427
[19] Sasano M et al 2012 *Phys. Rev. Lett.* **107** 202501
[20] Iwamoto K et al 1999 *Astrophys. J. Suppl.* **125** 439
[21] Suzuki T and Otsuka T 2008 *Phy. Rev.* C **78** 061302
[22] Suzuki D et al 2008 *Phys. Lett.* **B666** 222
[23] Iwasaki H, 2014 private communication
[24] Warburton E K and Brown B A 1992 *Phys. Rev.* C **46** 923
[25] Fujita J and Miyazawa H 1957 *Prog. Theor. Phys.* **17** 360
[26] Otsuka T, Suzuki T, Holt J D, Schwenk A and Akaishi Y 2010 *Phys. Rev. Lett.* **105** 032501
[27] Hjorth-Jensen M, Kuo, T T S and Osnes E 1995 *Phys. Rep.* **386** 1
[28] Holt J D, Otsuka T, Schwenk A and Suzuki T 2012 *J. Phys.* G **39** 085111
[29] Steppenbeck D et al 2013 *Nature* **502** 207
[30] Steffen W et al 1983 *Nucl. Phys.* **A404** 413
[31] Takayanagi K 2011 *Nucl. Phys* **A852** 61; **A864** 91
[32] Tsunoda N, Takayanagi K, Hjorth-Jensen and Otsuka T 2014 *Phys. Rev.* C **89** 024313

Shell model studies for nuclear astrophysics

K. Langanke[1,2,3] and G. Martínez-Pinedo[2,1]

[1]GSI Helmholtzzentrum für Schwerionenforschung, Planckstr. 1, 64291 Darmstadt, Germany
[2]Institut für Kernphysik, TU Darmstadt, Schlossgartenstr. 9, 64291 Darmstadt, Germany
[3]Frankfurt Institute for Advanced Studies, Ruth-Moufang-Str. 1, 60438 Frankfurt, Germany

E-mail: k.langanke@gsi.de

Abstract. Shell model studies have contributed in recent years significantly to improve nuclear input required in simulations of the dynamics of astrophysical objects and their associated nucleosynthesis. This manuscript highlights a few examples like electron capture rates of importance for the evolution of core-collapse supernovae and the nucleosynthesis in thermonuclear supernovae, neutrino-nucleus cross sections with relevance to the supernova neutrino spectra and finally half lives of neutron-rich nuclei with magic neutron numbers which serve as waiting points in the mass flow of the astrophysical r-process.

1. Introduction

The advances of computer hardware, decisive progress in modelling and the development of powerful computer codes has strengthened the position of the interacting shell model as the method of choice to describe medium-mass nuclei [1, 2]. Such nuclei play often crucial roles for the dynamics of astrophysical objects and their associated nucleosynthesis. However, a direct experimental determination of the required input is often prohibited due to the extreme conditions of the astrophysical environment and hence the information has to be modelled. Here the shell model has become the method of choice due its ability to account for the relevant correlations among nucleons and to accurately reproduce low-energy spectra and electromagnetic transitions [3, 4].

This manuscript will highlight a few examples where shell model studies, exploiting the diagonalization and Shell Model Monte Carlo (SMMC) varieties of the approach, have contributed to a better understanding of astrophysical objects. Diagonalization shell model calculations have been performed to derive rates for weak interaction processes of nuclei up to the iron-nickel mass range [5, 6, 7]. In particular, the shell model rates for electron captures on nuclei have significant impact on the presupernova core evolution of massive stars [8, 9] and on the nucleosynthesis in thermonuclear supernovae [10]. For r-process nucleosynthesis shell model calculations of half lives for neutron-rich nuclei with magic neutron numbers (called waiting points) indicate that the r-process mass flow through these waiting points is faster than previously anticipated [11, 12, 13]. The SMMC approach allows to derive nuclear properties at finite temperature in extremely large model spaces taking the relevant nuclear correlations into account [14, 15]. The SMMC has been the base to derive electron capture rates for heavier neutron-rich nuclei for which cross-shell correlations are essential establishing capture on nuclei as the main weak interaction process for the dynamics of the core collapse of a massive star [16, 17, 18]. The SMMC also allows for the microscopic derivation of nuclear level densities

[19, 20] which is an important input for calculations of nuclear reaction rates in the framework of the statistical model. Such calculations are reviewed by Yoram Alhassid in this volume.

2. Electron captures in supernovae

A massive star ends its life in a supernova explosion triggered by the gravitational collapse of its inner core which is no longer supported by energy released in charged-particle reactions [21, 18]. Electron captures on nuclei have three important consequences during the collapse [22, 3]: i) they reduce the number of electrons and hence the pressure which the degenerate (relativistic) electron gas can stem against the gravitational contraction; ii) the neutrinos, generated by the capture process, leave the star mainly unhindered carrying away energy and keeping the entropy in the core low such that heavy nuclei survive during collapse, iii) electron capture changes a proton in the nucleus into a neutron driving the core composition to more neutron-rich (and heavier) nuclei.

At the stellar conditions, at which the core composition (described by Nuclear Statistical Equilibrium) is dominated by nuclei from the iron-nickel mass range (pf shell nuclei), the capture process is dominated by GT_+ transitions[1]. Thanks to advances in computer hardware and in nuclear modelling it has become possible to study the low-energy spectra and transitions of pf-shell nuclei in the respective model space [1]. In fact it turned out that, besides a constant renormalization of the Gamow-Teller operator [23, 24], such highly correlated wave functions are required to describe the strong fragmentation and total value of the GT_+ strength [25] as experimentally determined by charge-exchange experiments. The pioneering work to measure GT_+ strength functions has been performed at TRIUMF using (n,p) reactions, achieving, however, only modest energy resolutions [26]. This situation has been dramatically improved by the development of the (d,^{2}He) technique at the KVI Groningen which allowed to determine GT_+ distributions for many nuclei in the iron-nickel mass range with an energy resolution of about 150 keV [27].

We note that the strong energy dependence of phase space as well as the fact that the electron Fermi energy is of the same order as the Q-value of the abundant nuclei under presupernova conditions, makes a detailed and accurate description of the GT_+ distribution an important requirement for a reliable description of stellar electron capture during this phase of the collapse. That the diagonalization shell model is up to this task and indeed the method of choice to describe stellar weak-interaction rates during presupernova collapse has been recently demonstrated by Cole and collaborators [28]. These authors compared capture rates derived from experimental GT_+ data for all pf shell nuclei, for which data exist, with rates calculated within the shell model using two different residual interactions. As is shown in Fig. 1 the agreement is quite satisfactory at the conditions at which these nuclei are abundant and relevant for the core dynamics (right panel).

We note that in the stellar environment electron capture is reduced (approximately by a factor of order 2) due to screening effects [31]. As beta decays are enhanced by screening, stellar environment effects can alter the conditions at which URCA pairs operate. This is particularly important for the core collapse of $8-12 M_\odot$ stars, (electron capture supernovae) [32], where several URCA pairs influence the dynamics of the core evolution. Stars in this mass range develop degenerate ONe or ONeMg cores which are driven towards collapse by the loss of electron pressure support due to electron captures, mainly in the very abundant nuclear species ^{20}Ne and ^{24}Mg [32, 33]. Ref. [7] has evaluated the electron capture rates on ^{20}Ne, ^{20}F, ^{24}Mg, ^{24}Na and the β decay rates for ^{20}Na and ^{24}Na at temperature and density conditions relevant

[1] The subscript refers to the isospin component in the GT operator such that in GT_+ transitions a proton is changed into a neutron, in GT_- transitions, relevant for β decay of nuclei with neutron excess, a neutron is changed into a proton, and the GT_0 strength, important to describe low-energy inelastic neutrino-nucleus scattering, refers to transitions between proton states and neutron states, respectively

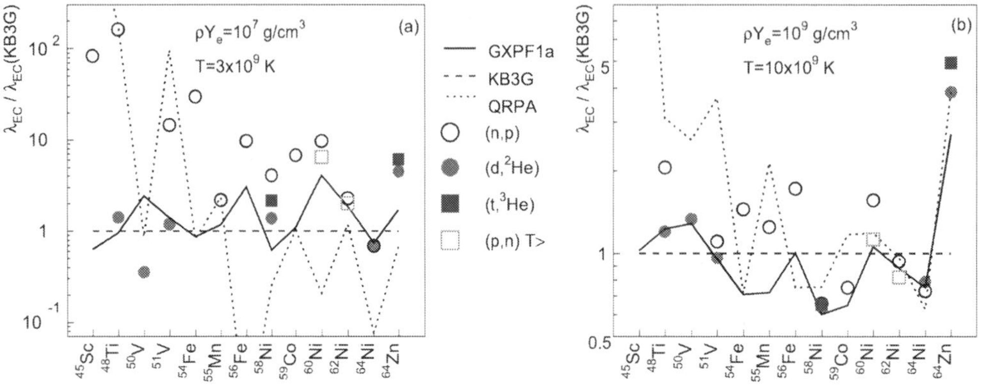

Figure 1. Comparison of electron captures rates as function of temperature calculated from experimental GT_+ data and distributions calculated within the large-scale shell model with two different interactions (KB3G [29] and GXPF1 [30]) and within the QRPA approach. The left and right panels correspond to conditions during the earlier stages of the collapse where the capture rates are sensitive to details of the GT_+ distribution. (from [28]).

for the late-evolution stages of stars with $M = 8 - 12M_\odot$, based on recent experimental data and large-scale shell model calculations.

In the later stage of the collapse of massive stars, the nuclei present in the core composition get heavier and more neutron-rich. The appropriate model space to describe electron capture for such nuclei is too large (requiring two major shells) to allow for shell model diagonalization calculations. The calculations are then based on a statistical shell model approach (Shell Model Monte Carlo [15]) which allows to determine nuclear properties at finite temperature and in large multi-shell model spaces taking the relevant nuclear correlations into account. Such correlations are particularly important for nuclei with proton number below and neutron number above an oscillator shell closure (like N=40). In such situations GT_+ transitions would be completely blocked by the Pauli principle in the Independent Particle Model (IPM) [34] suppressing electron capture on nuclei drastically. However, it has been shown in [16] that nuclear correlations induced by the residual interaction move nucleons across the shell gap enabling GT_+ transitions and making electron capture on nuclei the dominating weak interaction process during collapse [16, 46]. We add two remarks: The unblocking of the GT_+ strength across the $N = 40$ shell closure has been experimentally confirmed for ^{76}Se (with 34 protons and 42 neutrons) [35], in agreement with shell model studies [37]. Supplemented by transfer reaction experiments [36] the shell model study shows that the neutron occupation numbers in the $g_{9/2}$ orbital is about 6 (rather than 2 as in the IPM). Furthermore, shell model studies clearly show that the description of cross-shell correlations is a rather slowly converging process requiring the consideration of multi particle-multi hole configurations [38, 39, 37]. The diagonalization shell model electron capture rates agree reasonably well with those obtained by the SMMC+RPA approach at such astrophysical conditions at which nuclei with $Z < 40, N > 40$ dominate.

The modern shell model weak-interaction rates have significant impact on collapse simulations. In the presupernova phase ($\rho < 10^{10}$ g/cm^3) the captures proceed slower than assumed before and for a short period during silicon burning β-decays can compete [8, 9]. As a consequence, the core is cooler, more massive and less neutron rich before the final collapse. However, until recently simulations of this final collapse assumed that electron captures on nuclei are prohibited by the Pauli blocking mechanism, mentioned above (e.g. [21]). However, based on the SMMC calculations it has been shown in [16] that capture on nuclei dominates over capture on free protons. The changes compared to the previous simulations are significant [16, 17, 18]. Importantly the shock is now created at a smaller radius with more infalling material to traverse, but also the density, temperature and entropy profiles are strongly modified [17].

Finally we note that the shell model electron capture rates [6, 42], which are noticeably slower than the pioneering rates of Fuller *et al.* (FFN) for *pf* shell nuclei [40], have important consequences in nucleosynthesis studies for thermonuclear (type Ia) supernovae by less reducing the electron-to-nucleon ratio behing the burning front [10]. As a consequence very neutron-rich nuclei like ^{50}Ti and ^{54}Cr are significantly suppressed compared to calculations which use the FFN rates [41]. In fact, in calculations using the shell-model rates no nuclide is significantly overproduced compared to solar abundances [10].

3. Neutrino-nucleus scattering

The shell model has also been used to derive inelastic neutrino-nucleus cross sections for applications in supernova dynamics and nucleosynthesis. To validate this approach it has been exploited that the GT_0 strength, which is relevant for neutrino scattering on nuclei at supernova neutrino energies, is, in a very good approximation, proportional to the M1 strength of spherical nuclei [43]. Precision M1 data, obtained by inelastic electron scattering for such nuclei, are well reproduced by shell model calculations, which are also used to derive GT_0 distributions for excited nuclear states which can be thermally populated at finite supernova conditions [43, 44]. At higher neutrino energies forbidden transitions can contribute to the inelastic scattering cross section which have been derived by RPA calculations. Supernova simulations which incorporate inelastic neutrino-nucleus scattering indicate that this mode has a noticeable effect on the early neutrino spectra emitted from supernova [45]. Here nuclei act as obstacles to high-energy neutrinos which are down scattered in energy. This reduces significantly the tail of the neutrino spectra and hence also the predicted event rates for the observation of supernova neutrinos by earthbound detectors [45]. The simulations also show that inelastic neutrino-nucleus scattering contributes only slightly to the neutrino thermalization with matter during the collapse [18]. Recent studies also indicate that nuclear deexcitation by neutrino pair emission has no impact on the collapse dynamics [46]. It is, however, the dominating source of $\bar{\nu}_e$ and ν_μ, ν_τ neutrinos during collapse.

4. Half-lives of r-process waiting points

The r-process is the astrophysical origin of about half of the elements heavier than iron. It occurs in an astrophysical environment with extreme neutron densities [47], associated with neutron star mergers [48] and the neutrino-driven wind in a core-collapse supernova [49]. The r-process path in the nuclear chart runs through nuclei with such large neutron excesses that most of them have yet not been made in the laboratory and their properties have to be modelled. Of particular importance are the nuclei with magic neutron numbers. With their relatively long half-lives they act as obstacles in the r-process flow (so-called waiting points).

There have been a few crucial half-life measurements of waiting points with $N = 50$ and 82. These data turned out to be shorter than anticipated by global predictions based on QRPA models, likely pointing to the importance of correlations beyond such approaches. Such correlations have been considered in shell model studies performed for the waiting point nuclei. These studies, in some cases done before the measurement, achieved in general good agreement with the experimental half lives [11, 3]. Very recently the shell model calculations have been extended to include also contributions from forbidden transitions [12, 13] which turn out to reduce the half lives for the $N = 126$ waiting points significantly. This fact had already been noticed previously by Borzov [50]. By using recent half-life data for neutron-rich nuclei he also succeeded to improve the Fayan's density functional to achieve a rather accurate description of measured half lives [50] (see Fig. 2). Unfortunately Borzov's approach is currently limited

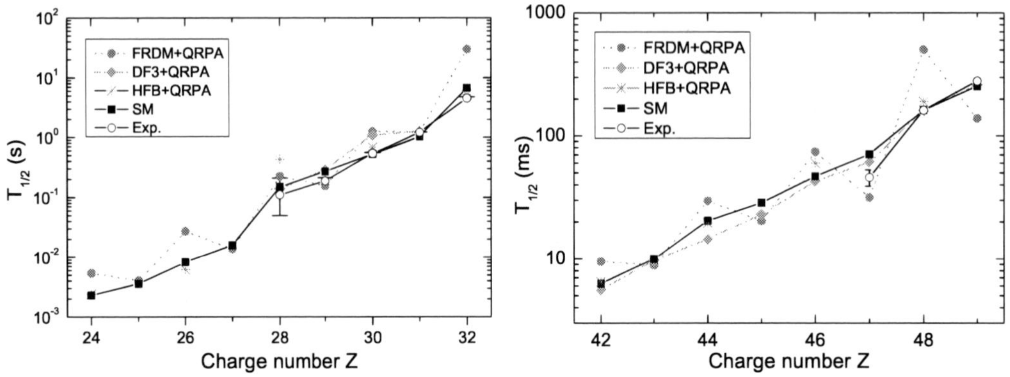

Figure 2. Comparison of shell model half lives for $N = 50$ (left) and $N = 82$ r-process waiting point nuclei with data and predictions from other models (from [13]).

to spherical nuclei, while the shell model studies are even more restricted to nuclei with magic neutron numbers. Nevertheless these calculations, together with the available data, indicate that the mass flow in the r-process to heavier nuclei is faster than anticipated from previous simulations. This can have interesting consequences for the competition of beta decays and neutron captures at freeze out [51]. In particular for r-process simulations within the neutron star merger scenario the faster half lives can lead to faster fission cycling influencing the supply of free neutrons after freeze out [52, 53]. Fission cycling is a characteristic of r-process simulations for neutron star mergers. Here the very extreme neutron densities support mass flow up to the mass range $A \sim 280 - 300$ where the nuclei decay by fission, producing nuclides around ^{132}Sn as fission fragments. Simulations thus have the interesting feature that the third peak is generated by neutron captures already before freeze-out, while the second peak at $A \sim 130$ and the lead peak develop as products of fission and alpha decays, respectively, at late stages of the process after freeze out. While the r-process abundance pattern depends somewhat on the fission yield distributions, it is relatively insensitive to the underlying neutron star models [48] and the adopted nuclear mass models [54]. Hence the neutron star merger scenario with its characteristic fission cycling might be able to describe the rather robust pattern of r-process abundances between the second and third peaks as it is observed in old stars [55].

Acknowledgements
The authors are grateful for a longterm collaboration with the members of the Strasbourg-Madrid shell model group: Etienne Caurier, Frederic Nowacki, Alfredo Poves, Kamila Sieja and Andres Zuker. Our work has also benefitted strongly from collaborations with many astrophysicists, most notably with Hans-Thomas Janka and Friedel Thielemann. This work was supported by the ExtreMe Matter Institute EMMI in the framework of the Helmholtz Alliance HA216/EMMI, the Deutsche Forschungsgemeinschaft through contract SFB 634, the Helmholtz International Center for FAIR within the framework of the LOEWE program launched by the state of Hesse, the Helmholtz Association through the Nuclear Astrophysics Virtual Institute (VH-VI-417).

References

[1] Caurier E, Martinez-Pinedo G, Nowacki F, Poves A , Zuker A P 2005, Rev. Mod. Phys. **77** 427
[2] Corragio L *et al.* 2009, Prog. Part. Nucl. Phys. **62** 135
[3] Langanke K, Martínez-Pinedo G 2003, Rev. Mod. Phys. **75** 819
[4] Grawe H, Langanke K and Martinez-Pinedo G 2007, Rep. Prog. Phys. **70** 1525

[5] Oda T, Hino M, Muto K, Takahara T and Sato K 1994, At. Data Nucl. Data Tables **56** 231
[6] Langanke K and Martinez-Pinedo G 2001, At. Data Nucl. Data Tables **79** 1
[7] Martínez-Pinedo G *et al.* 2014, Phys. Rev. **C89** 045806
[8] Heger A *et al.* 2001, Phys. Rev. Lett. **86** 1678
[9] Heger A *et al.* 2001, Astr. J. **560** 307
[10] Brachwitz F *et al.* 2000, Astrophys. J. **536** 934
[11] Martinez-Pinedo G and Langanke K 1999, Phys. Rev. Lett. **83** 4502
[12] Suzuki T *et al.* 2012, Phys. Rev. **C85** 015802
[13] Zhi Q *et al.* 2013, Phys. Rev. **C87** 025803
[14] Johnson C W, Koonin S E, Lang G H and Ormand W E 1992, Phys. Rev. Lett. **69** 3157
[15] Koonin S E, Dean D J and Langanke K 1997, Phys. Rep. **278** 2
[16] Langanke K *et al.* 2003, Phys. Rev. Lett. **90** 241102.
[17] Hix R W *et al.* 2003, Phys. Rev. Lett. **91** 210102
[18] Janka H.-Th. *et al.* 2007, Phys. Rep. **442** 38
[19] Nakada H and Alhassid Y 1997, Phys. Rev. Lett. **79** 2939
[20] Langanke K 1998, Phys. Lett. **438** 235
[21] Bethe H A 1990, Rev. Mod. Phys. **62**, 801
[22] Bethe H A, Brown G E, Applegate J and Lattimer J M 1979, Nucl. Phys. **A324** 487
[23] Langanke K *et al.* 1995, *Phys. Rev.* C **52** 718
[24] Martinez-Pinedo G *et al.* 1996, Phys. Rev. **C53** R2602
[25] Caurier E, Langanke K, Martínez-Pinedo G and Nowacki F 1999, *Nucl. Phys.* A **653** 439
[26] Vetterli M C *et al.* 1992, Phys. Rev. **C45** 997
[27] Frekers D 2006, Prog. Part. Nucl. Phys. **57**
[28] Cole A L *et al.* 2012, Phys. Rev. **86** 015809
[29] Poves A, Sánchez-Solano J, Caurier E , Nowacki F 2001, Nucl. Phys. A **694** 157.
[30] Honma M, Otsuka T, Brown B A and Mizusaki T 2004, Phys. Rev. **C69** 034335
[31] Juodagalvis A *et al* 2010, Nucl. Phys. **A848** 454
[32] Nomoto K 1987, Astrophys. J. **322** 206.
[33] Huedepohl L *et al.* 2010, Phys. Rev. Lett. **105** 249901
[34] Fuller G M 1982, Astr. J. **252** 741
[35] Grewe E W *et al.* 2008, Phys. Rev. **C78** 044301
[36] Kay B P *et al.* 2009, Phys. Rev. **C79** 021301
[37] Zhi Q *et al.* 2011, Nucl. Phys. A **A859** 172
[38] Dean D J *et al.* 1999, Phys. Rev. **C59** 2474
[39] Caurier E *et al.* 2001, Phys. Lett. **B522** 69
[40] Fuller G M, Fowler W A and Newman M J 1980, ApJS **42** 447, 1982 ApJS **48** 279
[41] Iwamoto K *et al.* 1999, Astrophys. J., Suppl. Ser. **125** 439
[42] Langanke K and Martínez-Pinedo G 2000, Nucl. Phys. **A673** 481
[43] Langanke K *et al.* 2004, Phys. Rev. Lett. **93** 202501
[44] Juodagalvis A *et al.* 2005, Nucl. Phys. **A747** 87
[45] Langanke K *et al.* 2008, Phys. Rev. Lett. **100** 011101
[46] Fischer T, Langanke K, Martínez-Pinedo G 2013, Phys. Rev. **C88** 065804.
[47] Thielemann F-K *et al.* 2011, Prog. Part. Nucl. Phys. **66** 346
[48] Korobkin O, Rosswog S, Arcones A and Winteler C 2012, Mon. Not. R. Astron. Soc. **426** 1940
[49] Woosley S E *et al.* 1994, Astr. J. **399** 229
[50] Borzov I N 2011, Phys. At. Nucl. **74** 1435
[51] Arcones A and Martinez-Pinedo G 2011, Phys. Rev. **83** 045809
[52] Eichler M *et al.*, submitted to Astr. and Astrophys.
[53] Caballero O L *et al.*, submitted to Phys. Rev. C
[54] de Jesus Mendoza-Temis J, Ph.D. thesis (Technical University Darmstadt, 2014)
[55] Sneden C, Cowan J J and Gallino R 2008, Ann. Rev. Astron. Astrophys. **46** 241

Journal of Physics: Conference Series **580** (2015) 012034 doi:10.1088/1742-6596/580/1/012034

The structure of proton rich nuclei in nuclear astrophysics

L .S. Ferreira

Centro de F´sica das Interacções Fundamentais, and Departamento de F´sica, Instituto Superior Técnico, Avenida Rovisco Pais, P1049-001 Lisbon, Portugal

Abstract. The properties of e otic proton rich nuclei are of great importance for nuclear astrophysics models. In the present work, we show how to address many nuclear structure properties of these nuclei at the e tremes of stability, from the analysis of proton radioactivity.

1. Introduction

Reaching the limits of bound nuclei is one of the major challenges for modern nuclear physics. Due to the appearance of new experimental facilities and technologies using radioactive ion beams, nuclei with an excess of protons or neutrons are being produced in the laboratory, mapping vast regions of the nuclear chart. The limits of stability for nucleon emission are established for proton rich nuclei almost up to charges equal to 83, and for neutron rich nuclei up to oxygen [1]. The development of new detection techniques and the use of storage rings, and Penning traps, allowed the measurement with high precision of masses, half-lives and branching ratios for decay, and even the energy spectrum in some cases. The availability of these data offered the possibility to test the quality of theoretical model predictions for fundamental properties of the nucleus at the extremes of isospin, where it becomes unstable against nucleon emission.

An overview of the main astrophysical processes shows that the vast majority of reactions encountered in the chains for the formation of the elements, involve quite unstable species. The various scenarios to explain the production of stable heavy nuclei, will go through combinations of (p,γ) and (n,γ) captures, or emission reactions (γ,n), (γ,p) and (γ, α) completed by β-decays and electron captures.

The path for the rapid proton capture processes might involve nuclei close to the proton drip-line[2], and in most cases, a stable p-nuclide will be synthesized through a quite unstable progenitor formed at the end stage of the evolution of massive stars, where there are enough seed nuclei, and high temperatures, for the reactions to take place.

Nuclear astrophysics models strongly depend on a very precise determination of the reaction rates and reaction energies. Since the nuclei involved are very unstable and short-lived, the cross sections for most reactions are unknown, but the energies can be obtained from nuclear masses. Particle-decay spectroscopy is for most of these nuclei the only way to determine their properties, like separation energies and identification of the decaying states. An interesting example[3] is provided by the indirect determination of the separation energy of a proton in ^{105}Sb, relevant for the rapid proton capture in the Sn-Sb-Te cycle at various Sn isotopes. It was achieved through the measurement of the Q-value for a very small α decay branch of the proton radioactive ^{109}I, and this new measurement, imposed strong constraints on the proton capture flow for the various isotopes.

For the calculation of the reaction rates, some of the nuclear ground state properties, like masses, and deformations, need to be known. Many reaction chains along the rp processes can go via direct or resonance capture of the proton, therefore, the properties of the excited states are also required. Moreover, the evaluation of the reverse (γ,p) reaction, is of key importance in the proton capture networks.

The previous discussion points out the strong link between astrophysical properties needed for explosive scenarios, and the properties of exotic proton rich nuclei, short lived and only known from their decay data. Nuclei at the proton drip line tend to be proton radioactive. The theoretical interpretation of this phenomena, can unveil relevant features of the structure of proton rich nuclei.

2. The spectra and shape of proton rich nuclei

The emission of a proton from the ground and excited states of exotic drip-line nuclei, with charges ranging from =50 up to =83 has been measured extensively in the last two decades [4]. The emission of two protons has been also observed in the decay of ^{45}Fe[5, 6], and afterwards for other lighter nuclei[7].

Theoretical models developed to explain this radioactive decay, provide consistent interpretation of the experimental data, helping to determine some nuclear properties. Realistic mean field models, can predict the nuclear shape parameters and quantum numbers of the decaying states of deformed nuclei[8, 9]. Fully selfconsistent calculations, using more fundamental interactions based on relativistic density functionals derived from meson exchange and point coupling models, were also able to account for the experimental data of proton radioactivity from spherical nuclei [10].

A model that reproduces the experimental half lives and branching ratios for decay to excited states, will also assign the spins and parities of the decaying states, and the nuclear shape parameters. Consequently, these decay modes can be used to probe the wave functions of the nuclear levels involved and to study also nuclear shell effects. The non-adiabatic-quasi-particle model, has been very successful in this context, and was able explain many aspects of the structure of odd-even nuclei, with an odd number of protons[11], as well as odd-odd nuclei[12], including the breaking of axial symmetry. We will discuss below a few cases which might be relevant in the nuclear astrophysics context.

It is di cult to determine experimentally the structure of very proton-rich nuclei due to the small production cross section. A quite recent experiment, provided very good data for the proton radioactive ^{151}Lu nucleus[13]. Using recoil proton tagging, the lifetime of the first excited state above the proton emitting ground state as well as higher-lying longer-lived states have been measured using γ ray coincidence techniques.

Theoretical calculations based on the non-adiabatic quasi-particle model, coupling the quasi-particles with the spectra of the daughter nucleus, suggest that the proton emitting ground state of ^{151}Lu is a h11 2^-, and has a mildly oblate quadrupole deformation of $\beta \approx$ - 0.11.The experimental value of the half-life for decay from the ground state is reproduced assuming this deformation, and is also consistent with the calculation of the life time of the first excited state, which decays by an electromagnetic γ-ray transition. These results, provide the best evidence to date for proton emission from an oblate nucleus.

The spectrum of ^{151}Lu is shown in Fig 1. The theoretical calculation of the energy levels were obtained by diagonalizing the interaction between the quasi-particle states and the spectrum of the daughter nucleus. For details see reference[13]. This example, is an interesting determination of shape parameters, and spectrum of drip-line nuclei, through the observation of proton decay.

Another important nucleus for astrophysical calculation is ^{18}Ne, since the breakout path from

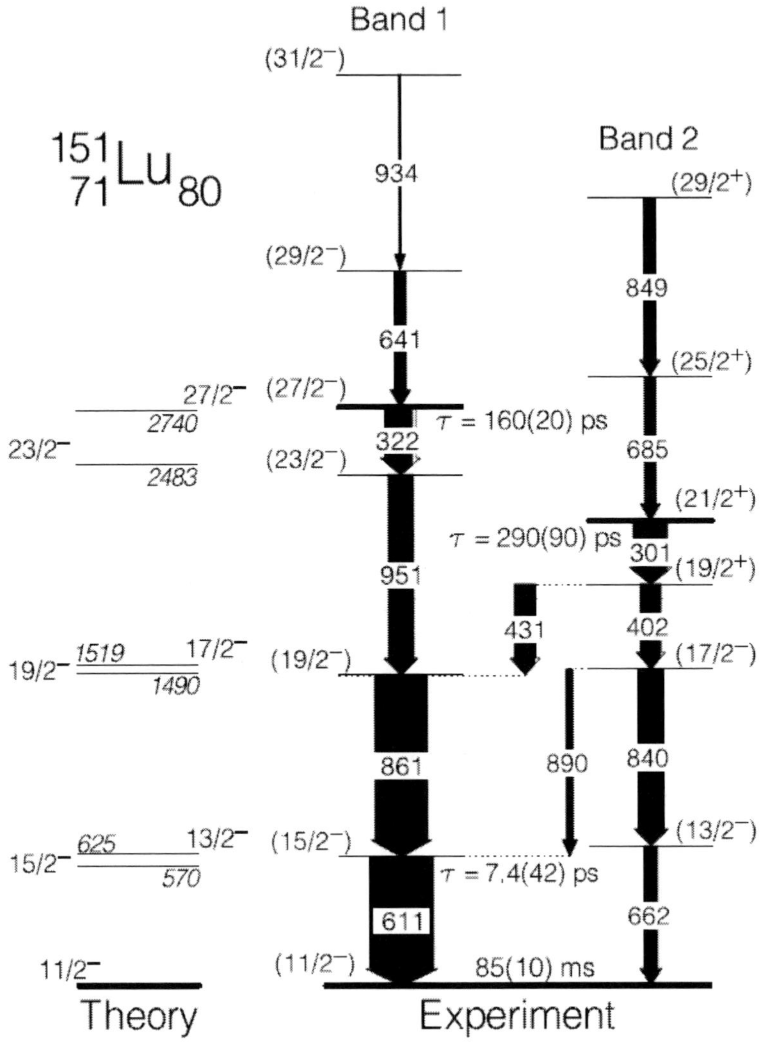

Figure 1. Theoretical and experimental excited states in ^{151}Lu from Ref[13].The widths of the arrows correspond to the intensities of each transition, with the white component of each arrow indicating the calculated internal conversion component.The life times of the states are also shown.

the hot CNO cycle involve sequences of alpha particle reactions that proceed through the ^{18}Ne resonances, above the alpha decay threshold. The determination of resonances in ^{18}Ne is then an important issue to obtain the reaction rates, but they are not easy to observe experimentally [14].

The reaction ^{17}F(p,γ)^{18}Ne is particularly important for the rp-process path, and it is just the inverse of proton decay from resonances in ^{18}Ne, which exhibits two proton radioactivity[15, 16]. The analysis of these experiments, beside finding the simultaneous two proton emission from a 1^- state at 6.15 MeV, that has been discussed together with the sequential decay through the ghost of the 1 2^+ in reference [17], has also shown at higher excitation energy a strong branch, $\approx$30%, for sequential emission of one proton after the other, going through excited states of ^{17}F

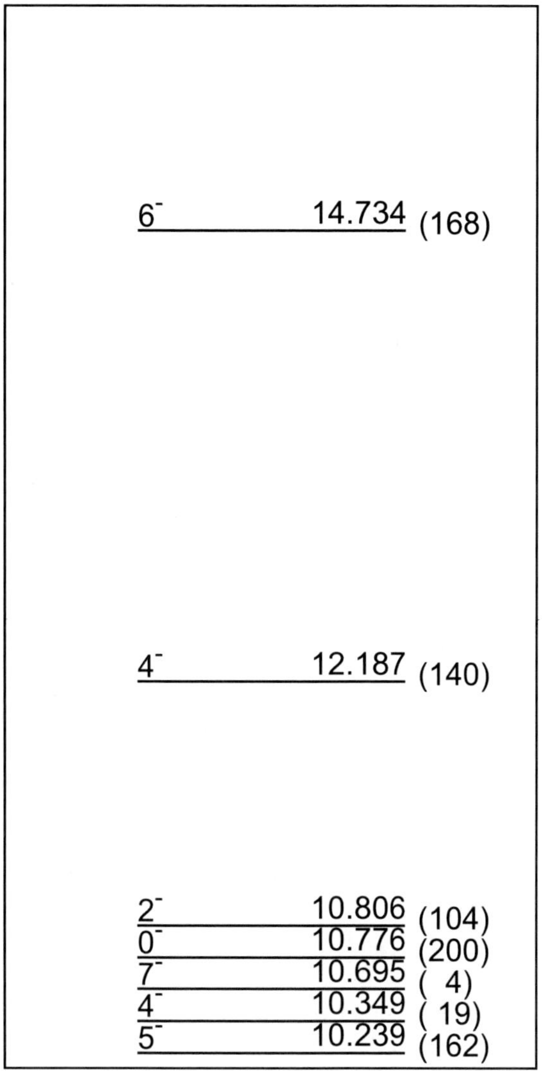

Figure 2. Calculated excited states in ^{18}Ne with a total width smaller than 200 keV and a branching ratio for decay to unbound states larger than 0.7. The spin and parity, and energy in MeV of the decaying level are given in the figure. The width of the resonances in KeV, are shown in parenthesis.

before reaching the final daughter nucleus ^{16}O in the ground state.

This is quite surprising, since simple barrier penetration considerations would favour a decay with the largest amount of energy being carried out by the first proton, thus to the ground state of ^{17}F that is bound, and consequently the second proton would not be emitted. A smaller percentage decays directly by simultaneous two proton emission. A theoretical analysis in terms of sequential decay, can lead to the identification of possible excited states, candidates for the emission of the second proton.

Since ^{18}Ne is a spherical nucleus, the half-life can be calculated according to scattering theory, from the knowledge of the proton state and the spectroscopic factor. The latter, can be determined from a standard shell model calculation with a realistic interaction. We have used the interaction of reference [18] which was fitted to the experimental excitation energies, and reproduces the experimental data for all sd nuclei.

The calculation shows that there are excited states of negative parity at quite high energies in ^{18}Ne, which are very narrow, and prefer to decay by one proton emission to the excited states than to the ground state of the daughter ^{17}F, thus becoming possible candidates for the emission of a second proton in a sequential two-proton decay process. Some were confirmed by the experimental studies of Raciti and collaborators[15]. The calculation of branching ratios for decay to unbound proton states in ^{17}F, shows meaningful values for negative parity states at quite high energy, and are presented in figure 2. This example shows the power of proton emission to identify the spectra of proton radioactive nuclei.

In the region of production of medium heavy isotopes, N= nuclei play an important role, since rp processes seem to follow the N= line up to the neighbourhood of the double shell closure N= =50 nucleus ^{100}Sn. The structure of these drip-line nuclei influences the rates at which the reactions can proceed, and the residual pairing proton neutron interaction can be quite influential.

Proton emission from odd-odd nuclei, can also be useful in this context, since the emission of one proton is quite dependent on the state of the unpaired neutron. The daughter nucleus in this case has an odd number of nucleons, and its angular momentum is determined by the Nilsson level occupied by the odd neutron. Different values of this angular momentum, will allow different values of the angular momentum of the escaping proton [19]. The theoretical interpretation of proton emission from odd-odd based on the adiabatic quasi-particle model, is also well established [12], and includes the proper treatment of residual pairing and neutron-proton interactions. It is possible to single out the angular momentum and parity of the ground state of these emitters and estimate the effect of the residual np interaction.

In conclusion, the theoretical interpretation of the experimental data on proton decay from proton rich nuclei, provides the possibility to predict the properties of the decaying state, the nuclear deformation, end the existence of isomeric states. This information, might be quite relevant for nuclear astrophysics studies.

Acknowledgments

This work was supported by the Fundação para a Ciência e a Tecnologia (Portugal), Project: CERN FP 123606 2011.

References

[1] Thoennessen M, Sherrill B M 2011 *Nature* **473** 25
[2] Arnould M, Goriely S 2003 *Phys. Rep.* **384** 184
[3] Mazzocchi C, *et al.* 2007 *Phys. Rev. Lett.* **98** 212501
[4] Sonzogni A 2002 *Nucl. Data Sheets* **95** 1
[5] Pfutzner M, et al. 2002 *Eur. Phys. J.* A **14** 279
[6] Giovinazzo J, *et al.* 2002 *Phys. Rev. Lett.* **89** 102501
[7] Blank B, Ploszajczak M 2008 *Rep. Prog. Phys.* **71** 046301
[8] Fiorin G, Maglione E and Ferreira L S 2003 *Phys. Rev.* C **67** 054302
[9] Arumugam P, Ferreira L S and Maglione E 2009 *Phys. Lett.* B **680** 443
[10] Ferreira L S, Maglione E and Ring P 2011 *Phys. Lett.* B **701** 508
[11] Arumugam P, Maglione E and Ferreira L S 2008 *Phys. Rev.* C **78** 041305
[12] Patial M, Arumugam P, Jain A K, Maglione E and Ferreira L S 2013 *Phys. Rev.* C **88** 054302
[13] Procter M G, *et al.* 2013 *Phys. Lett.* B **725** 7984
[14] Almaraz-Calderon S *et al.* 2012 *Phys. Rev.* C **86** 025801
[15] G. Raciti, *et al.* 2008 *Phys. Rev. Lett.* **100** 192503
[16] C. J. Lin, *et al.* 2011 *AIP Conf. Proc.* **98** 1409
[17] Brown B A, Barker F C , Millener D J 2002 *Phys. Rev.* C **65** 051309(R)
[18] M. Bouhelal *et al.* 2011 Nuc. Phys. **A** **864** 113
[19] Ferreira L S and Maglione E 2001 *Phys. Rev. Lett.* **86** 1721

Walking at the drip line

Angela Bonaccorso

Istituto Nazionale di Fisica Nucleare, Sez. di Pisa, Largo Pontecorvo 3, 56127 Pisa, Italy.

E-mail: bonac@df.unipi.it

Abstract. Among exotic nuclei those at the drip line which are unstable against neutron emission are particularly interesting because they convey information on the nuclear force in the most extreme situations. Strictly speaking they are not "nuclei" but they exist thanks to long living resonances between a neutron and a bound "core" nucleus. Adding one more neutron they become bound and are called "borromean". Being particularly exotic they have attracted much attention in past years, see for example Refs.[1, 2, 3]. One very challenging example is ^{13}Be whose level ordering has been discussed in a large number of papers in which it has been studied by transfer [4] and fragmentation experiments [5]-[11], or it has been discussed theoretically[12]-[19]. Although projectile fragmentation spectra show evident similarities, the interpretations of data all differ from each other. In this paper we argue that a way trough the problem could be to try to establish first, or at the same time, the quite elusive "nature" of the second s-state in the Beryllium isotopes with A=9-14. On the other hand there are other recent neutron removal experiments leading to nuclei unstable by one or more proton emissions [20], and thus somewhat mirror to borromean nuclei, performed with nuclei close to the proton drip line. It has been shown that by taking in coincidence all (charged) particles but the removed neutron, reconstructing the invariant mass and gating on the ground state peak, it is possible to obtain the longitudinal momentum distribution of the unbound "core". One can link it to the original wave function of the bound orbital and thus determine the initial neutron angular momentum from the shape of the distribution and the initial occupation probability from the absolute removal cross section. Then it is clear that modern experiments and theories are able to study unstable nuclei with the same degree of accuracy as stable nuclei. Such a line of research offers a great potential for numerous further studies beyond the drip line.

1. Introduction

Some time ago we thought we had established the level ordering of ^{13}Be and ^{14}Be on a firm basis using a time dependent projectile fragmentation model [3, 17], comparing to experimental data [9] and using structure inputs obtained from a semi-phenomenological core-vibration coupling model of two-neutron halo nuclei [18]. Thus from the shell ordering the neutron-core interaction was determined which is needed by any three body model of borromean nuclei. As a consistency test we showed that the energy spectra of some Beryllium isotopes obtained using pp-RPA predict the same shell ordering as extracted from the reaction model vs. data analysis [3, 17, 18].

Projectile fragmentation is used here to study a two neutron halo projectile ^{14}Be, as in Refs.[5]-[16]. The two-neutron emission is supposed to be due to the nuclear interaction with the target. The mechanism is such that one neutron is stripped by the target, taken away from the elastic channel and not measured in coincidence with the core. The second neutron however re-interacts with the core which, is going to be ^{12}Be in the case of the projectile fragmentation of ^{14}Be, since ^{13}Be is not bound. For a decaying s-state this mechanism has also been described by the sudden approximation in Ref.[19]. The sudden method leads to a transition probability independent of the impact parameter.

Another type of experiment consists in neutron removal leading to nuclei unstable by one or more proton emissions [20], performed with nuclei close to the proton drip line. It has been shown that by taking in coincidence all (charged) particles but the removed neutron, reconstructing the invariant mass and gating on the ground state peak, it is possible to obtain the longitudinal momentum distribution of the "unbound" core. One can link it to the original wave function of the bound orbital and thus determine the initial neutron angular momentum from the shape of the distribution and the initial occupation probability from the absolute removal cross section. We will show in the following some preliminary experimental spectra and the relative theoretical description. Comparison between experimental and theoretical cross sections can provide information on the single particle occupation in the initial state. These reactions will be described by the transfer-to-the-continuum model [21]-[23], using a recently developed n-^{9}Be optical potential [24] and a core-^{9}Be folding potential [25].

2. Projectile Fragmentation and Transfer-to-the-Continuum

We calculate breakup by a semiclassical approximation in which the dynamics is controlled by the three potentials describing nucleon-core, nucleon-target, and core-target interactions. The collision is described in terms of only the three-body variables of nucleon coordinate, projectile coordinate, and target coordinate. The projectile-target relative motion is treated semiclassically by using a trajectory of the center of the projectile relative to the center of the target. The formalism is then valid for incident energies above the Coulomb barrier. To first order such excitations can be described along this trajectory by a time dependent perturbation amplitude [21]-[26] for a transition from a nucleon state bound in the projectile, to a final continuum state. The radial part of the single particle initial state wave function is calculated in a potential $V_{WS}(r)$ which is fixed in space and chosen such as to reproduce the experimental valence particle separation energy. For the proton emitters nuclei of the second part of this paper our wave functions are also fitted to *ab initio* overlaps calculated with the Variational Montecarlo Method [28]. The role of the target is just to perturb the initial bound state wave function and to allow the transition to the continuum by transferring some momentum to the neutron. For the description of the neutron-core relative energy spectrum used to study ^{13}Be it is enough to choose a simplified form of the interaction, such as a delta-function potential $V(r) = v_2\delta(x)\delta(y)\delta(z)$. The value of the strength $v_2 \equiv$ [MeV fm^3] used in the calculation is taken equal to the volume integral of the appropriate neutron-target interaction. The coordinate system and all other details of the calculations can be found in Ref.[17]. On the other hand, the sudden approximation has been followed in Ref.[19] leading to a simplified formalism. Our formalism [17] reduces to that of [19] for a s-to-s transition in the sudden limit. However our approach is not restricted to transitions with conservation of the angular momentum.

For the inclusive transfer-to-the-continuum (TC) reactions which correspond to the measurement of the "core" longitudinal distribution a continuum final state in which the neutron interacts with the target via an optical potential [24] is taken and thus both elastic and inelastic breakup contribute. In the projectile fragmentation formalism instead only the elastic breakup contributes as neutron and core are taken in coincidence which means the neutron energy is not degraded by the interaction with the target.

3. Structure and Reactions of $^{13-14}$Be

The ground state of ^{14}Be has spin $J^\pi = 0^+$. In a simple model assuming two neutrons added to a ^{12}Be core in its ground state the wave function is: $|^{14}Be >= [b_1(2s_{1/2})^2 + b_2(1p_{1/2})^2 + b_3(1d_{5/2})^2] \otimes |^{12}Be, 0^+ >$. Then the bound neutron can be in a 2s, 1p$_{1/2}$ or 1d$_{5/2}$ state. However

Table 1. Scattering length of the 2s continuum state, energies and widths of the p- and d-resonances in ^{13}Be and corresponding strength parameters for the δV potential

	ε_{res} MeV	Γ_j MeV	a_s fm^{-1}	α MeV
$2s_{1/2}$			-0.8	8.0
$1p_{1/2}$	0.67	0.28		8.34
$1d_{5/2}$	2.0	0.40		-2.36

Table 2. Theoretical and experimental values of S_{2n} (MeV) in ^{12}Be and ^{14}Be, $\langle r^2\rangle^{1/2}$ (fm), $\langle\lambda^2\rangle^{1/2}$ (fm) and $\langle\rho^2\rangle^{1/2}$ (fm) in ^{14}Be in the two cases of non-inversion (first line) and inversion (second line) of the 2s and $1p_{1/2}$ shells.

	$\epsilon(1p_{1/2})$	$\epsilon(2s)$	$S_{2n}(^{12}Be)$	$S_{2n}(^{14}Be)$	$\langle r^2\rangle^{1/2}$	$\langle\lambda^2\rangle^{1/2}$	$\langle\rho^2\rangle^{1/2}$
Theory	-3.15	0.09	2.91	0.51	3.45	5.45	8.45
	0.48	-3.15	3.67	1.33	2.96	4.27	4.72
Exp.	-	-	3.67±0.01	1.34 ± 0.11	3.1±0.4	4.5±1.0	5.4±1.0

the situation is much more complicated [12]-[14] and in particular the calculations of Ref. [15] show that there is a large component $(2s_{1/2}, 1d_{5/2}) \otimes |^{12}Be, 2^+ >$ with the core in its low energy 2^+ state which can modify the neutron distribution.

To describe the valence neutron in the ^{13}Be continuum we assume that the single neutron hamiltonian with respect to ^{12}Be has the form $h = t + U$ where t is the kinetic energy and $U(r) = V_{WS} + \delta V$ is the real part of the neutron-core interaction. V_{WS} is a Woods-Saxon potential plus spin-orbit whose parameters are given in Ref.[17], and δV is a correction $\delta V(r) = 16\alpha e^{2(r-R)/a}/(1+e^{(r-R)/a})^4$ which originates from particle-vibration couplings [13] evaluated using Bohr and Mottelson collective model of the transition amplitudes between zero and one phonon states. Such couplings are relevant only for the low energy spectrum. Values for the parameter α are given in Table 1. Therefore our structure model is not a simple *single-particle in a potential* model but contains in it the full complexity of single-particle vs. collective couplings. If good fittings of experimental data will be obtained, then the parameters of a semi-phenomenological potential can be deduced and linked to a more microscopic model. Results obtained including the s, p and d initial states in ^{14}Be leading to unbound ^{13}Be states are shown in the LHS of Fig.1. Each curve corresponds to just one transition as indicated. The solid curve is the sum of all transitions from the s-bound state which gives the dominant contribution to the transition probability due to the long tail of the wave function. To make them visible some curves have been multiplied by a factor of five as indicated in the legenda. In the RHS figure we compare the s to s cross sections from our model with the results of the sudden formula used in [19]. Large differences are seen and the sudden formula seems to over predict the overall strength of the transitions. The s-threshold state in ^{13}Be has a very small scattering length and it is probably a complicated state coupled to an excited core. Finally to compare with available data, we show at the bottom of Fig. 1, the experimental points from H. Simon et al [9] for the reaction ^{14}Be+^{12}C $\rightarrow$ n+^{12}Be+X at 250 A.MeV and our calculations. More details in the legenda and in Refs. [17].

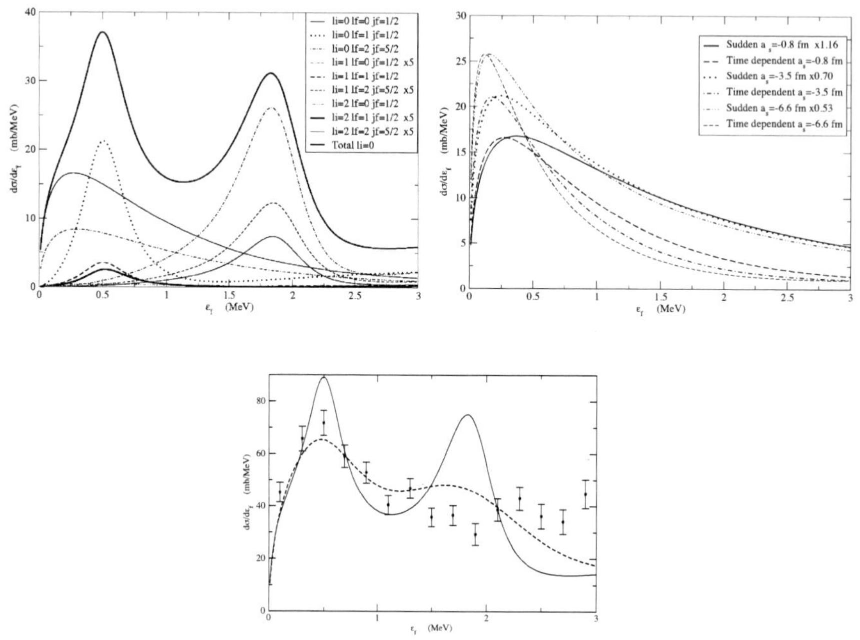

Figure 1. Top LHS: Individual transitions from the bound s,p,d components of the initial wave function to the unbound s,p,d states. Only the contribution from the initial s-orbital are summed to give the full curve. Top RHS: n-^{12}Be relative energy spectrum; comparison of fragmentation method for an s to s transition with results of the sudden formula using the same optical model phase shifts corresponding to scattering lengths as indicated. Bottom center: Sum of all transitions from the s initial state for the reaction [8] ^{14}Be+^{12}C $\rightarrow$ n+^{12}Be+X. Dashed line is the folding of the calculated spectrum with the experimental resolution curve.

On the other hand with the s and p single particle states given in Table 1, pp-RPA calculations have been performed [3] and [18]. Some of those results are reported here in Table 2 using two hypothesis for the level ordering. The best results are obtained for the s(bound)-p(unbound) inversion case, in particular notice the RPA amplitudes, given in Ref.[18] which are closely related to the spectroscopic factors and give a hint on the occupation probabilities of the orbitals. It is evident that the mixing of s-p-d states is very strong in both ^{12}Be and ^{14}Be and that these nuclei have a *dynamical* nature far from the traditional shell ordering and closure concepts. Furthermore, notice that ^{13}Be is the first nucleus in which a s-p inversion is shown to happen across threshold.

An alternative way of solving the problem of the level scheme of ^{13}Be would be to try to understand and establish consistently the "nature" and evolution of the second s-state in the Beryllium isotopes with A=9-14, see also Ref.[27]. As one can see from the level schemes of ^{9,10,11}Be [29] the position and contribution of the second s-state is either unknown or not well understood. In particular some recent efforts are illustrated in Fig. 2. We report: i) top LHS, a table from Ref.[30] with the parameters extracted for the possible $(1/2)^+$ resonant state in ^{9}Be from recent (γ,n) and (e,e$'$) experiments, where the authors tried to established whether the state is a single particle virtual state or a many-body resonance; ii) top RHS, the effective single particle energies for the 1s and 0d$_{5/2}$ states in the N=9 isotones presented in Ref. [32], in

TABLE I: R-matrix parameters ($\bar{E}_R$,Γ_R) extracted in recent experiments using real and virtual photons and the corresponding positions (E_{res}) and the widths (Γ) of the ^{9}Be $1/2^+$ resonant state.

Ref.	$\bar{E}_R$ (keV)	Γ_R (keV)	E_{res} (keV)	Γ (keV)
[2]	83(6)	274(8)	-31(6)	153(4)
[3]	70(3)	225(12)	-21	132
[4]	66(2)	213(6)	-21	124

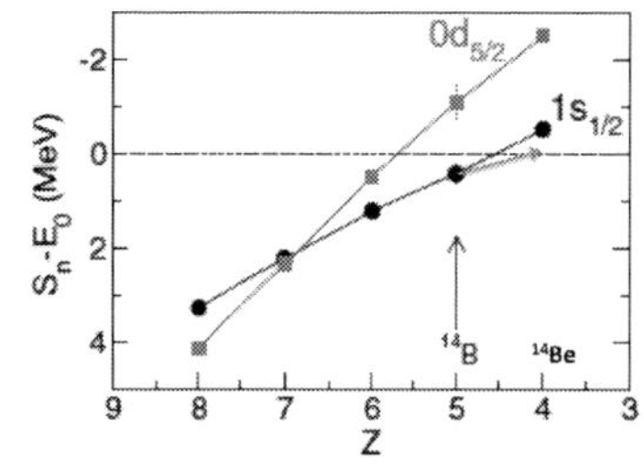

. (Color online) Effective single-particle binding energies of the $1s_{1/2}$ and $0d_{5/2}$ orbitals as a function of Z, for $N = 9$ isotones. The ^{14}B points are from the present measurement, and the lines are to guide the eye.

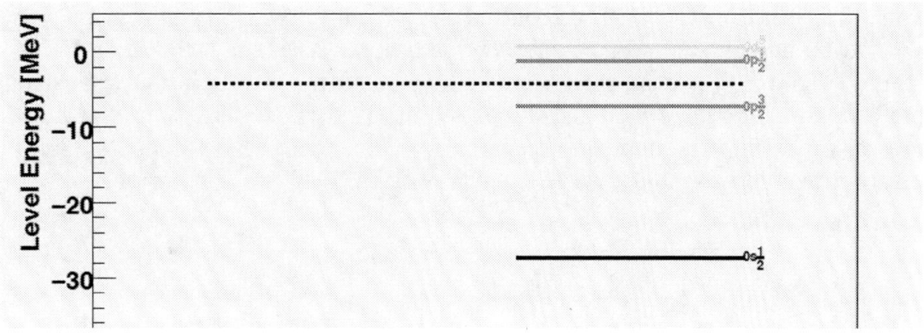

Figure 2. Top, from LHS to RHS: table from Ref.[30] with the parameters extracted for the possible $(1/2)^+$ resonant state in ^{9}Be from recent (γ,n) and (e,e′) experiments; effective single particle energies for the N=9 isotones [32], in blue our extrapolation for ^{13}Be. Bottom: bound states obtained by the dispersive optical model (DOM) for ^{9}Be [31].

blue our extrapolation for ^{13}Be; iii) at the bottom of the figure, the bound states obtained by the dispersive optical model (DOM) for ^{9}Be [31]; there is no hint of s-state in this case which suggests the possibility of a very wide resonance.

4. Proton emitters

As we mentioned in the Introduction we are now studying the possibility to obtain information on the shell structure and single particle properties of very light exotic nuclei which after one neutron knockout are unbound by one or more proton emissions. The relevant experiments have been performed by the HiRA collaboration at MSU and TEXAS AM laboratories. Details on the experiment will be given elsewhere [35]. Table 3 gives the relevant structure information on one of the cases under analysis. Such inputs were used in reaction calculations done with the TC model [21, 22, 23] and the optical potentials of Refs.[24, 25]. Fig.3 shows the invariant mass spectrum and parallel momentum distribution of the ^{9}Be(^{6}Li,^{5}Li)X reaction at 36.6A.MeV[20]. The calculated spectrum, obtained by summing both p-contributions, and convoluted with the experimental resolution function is shown by the green curve. It is in very good agreement with the data shown in black. In particular notice both the kinematical cut and the long tail at large

Table 3. Structure parameters for the indicated overlap from shell model [33] and VMC [34]. Given are the neutron (S_n) separation energy, the asymptotic normalization constants (ANC) and the spectroscopic factors (SF). VMC overlaps are normalized to the SF, shell model's to 1.

	S_n MeV	ANC_{WS} $fm^{-1/2}$	SF_{SM}	ANC_{VMC} $fm^{-1/2}$	SF_{VMC}
$\langle ^6Li \vert ^5Li \rangle$	5.66	2.85	$p_{1/2}$ 0.3301		$p_{1/2}$ 0.20463
		2.89	$p_{3/2}$ 0.3384		$p_{3/2}$ 0.30566

and small $P_{\parallel}$ respectively are perfectly reproduced by the reaction model. Also the calculated cross section for elastic and inelastic breakup is 38.5mb vs an experimental value of 38.1mb. The neutron wave function has been calculated in a Wood-Saxon plus spin orbit potential, fitting the separation energy such that the ANC would agree with the ANC of the VMC model. The core-target S-matrix has been parametrized by a smooth cutoff function with strong absorption radius R_s=5.8fm. We have excluded the possibility that the weakly bound proton (S_p=4.6MeV) would breakup before the neutron according to the model presented in Ref.[36], by calculating the proton breakup probability with the appropriate VMC wave function.

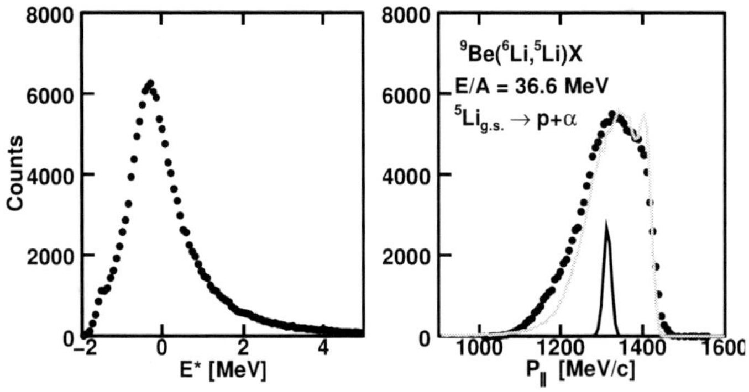

Figure 3. Invariant mass spectrum and parallel momentum distribution of the ^{9}Be(^{6}Li,^{5}Li)X reaction at 36.6A.MeV[20].

5. Conclusions and Outlook

From the structure point of view, in the search for the dripline position, a very important role is played by the study of nuclei unstable by neutron emission. Results of projectile fragmentation theory for elastic breakup have been presented together with structure calculations of $^{12-14}$Be. We have introduced also the study of neutron removal giving rise to an unbound core which emits protons in very light nuclei. They are somewhat mirror to borromean nuclei. For such nuclei one can use both the shell model and advanced *ab initio* model wave functions in reaction calculations and comparisons are in progress. This is one of the most important subjects of study within RIB physics which need to be addressed and developed further in the near future and for which some suggestions of possible structure and reaction formalisms applied to state-of-the-art experiments have been discussed.

Acknowledgements. I wish to acknowledge the collaboration with G. Blanchon, D. M. Brink, A. García-Camacho and the late N. Vinh Mau for the work presented in the first part of this contribution and with R. J. Charity and G. Salvioni for the work presented in the second part.

References

[1] Cho Adrian 2009 *Science Magazine*, **323** 1424, www.sciencemag.org
[2] Bonaccorso A 2013 *Phys. Scr.* T**152** 014019; Proc. Nobel Symposium NS152, Fahlander C, Jonson B eds.
[3] Blanchon G 2008 *Nuclear Reactions and Structure For Nuclei Far from Stability* Ph.D. thesis, University of Pisa, unpublished. http://www.infn.it/thesis/
[4] Korsheninnikov A A *et al* 1995 *Phys. Lett.* B **343** 53
[5] Marqués F M *et al* 2001 *Phys. Rev.* C **64** 061301(R); Orr N 2003 *Prog. Theor. Phys. Suppl.* **146** 201
[6] Lecouey J L 2004 *Few Body Syst* **34** 21
[7] Thoennessen M *et al* 1999 *Phys. Rev.* C **59** 111; 1999 *Phys. Rev.* C **60** 027303
[8] Labiche M *et al* 2001 *Phys. Rev. Lett.* **86** 600
[9] Simon H *et al* 2007 *Nucl. Phys.* A **791** 267
[10] Kondo Y *et al* 2010 *Phys. Lett.* B **690** 245
[11] Randisi G *et al* 2014 *Phys. Rev.* C **84** 034320
[12] Thompson I J and Zhukov M V 1996 *Phys. Rev.* C **53** 708
[13] Vinh Mau N and Pacheco J C 1996 *Nucl. Phys.* A **607** 163; 2002 *Phys. Rev.* C **65** 044004
[14] Descouvemont P 1994 *Phys. Lett.* B **331** 71 ; 1995 *Phys. Rev.* C **52** 704
[15] Tarutina T, Thompson I J, Tostevin J A 2004 *Nucl. Phys.* A **733** 53
[16] Labiche M, Marques F M, Sorlin O and Vinh Mau N 1999 *Phys. Rev.* C **60** 027303
[17] Blanchon G, Bonaccorso A, Brink D M, García-Camacho A and Vinh Mau N 2007 *Nucl. Phys.* A**784** 49
[18] Blanchon G, Vinh Mau N, Bonaccorso A, Dupuis M, Pillet N 2010 *Phys. Rev.* C **82** 034313
[19] Bertsch G F, Hencken K and Esbensen H 1998 *Phys. Rev.* C **57** 1366
[20] Charity R J and the HiRA collaboration 2013, private communication.
[21] Bonaccorso A and Brink D M 1988 *Phys. Rev.* C **38** 1776; 1991 *Phys. Rev.* C **43** 299; 1991 *Phys. Rev.* C **44** 1559
[22] Bonaccorso A 1999 *Phys. Rev.* C **60** 054604
[23] Salvioni G 2014 MSc Thesis, University of Pisa, unpublished
[24] Bonaccorso A and Charity R J 2014 *Phys. Rev.* C **89** 024619
[25] Bonaccorso A *et al* , in preparation
[26] Alder K and Winther A 1975 *Electromagnetic Excitation* North-Holland
[27] Hoffman C R, Kay B P, and Schiffer J P 2014 *Phys. Rev.* C **89** 061305(R)
[28] Pieper S C and Wiringa R B 2001 *Annu. Rev. Nucl Part. Sci.* **51** 53
[29] Tilley D R *et al* 2004 *Nucl. Phys.* A**745** 155
Kelly J H *et al* 2012 Nucl. Phys. A **880** 88
[30] Efros V D, von Neumann-Cosel P, Richter 2014 *Phys. Rev.* C **89** 027301
[31] Charity R J 2013, private communication.
[32] Bedoor S *et al* 2013 *Phys. Rev.* C **88** 01304
[33] Horoi M, private communication, calculations done in the p-shell using the P(5-16) effective interaction of Warburton E K 1992 *Phys. Rev.* C **46** 923
[34] Nollett K M and Wiringa R B 2011 *Phys. Rev.* C **83** 041001(R)
[35] Charity R J *et al* in preparation.
[36] Bonaccorso A 2014 *Acta Phys. Pol.* B **45** 397

Pseudospin Dynamical Symmetry in Nuclei

Joseph N Ginocchio
MS 283, Los Alamos National Laboratory, Los Alamos, NM, 87545,USA
E-mail: gino@lanl.gov

Abstract. Pseudospin symmetry has been useful in understanding atomic nuclei. We review the arguments that this symmetry is a relativistic symmetry. The condition for this symmetry is that the sum of the vector and scalar potentials in the Dirac Hamiltonian is a constant. We give the generators of pseudospin symmetry. We review some of the predictions that follow from this insight into the relativistic origins of pseudospin symmetry. Since in nuclei the sum of the scalar and vector potentials is not zero but is small, we discuss preliminary investigations into the conditions on the potentials to produce partial dynamic pseudospin symmetry. Finally we show that approximate pseudospin symmetry in nuclei predicts approximate spin symmetry in anti-nucleon scattering from nuclei.

1. Introduction

Fifty years ago Aldo and I first met at Rutgers University. We both were starting our first post-doctoral appointments. We have remained in touch primarily through the 10 seminars in nuclear physics that he has organized in different venues in and near the Bay of Naples for the last 28 years. These seminars have been enormously popular and successful. I have enjoyed knowing Aldo and Renata these last fifty years.

When Aldo and I started our careers nuclear physics was a very exciting field of research. About five years after our post-doctoral appointments at Rutgers, a quasi-degeneracy in the single nucleon states of spherical nuclei with quantum numbers $(n\ \ell_j, n'\ell'_{j'})$ was discovered [1, 2], where $n' = n - 1, \ell' = \ell + 2, j' = j + 1$ and n, ℓ, j are the radial, orbital angular momentum, and total angular momentum quantum numbers, respectively. These quasi-degeneracies persist in recent measurements in nuclei far from stability [3]. The authors realized that, if they define the average of the orbital angular momenta as a pseudo-orbital angular momentum $(\tilde{\ell})$ and then couple a pseudospin $(\tilde{s} = \frac{1}{2})$ to the pseudo-orbital angular momentum, they will get the total angular momenta $(j = \tilde{\ell} \pm \frac{1}{2})$. For example, for the $(1s_{\frac{1}{2}}, 0d_{\frac{3}{2}})$ orbits, $\tilde{\ell} = 1$, which gives the total angular momenta $j = \frac{1}{2}, \frac{3}{2}$. Subsequently pseudospin doublets in deformed nuclei were discovered [4]. Pseudospin symmetry was later revealed to be a symmetry of the Dirac Hamiltonian [5, 6].

2. Symmetries of the Dirac Hamiltonian

The Dirac Hamiltonian with a Lorentz scalar potential, $V_S(\vec{r})$, and a potential which is the fourth component of a Lorentz vector potential, $V_V(\vec{r})$, is

$$H = \vec{\alpha} \cdot \vec{p} + \beta(V_S(\vec{r}) + M) + V_V(\vec{r}), \tag{1}$$

where $\vec{\alpha}$, β are the Dirac matrices, $\vec{p}$ is the momentum, M is the mass, $\vec{r}$ is the radial coordinate, and the velocity of light is set equal to unity, c =1.

2.1. Spin Symmetry: A Symmetry of the Dirac Hamiltonian

The Dirac Hamiltonian has spin symmetry when the difference of the vector and scalar potentials in the Dirac Hamiltonian is a constant, $V_S(\vec{r}) - V_V(\vec{r}) = C_s$ [7]. Hadrons [8] and anti-nucleons in a nuclear environment have spin symmetry [9]. These are relativistic systems and normally, in such systems, we would expect large spin-orbit splittings, but, in this limit, spin doublets are degenerate. The generators for this SU(2) spin algebra, $\vec{S}$, which commute with the Dirac Hamiltonian with any potential $V(\vec{r})$, spherical or deformed, $[\,H\,,\,\vec{S}\,] = 0$, are given by [10]

$$\vec{S} = \begin{pmatrix} \vec{s} & 0 \\ 0 & U_p\,\vec{s}\,U_p \end{pmatrix}, \tag{2}$$

where $\vec{s} = \vec{\sigma}/2$ are the usual spin generators, $\vec{\sigma}$ the Pauli matrices, and $U_p = \boldsymbol{\vec{\sigma}} \cdot \boldsymbol{\hat{p}}$ is the helicity unitary operator introduced in [11] and $\hat{p} = \frac{\vec{p}}{p}$, the unit momentum. The generators are four by four matrices as appropriate for the Dirac Hamiltonian.

2.2. Pseudospin Symmetry: A Symmetry of the Dirac Hamiltonian

Another SU(2) symmetry of the Dirac Hamiltonian occurs when the sum of the vector and scalar potentials in the Dirac Hamiltonian is a constant, $V_S(\vec{r}) + V_V(\vec{r}) = C_{ps}$ [7]. The generators for this SU(2) algebra $\vec{\tilde{S}}$, which commute with the Dirac Hamiltonian with any potential $V(\vec{r})$, spherical or deformed, $[\,H\,,\,\vec{\tilde{S}}\,] = 0$, are given by [10]

$$\vec{\tilde{S}} = \begin{pmatrix} U_p\,\vec{s}\,U_p & 0 \\ 0 & \vec{s} \end{pmatrix}. \tag{3}$$

This symmetry was shown to be pseudospin symmetry [5]. The eigenfunctions of the Dirac Hamiltonian in this limit will have degenerate doublets of states, one of which has pseudospin aligned and the other with pseudospin unaligned. The "upper" matrix of the pseudospin generators in Eq (3), $U_p\,\vec{s}\,U_p$, have the spin intertwined with the momentum which enables the generators to connect the states in the doublet, which differ by two units of angular momentum. In finite nuclei both potentials go to zero at infinite distances. Hence $C_{ps} = 0$ and the scalar potential will be equal in magnitude but opposite in sign to the vector potential in the pseudospin limit. The approximate equality in the magnitude of the vector and scalar fields in nuclei and their opposite sign have been confirmed in relativistic mean field theories [6] and in QCD sum rules [12, 6].

3. Consequences of Relativistic Pseudospin Symmetry

One immediate consequence of pseudospin symmetry as a relativistic symmetry is that the "lower" matrix of the pseudospin generators in Eq (3), $\vec{s}$, does not change the radial wavefunction of the "lower" component of the Dirac eigenfunctions. Hence this symmetry predicts that the radial wavefunctions of the "lower" component are the same for the two states in the doublet. Previous to this discovery many relativistic mean field calculations of nuclear properties had been made. Hence this prediction was tested with existing calculations and, indeed, these wavefunctions are very similar for both spherical [13, 14] and deformed nuclei [15, 16]. Because of the momentum dependence of the "upper" matrix of the generators the relationship between the "upper" components involves a differential equation and these have also been tested in spherical [17] and deformed nuclei [16] with success.

Magnetic dipole and Gamow-Teller transitions between the two single-nucleon states in pseudospin doublets are forbidden non-relativistically because the states differ by two units of angular momentum. However, they are not forbidden relativistically which means that they are proportional to the lower component of the Dirac eigenfunction. This leads to a condition between the magnetic moments of the states and the magnetic dipole transtion between them because the radial amplitudes of the lower components of the two states in a pseudospin doublet are equal. Therefore the magnetic dipole transition between the two states in the doublet can be predicted if the magnetic moments of the states are known [18, 6]. Likewise pseudospin symmetry also predicts Gamow-Teller transitions between a state in a parent nucleus to the partner state in the daughter nucleus if the Gamow-Teller transition to the same states in the parent and daughter nucleus is known. We do not have space to discuss these relationships in detail but one example occurs in the mirror nuclei ${}^{39}_{19}K_{20}$ and ${}^{39}_{20}Ca_{19}$. The ground state and first excited state of ${}^{39}_{19}K_{20}$ are interpreted as a $0d_{3/2}$ and $1s_{1/2}$ proton hole respectively, while the ground state and first excited state of ${}^{39}_{20}Ca_{19}$ are interpreted as a $0d_{3/2}$ and $1s_{1/2}$ neutron hole respectively. These states are members of the $\tilde{n}_r = 1, \tilde{\ell} = 1$ pseudospin doublet. Using the magnetic moment of ${}^{39}Ca$ a transition rate is calculated which is only about 37 % larger than the measured value. However, the two states in the doublet are not pure single-particle states. A modification of these relations has been derived which take into account the fact that these states are not pure single particle states [19, 6]. The modified relations give a transition rate that agrees with the measured value to within experimental error. Again using the mass 39 nuclei, the Gamow - Teller transitions from the ground state of ${}^{39}Ca$ to the ground and first excited state of ${}^{39}K$ are known, which is enough information to predict the transition from the ground state to the excited state. In the non-relativistic shell model an effective tensor term $g_{eff}\ [Y_2\sigma]^{(1)}$ is added to the magnetic dipole operator and the Gamow-Teller operator to produce a transition, where g_{eff} is a calculated effective coupling constant. However, the magnetic dipole transition calculated between the same states is an order of magnitude lower than the experimental transition [20] although the calculated Gamow-Teller agrees with the experimental value within the limits of experimental and theoretical uncertainty. This inconsistency has been a puzzle for the non-relativistic shell model. On the other hand the relativistic single-nucleon model gives a consistent description of both of these transitions. A global prediction of magnetic dipole transitions throughout the periodic table has had reasonable success as well [19, 6]. However, a global prediction of Gamow-Teller transitions have not been done yet. Pseuodspin symmetry can also be used to relate quadrupole transitions between multiplets [6].

4. Pseudospin as a Partial Dynamical Symmetry

In the invariant pseudospin symmetry limit there are no bound states for nucleons above the Dirac sea [6]. Hence pseudospin symmetry can never be an invariant symmetry. Experimental evidence and theoretical calculations suggest that the symmetry is better preserved near the Fermi sea [6]. Also the "intruder" states, states that have zero nodes and have spin aligned with the orbital angular momentum, do not have bound pseudospin partners and therefore the intruder states definitely do not have pseudospin symmetry [21]. For this reason it is of interest to investigate whether pseudospin symmetry is a partial dynamical symmetry [22] . A nucleus with partial dynamical system is one in which some of the states in a nucleus may have the symmetry, while others do not. The question is then: "Are there Dirac Hamiltonians that have a partial dynamical pseudospin symmetry?".

We have started to address this question. We shall first consider spherically symmetric Dirac Hamiltonians. For such Hamiltonians the alignment of the spin with the orbital angular momentum is conserved. The conserved quantum number is κ and

$$\kappa = (j+1/2),\ j = \ell - 1/2 = \tilde{\ell} + 1/2$$
$$\kappa = -(j+1/2),\ j = \ell + 1/2 = \tilde{\ell} - 1/2. \tag{4}$$

The two states that are pseudospin doublets we label as κ_a (pseudospin aligned) and κ_u (pseudospin unaligned), where

$$\kappa_a = \tilde{\ell}+1,\ j = \tilde{\ell} + 1/2$$
$$\kappa_u = -\tilde{\ell},\ j = \tilde{\ell} - 1/2. \tag{5}$$

so that $\kappa_a + \kappa_u = 1$ for the two states in a pseudospin doublet. Denoting $G_{\kappa,j}(r)/r$ as the radial wavefunction of the upper component of the Dirac eigenfunction, the condition that these two states are members of a pseudospin doublet is

$$y_{\tilde{\ell}}(r) = G'_{\kappa_a,j}(r) + \frac{\kappa_a}{r}\, G_{\kappa_a,j}(r) = G'_{\kappa_u,j}(r) + \frac{\kappa_u}{r}\, G_{\kappa_u,j}(r). \tag{6}$$

The function $y_{\tilde{\ell}}(r)$ depends on $\tilde{\ell}$ but not on the two κ's of the states in the doublet. The Dirac equation implies a second order differential equation for $G_{\kappa,j}(r)$ which in turn implies the following second order differential equation for $y_{\tilde{\ell}}(r)$

$$y''_{\tilde{\ell}}(r) + C^{(1)}_{\kappa} y'_{\tilde{\ell}}(r) + C^{(0)}_{\kappa} y_{\tilde{\ell}}(r) = 0, \tag{7}$$

where

$$C^{(1)}_{\kappa} = -\left(2\frac{A'_{\kappa}(r)}{A_{\kappa}(r)} + \frac{B'_{\kappa}(r)}{B_{\kappa}(r)}\right),$$

$$C^{(0)}_{\kappa} = -\frac{\tilde{\ell}(\tilde{\ell}+1)}{r} - \frac{A''_{\kappa}(r)}{A_{\kappa}(r)} + 2\left(\frac{A'_{\kappa}(r)}{A_{\kappa}(r)}\right)^2 + \left(\frac{\kappa}{r} + \frac{A'_{\kappa}(r)}{A_{\kappa}(r)}\right)\frac{B'_{\kappa}(r)}{B_{\kappa}(r)} + A_{\kappa}(r)B_{\kappa}(r), \tag{8}$$

and

$$A_{\kappa}(r) = E_{\kappa} + M + V_S(r) - V_V(r),\ B_{\kappa}(r) = E_{\kappa} - M - V_S(r) - V_V(r), \tag{9}$$

where E_{κ} is the eigenenergy of the Dirac eigenfunction. The condition then for two states to be pseudospin doublets is $C^{(i)}_{\kappa_a} = C^{(i)}_{\kappa_u}$ where $i = 0, 1$. Not every eigenstate will satisfy these conditions. We are investigating the possible solutions of these conditions.

Another approach is to require that the commutator of the Dirac Hamiltonian with the pseudospin generators not be identically zero but zero only when acting on the states which have good pseudospin symmetry; i.e.,

$$[H\,,\,\vec{\tilde{S}}]\ \Psi_{\kappa,j}(\vec{r}) = 0, \tag{10}$$

for both the states in the doublet, $\kappa = \kappa_a, \kappa_u$, and where $\Psi_{\kappa,j}(\vec{r})$ is the two component Dirac eigenfunction. This leads to a condition only on the upper component

$$[V_S(r) + V_V(r)\,,\,U_p\,\hat{p}]\ G_{\kappa_{a,u},j}(r)/r\left[Y^{(\tilde{\ell}\pm 1)}\chi\right]^{j=\tilde{\ell}\pm\frac{1}{2}} = 0, \tag{11}$$

where $Y^{(\ell)}$ is the spherical harmonic of rank ℓ and χ is the spin function and $[\ ]^j$ means coupled to angular momentum j. These conditions are being studied to see if there are potentials which satisfy these conditions.

5. Anti-nucleon in a Nuclear Environment

Charge conjugation changes a nucleon into an anti-nucleon. Under charge conjugation the scalar potential of a nucleon remains invariant while the its vector potential changes sign. Hence the pseudospin condition $V_S(\vec{r}) + V_V(\vec{r}) \approx C_{ps}$ becomes the spin condition $\bar{V}_S(\vec{r}) - \bar{V}_V(\vec{r}) \approx C_{ps} = \bar{C}_s$ where $\bar{V}_{S,V}(\vec{r})$ are the scalar and vector potentials of an anti-nucleon in a nuclear environment respectively. Hence approximate pseudospin for nucleons predicts approximate spin symmetry for anti-nucleons in a nuclear environment. Since the potentials are also very deep we would expect approximate $U(3)$ symmetry as well [23].

Of course the anti-nucleons scattered from nuclei experience a strong annihilation potential while nucleons scattered from nuclei do not. However, since the nucleons do not feel an annihilation potential, this means that the sum of the scalar and vector annihilation potentials for nucleons must be zero which means that the difference of the anti-nucleon annihilation potentials is zero as well. Hence the anti-nucleon annihilation potentials conserve spin symmetry as well.

The sparse data on the scattering of polarized anti-protons on nuclei supports the prediction that spin symmetry is approximately conserved [24]. Perhaps more extensive data on the scattering of polarized anti-protons on nuclei will be forthcoming from GSI.

6. Conclusions

Most features of nuclei can be understood within the non-relativistic shell model. Therefore it comes as a surprise that pseudospin symmetry in nuclei is a relativistic symmetry. However, it can never be an invariant symmetry since no nucleon bound states exist when the sum of the scalar and vector potentials is a constant. Investigations have begun to determine if certain potentials produce a partial dynamic symmetry.

Approximate pseudospin symmetry in nuclei implies that spin symmetry will be approximately conserved in anti-nucleon scattering from nuclei. The new anti-proton facility at GSI could test this prediction by measuring the polarization of anti-protons scattered from nuclei. This prediction follows simply from charge conjugation and, if confirmed, will be another manifestation of the relativistic origin of pseudospin symmetry.

In addition, since QCD sum rules predict that the sum of the scalar and vector potentials is small in nuclear matter, perhaps there is a more fundamental rationale for pseudospin symmetry in terms of quark dynamics.

Acknowledgments

The author would like to thank Prof. Ami Leviatan for his collaboration on the preliminary research on pseudospin symmetry as a partial dynamic symmetry discussed in Section 4. This research was supported by the US Department of Energy, Contract No. W-7405-ENG-36.

References

[1] Arima A, Harvey M and Shimizu K 1969 *Phys. Lett. B* **30** 517
[2] Hecht K T and Adler A 1969 *Nucl. Phys. A* **137** 129
[3] Sorlin O and Porquet M G 2008 *Prog. Part. Nucl. Phys.* **61** 602
[4] Bohr A, Hamamoto I and Mottelson B R 1982 *Phys. Scr.* **26** 267
[5] Ginocchio J N 1997 *Phys. Rev. Lett.* **78** 436
[6] Ginocchio J N 2005 *Phys. Rep.* **414** 165
[7] Bell J S and Ruegg H 1997 *Nucl. Phys. B* **98** 151
[8] Page P R, Goldman T, and Ginocchio J N 2001 *Phys. Rev. Lett.* **86** 204
[9] Ginocchio J N 1999 *Phys. Rep.* **315** 231
[10] Ginocchio J N and Leviatan A 1998 *Phys. Lett. B* **425** 1
[11] Blokhin A L, Bahri C and Draayer J P 1995 *Phys. Rev. Lett.* **74** 4149
[12] Cohen T D *et al* 1995 *Prog. Part. Nucl. Phys.* **35** 221
[13] Ginocchio J N and Madland D G 1998 *Phys. Rev. C* **57** 1167
[14] Meng J, Sugawara-Tanabe K, Yamaji S, Ring P and Arima A, 1998 *Phys. Rev. C* **58** R628
[15] Sugawara-Tanabe K, Yamaji S and Arima A 2002 *Phys. Rev. C* **65** 054313

[16] Ginocchio J N, Leviatan A, Meng J and Zhou S G 2004 *Phys. Rev. C* **69** 03430
[17] Ginocchio J N 2002 *Phys. Rev. C* **66** 064312
[18] Ginocchio J N 1999 *Phys. Rev. C* **59** 2487
[19] von Neumann-Cosel P and Ginocchio J N 2000 *Phys. Rev. C* **62** 014308
[20] Hagberg E *et al* 1994 *Nucl. Phys. A* **571** 555
[21] Ginocchio J N and Leviatan A 2001 *Phys. Rev. Lett.* **87** 072502
[22] Leviatan A 1996 *Phys. Rev. Lett.* **77** 818
[23] Ginocchio J N 2005 *Phys. Rev. Lett.* **95** 252501
[24] Martin A *et al* 1988 *Nucl. Phys. A* **487** 563

High-spin shell model states in neutron-rich Sn isotopes

Ł W Iskra,[1] R Broda,[1] R V F Janssens,[2] J Wrzesiński,[1] B Szpak,[1] C J Chiara,[2,3] M P Carpenter,[2] B Fornal,[1] N Hoteling,[2,3] F G Kondev,[4] W Królas,[1] T Lauritsen,[2] T Pawłat,[1] D Seweryniak,[2] I Stefanescu,[2,3] W B Walters,[3] and S Zhu[2]

[1] Institute of Nuclear Physics, Polish Academy of Sciences, PL-31-342 Kraków, Poland
[2] Physics Division, Argonne National Laboratory, Argonne, Illinois 60439, USA
[3] Department of Chemistry and Biochemistry, University of Maryland, College Park, Maryland 20742, USA
[4] Nuclear Engineering Division, Argonne National Laboratory, Argonne, Illinois 60439, USA

E-mail: lukasz.iskra@ifj.edu.pl

Abstract. High-spin states with the seniority $v \geq 2$ have been investigated in the neutron-rich ^{118,120,122,124,126}Sn isotopes. They were produced in fusion-fission processes following ^{48}Ca + ^{208}Pb, ^{48}Ca + ^{238}U reactions and via fission of target nuclei in the ^{64}Ni + ^{238}U system. By employing techniques of delayed- and cross-coincidences, it was possible to establish level schemes up to an 8 MeV excitation energy. The 13^- and 15^- states were identified as being isomeric and their half-lives were determined. The reduced transition probabilities extracted for isomeric transitions behave very regularly with the mass number A. The spin-parity values assigned to or suggested for the identified states were supported by shell-model calculations and by systematics.

1. Introduction

Spectroscopic studies of Sn isotopes with the closed $Z = 50$ proton shell are important to verify predictions of the shell model for excitations which involve the orbitals filled by neutrons between the $N = 50$ and $N = 82$ major shells. In the neutron-rich region, the high-j $h_{11/2}$ orbital plays a dominant role in the structure of the yrast states. The maximum spin coupling of the $h_{11/2}$ neutrons leads to the formation of isomeric states starting from the ^{116}Sn isotope. The seniority $v = 2$ and 3 $(h_{11/2})^n$ 10^+ and $27/2^-$ isomers were established in a series of deep-inelastic heavy ion reactions [1-4] up to the ^{128}Sn nucleus. They exhibit an impressive regularity in the excitation energies and especially in the reduced transition probabilities of the isomeric transitions (*B(E2)* values). The latter reflect the filling of the $h_{11/2}$ neutron orbital and define the half-filling to occur for the ^{123}Sn isotope. The present report aims at a considerable extension of the study of the Sn isotopes, involving the identification of hitherto unknown states with high values of seniority. Investigations with the same goal have also been carried out by two other groups which reported results of their parallel studies in recent publications. Several transitions located above the 10^+ isomers in Sn isotopes have been identified by Fotiades *et al.* [5] and a more complete study revealing also structures in the odd Sn isotopes as well was presented by Astier *et al.* [6]. The detailed results of the present study were reported in a recent publication [7]. Apart from the comforting similarity of the results obtained in independent experimental efforts, the level schemes presented here and in [7] are often more detailed, especially in instances which may turn out to be critical for the interpretation of some of the observed structures. Additionally, shell-model calculations have been performed for isotopes down to the A = 122 mass number and the results convincingly confirm the experimental findings.

2. Experimental procedures and data analysis techniques

The analysis was based on gamma coincidence data collected during three experiments performed at the Argonne National Laboratory. In all cases, the Argonne Tandem Linear Accelerator System (ATLAS) and the GAMMASPHERE [8] germanium detector array, which consisted of 101 Compton-suppressed Ge detectors, were employed. During the first experiment, a 330-MeV ^{48}Ca beam was bombarding a 55 mg/cm^2 ^{238}U target. In the second one, the target was changed to a 50 mg/cm^2 ^{208}Pb foil. Finally, in the last experiment, the uranium target was bombarded by a 430-MeV ^{64}Ni beam. In the ^{48}Ca + ^{238}U and ^{48}Ca + ^{208}Pb experiments, the $^{118\text{-}126}$Sn isotopes were produced in fusion-fission reactions, whereas fission of ^{238}U target nuclei populated these isotopes predominantly in the ^{64}Ni + ^{238}U measurement. In all experiments, gamma-ray coincidence data were collected with a trigger requiring that three or more Compton-suppressed gamma quanta be detected in coincidence. The beams were pulsed with a 412 ns repetition rate to provide a clean separation between prompt and delayed transitions.

The identification of gamma rays in a specific Sn isotope was obtained using two techniques described in detail in Refs. [9,10]. In the first, the delayed gamma coincidence technique was used requiring a delayed time condition for gamma rays selected by gates placed on previously known transitions in the corresponding isomeric decays. In this way it was possible to identify states above isomers with half-lives of less than a few μs. In order to identify states above much longer-lived isomeric states, the method of gamma cross-coincidences had to be applied. In the case of the ^{48}Ca + ^{238}U data, for example, gamma rays associated with the complementary Sm isotopes were found in mutual coincidence with the prompt gamma rays emitted by Sn nuclei. The fairly complex analysis of such data allowed to unambiguously identify the most intense gamma transitions located above the states with long half-lives in the respective Sn isotopes. These also include transitions occurring in the decay of higher-lying isomers. The cross-coincidence technique also allows to determine the number of evaporated neutrons in fission of the compound nucleus and serves as an important observable that can be used in studies of the reaction mechanism [11]. Following the crucial identification of the most intense gamma transitions above the isomers, a standard analysis of the data involving prompt and delayed gamma coincidence techniques established fairly detailed level schemes for all the Sn isotopes of interest.

3. Results

The data analysis outlined above established many states which can be classified as seniority v = 4,6 excitations in even neutron-rich Sn isotopes in the mass range A = 118-126. In all the isotopes, the level schemes above the 10^+ isomers show rather similar features and extend up to about 8 MeV in excitation energy and I ~ 20 in spin value. The ^{122}Sn level scheme is displayed in Fig. 1 as a representative example of the results obtained in this work. In this case, the identification of the three most intense transitions above the 10^+ (62 μs) isomer required the use of the gamma cross-coincidence method. It turned out that all three transitions were found to be delayed and the analysis established the presence of two isomeric states with respective half-lives of 134 and 40 ns. Using the delayed coincidence technique, the gamma rays preceding in time the isomers could be safely identified, simplifying the construction of the level scheme above the new long-lived states. A detailed analysis has shown that the 4721-keV isomeric state decays in a complex way to the 10^+, 8^+ and 7^- levels. The consideration of the observed decay pathways strongly supported the 15^- and 13^- spin-parity assignments to the 4721- and 4479-keV isomers, respectively. Two of the observed branches involve four transitions connecting the upper isomer with the well-established 7^- level. Whereas the intermediate state lifetimes (< 3ns) exclude *M2* and higher multipolarities, all four gamma rays can be characterized as *E2* transitions settling spin-parity assignments to all the intermediate levels. Consideration of any spin-parity values lower than the 15^- assignment to the isomer would be in contradiction with the branching observed in the isomer decay. In addition, the 12^+-14^+-16^+ multiplet with a seniority 4 value is expected above the 10^+ isomer. These prompt transitions have been identified as bypassing the 4479- and 4721-keV isomeric states and the observed 666-keV decay to the 15^- isomer is fully consistent with the 16^+ assignment to the 5387-keV level depopulated by competing *E1* and *E2* branches.

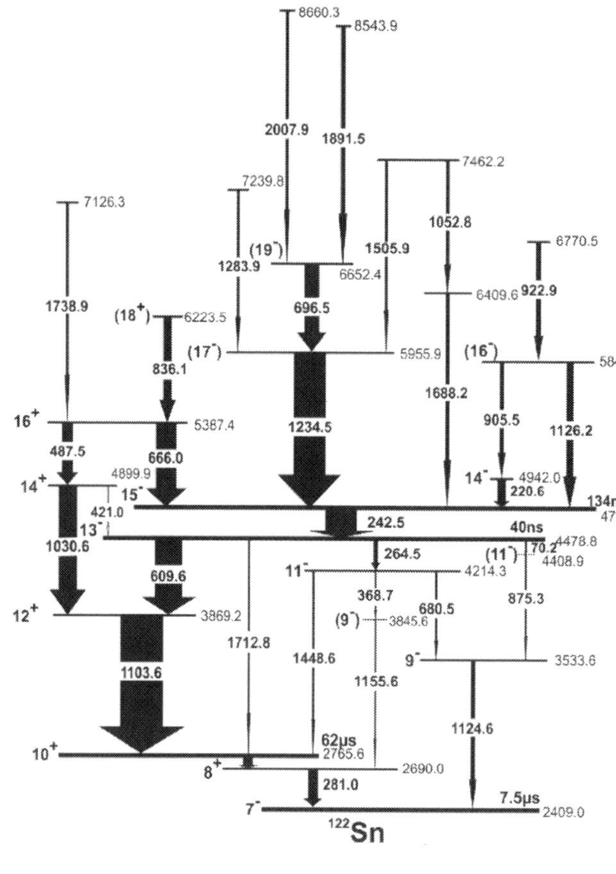

Figure 1. The ^{122}Sn level scheme established in the present study. The arrow widths reflect the observed transition intensities. Figure taken from Ref. [7].

Similar analyses and considerations established level schemes in all other even Sn isotopes and the deduced spin-parity assignments to the observed states were strongly supported by shell-model predictions. Calculations carried out for the tin isotopes with *A* from 122 to 130, with the OXBASH code [12], assumed ^{132}Sn as the closed core with the neutron holes occupying the $g_{7/2}$, $d_{5/2}$, $d_{3/2}$, $s_{1/2}$, and $h_{11/2}$ orbitals. The respective single-particle energies were selected to reproduce the experimental level energies in the ^{131}Sn isotope. A set of two-body matrix elements was computed using the shell-model interaction described in detail in Ref. [13]. Comparison of the experimental and calculated energies for the yrast and near yrast levels in ^{122}Sn is presented in Fig. 2. Similar comparisons of experimental and calculation results were made also for the heavier Sn isotopes. In general, the calculations reproduce the experimental levels rather well. However, some systematic deviations indicate that further adjustments of shell-model interactions might improve the agreement. The calculated wave functions indicate the high complexity of the intrinsic structure of many observed states. Nevertheless, the results of the calculations fully confirmed the spin-parity assignments deduced from experiment.

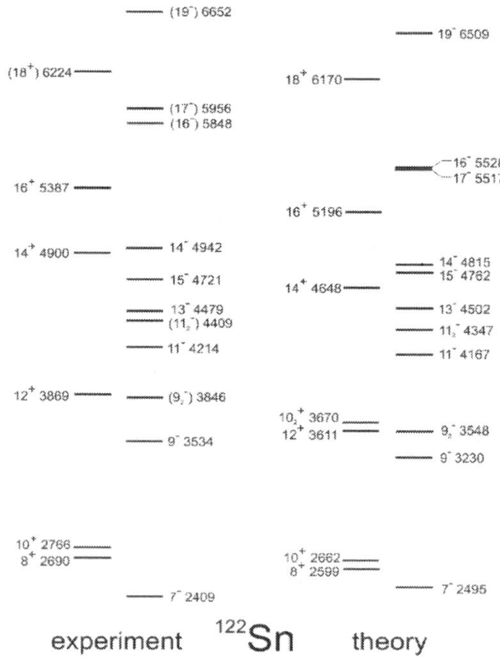

Figure 2. Comparison of the level energies established in the experiments with the theoretical shell-model calculations for ^{122}Sn. Figure taken from Ref. [7].

Furthermore, the systematics of the level energies in the isotopic chain of the even neutron-rich Sn nuclei (Fig. 3) looks consistent with the identified structures. The 2^+ state energies are very stable for all of the isotopes up to ^{130}Sn, with a shallow minimum in the vicinity of ^{124}Sn which is the result of the half-filling the $h_{11/2}$ subshell. However, higher lying states exhibit a regular pattern as well with a smooth decrease of level energies for heavier isotopes. This behavior is also consistent with trends predicted by the shell-model calculations.

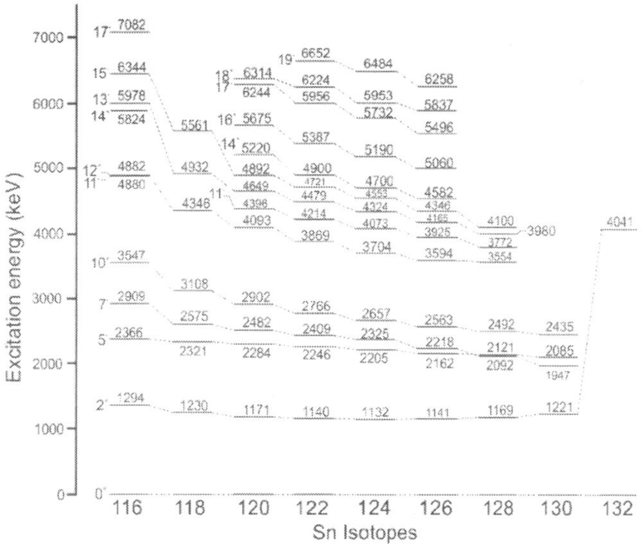

Figure 3. Systematics of the energies of yrast levels established in neutron-rich even Sn isotopes. Figure taken from Ref. [7].

The precise half-life determinations for the isomeric states in the ^{120,122,124,126}Sn isotopes assigned as 13^- and 15^- [7] and the isomeric *E2* transition energies allowed to extract the reduced transition probabilities *B(E2)* as presented in Fig. 4. At the top of this figure, *B(E2)* amplitudes for these 13^- and 15^- isomeric decays are shown, including also the value for the ^{128}Sn isotope, where a similar 15^- isomer had been identified previously [14]. In the lower part, the reduced transition probabilities *B(E2)* for the low-lying 10^+ and $27/2^-$ isomers are displayed showing a remarkable similarity in the *B(E2)* behavior. The smooth variation with mass number *A* of the *B*(*E*2) reduced transition probabilities, extracted for the isomeric transitions, reflects the filling of the neutron $h_{11/2}$ subshell. The same phenomenon was studied earlier for protons in proton-rich region of the chart of nuclides [15]. In this case an analogous behavior of the *B(E2)* values was obtained and the half filling of the proton $h_{11/2}$ subshell was determined to occur for the $Z = 71$, ^{153}Lu nucleus. Surprisingly, also the *B(E1)* values calculated for the observed *E1* decays of the 13^- isomers in the Sn isotopes exhibit a similar regularity. It should be pointed out that the convention introduced in Refs. [1-4] was adopted here as well; i.e., *B*(*E*2) amplitudes are shown rather than the *B*(*E*2) values, and arbitrarily the positive sign is assigned to the amplitudes of the light Sn isotopes with a change to negative values when the *B*(*E*2) = 0 line is crossed.

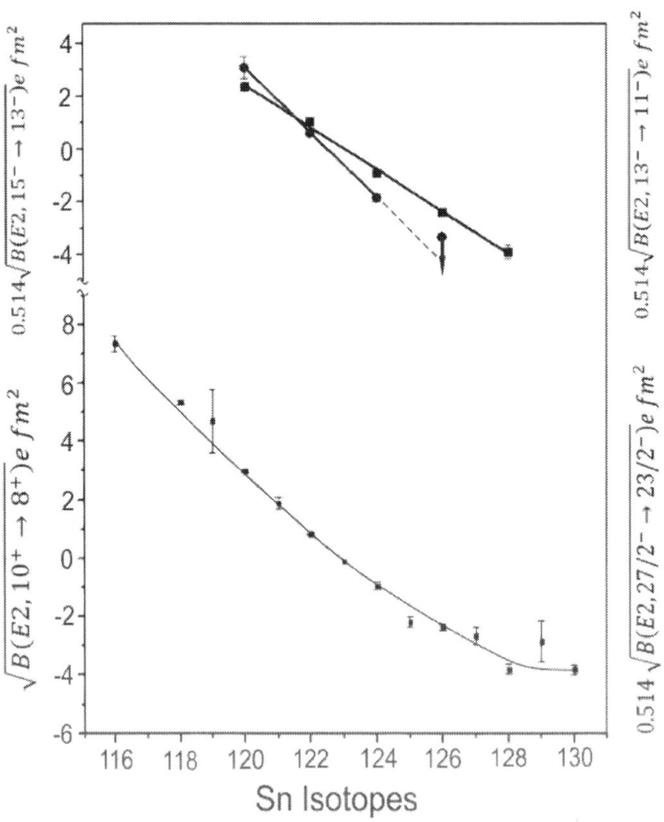

Figure 4. Transition amplitudes for isomeric decays observed in the Sn isotopes as calculated from the reduced transition rates *B(E2)*. On the bottom of the graph, the mass dependence related to the decay of the 10^+ and $27/2^-$ isomers is shown [16]. On the top, the corresponding values for the 13^- (circles) and 15^- (squares) established in the present study are given. In the ^{126}Sn isotope, only a limit could be determined. Figure taken from Ref. [7].

4. Conclusions

In the present work, extended level structures located above the 10^+ and 7^- isomers were established in the even $^{118\text{-}126}$Sn isotopes. It was demonstrated that with complex fusion-fission reactions such an identification is possible across the full range of neutron-rich isotopes and allows to construct level schemes up to relatively high spins. In the ^{120,122,124,126}Sn isotopes, the presence of 13^- and 15^- isomeric states could be established, yielding reduced transition probability *B(E2)* values for isomeric transitions which resemble closely the *B(E2)* values and mass dependence observed earlier in the 10^+ and $27/2^-$ isomeric decays. A number of states located above these 15^- isomers could also be identified. They correspond to seniority $v = 6$ excitations. Theoretical shell-model calculations performed for all isotopes down to $A = 122$ reproduced reasonably well the experimental levels. The results obtained from the same experiments for the neutron-rich odd Sn isotopes will be presented in a forthcoming publication [17].

Acknowledgments:

This work was supported by the Polish National Science Center, Projects No. UMO-2012/07/N/ST2/02861 and No. NN202-008640, the U.S. Department of Energy, Office of Nuclear Physics, under Contract No. DE-AC02-06CH11357 (ANL) and Grant No. DE-FG02-94ER40834 (UM), as well as by the Marian Smoluchowski Krakow Research Consortium "Matter-Energy-Future" as a Leading National Center (KNOW).

References

[1] Lunardi S et al. 1987 *Z. Phys. A* **328** 487
[2] Broda R et al. 1992 *Phys. Rev. Lett.* **68** 1671
[3] Mayer R H et al. 1994 *Phys. Lett. B* **336** 308
[4] Zhang C T et al. 2000 *Phys. Rev.* ***C* 62** 057305
[5] Fotiades N et al. 2011 *Phys. Rev.* ***C* 84** 054310
[6] Astier A et al. 2012 *Phys. Rev.* ***C* 85** 054316
[7] Iskra Ł W et al. 2014 *Phys. Rev.* **C 89** 044324
[8] Lee I Y 1990 *Nucl. Phys* ***A* 520** 641c
[9] Broda R 2006 *J. Phys G: Nucl. Part. Phys.* **32** R151
[10] Broda R et al. 2012 *Phys. Rev.* ***C* 86** 064312
[11] Iskra Ł W et al. 2013 *Acta Phys. Pol. B* **44** 395
[12] Brown B A et al. 2004 *MSU-NSCL Report* 1289
[13] Brown B A et al. 2005 *Phys. Rev.* ***C* 71** 044317
[14] Pietri S et al. 2011 *Phys. Rev.* ***C* 83** 044328
[15] McNeill J H et al. 1989 *Phys. Rev. Lett.* **63** 860
[16] Lozeva R L et al. 2008 *Phys. Rev.* ***C*** 77 064313
[17] Iskra Ł W et al. unpublished.

Production and decay of the heaviest odd-Z nuclei in the ^{249}Bk + ^{48}Ca reaction

Yu Ts Oganessian[1], F Sh Abdullin[1], C Alexander[2], J Binder[2], R A Boll[2], S N Dmitriev[1], J Ezold[2], K Felker[2], J M Gostic[3], R K Grzywacz[2,4], J H Hamilton[5], R A Henderson[3], M G Itkis[1], K Miernik[2], D Miller[4], K J Moody[3], A N Polyakov[1], A V Ramayya,[5] J B Roberto[2] , M A Ryabinin[6], K P Rykaczewski[2], R N Sagaidak[1], D A Shaughnessy[3], I V Shirokovsky[1], M V Shumeiko[1], M A Stoyer[3], N J Stoyer[3], V G Subbotin[1], A M Sukhov[1], Yu S Tsyganov[1], V K Utyonkov[1], A A Voinov[1, a], and G K Vostokin[1]

[1]Joint Institute for Nuclear Research, RU-141980 Dubna, Russian Federation
[2]Oak Ridge National Laboratory, Oak Ridge, TN 37831, USA
[3]Lawrence Livermore National Laboratory, Livermore, CA 94551, USA
[4]University of Tennessee, Knoxville, TN 37996, USA
[5]Vanderbilt University, Nashville, TN 37235, USA
[6]Research Institute of Atomic Reactors, RU-433510 Dimitrovgrad, Russian Federation

Abstract. The reaction of ^{249}Bk with ^{48}Ca has been investigated with an aim of synthesizing and studying the decay properties of isotopes of the new element 117. The experiments were performed at five projectile energies (in two runs, in 2009-2010 and 2012) and with a total beam dose of ^{48}Ca ions of about 9×10^{19}. The experiments yielded data on α-decay characteristics and excitation functions of the produced nuclei that establish these to be ^{293}117 and ^{294}117 – the products of the 4*n*- and 3*n*-evaporation channels, respectively. In total, we have observed 20 decay chains of *Z*=117 nuclides. The cross sections were measured to be 1.1 pb for the 3*n* and 2.4 pb for the 4*n*-reaction channel. The new ^{289}115 events, populated by α decay of ^{293}117, demonstrate the same decay properties as those observed for ^{289}115 produced in the ^{243}Am(^{48}Ca,2*n*) reaction thus providing cross-bombardment evidence. In addition, a single decay of ^{294}118 was observed from the reaction with ^{249}Cf – a result of the in-growth of ^{249}Cf in the ^{249}Bk target. The observed decay chain of ^{294}118 is in good agreement with decay properties obtained in 2002-2005 in the experiments with the reaction ^{249}Cf(^{48}Ca,3*n*)^{294}118. The energies and half-lives of the odd-*Z* isotopes observed in the 117 decay chains together with the results obtained for lower-*Z* superheavy nuclei demonstrate enhancement of nuclear stability with increasing neutron number towards the predicted new magic number *N*=184.

1. Introduction

After many unsuccessful attempts to reach the predicted "island of stability" of the superheavy elements (SHE) by means of different methods (see, e.g., Review [1] and references therein), including usage of kinematic separators of the fusion-evaporation reaction products [2], the work aimed at achieving this goal has been resumed in the Flerov Laboratory of Nuclear Reactions (JINR, Dubna).

[a] voinov_2000@mail.ru

For the first time, in complete-fusion reactions ^{242,244}Pu, ^{245,248}Cm, and ^{249}Cf targets with ^{48}Ca projectiles we have synthesized nine isotopes of even-Z nuclei of the new elements 114 (Fl), 116 (Lv) and 118 employing the Dubna gas-filled recoil separator (DGFRS) [3]. The production cross-sections and decay properties of the isotopes of $^{286\text{-}289}$114 and ^{292,293}116 were recently confirmed in independent experiments performed at FLNR [4], LBNL [5,6], and GSI [7-9].

The next step towards the island of stability of the nuclei around N=184 and Z=114 (or 120-126) was the production of odd-Z nuclides in reactions with odd-Z targets (^{237}Np, ^{243}Am, and ^{249}Bk). Studies of odd-Z nuclei provide even more detailed information about the structure of the heaviest nuclides than studies of even-Z nuclei because of the longer odd-Z decay chains; this results from strong fission hindrance caused by the unpaired nucleons. For the first time, the nuclei ^{287}115 and ^{288}115 and their decay products including ^{283}113 and ^{284}113 isotopes were observed in 2003 [10] in the ^{243}Am+^{48}Ca reaction; then new isotopes ^{289}115 and ^{285}113 were observed in 2010-2012 [11]. In 2006 a lighter isotope ^{282}113 was synthesized in the ^{237}Np(^{48}Ca,3n) reaction [12]. In addition, three decay chains of the lightest isotope ^{278}113 were observed in the reaction ^{209}Bi(^{70}Zn,n) in 2004, 2005, and 2012 [13].

The discovery of element 117 in the ^{249}Bk+^{48}Ca reaction has been reported in 2010 [14]. A relatively high stability of all these odd-Z nuclei is caused by the influence of presumably spherical nuclear shells at Z=114–126 and N=184. In 2012, we have performed a new series of experiments to obtain a more detailed information on the decay properties of odd-Z nuclei, to measure the excitation function of the ^{249}Bk+^{48}Ca reaction in a more extended range of projectile energies, and to make a cross-bombardment consistency check of the reported discoveries of elements 113, 115 and 117 [11,15].

2. Experiment

For the synthesis and identification of these odd-Z nuclei, we used the Dubna gas-filled recoil separator that selects evaporation products of complete-fusion reactions that are strongly forward peaked and suppresses the products of transfer reactions and reactions with emission of charged particles (pxn, αxn, etc.). The evaporation residues were separated in flight, then passed through a time-of-flight system and were implanted in a focal-plane silicon detector, with an estimated transmission efficiency of about 35% for Z=117 nuclei. Prior to 2011 in our work we used 4×12 cm^2 Si-detector array with 12 vertical position-sensitive strips surrounded by eight 4×4 cm^2 detectors without position sensitivity (see Refs. [3,10-12,14], and references therein). The detection system was calibrated by registering the recoil nuclei and decays (α or SF) of the known isotopes of No and Th and their descendants [16] produced in the reactions ^{206}Pb(^{48}Ca,2n) and ^{nat}Yb(^{48}Ca,3-5n), respectively. In the above experiments, the full-width-at-half-maximum (FWHM) energy resolution of α particles implanted in the focal-plane detectors was 60–140 keV, depending on the strip and the position within the strip. The FWHM position resolutions of the signals of correlated decays of nuclei implanted in the detectors were 1.1–1.3 mm for ER-α signals and 0.4–0.8 mm for ER-SF signals [14].

In 2011, before performing the new run for additional study of the excitation function of the ^{249}Bk+^{48}Ca reaction and decay properties of Z=117 isotopes and descendant nuclei, the detection system was modified to increase the position granularity of the detectors, which reduces the probability of observing sequences of random events that mimic decay chains of synthesized nuclei. The new focal-plane detectors consisted of two 6×6 cm^2 detectors each having 16 strips surrounded by six 6×6 cm^2 side detectors. The appropriate multi-channel data acquisition system [17] was developed by the DGFRS research crew on the basis of POLON spectroscopic modules and products of TekhInvest Ltd. (Dubna, Russia). This allowed us to get for the focal-plane detector the FWHM energy resolution of 34–73 keV, while the sum signals recorded by the side and focal-plane detectors had an energy resolution of about 83 to 117 keV. The FWHM position resolutions of the implantation detector were 1.1–1.8 mm for ER-α signals and 0.5–1.2 mm for ER-SF signals [15]. Other experimental conditions were the same as in Refs. [3,10-12,14]. The summary conditions of the ^{249}Bk+^{48}Ca reaction studied at the DGFRS are shown in Table 1. In order to reduce the background rate in the detector, the beam was switched off (like it was implemented in all the previous experiments) after a recoil signal was recorded being followed by an α-like signal in the focal-plane detector within energy and time intervals corresponding to decays of parent and/or daughter nuclei, in the same strip and in close position.

Table 1. The ^{249}Bk target thickness [a], lab-frame beam energies in the middle of the target layers, resulting excitation-energy intervals, total beam doses, and numbers of observed decay chains assigned to the parent nuclei ^{293}117 (4*n*) and ^{294}117 (3*n*) are listed.

Target thickness (mg/cm^2)	E_{lab} (MeV)	E_{exc} (MeV)	Beam dose $\times 10^{18}$	Number of chains 4*n* / 3*n*	Ref.
0.33	243.7	30.4–34.7	9.4	0 / 1	[15]
0.31	247.0	33.2–37.5	20.	0 / 1	[14]
0.33	246.8	32.8–37.5	5.4	0 / 2	[15]
0.31	251.7	37.2–41.4	24.	5 / 0	[14]
0.33	251.7	37.0–41.9	14.1	5 / 0	[15]
0.33	255.7	40.3–44.8	9.2	3 / 0	[15]
0.33	259.8	43.8–48.3	11.9	3 / 0	[15]

3. Results

In the experiments with the ^{243}Am+^{48}Ca reaction in 2010-2012, at the lowest excitation energies E*=31.1–36.4 MeV four decay chains of the isotope ^{289}115, the product of the 2*n* channel, were observed [11] (see Figure 1). Because of the mass difference between ^{243}Am and ^{249}Bk (α+2*n*) and lower yields of the 1*n* and 5*n* channels compared with the 2-4*n* channels, one and the same isotope of element 115, ^{289}115, can be produced only in the 2*n*- and 4*n*-evaporation channels of the ^{243}Am+^{48}Ca and ^{249}Bk+^{48}Ca reactions, respectively. Indeed, 16 decay chains of the parent isotope ^{293}117 were observed in the ^{249}Bk+^{48}Ca reaction at higher excitation energies E*=37.0–48.3 MeV. The radioactive decay properties of ^{293}117 and all the descendant nuclei discovered in 2010 [14] were confirmed by registration of 11 new decay chains in this new series of experiments [15]. One can see in Figure 2 that decay characteristics of ^{289}115, ^{285}113, and ^{281}Rg, α-decay products of ^{293}117, in the five events observed in the first experiment [14] and four events originating from ^{289}115 and produced in cross reaction with ^{243}Am [11] are in good agreement with the recent data. Thus, the isotope ^{289}115 was produced in two reactions with target nuclei ^{243}Am and ^{249}Bk that provide cross-bombardment evidence for the discovery of elements 117, 115, and 113.

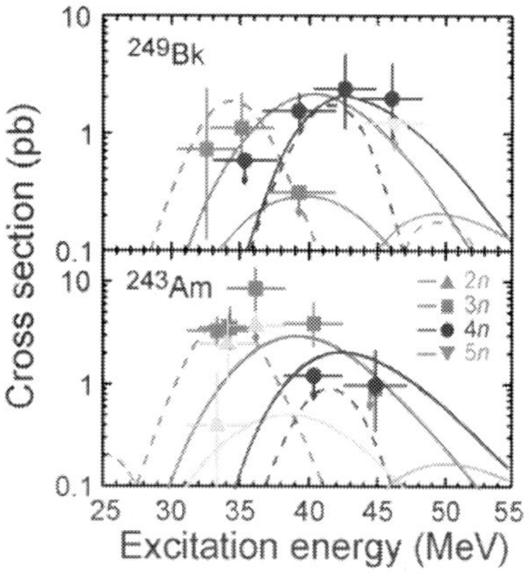

Figure 1. Measured cross sections of the x*n*-evaporation channels of the ^{249}Bk+^{48}Ca and ^{243}Am+Ca reactions. Vertical error bars correspond to statistical uncertainties [18]. Horizontal error bars represent the range of excitation energies populated at given beam energy. Symbols with arrows show upper cross-section limits. The results of theoretical calculations are shown by solid [19] and dashed [20] lines.

The heaviest isotope ^{294}117, product of the ^{249}Bk(^{48}Ca,3*n*) reaction, was synthesized at lower excitation energies of 30.4–37.5 MeV. The decay properties of all the nuclei determined in the four decay chains originating from parent nucleus ^{294}117 are shown in Figure 3. The properties of the nuclei in the new decay chains point to the same activities

[a] As in [14], the ^{249}Bk was produced at ORNL at the High Flux Isotope Reactor. The Bk fraction was chemically separated and purified at the Radiochemical Engineering Development Center at ORNL. Six arc-shaped targets were made at RIAR.

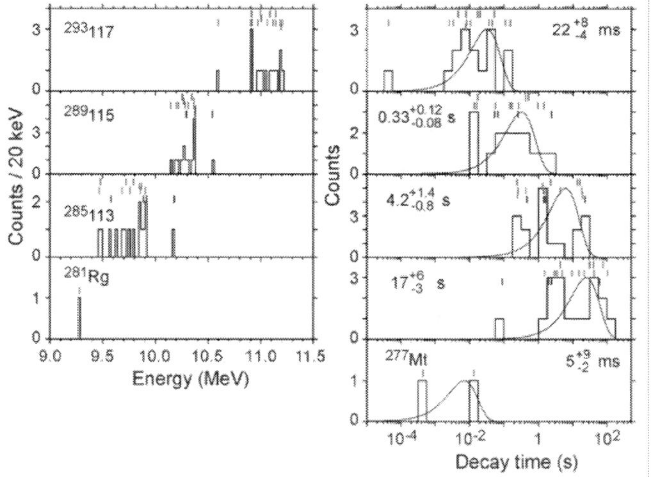

Figure 2. α-particle energy spectra registered by the focal-plane detector only or together with the side one (left-hand panel) and decay-time distributions on a logarithmic scale (right-hand panel) for isotopes originating from ^{293}117. The events originating from ^{293}117 observed in the first [14] and in the second experiment with ^{249}Bk [15] and in the reaction with ^{243}Am [11] are shown in the histograms by red (top), green (middle), and blue (bottom) lines, respectively.

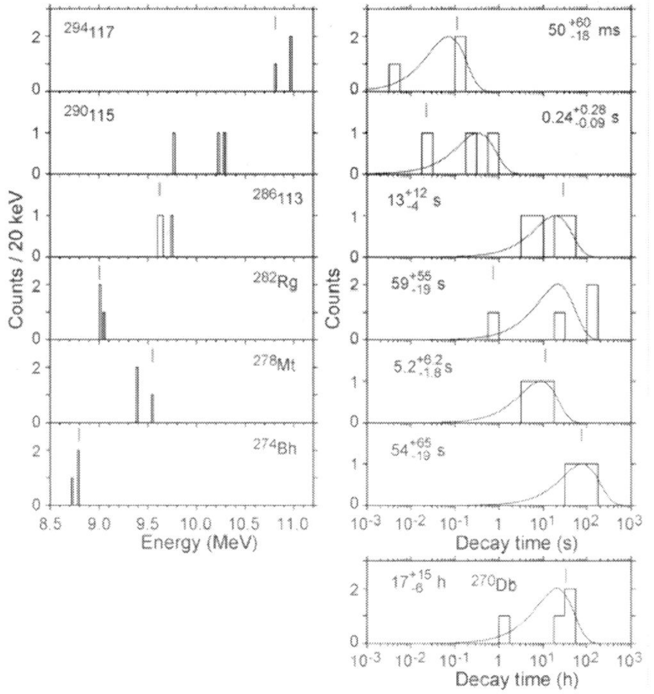

Figure 3. The same as Figure 1 but for ^{294}117. The vertical lines show the energies and decay times of the events obtained in the first experiment [14].

arising from ^{294}117 detected in the two experiments using the ^{249}Bk target [14,15].

The maximum cross sections for the 3*n* and 4*n* evaporation channels, were measured to be $\sigma_{3n}=1.1^{+1.2}_{-0.6}$ pb and $\sigma_{4n}=2.4^{+3.3}_{-1.4}$ pb, at $E^*=35$ and 43 MeV respectively (see Figure 1).

The target isotope ^{249}Bk with a half-life of 330 d decays into ^{249}Cf. During a long experiment, this creates an opportunity to produce $Z=118$ isotopes in the ^{249}Cf+^{48}Ca reaction [21]. Indeed, with 247-MeV ^{48}Ca we observed one more decay chain of nuclei whose radioactive properties agree well with those determined for ^{294}118 and its descendant nuclei ^{290}Lv and ^{286}Fl (fission branch of about 50%) [3,21] (see Figure 4b). Taking into account the buildup of ^{249}Cf in the preceding [14] and present [15] experiments, the detected decay chain of ^{294}118 corresponds to $0.3^{+0.7}_{-0.26}$ pb for the total excitation energy interval of 26.6–37.5 MeV of the compound nucleus ^{297}118, in good agreement with cross sections already measured in this reaction [3,21].

Therefore, the discovery of the chemical elements with atomic numbers 113, 115, and 117 that were synthesized for the first time in 2003 [10] and 2010 [14] has now been corroborated through the observation of additional decay chains in the reaction ^{249}Bk+^{48}Ca. In total, 59 decay chains originating from odd-*Z* parent nuclei ^{282}113 [12], $^{287-289}$115 [10,11], and ^{293,294}117 [14,15] were observed and radioactive decay properties of 29 new isotopes were determined (see Figure 4a). The measured α-particle energies for all the isotopes of odd-*Z* elements produced in the reactions ^{243}Am+^{48}Ca and ^{249}Bk+^{48}Ca agree well with the systematics of the α-decay energies of the heavy nuclei and have intermediate values between

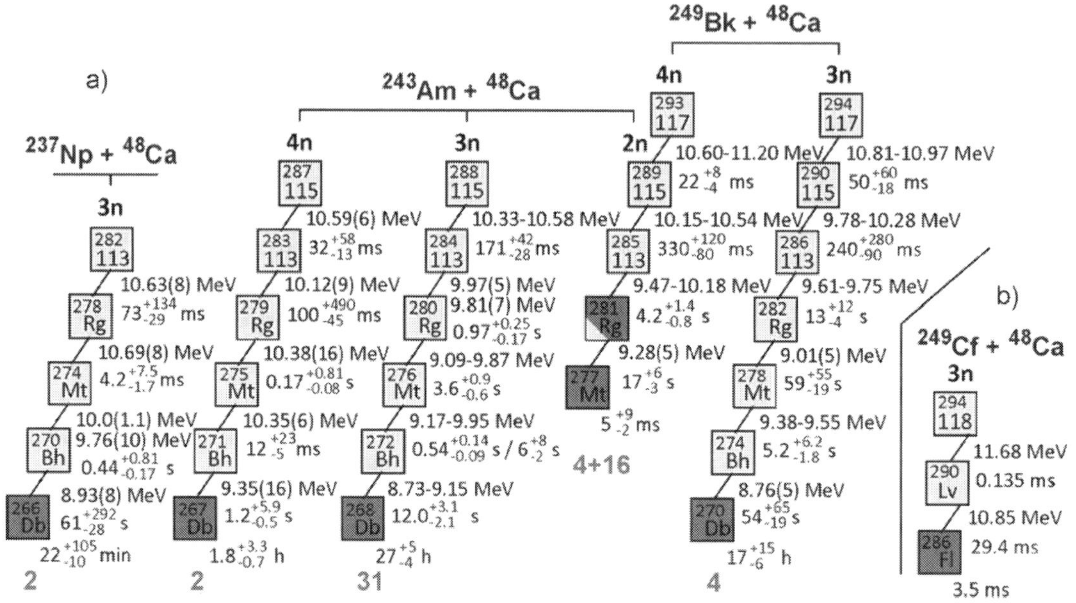

Figure 4. (a) Summary of the decay properties of the isotopes of elements 113, 115 and 117 observed for the first time among the reaction products of ^{48}Ca beam with ^{237}Np, ^{243}Am and ^{249}Bk targets. The number of the detected decay events of a given isotope is shown at the bottom of the chains. (b) Decay properties of ^{294}118–^{286}Fl measured in the ^{249}Cf(^{48}Ca,3*n*) reaction [14].

neighboring even-*Z* nuclei. The measured α-particle energies of the *Z*=107 and *Z*=109 isotopes as well as their behavior *vs.* neutron number are in full agreement with what is observed for the previously known neighboring lighter nuclei. Moreover, ^{268}Db, the descendant nucleus of ^{288}115 and ^{284}113, was identified in independent chemistry experiments [22,23].

The experimental half-lives of the isotopes of odd-*Z* elements are shown in Figure 5. The increase of neutron number in nuclei with $N \geq 165$ results in considerable rise of their half-lives. An especially strong growth of lifetimes with increasing *N* is observed for the isotopes of elements 109, 111, and 113. The half-lives, as well as α-particle energies of the odd-*Z* isotopes observed in the *Z*=117 decay chains, together with the results obtained for lower-*Z* superheavy nuclei, demonstrate the decisive role of nuclear shell effects resulting in enhancement of stability with increasing neutron number towards the predicted new magic number *N*=184.

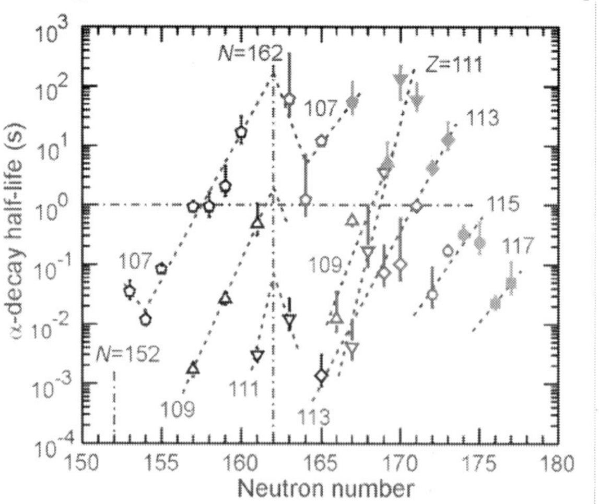

Figure 5. Half-lives *vs.* neutron number for the isotopes of odd-*Z* elements (results from ^{249}Bk+^{48}Ca reactions are shown by full red diamonds).

Experimental study of the properties of odd-*Z* nuclei was recently performed at the gas-filled recoil separator TASCA at GSI in fusion-evaporation reactions ^{243}Am+^{48}Ca [24] and ^{249}Bk+^{48}Ca [25]. Authors of Ref. [24] announced observation of 23 correlated α-decay chains starting from ^{287,288}115 isotopes; their observed decay pattern and properties are in good agreement with the DGFRS data [10,11]. Two decay chains of ^{294}117 and descendant nuclei observed in the ^{249}Bk+^{48}Ca reaction have been reported in [25]. The production cross section of ^{294}117 nucleus and decay properties of the observed descendant nuclei confirm previously reported DGFRS data [14,15].

Acknowledgments
We are grateful to the JINR directorate and U400 cyclotron and ion source crews for their continuous support of the experiments. We acknowledge the support of the RFBR Grants No. 13-02-12052 and 13-03-12205. Research at ORNL was supported by the U.S. DOE Office of Nuclear Physics under DOE Contract No. DE-AC05-00OR22725 with UT-Battelle, LLC. Research at LLNL was supported by LDRD Program Project No. 08-ERD-030, under DOE Contract No. DEAC52-07NA27344 with Lawrence Livermore National Security, LLC. This work was also supported by the U.S. DOE through a Grant No. DE-FG-05-88ER40407 (Vanderbilt University). These studies were performed in the framework of the Russian Federation/U.S. Joint Coordinating Committee for Research on Fundamental Properties of Matter.

References

[1] Flerov G N and Ter-Akopian 1983 *Rep. Prog. Phys.* **46**, 817.
[2] Armbruster P *et al.* 1985 *Phys. Rev. Lett.* **54**, 406.
[3] Oganessian Yu Ts 2007 *J. Phys. G: Nucl. Part. Phys.* **34**, R165; 2011 *Radiochim. Acta* **99**, 429.
[4] Eichler R *et al.* 2010 *Radiochim. Acta* **98**, 133.
[5] Stavsetra L *et al.* 2009 *Phys. Rev. Lett.* **103**, 132502.
[6] Ellison P A *et al.* 2010 *Phys. Rev. Lett.* **105**, 182701.
[7] Hofmann S *et al.* 2007 *Eur. Phys. J.* A **32**, 251.
[8] Düllmann Ch E *et al.* 2010 *Phys. Rev. Lett.* **104**, 252701.
[9] Hofmann S *et al.* 2012 *Eur. Phys. J.* A **48**, 62.
[10] Oganessian Yu Ts *et al.* 2004 *Phys. Rev. C* **69**, 021601(R); 2005 *Phys. Rev. C* **72**, 034611.
[11] Oganessian Yu Ts *et al.* 2012 *Phys. Rev. Lett.* **108**, 022502; 2013 *Phys. Rev. C* **87**, 014302.
[12] Oganessian Yu Ts *et al.* 2007 *Phys. Rev. C* **76**, 011601(R).
[13] Morita K *et al.* 2004 *J. Phys. Soc. Jpn.* **73**, 2593; 2007 *ibid.* **77**, 045001; 2012 *ibid.* **81**, 105201.
[14] Oganessian Yu Ts *et al.* 2010 *Phys. Rev. Lett.* **104**, 142502; 2011 *Phys. Rev. C* **83**, 054315.
[15] Oganessian Yu Ts *et al.* 2012 *Phys. Rev. Lett.* **109**, 162501; 2013 *Phys. Rev. C* **87**, 054621.
[16] Audi G *et al.* 2012 *Chinese Phys.* C **36**, 1287.
[17] Tsyganov Yu S, Polyakov A N, Voinov A A, Shumeyko M V 2013 *Proc. Int. Symp. on Nuclear Electronics and Computing (Varna, Bulgaria, 9-16 September 2013)* (Dubna: Russian Federation/JINR) p 250.
[18] Schmidt K-H *et al.* 1984 *Z. Phys. A* **316**, 19.
[18] Zagrebaev V and Greiner W 2008 *Phys. Rev. C* **78**, 034610.
[20] Siwek-Wilczyńska K *et al.* 2012 *Phys. Rev. C* **86**, 014611.
[21] Oganessian Yu Ts *et al.* 2006 *Phys. Rev. C* **74**, 044602.
[22] Dmitriev S N *et al.* 2005 *Mendeleev Commun.* **15**, 1.
[23] Stoyer N J *et al.* 2007 *Nucl. Phys. A* **787**, 388c.
[24] Rudolph D *et al.* 2013 *Phys. Rev. Lett.* **111**, 112502.
[25] Khuyagbaatar J *et al.* 2014 *Phys. Rev. Lett.* **112**, 172501.

Journal of Physics: Conference Series **580** (2015) 012039 doi:10.1088/1742-6596/580/1/012039

Production of evaporation residues in the vicinity of the N=126 closure

Roman Sagaidak

Flerov Laboratory of Nuclear Reactions, Joint Institute for Nuclear Research, 141980, Dubna, Moscow region, Russian Federation

E-mail: sagaidak@nrmail.jinr.ru

Abstract. The shell structure and fission-barrier heights of nuclei involved in the compound-nucleus (CN) de-excitation chains leading to observable evaporation residues (ER) determine essentially fusion-evaporation cross sections for heavy fissile nuclei produced in the vicinity of the N=126 neutron shell in heavy ion (HI) reactions. The entrance-channel effect of quasi-fission (QF) lowering the fusion cross section in reactions with massive nuclei and the effect of the collective enhancement in the nuclear level density (CENLD), corresponding to the exit channel, should strongly reduce production cross sections for ER and increase the observed fission cross section in the vicinity of N=126. These effects were not evidently manifested in the analysis of cross sections obtained in different CN reactions leading to Po isotopes produced in a wide region of N. This analysis has been performed in the framework of standard statistical model (SSM) approximations. As a whole, production cross sections for neutron deficient Po nuclei produced in HI reactions can be reproduced in the framework of SSM with the reduced liquid-drop fission barriers [1]. We present a similar analysis of ER and fission cross section data obtained in the reactions induced by ^{3,4}He on Pb and Bi targets, which lead to Po and At compound nuclei [2], which proved to be a good test of the preceding results [1]. Such reactions are characterized by much lower angular momenta transferred to a CN that allows to suggest more pronounced manifestation of CENLD effects in the CN decay. These reactions also do not reveal any indications of QF. This simplifies the data analysis and makes it more unambiguous. The results of the analysis of the ^{3,4}He reactions performed in the present work demonstrate the same changes in the fission barrier heights for Po and At nuclei as obtained earlier with HI reactions data [1].

1. Introduction

A comparison of fission barrier heights calculated in a region between two neutron shell numbers ($126 \lesssim N \lesssim 184$) reveals a large spread in their predictions for exotic neutron-rich nuclei [3]. At the same time, reasonable description of the fission barrier heights close to the stability line (where the model parameters were tuned to the experimental data) is accomplished [4]. Large deviations between predictions of different models and the fission barrier heights derived from nuclear reaction data were pointed out earlier for neutron-deficient nuclei [5]. The difference between calculations for extremely neutron-rich (hypothetical) nuclei reaches more than 30 MeV. Such uncertainty results in completely different scenarios as far as the r-process termination by fission is considered [6]. In macroscopic calculations of fission barriers, the height of the barrier is proportional to the nuclear surface energy which, in turn, depends on the charge asymmetry parameter $I = (N - Z)/A$. Therefore, the study of the isotopic dependence of fission-barrier heights for nuclei far from the stability line is one of the most important tasks in modern fission studies. Unfortunately, these studies for neutron-rich nuclei are very difficult due to their very low production rates as it follows, e. g., from the results obtained in relativistic HI collisions [7] and as one may expect from calculations exploring nuclear transfer reactions at Coulomb energies [8]. At the same time, the isotopic dependence of fission barrier heights can be presently studied in the more readily accessible region of neutron-deficient nuclei produced in HI fusion-evaporation reactions.

2. Statistical model considerations of ER production in HI fusion reactions

Production cross sections for the neutron-deficient Po to Th nuclei, obtained in very asymmetric projectile-target fusion-evaporation reactions are mainly determined by level density parameters and the fission-barrier heights for nuclei involved in the CN de-excitation chains. Analysis of the measured ER cross sections is usually performed in the framework of the SSM approximations, e.g., with HIVAP [9] using the liquid-drop (LD) fission barrier heights with scaling of the form: $B_f(L) = k_f B_f^{LD}(L) - \Delta W_{gs}$ (k_f is a scaling factor at the rotating LD barriers $B_f^{LD}(L)$ [10] and ΔW_{gs} is the ground state shell correction). The macroscopic components of the barriers $B_f^m = k_f B_f^{LD}$ derived from such analysis of the measured cross sections for Fr and Ra nuclei production [11]–[14] are compared to various predictions in Fig. 1. The derived values are lower than any calculations [15] (with the exception of [16]) and are a relatively smooth function of N, e.g., for Fr nuclei, k_f is gradually reduced from 0.85 to 0.8, in going from N=131 to 117. Prior analysis of ER and fission cross sections for Po nuclei in a wide region of $105 \leqslant N \leqslant 126$ [1] shows a sizable decrease in B_f^m as compared to any predictions (see Fig. 1). This analysis includes ER cross section data obtained in nearly symmetric and asymmetric reactions leading to the most neutron-deficient Po nuclei. The possibility of QF effects leading to the fusion probability $P_{fus} < 1$ cannot be ruled out in these reactions bearing in mind, e.g., the comparison of the ER production cross sections obtained in the $^{16}O+^{186}W$ and $^{48}Ca+^{154}Sm$ reactions leading to the $^{202}Pb^*$ CN [18]. Such comparison indicates the $P_{fus} < 1$ values [19], which are noticeably differed from similar values obtained with measured fission cross section and in calculations.

Note that the macro-microscopic models [3, 15] imply a smooth behavior of B_f^m observed for Fr and Ra (Fig. 1). It has to be also mentioned that any manifestations of the CENLD effects do not imply the data derived in Fig. 1. In particular, these effects are expected to be strongly pronounced in the vicinity of the N=126 spherical shell as the drop of ER production cross sections as it follows, e.g., from the model presented in [7]. Such effect is not observed in the cross section data of fusion experiments and, respectively, it is not revealed with the SSM analysis as an expected relative reduction in the B_f^m values (HIVAP does not take into account the CENLD effects).

Among other peculiarities in the ER production cross sections, the effect of the energy losses of a massive HI inside the target of finite thickness deserves attention. The effect leads to visible broadening of excitation functions along with lowering of their maxima. It has to be taken into account when the calculated excitation functions are adjusted to the measured one using variations in the macroscopic fission barrier heights [20]. Lastly, the reliability of the ER production cross sections measured in very asymmetric combinations implying a negligible fusion suppression may be of great importance for the derived B_f^m values. For example, available ER cross section data obtained in the $^{197}Au(^{12}C,xn)^{209-x}At$ reactions (see Refs in the most recent works [21]) demonstrate a difference in the values corresponding to a factor of 5–6 that leads to the difference in the "adjusted" k_f values in the range between 0.5 and 1.0 as it follows from the analysis with HIVAP.

3. Analysis of cross section data obtained in ^{3,4}He+Pb,Bi reactions

The He induced reactions on Pb and Bi targets, which lead to Po and At compound nuclei, can be particularly suited for this study considering the above mentioned difficulties inherent in the analysis of HI reaction cross sections. These reactions have at least two advantages over HI ones: a) much lower angular momenta are transferred to compound nuclei formed in the reactions ensuring, along with relatively high excitation energies, more favorable conditions for the manifestation of the CENLD effects as compared to the HI ones [1] associated with very high rotation energies and b) these reactions reveal no indications of QF due to the ultimate difference in the masses of projectiles and targets.

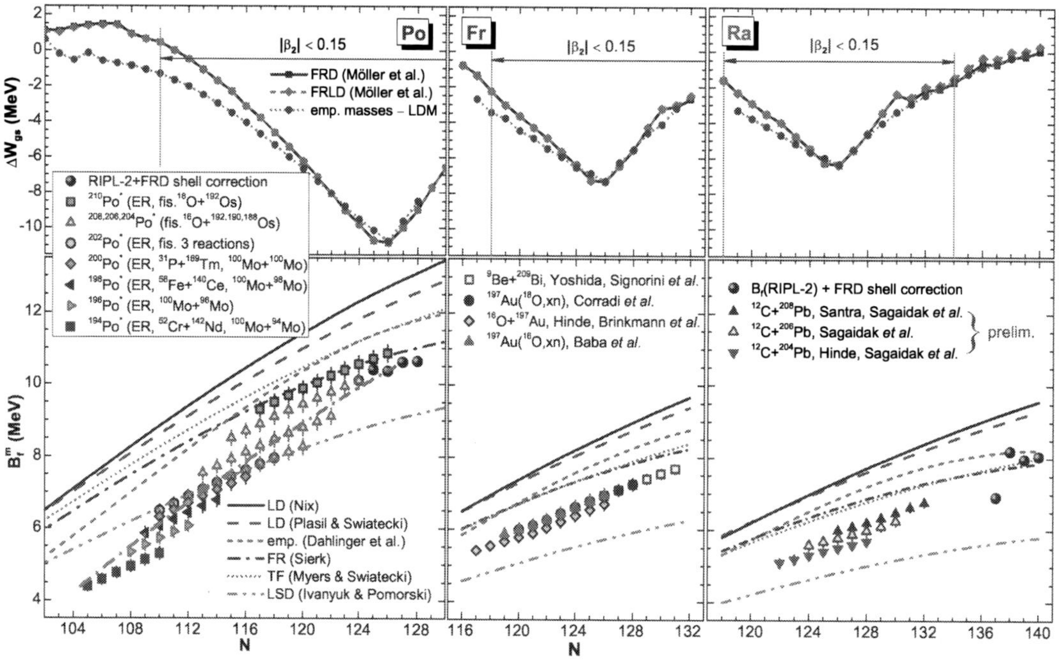

Figure 1. Bottom panels: the macroscopic fission barriers B_f^m for Po [1], Fr and Ra nuclei derived with the SSM [9] analysis of ER and fission excitation functions obtained in HI fusion reactions (symbols) in comparison to the different model predictions [5, 15, 16] (lines). Upper panels: the ground-state shell corrections ΔW_{gs} with the indicated regions of nearly spherical nuclei around N=126, corresponding to $|\beta_2| < 0.15$ [17].

Earlier, the fission barrier heights of Po and At nuclei (along with other parameters which determined the CN statistical decay) were derived using the only fission reactions induced by light charged particles (including He) on corresponding target nuclei [22]–[25]. At present, available experimental data on the production cross sections of Po and At ER produced in (^{3,4}He, xn) reactions [2] allow simultaneous analysis of the evaporation and fission cross section data as was done earlier [1] for some HI data. The main difference of the present approach from the previous fission data analysis [22]–[25] is the derivation of the nuclear potential parameters which are sensitive to the ER cross section data at barrier and sub-barrier energies, whereas fission data are more sensitive to the CN statistical decay parameters. Earlier, the fission data analysis [23]–[25] was performed using fusion cross sections *calculated* with the optical model.

The most representative experimental data were obtained in the ^{4}He+^{208}Pb reaction as shown in Fig. 2 for the $1n$ to $4n$ evaporation channels. The $1n$ and $2n$ evaporations allow choosing definitely the main parameter values of the nuclear exponential potential (V_0, r_0 and $\sigma(r_0)/r_0$)

which is used in the previous analysis of the HI Po data [1]. The evaporation data indicate importance of γ-emission at the CN de-excitation. Fission data (Fig. 3) are more sensitive to the level-density and fission barrier scaling parameters (a_ν, a_f/a_ν and k_f) of SSM. At high energies the critical angular momentum for fusion L_{cr} becomes a crucial value determining cross sections calculated for fission and for the $6n$ to $10n$ evaporations.

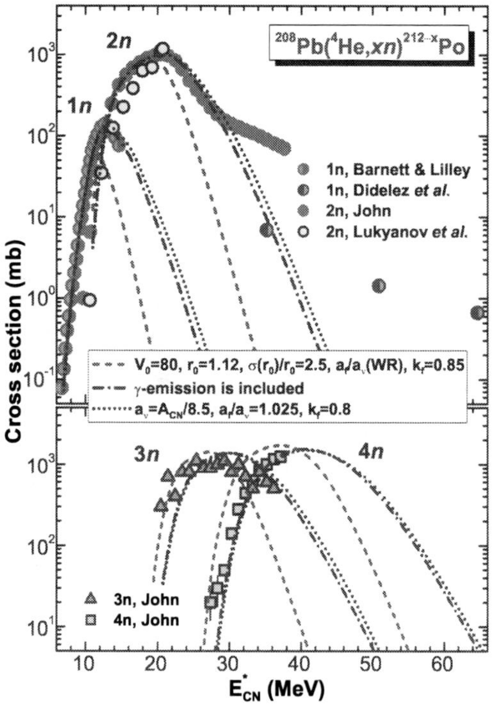

Figure 2. The $^{211-208}$Po isotopes production cross sections measured in the ^{208}Pb(^{4}He, $1n$–$4n$) reactions by different groups are shown by symbols (see Refs. in [2]). The corresponding excitation functions obtained with the SSM calculations using the best choice of the nuclear potential parameters and different sets of the CN-decay parameters are shown by lines.

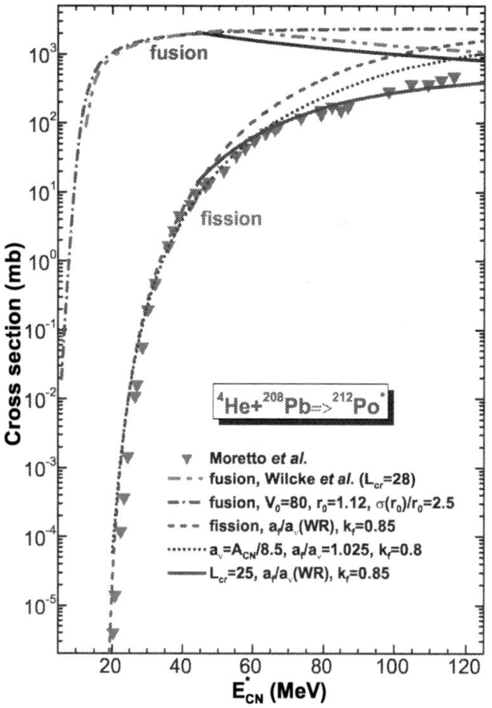

Figure 3. Fission cross sections measured in the ^{4}He+^{208}Pb reaction [24] (triangles) are shown together with the fission excitation functions calculated in the same as in Fig. 2 and with use of $L_{cr} = 25$ (lines). Fusion cross sections tabulated in [26] as well as calculated with the nuclear exponential potential are also shown for the reference.

In Table 1, the main parameters of the nuclear exponential potential and of the statistical decay of Po and At compound nuclei (see more details in [1]) are shown. Their values were varied within the SSM [9] data analysis and that corresponded to the best fit of calculations to the data [2] are indicated in Table 1. In the SSM description of the CN decay the preference was given to the values of the level density parameters calculated according to [9] and designated as a_f/a_ν(WR). The same approach was used earlier [1], although excitation functions calculated with $a_\nu = 8.5/A$ and $a_f/a_\nu = 1.05$ are very close to those corresponding to a_f/a_ν(WR).

In Table 2, as examples, the values of maxima and their positions for the xn-evaporation cross sections obtained in the most representative ^{4}He+Pb,Bi data [2] and in the corresponding SSM calculations with the parameter values listed in Table 1 are compared. Note that these calculations show a good fit to the measured fission cross sections (see Fig. 3).

In Fig. 4 the Po and At macroscopic fission barriers derived with the present analysis of

Table 1. The nuclear exponential potential and CN decay main parameters which were varied within the present SSM [9] data analysis. The values correspond to the best fit to the data [2].

Projectile	Target	V_0 (MeV/fm)	r_0 (fm)	$\sigma(r_0)/r_0$ (%)	L_{cr} ($\hbar$)	a_f/a_ν[9]	k_f	L_{cr} [26] ($\hbar$)
^{4}He	$^{208-206,204}$Pb	80	1.12	2.5	25	1.068–1.066	0.85	28
	^{209}Bi			3.5	26	1.0621		
^{3}He	^{208,207}Pb	60	1.25	8.5	16	1.068, 1.066		18
	^{209}Bi					1.0615		

Table 2. The values of maxima and their positions corresponding to the CN excitation energy for the (^{4}He,xn)-excitation functions obtained in experiments [2] (two upper lines) and in the present calculations using [9] with parameter values listed in Table 1 (next two bottom lines).

		Cross section maximum (mb) / Position of maximum (MeV)									
Target		1n	2n	3n	4n	5n	6n	7n	8n	9n	10n
^{208}Pb	Expt	135	1010	~1100	>1250		830	930	~400	242	142
		12.6	20.5	~30	>37		60.7	75.4	~87	102.4	111.7
	Calc	130	1040	1460	1610	1290	1020	730	550	377	256
		13	21	30	40	52	63	75	86	100	113
^{206}Pb	Expt		1050	>1380	>1270	1020	521	340	320	305	
			25	>30	>41	58.4	76.0	80.9	93.7	112.8	
	Calc	205	1020	1440	1560	1270	935	710	490		
		15	22	32	42	53	65	77	89		
^{209}Bi	Expt	144	954	1300	1090	736	603				
		12.3	21.2	27.7	40.5	55.0	63.4				
	Calc	202	1100	1400	1530	1280	980	640			
		13	22	31	42	53	65	76			

ER and fission excitation functions obtained in the ^{3,4}He fusion reactions are shown. As in the similar case of the HI reaction data analysis [1], the heights of the barrier in the region of $114 \leqslant N \leqslant 128$ are lower than any calculations [15] (with the exception of calculations in the framework of the LSD-model [16]).

4. Summary

ER and fission cross sections measured for ^{3,4}He reactions leading to the Po and At compound nuclei are analyzed in the framework of the Standard Statistical Model. This latter allows to reproduce reasonable well the cross sections in both channels. The macroscopic fission barriers derived for these nuclei are found lower than the calculated ones with different models, with the exception of those predicted by the LSD model [16] which underestimates the data. The same behavior has been observed earlier with HI- reaction data. Furthermore, the fission barriers for Po nuclei are in satisfactory agreement with those extracted from other authors with a similar SSM data analysis of very asymmetric HI-fusion reactions. These findings indicate that ^{3,4}He reactions can be used to verify the results of the HI data analysis using a similar approach, and that the derived macroscopic fission barriers (along with other SSM parameters) can provide an estimate of the fusion probability in more symmetric reactions

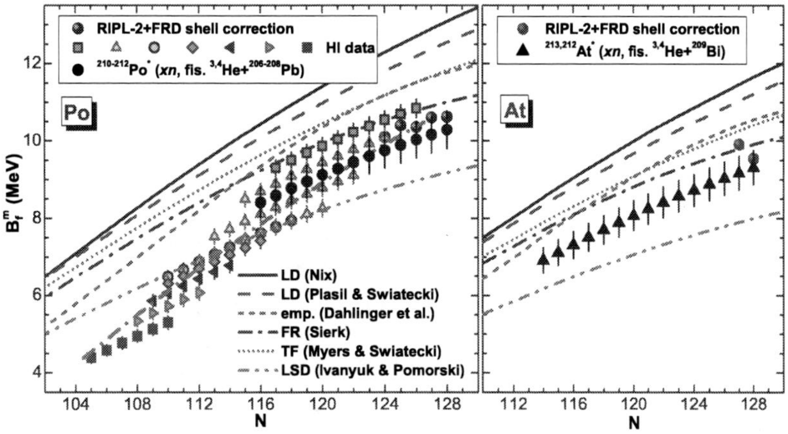

Figure 4. The same as in Fig 1 but for the Po and At macroscopic barriers derived with the analysis of ER and fission excitation functions obtained in ^{3,4}He fusion reactions (black symbols).

Acknowledgments
This work was partially supported by the RFBR Grant 13-02-12052.

References
[1] Sagaidak R N and Andreyev A N 2009 *Phys. Rev.* C **79** 054613
[2] Experimental Nuclear Reaction Data (EXFOR) `http://www.nndc.bnl.gov/exfor/exfor.htm`
[3] Mamdouh A, Pearson J M, Rayet M and Tondeur F 2001 *Nucl. Phys.* A **679** 337
[4] Belgya T *et al* 2006 *Handbook for calculations of nuclear reaction data, RIPL-2* IAEA-TECDOC-1506, `https://www-nds.iaea.org/ripl-2/`
[5] Dahlinger M, Vermeulen D and Schmidt K-H 1982 *Nucl. Phys.* A **376** 94
[6] Panov I V, Kolbea E, Pfeiffer B, Rauscher T, Kratz K-L and Thielemann F-K 2005 *Nucl. Phys.* A **747** 633
[7] Junghans A R, de Jong M, Clerc H-G, Ignatyuk A V, Kudyaev G A and Schmidt K H 1998 *Nucl. Phys.* A **629** 635
[8] Zagrebaev V I and Greiner Walter 2013 *Phys. Rev.* C **87** 034608
[9] Reisdorf W 1981 *Z. Phys.* A **300** 277; Reisdorf W and Schädel M 1992 *Z. Phys.* A **343** 47
[10] Cohen S, Plasil F and Swiatecki W J 1974 *Ann. of Phys.* **82** 557
[11] Corradi L. *et al* 2005 *Phys. Rev.* C **71** 014609
[12] Hinde D J, Berriman A C, Butt R D, Dasgupta M, Gontchar I I, Morton C R, Mukherjee A and Newton J O 2002 *J. Nucl. Rad. Sci.* **3** 31.
[13] Sagaidak R N *et al* 2003 *Phys. Rev.* C **68** 014603.
[14] Santra S, Singh P, Kailas S, Shrivastava A and Mahata K 2001 *Phys. Rev.* C **64** 024602
[15] Nix J R 1969 *Nucl. Phys.* A **130** 241; Plasil F and Swiatecki W J 1973 in Vandenbosch R and Huizenga J R *Nuclear fission* (New York: Academic Press) p 246; Sierk A J 1986 *Phys. Rev.* C **33** 2039; Myers W D and Świątecki W J 1999 *Phys. Rev.* C **60** 014606
[16] Ivanyuk F A and Pomorski K 2009 *Phys. Rev.* C **79** 054327
[17] Möller P, Nix J R, Myers W D and Swiatecki W J 1995 *At. Data Nucl. Data Tables* **59** 185
[18] Stefanini A M *et al* 2005 *Eur. Phys. J.* A **23** 473
[19] Sagaidak R N 2012 *EPJ Web of Conferences* **21** 06001
[20] Sagaidak Roman and Andreyev Andrei 2014 *Int. J. Modern Phys.* E **23** 1450001
[21] Cavinato M *et al* 1995 *Phys. Rev.* C **52** 2577; Joosten R *et al* 2000 *Phys. Rev. Lett.* **84** 5066
[22] Becchetti F D, Hicks K H, Fields C A, Peterson R J, Raymond R S, Ristinen R A, Ullmann J L and Zaidins C S 1983 *Phys. Rev.* C **28** 1217
[23] Ignatyuk A V, Smirenkin G N, Itkis M G, Mulgin S I, and Okolovivh V N 1985 *Sov. J. Part. Nucl.* **16** 307
[24] Moretto L G, Jing K X, Gatti R, Wozniak G J and Schmitt R P 1995 *Phys. Rev. Lett* **75**, 4186
[25] Rubehn Th, Jing K X, Moretto L G, Phair L, Tso K, and Wozniak G J 1996 *Phys. Rev.* C **54** 3062
[26] Wilcke W W, Birkelund J R, Wollersheim H J, Hoover A D, Huizenga J R, Schröder W U and Tubbs L E 1980 *At. Data Nucl. Data Tables* **25** 391

Search for the Cosmic Neutrino Background

A Faessler[1], R Hodak[2], S Kovalenko[3], F Simkovic[4,5]

[1] Institute für Theoretische Physik der Universität Tübingen,
D-72076 Tübingen, Germany; e-mail: faessler@uni-tuebingen.de;
[2] Institute of Experimental and Applied Physics, Czech Tecnical University,
Horská 3a/22, 12800 Prague, Czech Republic.
[3] Universidad Técnica Federico Santa María,
Centro-Científico-Tecnológico de Valparaíso,
Casilla 110-V, Valparaíso, Chile
[4] JINR, 141980 Dubna, Moscow Region, Russia
[5] Comenius University, Physics Dept., SK-842 15 Bratislava, Slovakia.

E-mail: faessler@uni-tuebingen.de

Abstract.
One expects three Cosmic Backgrounds: (1) The Cosmic Microwave Background (CMB) originated 380000 years after the Big Bang (BB). (2) The Neutrino Background decoupled about one second after the BB, while (3) the Cosmic Gravitational Wave Background created by the inflationary expansion decoupled directly after the BB. Only the Cosmic Microwave Background (CMB) has been detected and is well studied. Its spectrum follows Planck's black body radiation formula and shows a remarkable constant temperature of $T_{0\gamma} \approx 2.7$ K independent of the direction. The present photon density is about 370 photons per cm^3. The size of the hot spots, which deviates only in the fifth decimal of the temperature from the average value, tells us, that the universe is flat. About 380 000 years after the Big Bang at a temperature of $T_{0\gamma} = 3000$ K already in the matter dominated era the electrons combine with the protons and ^4He and the photons move freely in the neutral universe and form the CMB. So the temperature and distribution of the photons give us information of the universe 380 000 years after the Big Bang. The Cosmic Neutrino Background ($C\nu B$) decoupled from matter already one second after the BB at a temperature of about 10^{10} K. Today their temperature is $\sim$ 1.95 K and the average density is 56 electron-neutrinos and the total density of all neutrinos about 336 per cm^3. Measurement of these neutrinos is an extremely challenging experimental problem which can hardly be solved with the present technologies. On the other hand it represents a tempting opportunity to check one of the key elements of the Big Bang Cosmology and to probe the early stages of the universe. The search for the $C\nu B$ with the induced beta decay $\nu_e + ^3H \rightarrow ^3He + e^-$ using KATRIN (KArlsruhe TRItium Neutrino experiment) is the topic of this contribution.

1. Introduction

Penzias and Wilson [1] detected the Cosmic Microwave Background (CMB) in 1964 (Nobel Prize 1978) as byproduct of their search for possible perturbations of the communication with satellites. The frequency and wave length distribution follows exactly Planck's black body formula and yields a surprisingly identical temperature of four digits independent of the direction

($T_{0\gamma}$ = 2.7255(6) Kelvin). This led to the 'inflationary' model. The satellite observations [2, 3, 4] show deviations of the constant temperature of the background photons in the fifth digit.

Recently the BICEP2 collaboration [5] claimed to have seen in the fluctuations and the polarization of the CMB fingerprints of the Gravitational Wave Background originating from the Inflationary Expansion during the BB.

The neutrinos due to their weak interaction decouple earlier from matter about 1 second after the BB at a temperature of $T_{0\nu} \approx 1MeV \approx 10^{10}\,Kelvin$. The Cosmic Neutrino Background ($C\nu B$) of today $T_{0\nu} \approx 1.95$ K contains therefore information of the universe about ~ 1 second after the BB. The detection of this $C\nu B$ seems to be hardly possible due to the weak interaction of neutrinos with matter and due to their low energy ($T_\nu = 1.95\text{ K} \approx 2 \cdot 10^{-4}[eV]$). Nevertheless several methods have been discussed in the literature to search for these relic neutrinos [6, 7, 8, 9, 10, 11, 12, 13, 14]. The search for the $C\nu B$ with the induced beta decay [6, 7] seems to be the most promising.

$$\nu_e +^3 H \rightarrow^3 He + e^-. \tag{1}$$

The signal would show up by a peak in the electron spectrum with an energy of the neutrino mass above the Q value. We discuss prospects of searches for the reaction (1) with the KATRIN experiment [16].

2. The Microwave Background

The detection and measurement of the Cosmic Background Radiation in 1964 by Penzias and Wilson [1] surprised, because the radiation followed very accurately Planck's black body radiation law:

$$\epsilon(f)df = \frac{8\pi h}{c^3} \cdot \frac{f^3 df}{exp(hf/k_B T_0) - 1}\,[Energy/Volume] \tag{2}$$

The temperature parameter T_0 fitted by (2) is independent of the observation direction up to the fourth digit. The average temperature of the photons in the CMB is about $3 \cdot T_0$.

$$T_{0,rad} \equiv T_{0,\gamma} = 2.7255 \pm 0.0006\,[Kelvin] \tag{3}$$

Integration over all frequencies f yields the Stefan-Boltzmann law for the energy density.

$$\epsilon_{rad} = \varrho_{rad} c^2 = \alpha T_0^4; \tag{4}$$

The photons decouple from matter as soon as the electrons get bound to the protons and to the ^4He nuclei and the universe is electrically neutral.

3. Decoupling of the Cosmic Neutrino Background

The neutrinos decouple much earlier from matter than the photons due to their very weak interaction. The competition for decoupling of the neutrinos is between the expansion rate of the universe given by the Hubble constant

$$H = \frac{\dot{a}}{a} = \sqrt{\frac{8\pi G}{3}\varrho_{total}} = \sqrt{\frac{8\pi\varrho}{3M_{Planck}^2}} \propto T_0^2 \propto a^{-2}; \tag{5}$$

Journal of Physics: Conference Series **580** (2015) 012040 doi:10.1088/1742-6596/580/1/012040

and the reaction rate for relativistic neutrinos:

$$\Gamma = n_\nu < \sigma v > \approx T_0^3 G_{Fermi} T_0^2 = G_F T_0^5 \propto a^{-5}; \qquad v \approx c = 1. \tag{6}$$

For decreasing temperature T the neutrinos decouple, when the Hubble expansion rate (5) is about equal to the neutrino reaction rate (6):

$$\text{in the radiation dominated era:} \quad T_0 = \left(\frac{45\hbar^3 c^5}{32\pi^3 G}\right)^{1/4} \cdot \frac{1}{t^{1/2}} \equiv 1.3\left(\frac{t}{sec}\right)^{-1/2} [MeV];$$

This corresponds to the time after the Big Bang of about 1 *second* and a temperature of 1 MeV. Today the temperature is 1.95 Kelvin.

4. Search for the Cosmic Neutrino Background with KATRIN.

The tritium beta decay probability is according to Fermi's Golden Rule given by:

$$\Gamma^\beta_{decay}(^3H) = \frac{1}{2\pi^3} \cdot \sum \int | <^3 He|T|^3 H > |^2 \cdot 2\pi\delta(E_\nu + E_e + E_f - E_i) \frac{d\vec{p}_e}{2\pi^3} \cdot \frac{d\vec{p}_\nu}{2\pi^3}; \tag{7}$$

The half life is then:

$$Theory: \quad T^\beta_{1/2} = \frac{\ln 2}{\Gamma^\beta_{decay}(^3H)} = 12.32 \ years; \qquad Experiment: T^\beta_{1/2} = 12.33 \ years; \tag{8}$$

The same reduced Fermi and GT transitions are needed for the determination of the induced relic neutrino capture reaction (1).

$$\begin{aligned} \Gamma^\beta_{capture}(^3H) &= \frac{1}{\pi}(G_F \cos(\vartheta_C))^2 F_0(Z+1, T_e)[B_F(^3H) + B_{GT}(^3H)] p_e T_e \cdot < n_{\nu,e} > \frac{n_{\nu,e}}{< n_{\nu,e} >} \\ &= 4.2 \cdot 10^{-25} \frac{n_{\nu,e}}{< n_{\nu,e} >} [\text{for 1 tritium atom/year}]; \\ & with: \ < n_{\nu,e} > = 56 \ cm^{-3}; \end{aligned} \tag{9}$$

The Beta capture expression contains Fermi and Gamow-Teller contributions. $F_0(Z+1, T_e)$ is the Fermi function [17], which takes into account the Coulomb distortion of the outgoing s-electron with the asymptotic kinetic energy T_e in the final nucleus with the charge $Z+1$. The matrix elements can be calculated analytically by angular momentum and isospin algebra [6].

Different values are given in the literature for this effective strength of the Tritium source [18] and [6]. Drexlin [19] told us, that the correct value is $20\mu g$. This means $2 \cdot 10^{18}$ $Tritium_2$ molecules. The capture rate of relic neutrinos is then:

$$\text{Capture rate at KATRIN: } N_\nu(KATRIN) = 1.7 \cdot 10^{-6} \cdot \frac{n_{\nu,e}}{< n_{\nu,e} >}; \tag{10}$$

For the average relic neutrino number density $< n_{\nu,e} > = 56 \, cm^{-3}$ this corresponds to every 590 000 years a count. So the critical question is how much is the number density of the relic neutrinos increased by gravitational clustering in the solar system or better in our galaxy. If one assumes the result of Ringwald and Wong [8], that relic neutrinos can cluster on the scale of a single galaxy and their halo and uses the proportionality to the baryon overdensity of Lazauskas et al. [7], then one can expect very optimistically overdensities up to $n_{\nu,e}/<n_{\nu,e}>\leq 10^6$ in our neighbourhood. With this optimistic estimate of the upper limit for the relic neutrino overdensity of 10^6 one obtains from equation (10) :

$$N_\nu(KATRIN) = 1.7 \cdot 10^{-6} \cdot \frac{n_{\nu,e}}{<n_{\nu,e}>}\,[year^{-1}] \approx 1.7\,[counts\,per\,year]; \tag{11}$$

This seems not possible to measure for the moment. One way out would be to increase the effective activity of the tritium source. An effective mass of 2 milligrams Tritium would mean with the above optimistic estimate of the relic neutrino number overdensity $n_{\nu,e}/<n_{\nu,e}>\approx 10^6$ about 170 counts per year, which should be feasible.
The possibility to increase the the Tritium source strength can be seen from figure 15 of the KATRIN Design Report [20]. The increase of the tritium source strength is limited by the scattering of the emitted electrons by the tritium gas. After the mean free path

$$\lambda_{mean\ free\ path} = \frac{1}{\rho \cdot \sigma(electron - tritium)} = d_{free} \tag{12}$$

only about 37% decay electrons have not yet scattered. All the others including also electrons with the maximum energy, which contain the information on the neutrino mass and on the relic neutrino capture by tritium, are lost for the measurement. A detailed analysis [20] shows, that the maximum number of unscattered decay electrons can escape the source within the last area of the tritium gas of a width of half the mean free path. Thus to increase the column density (number of tritium atoms in a column with the base area $1cm^2$ and a specific length d of the column) of the tritium gas source beyond this value does not increase the the effective source strength, but yields only more background. Thus an increase of the tritium source strength with the present geometry a source with the area of about 50 cm^2; 8 cm in diameter is not able to increase the source strength. The increase of the source area to 5000 cm^2, 80 cm $diameter$ looks like a way out. But also this seems to be impossible.
The magnetic flux at the present source is given by

$$MagneticFlux(source) = 53\ cm^2 \cdot 3.6\ Tesla \approx 190\ Tesla \cdot cm^2, \tag{13}$$

which must be the same in the spectrometer.

$$MagneticFlux(spectrometer) = 63.6\ m^2 \cdot 3\ Gauss \approx 190\ Tesla \cdot cm^2, \tag{14}$$

The low magnetic field in the spectrometer of only 3 $Gauss$ is needed to transform the electron momenta, which are at the source almost perpendicular to the beam direction due to the cyclotron motion, into almost a translational direction (see figure 9 of the KATRIN design report [20]), to reject with an electric opposing field all electrons almost up to the tritium Q-value ($Q = 18.562\ keV$). To increase the source strength by a factor 100 with a corresponding increase of the source area to 5000 cm^2, 80 cm diameter, one needs also to increase the spectrometer cross section by a factor 100 or the diameter to about 90 m, which is not possible.

Till now we did not include the requirement for an energy resolution of the spectrometer of about $\Delta E \approx 1.0\,eV$ (see ref. [16]). The energy resolution of a KATRIN type spectrometer is determined by the perpendicular energy of the decay electrons in the spectrometer in the cyclotron motion $E_{f\perp}$. The electrons with a relatively large perpendicular energy in the spectrometer with longitudinal energy just below the Q-value cant be rejected by the opposing electric field and therefore arrive all in the detector. The angular momentum in the circular cyclotron motion of an electron and also the corresponding magnetic moment of this ring current conserve the ratio of the perpendicular energy over the magnetic field of the electrons:

$$|\vec{L}| = |\vec{r} \times \vec{p}| \propto \mu = const \propto \frac{E_{i\perp}}{B_i} = \frac{E_{f\perp}}{B_f} \tag{15}$$

An energy resolution of about $\Delta E = 1\ eV$ thus requires:

$$\Delta E = 1\ eV = E_{f\perp} = \frac{B_f}{B_i} \cdot E_{i\perp} = \frac{3\ Gauss}{360\ Gauss} E_{i\perp} \tag{16}$$

Thus at the Tritium source one can have only $E_{i\perp} = 120\ eV$ in the perpendicular motion. The rest of the Q-value of $18.5\ keV$ must be in the longitudinal direction at the source. (Only the momentum of the electrons is a vector and can be decomposed in a longitudinal part and a transversal part. The longitudinal and the transversal energy do not add up to the Q-value.) This allows only a cone with a small opening of $\vartheta_i = 5.7°$ relative to the beam axis. The accepted space angle of the emitted electrons is therefore reduced to $\Delta\Omega/(2 \cdot \pi) = 0.005 \rightarrow 0.5\ \%$. Thus this requirement for the energy resolution reduces the electron beam intensity by a factor 1/200. As a whole with an increase of 100 from the size of the source and a decrease of 1/200 due to the required energy resolution one looses intensity.
With the KATRIN spectrometer and the resolution of $\Delta E = 0.93\ eV$ and the background one hopes to reduce the upper limit of the electron neutrino mass to about $0.2eV$ (90% $C.L.$). Fitting at the upper end of the Kurie plot at $Q - m_{\nu,e}$ the electron spectrum the KATRIN collaboration hopes to determine the Q-value and the neutrino mass. The electron peak due to the capture of the relic neutrinos lies at $Q + m_{\nu,e}$. The neutrino mass and the energy resolution and the background remain the same as for the determination of the neutrino mass. One has only one additional fit parameter more (or two, if one counts the width of this peak,), the counts in the peak at $Q + m_{\nu,e}$. At the moment it does not seem possible to detect with a KATRIN type spectrometer the Cosmic Neutrino Background. But one should be able to give an upper limit for the local relic neutrino overdensity $n_{\nu,e}/ < n_{\nu,e} >$ in our Galaxy.

5. Summary.

- With the average relic electron neutrino number density of $< n_{\nu,e} > = 56\ cm^{-3}$ KATRIN could measure only every 590 000 years a count. So the hope is with the local overdensity due to gravitational clustering of the neutrinos in our galaxy. Estimates for this overdensity $n_{\nu,e}/ < n_{\nu,e} >$ vary widely from about 10^2 to 10^6. With the optimistic estimate of a local overdensity of 10^6 one obtains with KATRIN 1.7 counts per year. If one could increase the effective mass of the tritium source from 20 micrograms to 2 milligrams, this optimistic estimate of the overdensity would mean 170 counts per year or every second day a count. Detection would be possible.
- The second problem could be the energy resolution of the KATRIN spectrometer of $\Delta E = 0.93\ \ eV$. With this resolution one expects to extract from the electron spectrum at the upper end of the Kurie plot at $Q - m_{\nu,e}$ a very accurate Q-value of the accuracy of milli-eV and an upper limit of the electron neutrino mass of about $m_{\nu,e} \leq 0.2\ \ [eV]\ \ 90\%\ \ C.L.$ with an energy resolution of about $1.0\ eV$ and the background. If one can fit the Q value and the electron neutrino mass accurately enough, the position of the electron peak from the induced capture of the relic neutrinos is known to be at an electron energy of $Q + m_{\nu,e}$. The energy resolution and the background is the same at $Q - m_{\nu,e}$ and at $Q + m_{\nu,e}$. One should with KATRIN at least be able to determine an upper limit for the local relic neutrino density.

Acknowledgments:

We thank Christian Weinheimer and Guido Drexlin for discussions about the KATRIN spectrometer, the Deutsche Forschungsgemeinschaft for support within the project "Nuclear matrix elements of Neutrino Physics and Cosmology" FA67/40-1 and by the grant of the Ministry of Education and Science of the Russian Federation (contract 12.741.12.0150). F. S. acknowledges the support by the VEGA Grant agency of the Slovak Republic under the contract No. 1/0876/12 and by the Ministry of Education, Youth and Sports of the Czech Republic under contract LM2011027. S. K. thanks for support by the FONDECYT grant 1100582 and the Centro Cientifico-Tecnológico de Valparaíso PBCT ACT-028.

6. References

[1] Penzias A A, Wilson R W 1965 *Astrophys. J.* **142** 419
[2] Boggess N W, et al. 1992 *Astrophys. J.* **397** 420
[3] Bennett C L 2012 **arXiv:1212.5225** [astro-ph]
[4] Planck collaboration 2013 *submitted to Astronomy and Astrophysics* (March 21, 2013)
[5] BICEP2 Collaboration **arXiv: 1403.3985v2 [astro-ph]**, March 18. 2014
[6] Faessler A, Hodak R, Kovalenko S, Simkovic F 2011 *J. Phys. G.* **38** 075202 and **arXiv:1102.1799** [hep-ph]
[7] Lazauskas R, Vogel P, Volpe C 2008 *J. Phys. G.***35** 1
[8] Ringwald A, Wong Y 2004 *J. Cosmology and Astroparticle Phys.* **0412** 005
[9] Wigmans R 2003 *Astropart. Phys.* **19** 379
[10] Ringwald A 2003 *arXiv:0301157* [hep-ph]
[11] Weiler T 1982 *Phys. Rev. Lett.* **49** 234
[12] de Vega H J, Moreno O, Moya de Guerra E, Ramon Medrano M, Sanchez N G 2012 *Nucl. Phys.* **B866** 177 and **arXiv: 1109.3452** [hep-ph]
[13] Munyaneza F, Biermann P L 2006 *Astronomy and Astrophysics* **458** L9
[14] Päs H, Weiler T J 2001 *Phys. Rev. D* **63** 113015.
[15] Hagmann C 1998 *COSMO 98, Asilomar* **arXiv:9902102** [astro-ph]
[16] Drexlin G, Hannen V, Mertens S, Weinheimer C 2013 *Advances in High Energy Physics* **2013** 293986
[17] Doi M, Kotani D, Takasugi E 1985 *Prog. Theor. Phys. (Supp.)* **83** 1
[18] Kaboth A, Formaggio J A, Monreal B 2010 *Phys. Rev. D* **82** 062001
[19] Drexlin G, Weinheimer Ch, private communication 2013
[20] KATRIN Design Report 2004, FZKA Scientific Report**7090**; MS-KP-0501; http://www-ik.fzk.de/ katrin/publications/documents/DesignReport2004-12Jan2005.pdf

The role of symmetry in nuclear physics

Francesco Iachello

Center for Theoretical Physics, Sloane Laboratory, Yale University,
New Haven, Connecticut 06520-8120, USA

E-mail: francesco.iachello@yale.edu

Abstract. The role of discrete symmetries in nuclear physics is briefly reviewed within the context of the algebraic cluster model (ACM). The symmetries D_3 (triangle) for 3α and T_d (tetrahedron) for 4α are discussed and evidence shown for their occurrence in ^{12}C (D_3) and ^{16}O (T_d).

1. Continuous symmetries

Continuous symmetries were introduced in nuclear physics as early as the 1930's. In 1932 Heisenberg introduced isospin, $SU_T(2)$, in 1937 Wigner combined isospin with spin to $SU(4) \supset SU_T(2) \otimes SU_S(2)$, in the 1940's Racah developed the group theory of the shell model in ℓs and jj coupling, $U\left((2\ell+1)(2s+1)\right)$ and $U(2j+1)$, and in 1958 Elliott introduced the symmetry of mixed configurations, in particular the symmetry of the *sd* and *pf* shells. The use of continuous symmetries culminated in 1974 with the introduction of the interacting boson model, Arima-Iachello $U(6)$. The role of these symmetries in nuclear physics is well documented and will not be discussed here.

2. Discrete symmetries

Discrete symmetries were also introduced in the early days of nuclear physics, mostly within the framework of the α-particle model, by Wheeler in 1937 [1] and Dennison in 1954 [2] and later exploited by Brink [3], Robson [4] and others. We have recently readdressed this problem within an algebraic description of clustering, the Algebraic Cluster Model (ACM), and in this contribution some recent results will be reported.

The algebraic cluster model is a description of cluster states as representations of $U(3k-2)$, where k is the number of constituents. The elements of the algebra are the bilinear products of boson creation and annihilation operators which are a bosonic quantization of the Jacobi variables and their associated momenta plus an additional s boson. For $k=2$, there is only one Jacobi vector, $\boldsymbol{\rho} = \mathbf{r}_1 - \mathbf{r}_2$, for k=3 there are two vectors, $\boldsymbol{\rho} = (\mathbf{r}_1 - \mathbf{r}_2)/\sqrt{2}$, $\boldsymbol{\lambda} = (\mathbf{r}_1 + \mathbf{r}_2 - 2\mathbf{r}_3)/\sqrt{6}$, and for k=4 there are three vectors, $\boldsymbol{\rho} = (\mathbf{r}_1 - \mathbf{r}_2)/\sqrt{2}$, $\boldsymbol{\lambda} = (\mathbf{r}_1 + \mathbf{r}_2 - 2\mathbf{r}_3)/\sqrt{6}$, $\boldsymbol{\eta} = (\mathbf{r}_1 + \mathbf{r}_2 + \mathbf{r}_3 - 3\mathbf{r}_4)/\sqrt{12}$, where $\mathbf{r}_i$ are the coordinates of the constituent particles. For two-body clusters, ACM was introduced years ago for applications to molecules [5] and nuclear molecules [6]. For three-body clusters ACM was introduced in [7], and discussed in [8], and for four-body clusters was introduced in [9], [10] and discussed, very recently, in [11], [12]. The discrete symmetries of the α-particle model discussed in those articles and here are shown in Table 1.

k	Nucleus	$U(3k-2)$	Discrete symmetry	Jacobi variables
2	^{8}Be	$U(4)$	Z_2	$\boldsymbol{\rho}$
3	^{12}C	$U(7)$	D_3	$\boldsymbol{\rho}, \boldsymbol{\lambda}$
4	^{16}O	$U(10)$	T_d	$\boldsymbol{\rho}, \boldsymbol{\lambda}, \boldsymbol{\eta}$

Table 1. Discrete symmetries of the α-particle model considered in this article.

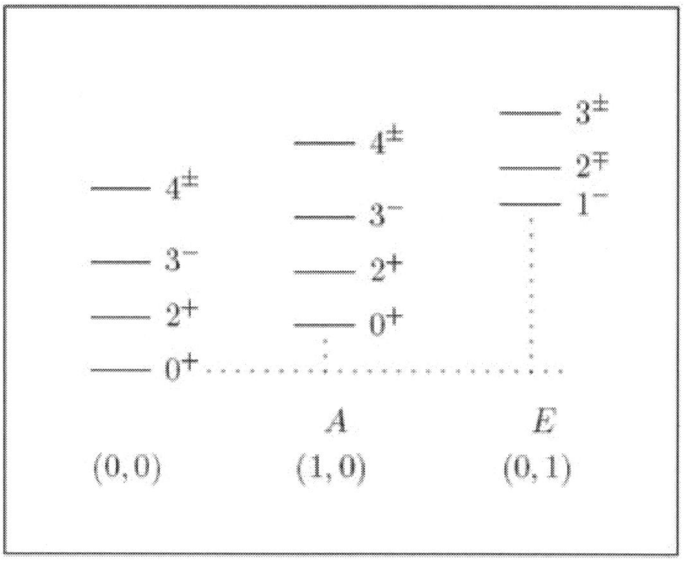

Figure 1. States of 3α with D_3 symmetry.

Within the algebraic cluster model (ACM) it is possible to derive many analytical results. In particular, for a rigid roto-vibrator one has explicit expressions for the energy levels in terms of the angular momentum L and the vibrational quantum numbers $v_i (i = 1, 2...)$. These are:

$$\begin{array}{ll} 2\alpha(Z_2) & E(v,L) = E_0 + \omega\left(v+\frac{1}{2}\right) + \kappa L(L+1) \\ 3\alpha(D_3) & E(v_1,v_2,L) = E_0 + \omega_1\left(v_1+\frac{1}{2}\right) + \omega_2\left(v_2+1\right) \\ & \quad +\kappa L(L+1) \\ 4\alpha(T_d) & E(v_1,v_2,v_3,L) = E_0 + \omega_1\left(v_1+\frac{1}{2}\right) + \omega_2\left(v_2+1\right) \\ & \quad +\omega_3\left(v_3+\frac{3}{2}\right) + \kappa L(L+1) \end{array} . \tag{1}$$

Spectra are characterized by the representations of the discrete group G and consist in a set of rotation-vibration bands, with specific values of the angular momentum and parity. Representations can be labeled either by G or by the isomorphic group S_n (the permutation group). The conversion from G to S_n is: $G = Z_2 \sim S_2 \sim P$, $A \equiv [2]$; $G = D_3 \sim S_3$, $A \equiv [3]$, $E \equiv [21]$; $G = T_d \sim S_4$, $A \equiv [4]$, $F \equiv [31]$, $E \equiv [22]$. Fig.1 shows the expected rotation-vibration spectra of states for 3α with D_3 symmetry. For A representations, the rotational band has $L^P = 0^+, 2^+, 3^-, 4^\pm, ...$, while for E representations the rotational band has $L^P = 1^-, 2^\mp, 3^\pm, ...$. Note the unusual angular momentum and parity content of the rotational bands. This content is not the same as that of the rotational bands of a rotating ellipsoidal shape. [See, for example, the $SU(3)$ limit of the IBM].

In Fig.2, the expected spectrum of a 4α configuration with T_d symmetry is shown. For

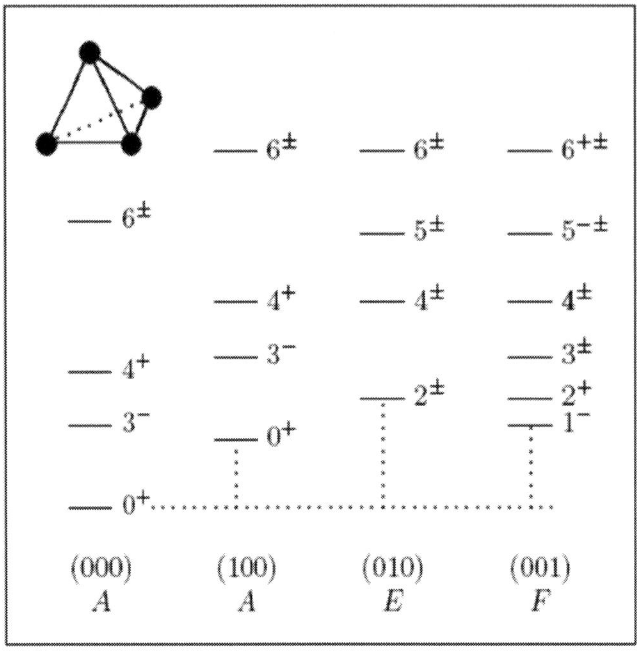

Figure 2. States of a 4α configuration with T_d symmetry.

A representations, the spectrum consists of a rotational band $L^P = 0^+, 3^-, 4^+, 6^\pm, ...$, for E representations $L^P = 2^\pm, 4^\pm, 5^\pm, ...$, and for F representations $L^P = 1^-, 2^+, 3^\pm, ...$

The breaking of representations of $U(3k-2)$ into those of S_n and thus the determination of the angular momentum content of each band is one of the novel results of ACM.

The occurrence of D_3 symmetry in ^{12}C has been confirmed by a very recent experiment [13], [14] and its evidence is shown in Figs. 3 and 4.

The occurrence of T_d symmetry in ^{16}O was discussed long ago by Robson [4], and it has been emphasized recently in [11].The evidence is shown in Figs. 5 and 6.

In addition to energy spectra one can also derive within the ACM, analytic expressions for other observables, most notably electromagnetic transition rates. For example, for 4α configurations with T_d symmetry, we have

$$B(EL; 0^+ \to L^P) = \left(\frac{Ze\beta^L}{4}\right)^2 \frac{(2L+1)}{4\pi}\left[4 + 12P_L\left(-\frac{1}{3}\right)\right] \tag{2}$$

and the form factors in electron scattering are

$$F_L\left(0^+ \to L^P\right) = c_L j_L\left(q\beta\right) \tag{3}$$

with $c_0^2 = 1; c_3^2 = \frac{35}{9}; c_4^2 = \frac{7}{3}; c_6^2 = \frac{32}{81}$.

The occurrence of T_d symetry in ^{16}O is confirmed by the $B(EL)$ values given in Table 2.

3. Conclusion

In conclusion, the role of discrete symmetry in cluster physics is that of providing benchmarks for energy levels and other obeservables that can be used to analyze data. In particular, by

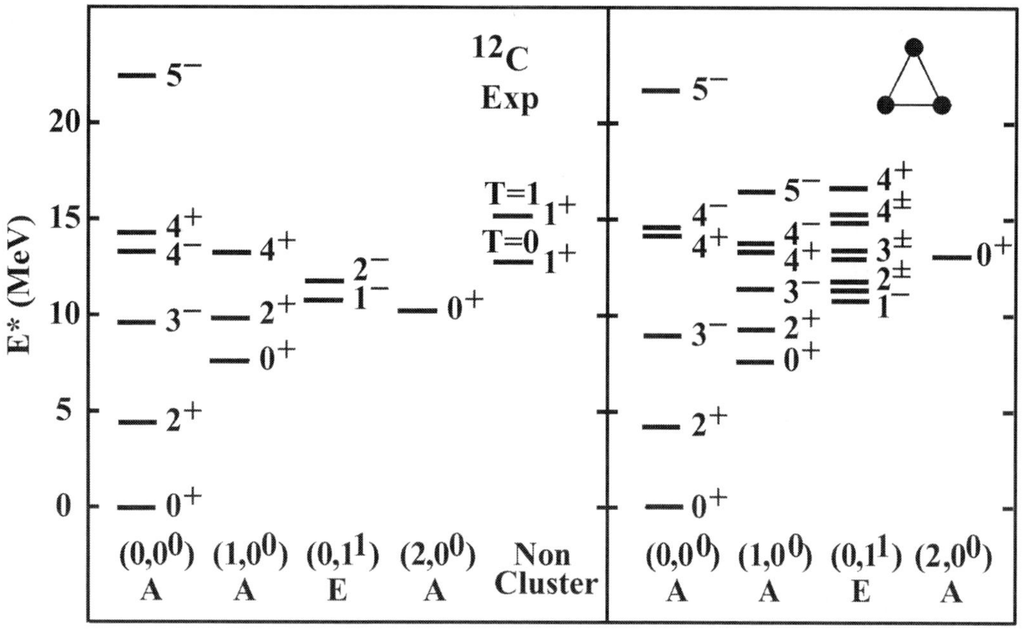

Figure 3. The experimental spectrum of ^{12}C(left) and its comparison with the theoretical spectrum with D_3 symmetry (right). The lowest non-cluster states are also shown. Reproduced with permission from [13].

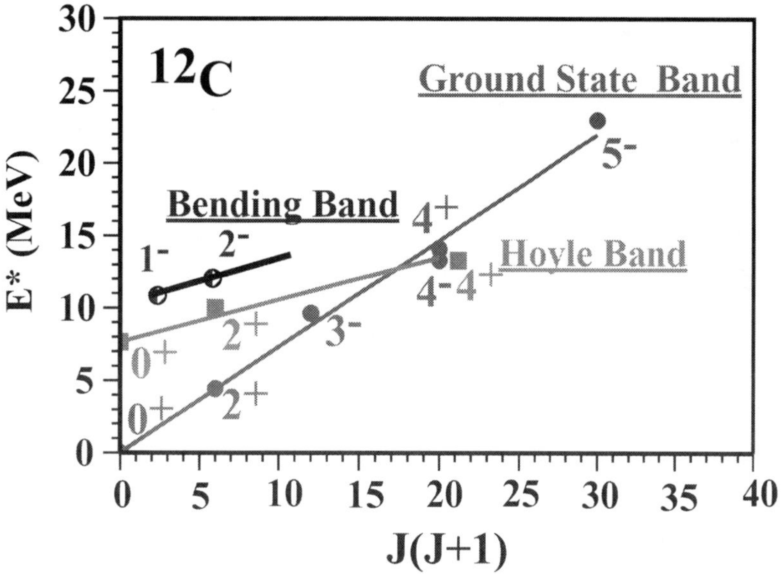

Figure 4. Observed cluster rotational bands in ^{12}C. Reproduced with permission from [13].

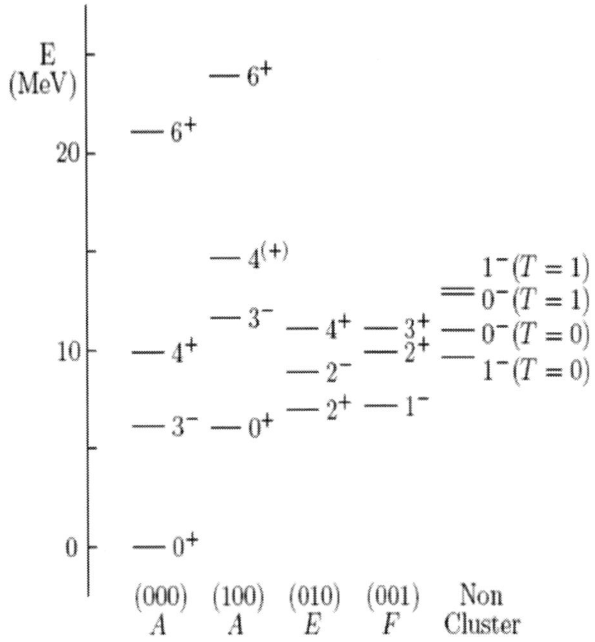

Figure 5. The experimental spectrum of ^{16}O showing evidence for T_d symmetry. The lowest non-cluster states are also shown.

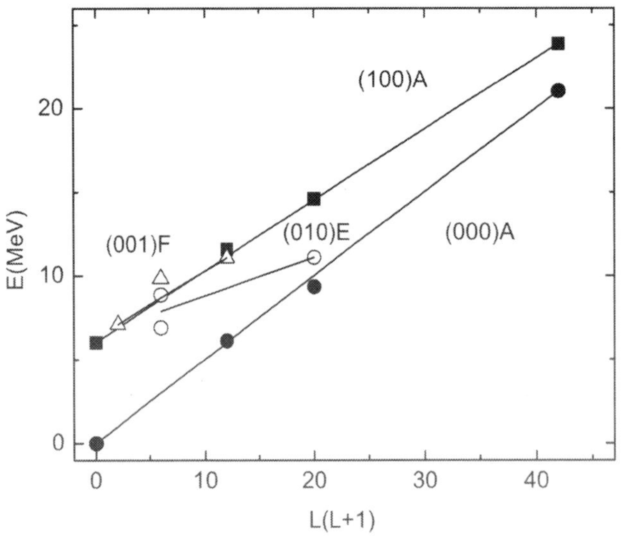

Figure 6. Observed cluster rotational bands in ^{16}O.

$B(EL;L^P \to 0^+)$	Th	Exp	$E(L^P)$	Th	Exp
$B(E3;3^- \to 0^+)$	181	205±10	$E(3^-)$	6132	6130
$B(E4;4^+ \to 0^+)$	338	378±133	$E(4^+)$	10220	10356
$B(E6;6^+ \to 0^+)$	8245		$E(6^+)$	21462	21052

Table 2. $B(EL)$ values and energies in ^{16}O compared with those expected from a T_d symmetry. $B(EL)$ values in e^2fm^{2L} and E in keV. The theoretical energies in column 5 are calculated from $E(keV) = 511\ L(L+1)$. The value of β in Eq.(2) is extracted from the elastic form factor measured in electron scattering, $\beta = 2.0$fm.

providing analytic expression for rotation-vibration spectra as well as electromagnetic transition rates and form factors. Strong evidence for the occurrence of D_3 symmetry in ^{12}C and of T_d symmetry in ^{16}O has been presented. [There is evidence for Z_2 symmetry in ^{8}Be but this has not been presented here].

The occurrence of D_3 and T_d symmetry is also confirmed by microscopic calculations in the (i) molecular orbital method [15] and in (ii) lattice calculations [16].

Cluster states are instead very difficult to describe within the framework of the one-center shell model, where one needs multiparticle-multihole configurations and a model space with several $\hbar\omega$ quanta of excitation.

Acknowledgements

This work was supported in part by U.S.D.O.E. Grant DF-FG02-91ER40608. It is dedicated to Aldo Covello on the occasion of his retirement.

References

[1] Wheeler JA 1937 *Phys. Rev.* **52** 1083
[2] Dennison DM 1954 *Phys. Rev.* **96** 378
[3] Brink DM 1965 Proceedings of the International School of Physics "Enrico Fermi", Course XXXVI, 247
[4] Robson D 1978 *Nucl. Phys.* **A308** 381; Robson D 1979 *Phys. Rev. Lett.* **42** 876; Robson D 1982 *Phys. Rev. C* **25** 1108(R)
[5] Iachello F 1981 *Chem. Phys. Lett.* **78** 581
[6] Iachello F 1981 *Phys. Rev. C* **23** 2778(R)
[7] Bijker R and Iachello F 2000 *Phys. Rev. C***61** 067305
[8] Bijker R and Iachello F 2002 *Ann. Phys. (N.Y.)* **298** 334
[9] Bijker R 2010 *AIP Conf. Proc.* **1323** 28
[10] Bijker R 2012 *J. Phys.: Conf. Ser.* **380** 012003
[11] Bijker R and Iachello F 2014 *Phys. Rev. Lett.* **112** 152501
[12] Bijker R and Iachello F 2014 in preparation
[13] Marin-Lambarri DJ *et al* 2014 *Phys. Rev. Lett.* **113** 012502
[14] Gai M 2014 These Proceedings
[15] Feldmeier H and Neff T 2008 Proceedings of the International School "Enrico Fermi", Course CLXIX, 185
[16] Epelbaum E *et al* 2014 *Phys. Rev. Lett.* **112** 102501

Electroweak structure of light nuclei

Laura E Marcucci

Department of Physics "E. Fermi", University of Pisa, and INFN-Pisa, 56127 Pisa, Italy

E-mail: `laura.marcucci@df.unipi.it`

Abstract. We review the most recent studies performed within the chiral effective field theory framework of electroweak processes, among which, in particular, the electromagnetic structure of light nuclei, the weak muon capture on deuteron and ^{3}He, and the weak proton-proton capture reaction at energies of astrophysical interest. The presented results will be compared with the available experimental data and with those obtained within the conventional phenomenological approach.

1. Introduction

The traditional picture of a nucleus consisting of nucleons interacting among themselves and with external electroweak (EW) probes can be extensively tested by comparing theoretical and experimental results for many observables as, for instance, form factors, magnetic moments, and capture cross sections. Here we focus on the $A = 2, 3$ charge and magnetic form factors and static properties, the muon capture on deuteron and ^{3}He, and the proton-proton (pp) weak capture reaction ($p + p \rightarrow d + e^+ + \nu_e$). The conventional approach, where realistic models for the nuclear interactions and currents are constructed phenomenologically (for a review see Ref. [1]), has proven to be extremely successful in reproducing the many experimental data, both in the electromagnetic and weak sector (for the observables of interest here see Refs. [2, 3, 4, 5]). However, it presents an "original sin", i.e., no connection with the underlying theory of strong interaction, quantum chromodynamics (QCD), is clearly visible. An alternative approach now available is the so-called chiral effective field theory (χEFT) one. In a very schematic view, χEFT can be seen as a formulation of QCD in terms of effective degrees of freedom suitable for low-energy nuclear physics: pions and nucleons. The symmetries of QCD, in particular its (spontaneously broken) chiral symmetry, severely restrict the form of the interactions of nucleons and pions among themselves and with external EW fields, and make it possible to expand the Lagrangian describing these interactions in powers of Q/Λ_χ, where Q is pion momentum and $\Lambda_\chi \sim 700$ MeV is the chiral-symmetry-breaking scale. As a consequence, classes of Lagrangians emerge, each characterized by a given power of Q/Λ_χ and each involving a certain number of unknown coefficients, the so-called low-energy constants (LECs). While these LECs could in principle be determined by theory (for instance, in lattice QCD calculations), they are in practice constrained by fits to experimental data. The potentials and currents derived within this framework have power-law behavior for large momenta, and need to be regularized. This is accomplished in practice by introducing a momentum-cutoff function. Among the great advantages of the χEFT framework, the two following ones are here of interest: (i) the possibility of deriving nuclear EW currents consistently with the nuclear interaction, and (ii) the possibility of setting a hierarchy among the different contributions, both for the interactions and the currents. In fact, it is well known that χEFT can justify *a priori* the empirical observation that the contribution of three-nucleon interactions (TNIs) to nuclear structure is far less significant than that of the two-nucleon (NN) force. Furthermore, the χEFT power counting allows to estimate and in principle improve the theoretical uncertainty of the calculation.

The idea of using χEFT to derive the nuclear EW transition operators was first implemented by Park *et al* [6] in the nineties for the electromagnetic (EM) current and charge operators and few years later [7] for the weak axial current and charge operators. The so-called heavy-baryon chiral perturbation theory (HBχPT) approach was used, where the baryons are treated as heavy static sources, and the perturbative expansion is performed in terms of the involved momenta over the baryon mass. Since at that time the χEFT nuclear potentials were not yet as accurate as the conventional phenomenological ones, the first χEFT calculations of EW observables were performed within a "hybrid" approach, in which nuclear wave functions were obtained from phenomenological potentials [5, 7]. Only very recently [8, 9, 10], these χEFT weak operators have been used to study weak processes which involve few-nucleon systems in conjunction with modern accurate χEFT potentials, in particular the NN potential derived at next-to-next-to-next-to leading order (N3LO) by Entem and Machleidt [11], augmented, when needed, by the TNI obtained at next-to-next-to leading order (N2LO), in the version of Ref. [12]. The cutoff Λ is varied between 500 and 600 MeV.

Few years ago, the problem of deriving the EM current and charge operators in χEFT has been revisited by Pastore *et al* [13] and, in parallel, by Kölling *et al* [14]. Pastore *et al* have used time-ordered perturbation theory (TOPT) to calculate the EM transition amplitudes, which allows for an easier treatment of the so-called reducible diagrams than the HBχPT approach. On the other hand, Kölling *et al* have used the method of unitary transformation, the same one used to derive the chiral potentials mentioned above. We will focus here only on the work of Pastore *et al*, but we would like to remark that the results obtained by these two groups, although with different methods, are in good agreement with each other. To be noted that TOPT has not yet been applied to derive the weak axial current and charge operators at N3LO, but work along this line is currently underway. Therefore, the results presented here have been obtained within the model of Pastore *et al* for the EM observables and that of Park *et al* for the weak sector. A summary description of the two models can be found in Ref. [15]. Here we only notice that (i) the leading-order (LO) contribution is that due to the one-body current operator, obtained from a non-relativistic reduction of the covariant single-nucleon current. (ii) The NN EM currents derived by Pastore *et al* have been found significantly different from those of Park *et al.* The largest differences arise at N3LO for the box diagrams and the contact terms. In particular, the contact terms of Park *et al* are much simpler than those of Pastore *et al* and can be written as sum of two terms, one isoscalar and one isovector, with two different LECs in front, usually fitted to reproduce ths isoscalar and isovector combination of the $A = 3$ magnetic moments [8, 9]. We refer to Ref. [10] for the LECs fitting procedure of Pastore *et al.* (iii) At N2LO, the weak axial current operator presents a contact three-nucleon contribution whose LEC can be related with the LEC c_D entering one of the two contact terms present in the TNI at N2LO, via the relation $d_R = \frac{m}{\Lambda_\chi g_A} c_D + \frac{1}{3} m(c_3 + 2c_4) + \frac{1}{6}$, where g_A is the single-nucleon axial coupling constant, c_3 and c_4 are LECs of the πN Lagrangian, already part of the chiral NN potential at NLO, and Λ_χ MeV is the the chiral-symmetry-breaking scale mentioned above. Therefore, it has become common practice to fit c_D (and c_E— the other LEC entering the N2LO TNI) to the triton binding energy and the Gamow-Teller matrix element in tritium β-decay. The values obtained in this way for c_D and c_E are listed in Refs. [8, 16], and they have been used in Ref. [8] to study the muon capture on deuteron and ^{3}He, in Ref. [9] to investigate the *pp* weak capture, in Ref. [17] to study $A = 3$ and 4 elastic scattering observables, as cross sections and analyzing powers, and in Ref. [16] to calculate the nuclear matter equation of state in many-body perturbation theory.

2. Results

In this section we review some of the results on the electromagnetic structure of $A = 2, 3$ nuclei, extensively studied in Ref. [10]. Then we report the results for the muon capture on deuteron and ^{3}He [8], and for the *pp* reaction [9].

2.1. *The electromagnetic structure of $A = 2, 3$ nuclei*

The static properties of $A = 2, 3$ nuclei are summarized in table 1, where we present the χEFT results for the deuteron r.m.s. radius and quadrupole moment, and the charge and magnetic radii for the $A = 3$ nuclei. The experimental data are also reported. To be noticed that the deuteron and $A = 3$ magnetic moments results are not listed, since these observables are used to fit the LECs entering the EM current operators at N3LO [10]. The theoretical uncertainties in table 1 are due to the cutoff dependence and the fitting procedure. By inspection of the table, we can conclude that the static properties of the $A = 2, 3$ nuclei are nicely reproduced. It should be noticed that within the phenomenological approach, based on the Argonne v_{18} potential [18] (AV18), the quadrupole moment, calculated to be 0.275 fm^2, is found in significant disagreement with the experimental value.

Table 1. Deuteron r.m.s. radius (r_d) and quadrupole moment (Q_d), and ^{3}H and ^{3}He charge (r_c) and magnetic (r_m) radii. The corresponding experimental values are also reported.

	Theory	Experiment
r_d [fm]	1.972 $\pm$ 0.004	1.9733 $\pm$ 0.0044
Q_d [fm^2]	0.2836 $\pm$ 0.0016	0.2859 $\pm$ 0.0003
$r_c(^3\text{He})$ [fm]	1.962 $\pm$0.004	1.959 $\pm$0.030
$r_c(^3\text{H})$ [fm]	1.756 $\pm$0.006	1.755 $\pm$0.086
$r_m(^3\text{He})$ [fm]	1.905 $\pm$0.022	1.965 $\pm$0.153
$r_m(^3\text{H})$ [fm]	1.791 $\pm$0.018	1.840 $\pm$0.181

The deuteron $B(q)$ structure function and magnetic form factor, together with the $A = 3$ magnetic form factors, and their isoscalar and isovector combinations, are reported in figure 1. Besides the fully consistent χEFT results, we present also the results obtained within the "hybrid" χEFT approach, where the nuclear wave functions are obtained from the AV18 potential for the deuteron and the AV18 augmented with the Urbana IX (UIX) TNI [19] for the $A = 3$ nuclei. From the inspection of the figure, we can conclude that the χEFT calculation is in agreement with the experimental data in a range of momentum transfer q much larger than one would naively expect (up to 1-2 times the pion mass). On the other hand, this χEFT calculation is unable to reproduce the first diffraction region of the $A = 3$ magnetic form factors. This problem is also present in the hybrid calculation, as well as in the phenomenological one (see for instance Ref. [2, 3]).

2.2. *Muon capture on $A = 2, 3$ nuclei*

The muon capture reactions here under consideration are $\mu^- + d \to n + n + \nu_\mu$ and $\mu^- + {}^3\text{He} \to {}^3\text{H} + \nu_\mu$, for which we are interested in the capture rate in the doublet hyperfine initial state (Γ^D) and in the total capture rate (Γ_0), respectively. The χEFT results of Ref. [8] are summarized in table 2, where the value for Γ^D obtained retaining in the final *nn* scattering state only the $^1S_0$ partial wave is also shown. The results of the table can be summarized as $\Gamma^D = (399 \pm 3)$ sec^{-1} and $\Gamma_0 = (1494 \pm 21)$ sec^{-1}. The errors are due to cutoff dependence, the uncertainty

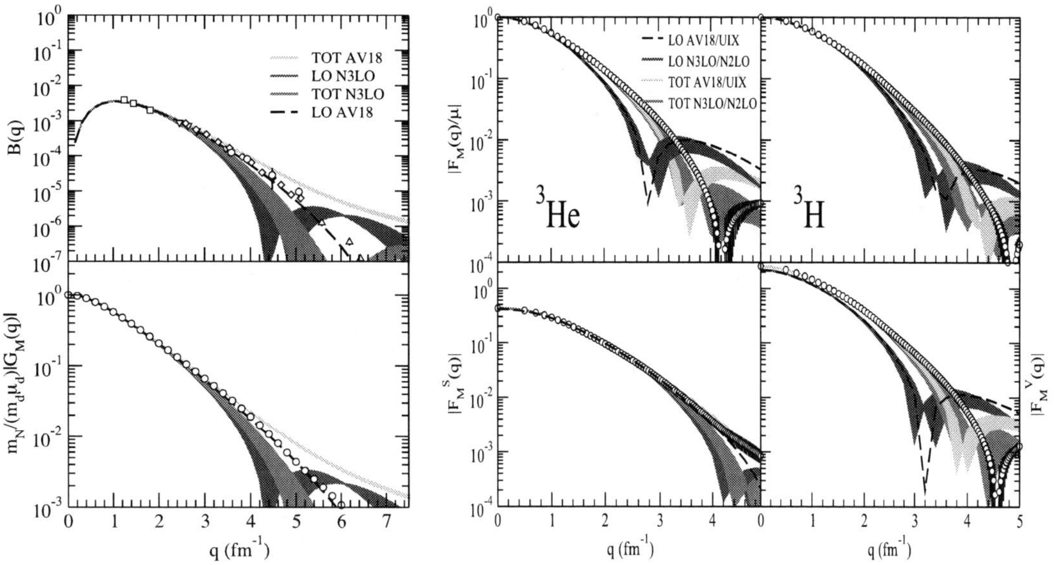

Figure 1. (Color online) The deuteron $B(q)$ structure function and magnetic form factor (on the left), and the $A = 3$ magnetic form factors, together with their isoscalar and isovector combinations (on the right), obtained at leading order (LO) and with inclusion of current operators up to N3LO (TOT), compared with the experimental data. The curves labelled "AV18" (or "AV18/UIX") have been obtained within the hybrid χEFT approach using the AV18 (and Urbana IX [19] TNI for $A = 3$) phenomenological potentials. The curves labelled "N3LO" (or "N3LO/N2LO") have been obtained instead using the N3LO NN potential [11] (and N2LO [12] TNI for $A = 3$).

inherent the fitting procedure of the LEC d_R, and radiative corrections [8]. The experimental data for Γ^D are affected at present by quite large uncertainties and no significant comparison between theory and experiment can be made. However, the ongoing experiment conducted by the MuSun collaboration at the Paul Scherrer Institute (PSI) is expected to measure Γ^D with a 1% precision. The theoretical prediction for Γ_0, instead, is in very nice agreement with the remarkably precise experimental determination of Ref. [20], (1496 ± 4) sec^{-1}. To be noticed that muon capture on ^{3}He can proceed also through the two- and three-body breakup channels, i.e. $\mu^- + ^3\text{He} \to n + d + \nu_\mu$ and $\mu^- + ^3\text{He} \to n + n + p + \nu_\mu$, with a 20% and 10% of branching ratio, respectively. No χEFT calculation for these processes has ever been performed, but the first study within the conventional phenomenological approach has been performed in Ref. [21], although using only the one-body current operator. Work on these reactions with a model for the weak axial current which includes also NN contributions is currently underway.

2.3. Weak proton-proton capture

The astrophysical S-factor for pp weak capture is defined as $S(E) = E \exp(2\pi\,\eta)\,\sigma(E)$, where $\eta = \alpha/v_{\rm rel}$, α being the fine structure constant and $v_{\rm rel}$ the pp relative velocity, and $\sigma(E)$ is the pp cross section at the center-of-mass energy E. Typically, $S(E)$ is given as a Taylor expansion around $E = 0$, and the coefficients of the expansion, $S(0)$, $S'(0)$ and $S''(0)$, are the quantities of interest [22]. Alternatively, the energy dependence of $S(E)$ can be made explicit by calculating

Table 2. Doublet capture rate for muon capture on deuteron and total capture rate for muon capture on ^{3}He obtained at leading order (LO) and with inclusion of current operators up to N2LO (TOT). The results calculated with the different values of two cutoff Λ are reported. The theoretical uncertainties are due to the fitting procedure of the LEC d_R.

	$\Gamma^D(^1S_0)$ [sec^{-1}]	Γ^D [sec^{-1}]	Γ_0 [sec^{-1}]
LO - $\Lambda = 500$ MeV	238.8	381.7	1362
LO - $\Lambda = 600$ MeV	238.7	380.8	1360
TOT - $\Lambda = 500$ MeV	254.4±9	399.2±9	1488±9
TOT - $\Lambda = 600$ MeV	255±1	399±1	1499±9

it directly. Note that the Gamow peak for the *pp* reaction is at $E = 6$ keV in the Sun, and it becomes of about 15 keV in larger mass stars. Therefore, we have studied the *pp* reaction in the energy range $E = 3 - 100$ keV. Two ingredients are essential in the calculation: (i) the initial *pp* scattering state is calculated using the χEFT N3LO NN potential [11], augmented not only by the Coulomb interaction, but also by the higher order electromagnetic terms, due to two-photon exchange and vacuum polarization. The additional distortion of the *pp* wave function, induced primarily by vacuum polarization, reduces $S(0)$ by $\sim 1\%$ [22, 9]. (ii) To have the correct energy-dependence of the S-factor up to $E = 100$ keV, we have included, in addition to the S-wave (the $^1S_0$ partial wave), all the P-wave channels ($^3P_0$, $^3P_1$, and $^3P_2$), and we have retained the explicit dependence on the momentum transfer via a standard multipole expansion. More details can be found in Ref. [9]. Finally, we recall that the model for the weak current is the same used in the successful studies of the muon capture reactions presented above. The S-factor at zero energy is found to be $S(0) = (4.030 \pm 0.006) \times 10^{-23}$ MeV fm^2, with a P-wave contribution of 0.020×10^{-23} MeV fm^2. The theoretical uncertainty is due again to the fitting procedure of the LECs and to the cutoff dependence. This value is $\simeq 1\%$ larger than the value reported in the literature [22], exactly due to the P-wave contributions. Note that within the potential model approach, using the AV18 potential, we have found $S(0) = (4.033 \pm 0.003) \times 10^{-23}$ MeV fm^2, with a P-wave contribution of 0.033×10^{-23} MeV fm^2, in nice agreement with the χEFT predictions, and with the value presented in Ref. [22] when the P-wave contributions are removed.

The energy dependence of $S(E)$ is shown in figure 2. The S- and $(S + P)$-wave contributions are displayed separately, and the theoretical uncertainty is included—the curves are in fact very narrow bands. As expected, the P-wave contributions become significant at higher values of E. The study of the effects of these P-wave contributions, as well as the new determination of $S(0)$ on the stellar structure and evolution is currently underway.

3. Conclusions and outlook

We have presented the most recent studies of EW observables involving light nuclei within the χEFT framework. The considered observables are the $A = 2, 3$ electromagnetic form factors and static properties, the muon capture on deuteron and ^{3}He and the proton weak capture by proton. The theoretical predictions have been found in a good agreement with the available experimental data, except for the $A = 3$ magnetic form factors at high values of momentum transfer, in a region, though, well beyond the applicability range of χEFT. The calculation of the *pp* astrophysical S-factor has been performed, for the first time, retaining the P-wave contributions, which have been found of the order of ~ 1 %. The theoretical uncertainty estimated within the χEFT framework has been found smaller than 1 %. To be noticed that the radiative proton-deuteron and the weak proton-^{3}He capture reactions, also relevant in astrophysics, are at reach within the present framework, and work along these lines is currently underway.

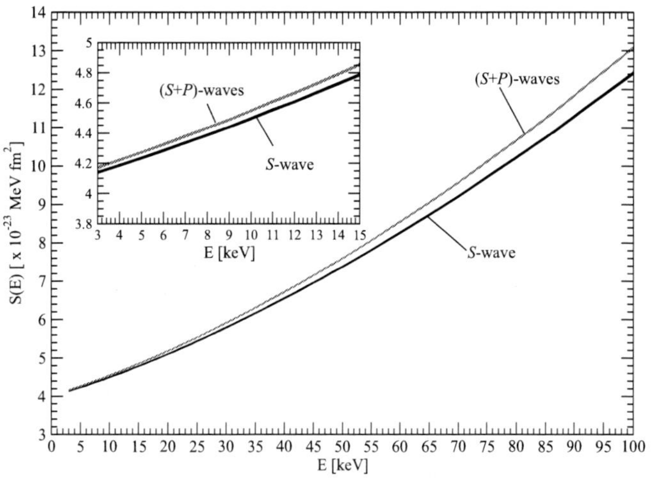

Figure 2. (Color online) The astrophysical S-factor for $E = 3 - 100$ keV. The S- and $(S+P)$-wave contributions are displayed separately. In the inset, $S(E)$ is shown in the range 3–15 keV.

References

[1] Carlson J and Schiavilla R 1998 *Rev. Mod. Phys.* **70** 743
[2] Marcucci L E, Riska D O, and Schiavilla R 1998 *Phys. Rev. C* **58** 3069
[3] Marcucci L E *et al* 2005 *Phys. Rev. C* **72** 014001
[4] Marcucci L E *et al* 2002 *Phys. Rev. C* **66** 054003
[5] Marcucci L E *et al* 2011 *Phys. Rev. C* **83** 014002
[6] Park T S *et al* 1996 *Nucl. Phys. A* **596** 515
[7] Park T S *et al* 2003 *Phys. Rev. C* **67** 055206
[8] Marcucci L E *et al* 2012 *Phys. Rev. Lett.* **108** 052502
[9] Marcucci L E *et al* 2013 *Phys. Rev. Lett.* **110** 192503
[10] Piarulli M *et al* 2013 *Phys. Rev. C* **87** 014006
[11] Entem D R and Machleidt R 2003 *Phys. Rev. C* **68** 041001
[12] Navràtil P 2007 *Few-Body Syst.* **41** 117
[13] Pastore S *et al* 2009 *Phys. Rev. C* **80** 034004; Pastore S *et al* 2011 *Phys. Rev. C* **84** 024001
[14] Kölling S *et al* 2009 *Phys. Rev. C* **80** 045502; Kölling S *et al* 2011 *Phys. Rev. C* **84** 054008
[15] Marcucci L E 2014 *Few-Body Syst.* **55** 615
[16] Coraggio L *et al* 2014 *Phys. Rev. C* **89** 044321
[17] Viviani M *et al* 2013 *Few-Body Syst.* **54** 885; Viviani M *et al* 2013 *Phys. Rev. Lett.* **111** 172302
[18] Wiringa R B, Stoks V G J, and Schiavilla R 1995 *Phys. Rev. C* **51** 38
[19] Pudliner B S *et al* 1995 *Phys. Rev. Lett.* **74** 4396
[20] Ackerbauer P *et al* 1998 *Phys. Lett. B* **417** 224
[21] Golak J *et al* 2014 *Phys. Rev. C* **90** 024001
[22] Adelberger E G *et al* 2011 *Rev. Mod. Phys.* **83** 195

Triangular $\mathcal{D}_{3h}$ Symmetry in the Rotation-Vibration Spectrum of ^{12}C

Moshe Gai
LNS at Avery Point,University of Connecticut , Groton, CT 06340-6097, USA
and Wright Lab, Dept of Physics, Yale University, New Haven, CT 06520-8124
E-mail: moshe.gai@yale.edu, http://astro.uconn.edu

Abstract. Our recent measurements of new states in ^{12}C including the second 2^+ at 10 MeV and the high spin 5^- state at 22.4 MeV allow us to study the Rotation-Vibration spectrum of ^{12}C from which evidence for a new ($\mathcal{D}_{3h}$) geometrical symmetry emerges. The data fit very well to the predicted (ground state) rotational band of an oblate equilateral triangular spinning top with a $\mathcal{D}_{3h}$ symmetry characterized by the sequence of states: $0^+, 2^+, 3^-, 4^{\pm}, 5^-$ with almost degenerate 4^+ and 4^- (parity doublet) states. Such a $\mathcal{D}_{3h}$ symmetry was observed in triatomic molecules, and it is observed in ^{12}C for the first time in nuclear physics. The triatomic like structure in nuclei is reminiscent of the discovery of diatomic α+^{14}C structure in ^{18}O. We discuss a classification of other rotation-vibration bands in ^{12}C such as the (0^+) Hoyle band and the (1^-) bending mode band and suggest measurements in search of the predicted ("missing") states that may shed new light on clustering in ^{12}C and light nuclei. In particular, the observation (or non observation) of the predicted ("missing") states in the Hoyle band will allow us to conclude the geometrical arrangement of the three alpha particles composing the Hoyle state at 7.654 MeV in ^{12}C.

1. Introduction

Molecular states are characterized by the separation vector connecting the constituent objects. For diatomic like object one single (separation) vector is required leading to the predicted U(4) symmetry [1] that was observed in ^{18}O via the characteristic enhanced E1 decays [2]. Triatomic symmetric spinning tops are characterized by the two Jacobbi vectors shown in figure 1 leading to the predicted U(7) with the geometrical $\mathcal{D}_{3h}$ symmetry [3, 4].

The recent observation of the 2^+ Hoyle rotational excitation in ^{12}C [5] allows for the first time the study of the Rotation-Vibration Structure of ^{12}C. These new data and the recently discovered 4^- [6, 7] and 4^+ [8] states in ^{12}C are in agreement with the predicted spectrum of an oblate spinning top with a $\mathcal{D}_{3h}$ symmetry [3, 4]. It was predicted [3, 4] that the three alpha-particle system of ^{12}C leads to the ground state rotational band including the most unusual sequence of states: $0^+, 2^+, 3^-, 4^{\pm}, 5^-$. The new high spin 5^- state [9] as well the previously published 4^- state [6, 7] lead to a J(J+1) trajectory as predicted by this U(7) model [9, 10] including the nearly degenerated 4^- and 4^+ states as shown in figure 2.

2. The Algebraic Cluster Model

The spectrum of ^{12}C predicted by the Algebraic Cluster Model (ACM) [3, 4] is shown in figure 3 where it is also compared to the measured spectrum of ^{12}C [9]. In addition to the ground state rotational band this U(7) model [3, 4] predicts the Hoyle state at 7.65 MeV in ^{12}C to be the first vibrational breathing mode of the three alpha-particle equilateral configuration leading to

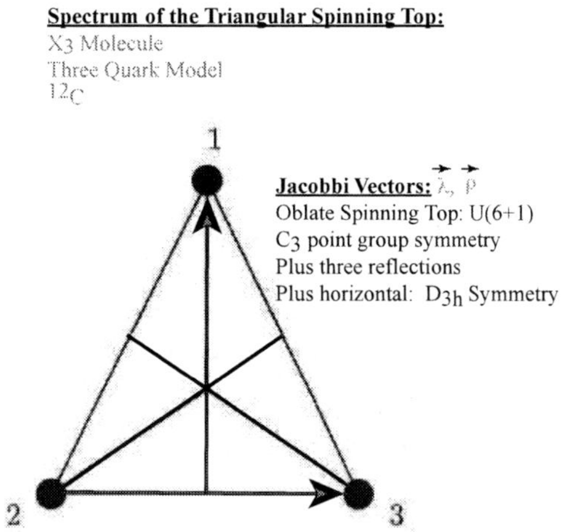

Figure 1. The Jacobbi vectors characterizing the triatomic symmetric spinning tops

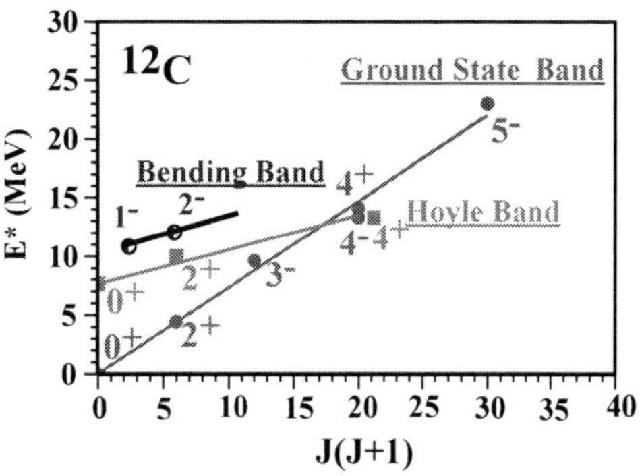

Figure 2. Rotational band structure of the ground-state band, the Hoyle band and the bending vibration band in ^{12}C. The spin-parity of the ground-state band are indicated (in blue), the Hoyle band (in red) and the bending mode band (in black). For each band the straight line fit to a J(J+1) curve is shown. Note the predicted "missing" 3^- and 4^- states in the Hoyle band shown in figure 3.

the same rotational structure albeit with a larger moment of inertia (by a factor of 2). Recent measurements revealed the 2^+ [5] and 4^+ [8] members of the predicted Hoyle rotational band and we are currently searching [11] for the 4^- predicted by the ACM to be nearly degenerated with the 4^+ state and the 3^- (broad) state that was suggested to lie between 11 and 14 MeV [6]. The observation (or lack there) of these "missing" states will allow us to determine wether the Hoyle state is composed of three alpha-particles in an equilateral triangle arrangement [9, 12, 13] or an obtuse triangle [14] or whether it is better described as vibrational excitation of a "diffuse gas" of three alpha-particles [15].

The U(7) model also predicts the 1^- state at 10.84 MeV to be the vibrational bending mode with a rotational band including the 1^- and a degenerate $2^{\pm}$ states. We are searching [11] for the third 2^+ of ^{12}C that is predicted by the U(7) model to lie near the observed 2^- state at 11.8 MeV.

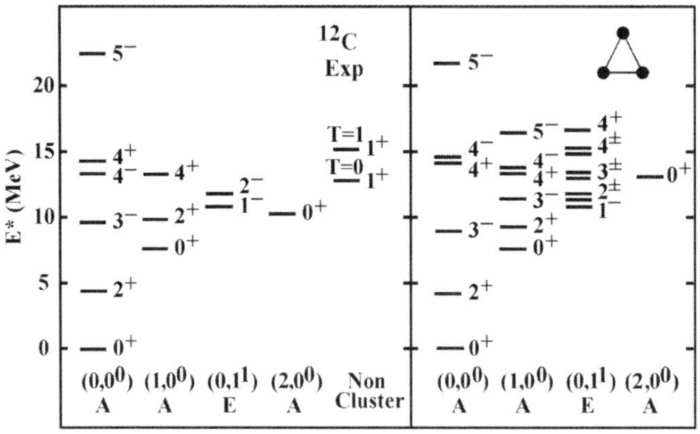

Figure 3. Comparison between the low-lying experimental spectrum of ^{12}C and the energies of the oblate symmetric top. The levels are organized in columns corresponding to the ground state band and the vibrational bands with A and E symmetry [3, 4] of an oblate top with triangular symmetry. The last column on the left-hand side, shows the lowest observed non-cluster (1^+) levels. The levels are characterized by angular momentum and parity L^P, and the vibrational labels (v_1, v_2^ℓ) as discussed in [3].

In conclusion the ACM appears to open a new chapter in cluster physics of light nuclei and it presents an opportunity for further experimental investigation of light nuclei.

Acknowledgments

This material is based upon work supported by the U.S. Department of Energy, Office of Science, Office of Nuclear Physics, under Award Number DE-FG02-94ER40870.

References

[1] Iachello F 1981 *Phys. Rev. C* **23**, 2778
[2] Gai M, Ruscev M, Hayes A C, Ennis J F, Keddy R, Schloemer E C, Sterbenz S M and Bromley D A,1983 *Phys. Rev. Lett.* **50** 239
[3] Bijker B and Iachello F 2000 *Phys. Rev. C* **61** 067305
[4] Bijker R and Iachello F 2002 *Ann. Phys.* **298** 334
[5] Itoh M *et al.* 2004 *Nucl. Phys.* **A738** 268; Itoh M *et al.* 2011 *Phys. Rev. C* **84** 054308; Freer M *et al.* 2009 *Phys. Rev. C* **80** 041303(R); Zimmerman W R *et al.* 2011 *Phys. Rev. C* **84** 027304(BR); Zimmerman W R *et al.* 2013 *Phys. Rev. Lett.* **110** 152502
[6] Freer M *et al.* 2007 *Phys. Rev. C* **76** 034320
[7] Kirsebom O S *et al.* 2010 *Phys. Rev. C* **81** 064313
[8] Freer M *et al.* 2011 *Phys. Rev. C* **83** 034314
[9] Marin-Lambarri D J, Bijker R, Freer M, Gai M, Kokalova Tz, C. Wheldon, 2014 *Phys. Rev. Lett.* **113** 012502.
[10] Popular press of Ref. [9]: Kevin Dusling, Physics, Synopsis: http://physics.aps.org/synopsis-for/10.1103/PhysRevLett.113.012502, Hamish Johnston, Physicsworld.com: http://physicsworld.com/cws/article/news/2014/jul/08/carbon-nu.
[11] A 12C(e,e') experiment in progress at Darmstadt, P. von Neumann-Cosel, private communication, 2013.
[12] Chernykh M, Feldmeier H, Neff T, von Neumann-Cosel P, and Richter A 2007 *Phys. Rev. Lett.* **98** 032501
[13] Yoshiko Kanada-Enyo 2007 *Prog. of Theo. Phys.* **117** 655
[14] Epelbaum E, Krebs H, Lee D, Meissner Ulf-G 2012 *Phys. Rev. Lett.* **106** 192501; Epelbaum E, Krebs H, Lahde T, Lee D, and Meissner Ulf-G 2012 *Phys. Rev. Lett.* **109** 252501
[15] Funaki Y, Horiuchi H, von Oertzen W, Ropke G, Schuck P, Tohsaki A, and Yamada T 2009 *Phys. Rev. C* **80** 64326 and references therein

HPC-enabled Nuclear Structure Studies – Description and Applications of the Symmetry-adapted No-Core Shell Model

J P Draayer[1], T Dytrych[1], K D Launey[1], A C Dreyfuss[1], and D Langr[2]
[1]Department of Physics and Astronomy, Louisiana State University, Baton Rouge, LA 70803, USA
[2] Department of Computer Systems, Czech Technical University in Prague, Prague, Czech Republic

Abstract. By exploiting symmetries that enable the accounting of vital collective correlations in nuclei, we achieve significantly reduced dimensions for equivalent ultra-large model spaces, and hence resolve the scale explosion problem in nuclear structure calculations, *i.e.*, the explosive growth in computational resource demands with increasing number of particles and size of the spaces in which they reside. As a result, we provide – with the help of High Performance Computing (HPC) resources – first solutions for selected benchmark calculations with remarkable findings of large-deformation and low-spin dominance in low-lying nuclear states. In the framework of a complementary symmetry-adapted study, one is able, facilitated by symmetry-preserving pieces of the inter-nucleon interaction, to accommodate unprecedented shell-model spaces critical to capture the physics governing the Hoyle state in ^{12}C, thereby addressing a 60-year-old puzzle on the emergence of cluster substructures within a no-core shell model framework. All of these findings underline the key role of symmetries in nuclear structure studies.

1. Introduction

The *ab initio* symmetry-adapted no-core shell model (SA-NCSM) [1], which capitalizes on exact as well as partial symmetries that underpin the structure of nuclei, provides remarkable insight into how simple symmetry patterns emerge in the many-body nuclear dynamics from first principles. Furthermore, by recognizing that deformed configurations often dominate the low-lying nuclear states, the SA-NCSM provides a strategy – a symmetry-guided concept – for determining the nature of bound states of nuclei in terms of a relatively small subspace of the symmetry-reorganized complete model space, which opens new domains of nuclei for *ab initio* investigations, namely, the intermediate-mass region, including isotopes of Ne, Mg, and Si. This innovative symmetry-guided concept is discussed with a focus on emergent patterns in complex nuclei, in particular, in a fully microscopic no-core symplectic shell-model (NCSpM) study of the challenging ^{12}C Hoyle state. This state was predicted based on observed abundances of heavy elements in the universe [2] and has attracted much recent attention both in theory (e.g., see [3–5]) and experiment [6–13]. The NCSpM findings inform key features of the primary physics responsible for the emergent phenomena of large deformation and alpha-cluster substructures in ^{8}Be and ^{12}C [14, 15], as well as enhanced collectivity in intermediate-mass nuclei [16].

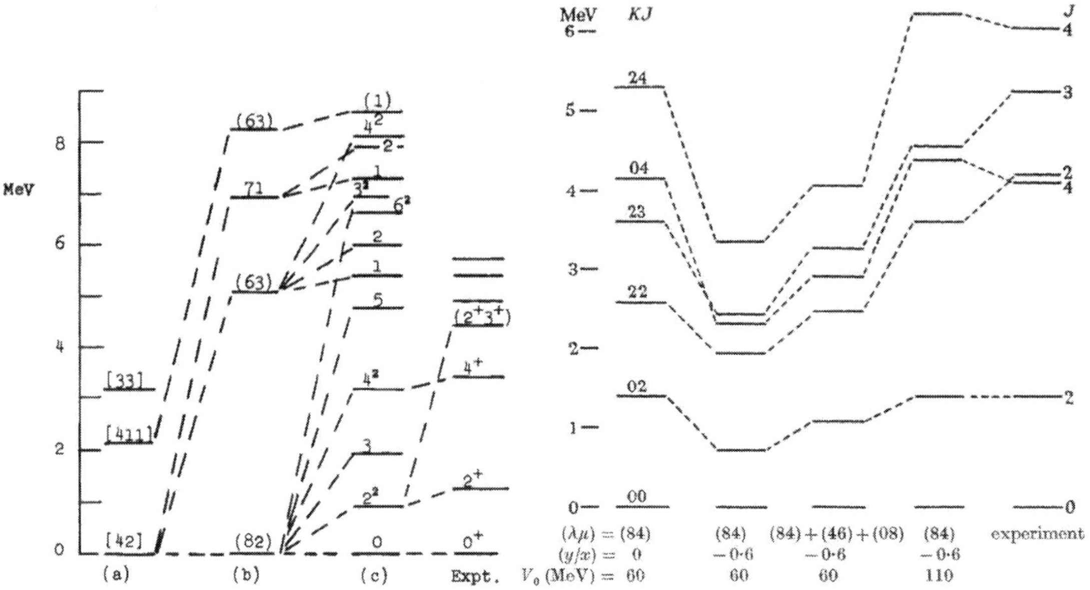

Figure 1. Elliott's SU(3) model applied to *sd*-shell nuclei. Left panel: Spectrum of ^{22}Ne (or ^{22}Mg) (a) with a Majorana potential, (b) with the addition of the second-order SU(3) Casimir invariant, C_2^{su3}, and (c) with the Majorana potential plus an attractive $Q \cdot Q$ interaction [or (b) with the addition of L^2]. Figure taken from [27]. Right panel: Spectrum of ^{24}Mg with a Gaussian central force. Figure taken from [20]. The vertical axis in both figures represents energy in MeV. Note the importance of the most deformed SU(3) configuration (8 2) in ^{22}Ne and (8 4) in ^{24}Mg for reproducing the experimental low-lying states.

Specifically, the *ab initio* SA-NCSM, with results that corroborate and are complementary to those enabled within the framework of the no-core shell model (NCSM) [17], and which can be used to facilitate *ab initio* applications to challenging lower *sd*-shell nuclei, reveal that bound states of light nuclei are dominated by high-deformation and low-spin configurations [1]. The applicable symmetries reveal the nature of collectivity in such nuclei and provide a description of bound states in terms of a relatively small fraction of the complete space when the latter is expressed in an $(LS)J$ coupling scheme with the spatial configurations further organized into irreducible representations (irreps) of SU(3). That SU(3) plays a key role tracks with the seminal work of Elliott [18–20], see figure 1. The SU(3)-symmetry dominance has been also observed in heavier nuclei, where pseudo-spin symmetry and its pseudo-SU(3) complement have been shown to play a similar role in accounting for deformation in the upper *pf* and lower *sdg* shells, and in particular, in strongly deformed nuclei of the rare-earth and actinide regions [21], as well as in many other studies (e.g., see [22]). This is further reinforced by the fact that SU(3) also underpins the microscopic symplectic model [23, 24], developed by Rowe and Rosensteel, which provides a theoretical framework for understanding deformation-dominated collective phenomena in nuclei [24–26].

One of the most successful particle-driven models is the no-core shell model (NCSM), which, in principle, can straightforwardly accommodate any type of inter-nucleon interaction [17,28,29]. It adopts the harmonic oscillator (HO) single-particle basis characterized by the $\hbar\Omega$ oscillator strength and retains many-body basis states of a fixed parity, consistent with the Pauli principle, and limited by a many-body basis cutoff $N_{\max}$. The $N_{\max}$ cutoff is defined as the maximum number of HO quanta allowed in a many-body basis state above the minimum for a given nucleus.

It divides the space in "horizontal" HO shells and is dictated by particle-hole excitations (this is complementary to the microscopic symplectic model, which divides the space in vertical slices selected by collectivity-driven rules). The NCSM has achieved remarkable descriptions of low-lying states from the lightest s-shell nuclei up through ^{12}C, ^{16}O, and ^{14}F, and is further augmented by several techniques, such as NCSM/RGM [30], Importance Truncation NCSM [31] and Monte Carlo NCSM [32]. This supports and complements results of other first-principle approaches, such as Green's function Monte Carlo (GFMC) [33], Coupled-cluster (CC) method [34], In-medium SRG [35], and Lattice Effective Field Theory (EFT) [5].

The next-generation *ab initio* symmetry-adapted no-core shell model (SA-NCSM) [1] combines the first-principle concept of the NCSM with symmetry-guided considerations of the collectivity-driven models. The model and its recent findings are described in the next section.

2. Symmetry-guided concept

2.1. *Simple pattern formation from first principles*

The *ab initio* symmetry-adapted no-core shell model (SA-NCSM) [1] adopts the first-principle concept and utilizes a many-particle basis that is reduced with respect to the physically relevant, deformation-related SU(3)⊃SO(3) subgroup chain (for a review, see [36]). This allows the full model space to be down-selected to the physically relevant space.

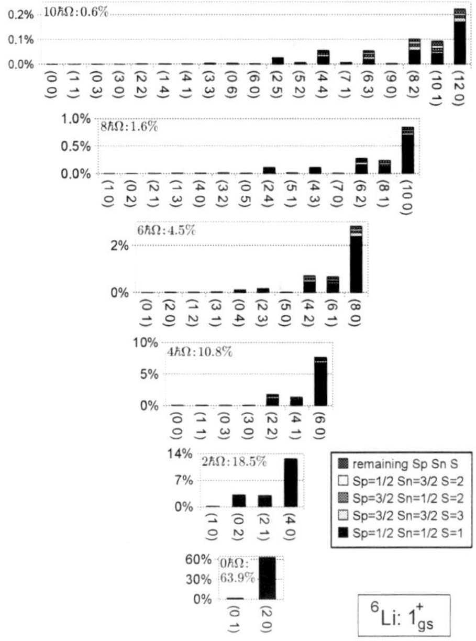

Figure 2. Probability distributions for proton, neutron, and total intrinsic spin components (S_pS_nS) across the Pauli-allowed deformation-related $(\lambda\,\mu)$ values and for $N\hbar\Omega$ subspaces for the calculated 1^+ ground state of ^{6}Li obtained for $N_{\rm max}=10$ and $\hbar\Omega=20$ MeV with the JISP16 bare interaction. The concentration of strengths to the far right demonstrates the dominance of collectivity in the calculated eigenstates.

The basis states of the SA-NCSM are based on HO single-particle states and for a given $N_{\rm max}$, are constructed in the proton-neutron formalism using an efficient construction based on powerful group-theoretical methods. The SA-NCSM basis states are related to the NCSM basis states through a unitary transformation (hence, the SA-NCSM results obtained in a complete $N_{\rm max}$ space are equivalent to the $N_{\rm max}$-NCSM results). They are labeled by the SU(3)⊃SO(3) subgroup chain quantum numbers $(\lambda\,\mu)\kappa L$, together with proton, neutron, and total intrinsic spins S_p, S_n, and S. The orbital angular momentum L is coupled with S to the total orbital momentum J and its projection M_J. Each basis state in this scheme is labeled schematically as $|\vec{\gamma}\,(\lambda\,\mu)\kappa L;(S_pS_n)S;JM_J\rangle$. The label κ distinguishes multiple occurrences of the same L value in the parent irrep $(\lambda\,\mu)$, and $\vec{\gamma}$ distinguishes among configurations carrying the same $(\lambda\,\mu)$ and $(S_pS_n)S$ labels.

The *ab initio* SA-NCSM results for p-shell nuclei reveal a dominance of configurations of large deformation in the $0\hbar\Omega$ subspace. For example, the *ab initio* $N_{\rm max} = 10$ SA-NCSM results with the bare JISP16 realistic interaction [37] (similarly, for the bare N^{3}LO realistic interaction [38]) for the 1^+ ground state (*g.st.*) and its rotational band for ^{6}Li reveal the dominance of the $0\hbar\Omega$ component with the foremost contribution coming from the leading $(2\,0)$ $S = 1$ irrep (figure 2). Furthermore, we find that important SU(3) configurations are then organized into structures with Sp(3, $\mathbb{R}$) symplectic symmetry, that is, the $(2\,0)$ symplectic irrep gives rise to $(4\,0)$, $(2\,1)$, and $(0\,2)$ configurations in the $2\hbar\Omega$ subspace and so on, and those configurations indeed realize the major components of the wavefunction in this subspace. Similar results are observed for other p-shell nuclei, such as ^{6}He, ^{8}Be, and ^{12}C.

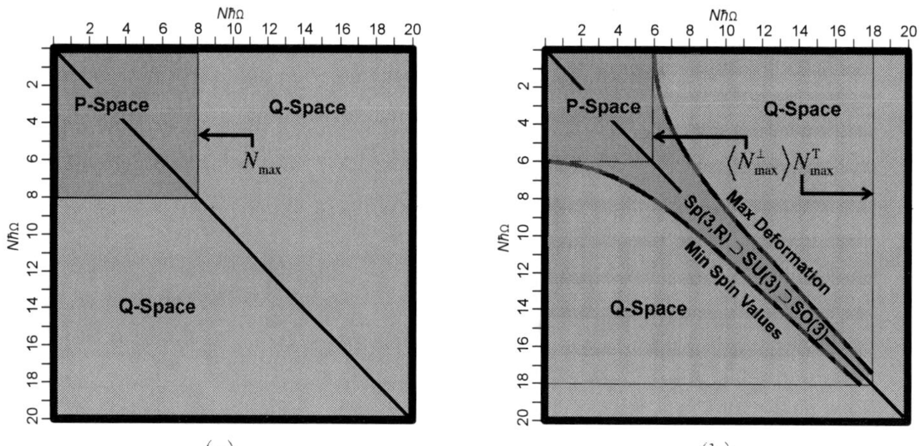

Figure 3. Schematic illustration of the symmetry-guided concept for a model-space selection (P-space, shaded in magenta) and associated interaction renormalization. (a) Conventional NCSM in $N_{\rm max}$ model spaces specified by HO $N\hbar\Omega$ many-body configurations, and (b) SA-NCSM in a $\langle N^{\perp}_{\rm max}\rangle N^{\top}_{\rm max}$ model space reorganized in terms of SU(3)-coupled $N\hbar\Omega$ many-body configurations; P-space includes all configurations up to $N^{\perp}_{\rm max}$ (small square), and large deformation/low-spin configurations at each $N\hbar\Omega$ subspace together with symplectic excitations built upon largely deformed SU(3) configurations in lower $N\hbar\Omega$ subspaces (intersection of horizontal and vertical light-magenta stripes). Typically, $N^{\perp}_{\rm max} \leq N_{\rm max}$, while $N^{\top}_{\rm max}$ extends much beyond $N_{\rm max}$. As indicated along the diagonal, the percentage (tremendously enhanced for illustration) of selected configurations relative to the corresponding complete subspace decreases for higher $N\hbar\Omega$.

The outcome points to a remarkable feature common to the low-lying states of nuclei that has heretofore gone unrecognized in other first-principle studies; namely, the emergence, without a priori constraints, of simple orderly patterns that favor strongly deformed configurations and low spin values. This feature confirms the dominant role the SU(3) and Sp(3, $\mathbb{R}$) symmetries play in nuclear dynamics and is central to expanding the reach of first-principle studies to heavier nuclei [39]. In particular, this provides a strategy to determine a physically relevant P-space (that is, the model space used in calculations) that could be used for interaction renormalization and for SA-NCSM calculations (figure 3). Such a model space can thus accommodate important physics in low $N\hbar\Omega$ subspaces (all configurations up to $N^{\perp}_{\rm max}$) together with symplectic excitations thereof up through $N^{\top}_{\rm max}$ including symplectic irreps that start over largely deformed configurations in high $N\hbar\Omega$ subspaces. Hence, model spaces, e.g., such as $\langle 6\rangle 18$ for ^{12}C are feasible and en route to be employed. This allows the SA-NCSM to advance an extensible microscopic framework for studying nuclear structure and reactions that capitalizes on advances being made in *ab initio* methods while exploiting symmetries found to dominate the dynamics.

2.2. Applications to the elusive Hoyle state

The symmetry-guided concept shown in figure 3 is utilized and further understood in the framework of the no-core symplectic shell model (NCSpM) with applications to the low-lying nuclear structure of ^{12}C [15]. This is a fully microscopic no-core shell model employed in a model space beyond current NCSM limits, namely, up through $N^{\top}_{\max} = 20$, that uses a symplectic $\mathrm{Sp}(3,\mathbb{R})$ basis and $\mathrm{Sp}(3,\mathbb{R})$-preserving interactions, while a nonpreserving spin-orbit interaction is applied to $0\hbar\Omega$ configurations. For example, results for the $\langle N^{\perp}_{\max} = 0\rangle N^{\top}_{\max} = 20$ model space are presented in Ref. [40].

The NCSpM employed within a full model space up through a given $N_{\max}$ coincides with the NCSM for the same $N_{\max}$ cutoff. However, in the case of the NCSpM, the symplectic irreps divide the space into 'vertical slices' that are comprised of basis states of a definite deformation $(\lambda\,\mu)$. Hence, the model space can be reduced to only a few important configurations that are chosen among all possible $\mathrm{Sp}(3,\mathbb{R})$ irreps within the $N^{\top}_{\max}$ model space.

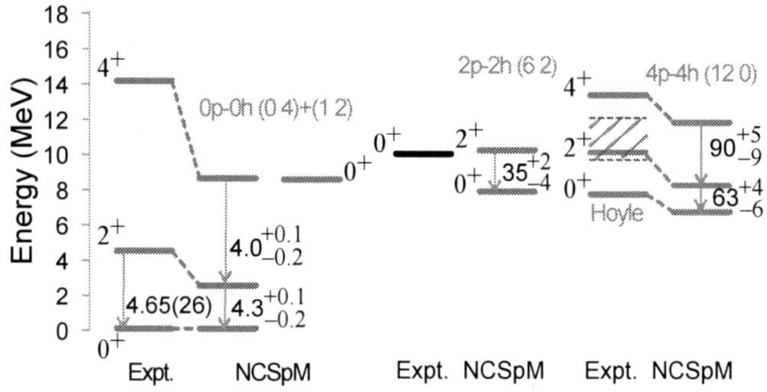

Figure 4. ^{12}C energy spectrum calculated by the NCSpM in the $N^{\top}_{\max} = 20$ model space of the $0\hbar\Omega$ (0 4) & (1 2), $2\hbar\Omega$ (6 2), and $4\hbar\Omega$ (12 0) symplectic irreps and compared to experiment ("Expt."). Figure taken from Ref. [15]. Results remain almost unchanged in the $\langle 0\rangle 20$ space, where the remaining $0\hbar\Omega$ $\mathrm{Sp}(3,\mathbb{R})$ irreps are included [40].

We employ a very simple Hamiltonian with an effective interaction derived from the long-range expansion of the two-body central nuclear force together with a spin-orbit term,

$$H_{\text{eff}} = H_0 + \frac{\chi}{2}\frac{\left(e^{-\gamma Q\cdot Q} - 1\right)}{\gamma} - \kappa\sum_{i=1}^{A} l_i \cdot s_i. \tag{1}$$

This includes the $\mathrm{Sp}(3,\mathbb{R})$-preserving terms: the spherical HO potential, which together with the kinetic energy yields the HO Hamiltonian, $H_0 = \sum_{i=1}^{A}\left(\frac{\mathbf{p}_i^2}{2m} + \frac{m\Omega^2\mathbf{r}_i^2}{2}\right)$, and the important $\frac{1}{2}Q\cdot Q = \frac{1}{2}\sum_s q_s \cdot (\sum_t q_t)$ interaction, which realizes the physically relevant interaction of each particle with the total quadrupole moment of the nuclear system; the nonpreserving spin-orbit term is turned on only among symplectic bandheads. The average contribution, $\langle Q\cdot Q\rangle_n$, of $Q\cdot Q$ within a subspace of n HO excitations is removed [41], that is, the trace of $Q\cdot Q$ divided by the space dimension for a fixed n. Hence, the large monopole contribution of the $Q\cdot Q$ interaction is removed, which, in turn, helps eliminate the considerable renormalization of the zero-point energy, while retaining the $Q\cdot Q$-driven behavior of the wavefunctions. This Hamiltonian in its zeroth-order approximation (for parameter $\gamma \to 0$) and for a valence shell goes back to the established Elliott model. We take the coupling constant χ to be proportional to $\hbar\Omega$ and, to leading order, to decrease with the total number of HO excitations, as shown by Rowe [42] based on self-consistent arguments. The success of such an effective nuclear interaction is not unexpected, as the spherical HO potential and the $Q.Q$ interaction directly follow from the second and third term, respectively, in the long-range expansion of any two-body central force, e.g., like the Yukawa radial dependence, $V^{(2)} = \sum_{i<j} V(r_{ij}/a) = \sum_{i<j}(\xi_0 + \xi_2 r_{ij}^2/a^2 + \xi_2 r_{ij}^4/a^4 + \ldots)$ [27], for a range parameter a.

As the interaction and the model space are carefully selected to reflect the most relevant physics, the outcome reveals a quite remarkable agreement with the experiment [15]. The low-lying energy spectrum and eigenstates for ^{12}C were calculated using the NCSpM with H of equation (1) for $\hbar\Omega = 18$ MeV given by the empirical estimate $\approx 41/A^{1/3} = 17.9$ MeV and for $\kappa \approx 20/A^{2/3} = 3.8$ MeV (e.g., see [43]). The results are shown for $N_{\max} = 20$, which we found sufficient to yield convergence. This $N_{\max}$ model space is further reduced by selecting the most relevant symplectic irreps, namely, the spin-zero ($S = 0$) $0\hbar\Omega$ 0p-0h (0 4), $2\hbar\Omega$ 2p-2h (6 2), and $4\hbar\Omega$ 4p-4h (12 0) symplectic bandheads together with the $S = 1$ $0\hbar\Omega$ 0p-0h (1 2) and all multiples thereof up through $N_{\max} = 20$ of total dimensionality of 6.6×10^3. In comparison to the experimental energy spectrum (figure 4), the outcome reveals that the lowest 0^+, 2^+, and 4^+ states of the 0p-0h symplectic slices calculated for $\gamma = -1.71 \times 10^{-4}$ closely reproduce the *g.st.* rotational band, while the calculated lowest 0^+ states of the $4\hbar\Omega$ 4p-4h (12 0) and the $2\hbar\Omega$ 2p-2h (6 2) slices are found to lie close to the Hoyle state and the 10-MeV 0^+ resonance (third 0^+ state), respectively. The model successfully reproduces other observables for ^{12}C that are informative of the state structure, such as mass rms radii, electric quadrupole moments and $B(E2)$ transition strengths (figure 4). The model we find is also applicable to the low-lying states of other p-shell nuclei, such as ^{8}Be, as well as sd-shell nuclei without any adjustable parameters [14, 16]. In particular, using the same $\gamma = -1.71 \times 10^{-4}$ as determined for ^{12}C, we describe selected low-lying states in ^{8}Be in an $N_{\max} = 24$ model space with only 3 spin-zero $0\hbar\Omega$ (4 0), $2\hbar\Omega$ (6 0), and $4\hbar\Omega$ (8 0) symplectic irreps. Furthermore, we have successfully applied the NCSpM without any adjustable parameters to the ground-state rotational band of heavier nuclei, such as ^{20}O, ^{20,22,24}Ne, ^{20,22}Mg, and ^{24}Si [16]. This suggests that the fully microscopic NCSpM model has indeed captured an important part of the physics that governs the low-energy nuclear dynamics and informs key features of the interaction and nuclear structure primarily responsible for the formation of such simple patterns.

In short, in the framework of the SA-NCSM, we show the emergence of orderly patterns from first principles. These patterns are linked to the SU(3) and symplectic Sp(3, $\mathbb{R}$) symmetries. This novel feature, in turn, enables the SA-NCSM – by using symmetry-dictated subspaces – to reach new domains of nuclear structure currently inaccessible by *ab initio* calculations, such as isotopes of Ne, Mg, and Si. Furthermore, the symmetry-guided concept is illustrated in the NCSpM. This study, in addition, reveals that, to explain emergent simple patterns of deformation and clustering, shell-model spaces well beyond the current limits (up through 22 major HO shells for the Hoyle state) are vital to accommodate particle excitations that appear critical to highly-deformed spatial structures and the convergence of associated observables. The outcome also points to the importance of simple many-body interactions and of the long-range part of the NN interaction, especially through its link to the HO potential and the interaction of individual particles with the total quadruple moment of the nuclear system.

Acknowledgments

We thank the PetaApps Collaboration, in particular, J. P. Vary, P. Maris, U. Catalyurek, E. Saule, and M. Sosonkina, as well as D. Rowe and G. Rosensteel for useful discussions. This work was supported by the U.S. NSF (OCI-0904874), the U.S. DOE (DE-SC0005248), and SURA. ACD acknowledges support by the U.S. NSF (grant 1004822) through the REU Site in Dept. of Physics & Astronomy at LSU. We also acknowledge DOE/NERSC, LSU/LONI, and NSF & U. of Illinois/Blue Waters for providing HPC resources.

References

[1] Dytrych T, Launey K D, Draayer J P, Maris P, Vary J P, Saule E, Catalyurek U, Sosonkina M, Langr D and Caprio M A 2013 *Phys. Rev. Lett.* **111** 252501
[2] Hoyle F 1954 *Astrophys. J. Suppl. Ser.* **1** 121
[3] Chernykh M, Feldmeier H, Neff T, von Neumann-Cosel P and Richter A 2007 *Phys. Rev. Lett.* **98** 032501
[4] Khoa D T, Cuonga D C and Kanada-En'yo Y 2011 *Phys. Letts. B* **695** 469
[5] Epelbaum E, Krebs H, Lee D and Meissner U G 2011 *Phys. Rev. Lett.* **106** 192501
[6] Fynbo H O U *et al.* 2005 *Nature* **433** 136
[7] Freer M *et al.* 2009 *Phys. Rev. C* **80** 041303
[8] Hyldegaard S *et al.* 2010 *Phys. Rev. C* **81** 024303
[9] Itoh M *et al.* 2011 *Phys. Rev. C* **84** 054308
[10] Zimmerman W R, Destefano N E, Freer M, Gai M and Smit F D 2011 *Phys. Rev. C* **84** 027304
[11] Raduta A R *et al.* 2011 *Phys. Letts. B* **705** 65
[12] Zimmerman W R *et al.* 2013 *Phys. Rev. Lett.* **110** 152502
[13] Marin-Lambarri D, Bijker R, Freer M, Gai M, Kokalova T, Parker D and Wheldon C 2014 *Phys. Rev. Lett.* **113** 012502
[14] Launey K D, Dytrych T, Draayer J P, Tobin G K, Ferriss M C, Langr D, Dreyfuss A C, Maris P, Vary J P and Bahri C 2013 *Proceedings of the 5th International Conference on Fission and properties of neutron-rich nuclei, ICFN5, November 4 - 10, 2012, Sanibel Island, Florida, edited by J. H. Hamilton and A. V. Ramayya* (World Scientific, Singapore) p 29
[15] Dreyfuss A C, Launey K D, Dytrych T, Draayer J P and Bahri C 2013 *Phys. Lett. B* **727** 511
[16] Tobin G K, Ferriss M C, Launey K D, Dytrych T, Draayer J P and Bahri C 2014 *Phys. Rev. C* **89** 034312
[17] Navrátil P, Vary J P and Barrett B R 2000 *Phys. Lett.* **84** 5728
[18] Elliott J P 1958 *Proc. Roy. Soc. A* **245** 128
[19] Elliott J P 1958 *Proc. Roy. Soc. A* **245** 562
[20] Elliott J P and Harvey M 1962 *Proc. Roy. Soc. A* **272** 557
[21] Bahri C, Draayer J and Moszkowski S 1992 *Phys. Rev. Lett.* **68** 2133
[22] Zuker A, Retamosa J, Poves A and Caurier E 1995 *Phys. Rev. C* **52** R1741
[23] Rosensteel G and Rowe D J 1977 *Phys. Rev. Lett.* **38** 10
[24] Rowe D J 1985 *Reports on Progr. in Phys.* **48** 1419
[25] Bahri C and Rowe D J 2000 *Nucl. Phys. A* **662** 125
[26] Draayer J, Weeks K and Rosensteel G 1984 *Nucl. Phys.* **A419** 1
[27] Harvey M 1968 *Adv. Nucl. Phys.* **1** 67
[28] Maris P, Shirokov A M and Vary J P 2010 *Phys. Rev. C* **81** 021301(R)
[29] BR Barrett P N and Vary J 2013 *Prog. Part. Nucl. Phys.* **69** 131
[30] Navrátil P, Quaglioni S, Stetcu I and Barrett B R 2009 *J. Phys. G: Nucl. Part.* **36** 083101
[31] Roth R and Navrátil P 2007 *Phys. Rev. Lett.* **99** 092501
[32] Abe T, Maris P, Otsuka T, Shimizu N, Utsuno Y and Vary J 2012 *Phys. Rev. C* **86** 054301
[33] Pieper S C, Varga K and Wiringa R B 2002 *Phys. Rev. C* **66** 044310
[34] Wloch M, Dean D J, Gour J R, Hjorth-Jensen M, Kowalski K, Papenbrock T and Piecuch P 2005 *Phys. Rev. Lett.* **94** 212501
[35] Tsukiyama K, Bogner S K and Schwenk A 2011 *Phys. Rev. Lett.* **106** 222502
[36] Dytrych T, Sviratcheva K D, Draayer J P, Bahri C and Vary J P 2008 *J. Phys. G: Nucl. Part. Phys.* **35** 123101
[37] Shirokov A, Vary J, Mazur A and Weber T 2007 *Phys. Lett. B* **644** 33
[38] Entem D R and Machleidt R 2003 *Phys. Rev. C* **68** 041001
[39] Dytrych T, Launey K D and Draayer J P 2014 *McGraw-Hill Yearbook of Science and Technology* YB140314
[40] Dreyfuss A C, Launey K D, Dytrych T, Draayer J P, Baker R, Deibel C and Bahri C 2014 *to be submitted to Phys. Rev. C*
[41] Rosensteel G and Draayer J P 1985 *Nucl. Phys. A* **436** 445
[42] Rowe D J 1967 *Phys. Rev.* **162** 866
[43] Bohr A and Mottelson B R 1969 *Nuclear Structure* vol 1 (Benjamin, New York)

Journal of Physics: Conference Series **580** (2015) 012045 doi:10.1088/1742-6596/580/1/012045

Shell-Model studies of chaos and statistical properties in nuclei

J. M. G. Gómez[1], E. Faleiro[2], L. Muñoz[1], R. A. Molina[3], A. Relaño[4]

[1] Departamento de Física Atómica, Molecular y Nuclear, Universidad Complutense de Madrid, E-28040 Madrid, Spain

[2] Departamento de Física Aplicada, E. U. I. T. Industrial, Universidad Politécnica de Madrid, E-28012 Madrid, Spain

[3] Instituto de Estructura de la Materia, CSIC, E-28006 Madrid, Spain

[4] Departamento de Física Aplicada I, Universidad Complutense de Madrid, E-28040 Madrid, Spain

E-mail: jmgomezg@ucm.es

Abstract. Shell-model calculations with realistic empirical interactions constitute an excellent tool to study statistical properties of nuclei. Using large-scale shell-model calculations in *pf*-shell nuclei, we study how the onset of chaos depends on different properties of the nuclear interaction and on excitation energy. We make use of classical random matrix theory and other theoretical developments based on information theory and time series analysis. We show that besides energy-level statistics, other statistical properties like the complexity of wave functions are fundamental for a proper determination of the dynamical regime of nuclei. Important deviations from GOE are observed in level fluctuations and in the complexity of wave functions.

1. Introduction

Beyond detailed nuclear spectroscopy, the statistical properties of nuclei are important not only in nuclear physics, but also in astrophysics. For example, statistical spectroscopy methods generated by many-body chaos and two-body random matrix ensembles provide useful information on Gamow-Teller strength sums and beta decay rates for stellar evolution and supernovae. For a recent review of these topics, as well as quantum chaos in Hamiltonian systems, see Gómez *et al.* [1]. We refer to this paper for references and a more detailed account of the topics of the present paper. In this work we discuss statistical properties related to chaos in nuclei and will try to give an answer to a simple question: *to what extent is nuclear motion chaotic?* We discuss experimental and theoretical results and conclude that there is no simple answer to this question. It depends on which nuclei and which excitation energy region we are talking about.

It is well known that real and complex systems are usually not fully ergodic and neither are they integrable. The chaotic orbits only occupy a certain part of its phase space. For quantum stationary systems with a classical analog, there is abundant evidence showing a clear relationship between the fluctuation properties of their energy levels and the large time scale behavior of the classical analog. The pioneering work of Berry and others, especially Bohigas *et al.* [2] lead to an important and concise statement known as the BGS conjecture: *the spectral properties of simple systems known to be ergodic in the classical limit follow very closely those of the Gaussian orthogonal ensemble (GOE) of random matrices.* During many years it has been mainly supported by many numerical calculations, and more recently an analytical proof was obtained as well by Heusler *et al.* [3]. The idea is generalized to quantum systems without a classical limit, and thus a quantum system is considered chaotic if its fluctuation properties agree with those of random matrix theory (RMT). In this case the distribution of the nearest-neighbor spacings is given to a good approximation by the Wigner surmise. It corresponds to the existence of linear repulsion between adjacent levels. On the contrary, integrable systems lead to level fluctuations that are well described by the Poisson distribution, i.e., levels behave as if they were uncorrelated.

The information on regular and chaotic nuclear motion available from experimental data is rather limited, because the analysis of energy levels requires the knowledge of sufficiently large pure sequences, i.e., consecutive level samples all with the same quantum numbers (J, π, T) in a given nucleus. The situation is clear above the one-nucleon emission threshold, where a large number of neutron and proton $J^{\pi} = 1/2^{+}$ resonances are identified. The agreement between this Nuclear Data Ensemble (NDE) [4] and the GOE predictions is excellent. In the low-energy domain, however, it is rather difficult (if not impossible) to get large enough pure sequences. For this reason, even if the data coming from different nuclei are conveniently scaled and gathered, the conclusions are less clear. There is some evidence that, at low energy, the spectral fluctuations are close to GOE predictions in spherical nuclei while they deviate towards the Poisson distribution in deformed nuclei.

In order to get a deeper understanding of what happens in the low-energy region we can use the interacting shell model. Using large-scale shell-model calculations in pf-shell nuclei with a realistic empirical interaction, we study how the onset of chaos depends on different properties of the nuclear interaction and on excitation energy. We make use of classical random matrix theory and other theoretical developments based on information theory and time series analysis. We compare the results of sd and pf nuclei and find that they are quite different in particular nuclei. We will also show that some simple statistics, like the nearest-neighbor spacing distribution, can sometimes become misleading by itself and one has to look at other energy-level statistics, as well as other statistical properties like the complexity of wave functions, for a proper determination of the dynamical regime of nuclei.

2. Statistics used to assess quantum chaos

The first step in the study of spectral fluctuations is the unfolding of energy levels. Fluctuations are somehow defined as the departure of the actual level density from a local uniform density. For this reason, and because of the exponential-like increase of the nuclear density, it is essential to eliminate this secular variation. It should be noted that unfolding the spectrum is a difficult task when the exact form of the smooth density is not known, which is often what happens in complex quantum systems. See Gómez *et al.* [5] for a discussion on this point.

We consider three spectral statistics for the unfolded energy levels: the nearest neighbor level-spacing distribution $P(s)$, the average spectral rigidity $\Delta_3(L)$ of Dyson and Mehta [6, 7], and the power spectrum P_k^{δ} of the δ_n statistic. Both $P(s)$ and $\Delta_3(L)$ are well known and have been extensively used as a measure of chaos in quantum systems. Therefore we will give only a brief description of δ_n.

Apart from the comparison with random matrix behavior, there is an alternative approach to the analysis of spectral fluctuations of quantum systems which is based on the power spectrum of the energy level fluctuations [8]. There is an analogy between a discrete time series and a quantum energy spectrum, if time t is replaced by the energy E of the quantum states. In time series analysis, fluctuations are usually studied by means of the power spectrum of the signal.

We define the statistic δ_n as a signal,

$$\delta_n = \sum_{i=1}^{n} (s_i - \langle s \rangle) = \varepsilon_{n+1} - \varepsilon_1 - n, \quad n = 1, 2, \ldots N-1, \tag{1}$$

where $s_i = \varepsilon_{i+1} - \varepsilon_i$, ε_i are the unfolded levels, and the average values $\langle s \rangle = 1$ and $\langle \varepsilon_n \rangle = n$ have been used.

Thus δ_n represents the deviation of the nth excited level from its expected mean value. The discrete power spectrum of this signal is given by

$$P_k^\delta = \left|\hat{\delta}_k\right|^2 = \left|\frac{1}{\sqrt{N}} \sum_{n=1}^{N} \delta_n \exp\left(\frac{-2\pi i k n}{N}\right)\right|^2 \tag{2}$$

where $\hat{\delta}_k$ is the Fourier transform of δ_n, and N is the size of the series.

The first quantum spectra analyzed by this method were the levels of several nuclei calculated using the interacting shell model. It turned out that P_k^δ behaved very accurately as $1/k$. Then the analysis was performed for the basic random matrix ensembles GOE, GUE and GSE [8] and for a quantum billiard with chaotic classical limit. In all the cases the result was the same, namely the power spectrum of δ_n behaved as $1/k$. This led to the conjecture that *quantum chaos is characterized by* $1/f$ *noise.* Thus, without need to refer to any other systems, like random matrices, there is an intrinsic property that characterizes quantum chaos: A quantum system is considered chaotic if its spectral fluctuations are characterized by $1/f$ noise [8]. An analytical proof of this behavior for all the classical random matrix ensembles, independently of the symmetries of the system, was given by Faleiro *et al.* [9]. On the contrary, regular quantum systems exhibit Poisson statistics and $1/f^2$ noise.

Finally, apart from energy spectra, one should pay attention to the wave functions. The complexity of states can be measured by means of the information entropy S^{inf}. For any state of the basis $S^{inf} = 0$, and it takes the maximum value, $\ln(N)$, for a uniformly distributed state. For GOE matrices $S^{inf} = \ln(0.48N)$. The information entropy measures, somehow, the degree of delocalization of a certain state with regard to the mean-field basis. However, to avoid the dependence on the dimension N, the normalized localization length $l_{loc} = \exp(S^{inf})/0.48N$ is often used instead of the information entropy.

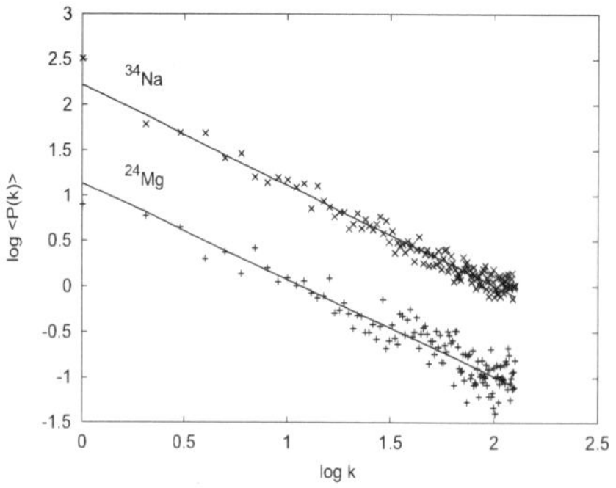

Figure 1. Average power spectrum of the δ_n function for ^{24}Mg and ^{34}Na, using 25 sets of 256 levels from the high level density region. The plots are displaced to avoid overlapping.

3. Shell-model calculations of chaotic properties in pf-shell nuclei

Large-scale shell-model calculations provide very long pure and complete sequences of energy levels. This makes possible the study of both short-range and long-range spectral correlations and in general the statistical analysis is quite reliable. Most shell-model calculations for statistical purposes have been done for nuclei in the *sd* shell and in the *pf* shell. We point out some of the most important theoretical results, focusing mostly on the *pf* shell, as these calculations are more recent and, because of the large dimensionalities of *pf* shell matrices, it becomes possible to study the dependence of the statistical properties on the excitation energy, angular momentum, pairing and isospin, for example. Results for the *sd* shell were summarized in the excellent review about shell model and many-body quantum chaos by Zelevinsky *et al.* [10]. The onset of the chaotic behavior in atomic nuclei is a question that has not been fully clarified experimentally as yet. We simply do not have sufficient experimental data. On the theoretical side this issue has been explored thoroughly using shell-model calculations. Early studies of level fluctuations in shell-model energy spectra showed a general agreement with GOE. This is indeed a remarkable result that has been known but has remained unexplained for many years. The many-body matrix elements of the nuclear shell-model Hamiltonian are certainly not random, as in GOE. Instead, all of them arise from the fairly well known nucleon-nucleon interaction. To understand the chaotic behavior of nuclei and to compare the results with RMT one needs to take into account the properties of realistic empirical shell-model interactions. See Section 6.2.1 of Gómez *et al.* [1] for a discussion on this point.

All theoretical work in the *sd* shell seems to indicate that nuclei present a strongly chaotic behavior, even in the proximity of the ground state. The spectral statistics $P(s)$ and $\Delta_3(L)$ follow closely the predictions of RMT [10]. Moreover, the power spectrum of δ_n for nuclei in this mass region clearly exhibits $1/f$ noise. This is illustrated by Fig. 1, which shows the results for a typical stable *sd*-shell nucleus, ^{24}Mg, with matrix dimensionalities up to $N \simeq 2000$; and for a very exotic nucleus, ^{34}Na, with N up to about 5000, in the *sd* proton and *pf* neutron shells. The shell-model calculations have been performed using the USD interaction [11] for ^{24}Mg, and the interaction described in [12] for ^{34}Na. Clearly, the power spectrum of δ_n follows closely a power law. We may assume the simple functional form $\langle P_k^\delta \rangle = 1/k^\alpha$.

A least squares fit to the data of Fig. 1 gives $\alpha = 1.11 \pm 0.03$ for ^{34}Na, and $\alpha = 1.06 \pm 0.05$ for ^{24}Mg [8]. Thus we clearly observe the existence of $1/f$ noise in these nuclei.

Turning now to pf-shell nuclei, we find a more diversified picture of spectral fluctuations. The shell-model calculations for pf-shell nuclei have been performed using the KB3 interaction [13]. Fig. 2 shows the $P(s)$ distribution for a very large number of levels in ^{52}Ca and ^{52}Sc [14]. After unfolding the spectrum for all the levels with the same $J^\pi T$, all the nearest neighbor level spacings for different $J^\pi T$ in a nucleus are gathered together. In ^{52}Ca all the J^π from 0^+ to 12^+, $T = T_z$ states are included, giving in total 11981 states. In ^{52}Sc the dimensionalities are much larger, and only the $J^\pi = 0^+$, 11^+ and 12^+, $T = T_z$ states were considered, giving a total of 11493 states. These large numbers of levels provide excellent statistics and we see that the agreement of $P(s)$ with GOE predictions is excellent, as could be expected. However, the $\Delta_3(L)$ behavior in the same nuclei is somewhat surprising, especially for the 0^+ states.

As Fig. 3 shows, in ^{52}Sc the $\Delta_3(L)$ statistic follows GOE up to large L values, but in ^{52}Ca it clearly deviates from GOE for stretches larger than $L \simeq 10$. Therefore, contrary to what the $P(s)$ results suggested, nuclear motion in ^{52}Ca cannot be considered to be very chaotic. As discussed in [14] and [1], several effects contribute to deviations from GOE in Ca isotopes. There are only active neutrons in the *pf* shell for Ca isotopes. For the 0^+ states, states of different seniority are included in the same sequence of levels, thus this approximate symmetry is not taken into account and that causes deviations towards Poisson, especially in the ground state region. More generally, for all J values there is a common feature in Ca isotopes. The residual n-n interaction is much weaker than the residual p-n interaction, and therefore the residual

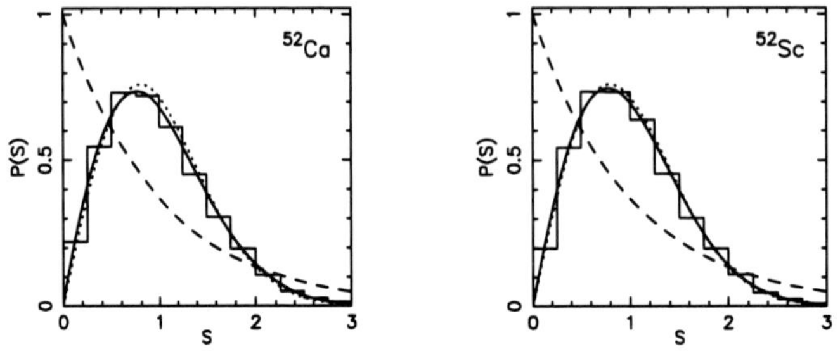

Figure 2. Nearest neighbor spacing distribution $P(s)$ for ^{52}Ca, $J^\pi = 0^+ - 12^+$, $T = 6$ states (left) and for ^{52}Sc, $J^\pi = 0^+, 11^+, 12^+$, $T = 4$ states (right). The dotted and dashed curves represent the GOE and Poisson $P(s)$ values, respectively.

interaction in Ca isotopes disturbs the ordered mean field motion much less than in other nuclei with valence protons and neutrons. In fact, a comparison of spectral statistics in the ^{46}Ca, ^{46}Sc and ^{46}Ti isobars shows strong deviations from GOE in ^{46}Ca, especially at low excitation energy. The simple presence of a proton in ^{46}Sc makes a great difference towards GOE, and this effect is enhanced in ^{46}Ti, in which the number of possible p-n configurations is even larger. Thus it can be concluded that chaos in nuclei has a strong isospin dependence [14, 1].

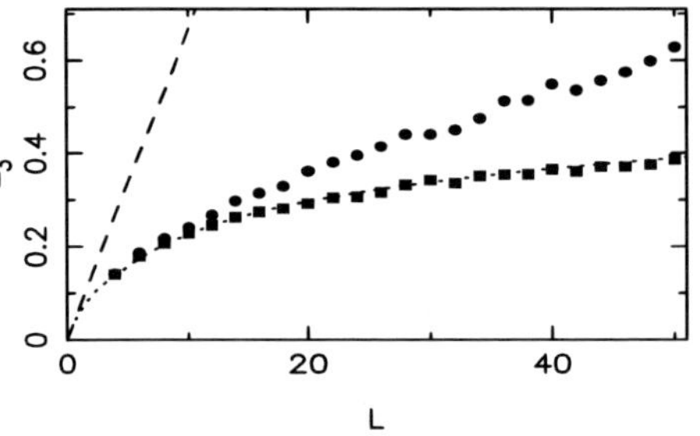

Figure 3. Average $\Delta_3(L)$ for all the $J^\pi = 5^+$, $T = T_z$ levels of ^{52}Ca (dots) and ^{52}Sc (squares). The dotted and dashed curves represent the GOE and Poisson $\Delta_3(L)$ values, respectively.

Finally we look at the complexity of wave functions. Fig. 4 compares the localization length l_{loc} for the 2^+ states in ^{28}Si [10] and in ^{46}Ti [1]. In both cases the number of levels is very large, providing good statistics. In both cases the behavior of $P(s)$ and $\Delta_3(L)$ are very close to GOE. We recall that $\exp(S^{inf})$ is the same as l_{loc}, except for a constant numerical factor. However, it is seen that in the region of largest density of states, the values in ^{28}Si are very close to the GOE limit, but in ^{46}Ti the maximum information entropy is about 65% of the GOE limit. Similarly, calculations for the $J^\pi T = 1^+2$ ($N = 2051$) states of ^{46}Sc and the $J^\pi T = 6^+5$ ($N = 2042$) states of ^{50}Ca, show that at the maximum density of states, l_{loc} is about 60% of the GOE value in ^{46}Sc, and about 40% in ^{50}Ca [14]. These results are consistent with the results for $\Delta_3(L)$, showing the existence of isospin dependence of chaos in nuclei, as well as the large deviations from GOE in nuclei like Ca isotopes in which all the valence particles are neutrons.

4. Conclusions

Large scale shell-model calculations with realistic empirical interactions provide long sequences of pure $J^\pi T$ states which facilitate reliable calculations of statistical nuclear properties. In

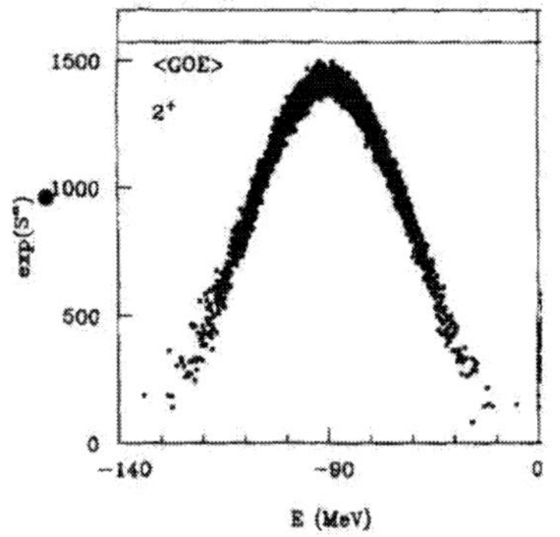

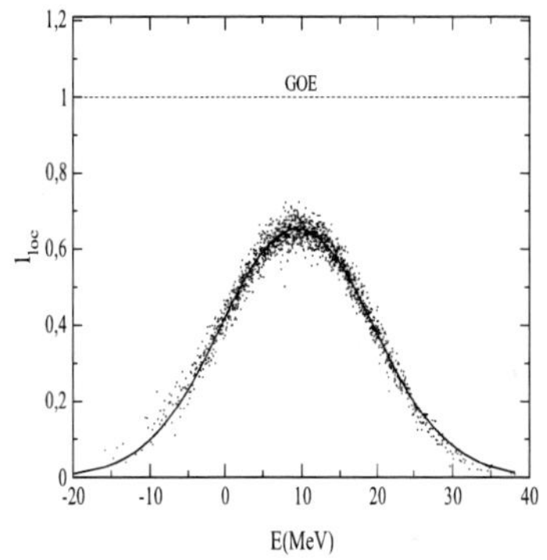

Figure 4. Comparison of the normalized localization length $l_{loc} = \exp(S^{inf})/0.48N$ as a function of excitation energy for the $J^\pi T = 2^+0$ states of ^{28}Si (left) [10] and for the $J^\pi T = 2^+1$ states of ^{46}Ti (right) [1]. In the figure on the left the information entropy S^{inf} is called S^α.

particular, in pf shell nuclei the sequences of states with the same $J^\pi T$ quantum numbers are very large and it has been possible to observe on the one hand the onset of chaos as the excitation energy increases, and on the other the onset of chaos as the strength of the residual interaction increases for fixed A, i. e., as the number of possible p-n configurations for the valence nucleons increases. In all the investigated nuclei, level fluctuations deviate from GOE in the lower part of the spectrum, and even when the full spectrum is taken into account for Ca isotopes the $\Delta_3(L)$ statistic clearly deviates from GOE for $L \geq 10$. The analysis of shell-model wave functions shows that the localization length l_{loc} in pf-shell nuclei is clearly smaller than the GOE value ($l_{loc}(\mathrm{GOE}) = 1$), and again a clear isospin dependence is observed for isobaric nuclei. We conclude that it is necessary to study the behavior of different spectral statistics, including short range and long range level correlations, as well as the complexity of wave functions, in order to get a clear picture and assess the degree of chaos in nuclei. In particular, it is clear that the results for the $P(s)$ distributions alone are not sufficient for a reliable assessment. Finally, we conclude that the deviations from GOE observed in our shell-model calculations in pf-shell nuclei make sense. After all, the nuclear interaction gives two-body matrix elements which are far from random, and the shell model configurations lead to nuclear wave functions with some components more important than others, especially, but not only, in the ground state region. On the contrary, in GOE the amplitudes of the wave function are more uniformly distributed.

References

[1] Gómez J M G, Kar K, Kota V K B, Molina R A, Relaño A and Retamosa J 2011 *Phys. Rep.* **499** 103-226.
[2] Bohigas O, Giannoni M J and Schmit C 1984 *Phys. Rev. Lett.* **52** 1.
[3] Müller S, Heusler S, Braun P, Haake F and Altland A 2004 *Phys. Rev. Lett.* **93** 014103.
[4] Haq R U, Pandey A and Bohigas O 1982 *Phys. Rev. Lett.* **48** 1086.
[5] Gómez J M G, Molina R A, Relaño A and Retamosa J 2002 *Phys. Rev.* E **66** 036209.
[6] Dyson F J and Mehta M L, 1963 *J. Math. Phys.* **4** 701.
[7] Mehta M L, 2004 *Random Matrices*, Academic Press, 3rd ed.
[8] Relaño A, Gómez J M G, Molina R A, Retamosa J and Faleiro E 2002 *Phys. Rev. Lett.* **89** 244102.
[9] Faleiro E, Gómez J M G, Molina R A, Muñoz L, Relaño A and Retamosa J 2004 *Phys. Rev. Lett.* **93** 244101.
[10] Zelevinsky V, Brown B A, Frazier N and Horoi M 1996 *Phys. Rep.* **276** 85.
[11] Wildenthal B H 1983 *Prog. Part. Nucl. Phys.* **11** 5.
[12] Poves A and Retamosa J 1994 *Nucl. Phys.* A **571** 221.
[13] Poves A and Zuker A 1981 *Phys. Rep.* **70** 4.
[14] Molina R A, Gómez J M G and Retamosa J 2000 *Phys. Rev.* C **63** 014311.

Ischia14 IOP Publishing
Journal of Physics: Conference Series **580** (2015) 012046 doi:10.1088/1742-6596/580/1/012046

On the relation of the shell, collective and cluster models

J. Cseh

Institute for Nuclear Research, Hungarian Academy of Sciences, Debrecen, Pf. 51, Hungary-4001

E-mail: cseh@atomki.mta.hu

Abstract. The intersection of the shell, collective and cluster models is described for multi-major-shell problems.

1. Introduction

The fundamental models of nuclear structure are based on different physical pictures. The shell model indicates that the atomic nucleus is something like a small atom, the cluster model suggests that it is similar to a molecule, while the collective model says that it is a microscopic liquid drop. Therefore, in order to understand the nuclear structure we need to study (among others) the interrelation of these models, find their common intersection, etc.

The basic connections were found in the fifties. Elliott [1] showed how the quadrupole deformation and collective rotation can be derived from the spherical shell model: the states belonging to a collective band are determined by their specific SU(3) symmetry. Wildermuth and Kanellopoulos [2] established the relation between the shell and cluster models. They proved that the Hamiltonians of the two models can be rewritten into each other exactly in the harmonic oscillator approximation. This relation results in a close connection between the corresponding eigenvectors, too: the wavefunction of one model is a linear combination of those of the other, which belong to the same energy. Later on this relation was interpreted by Bayman and Bohr [3] in terms of the SU(3) symmetry. As a consequence, the cluster states are also selected from the shell model space by their specific SU(3) symmetries. (In fact only one kind of cluster states have this feature, and there are other kinds, too, as discussed later.)

We will refer to this interrelation among the three basic structure models as the SU(3) connection. It was established in 1958 for a single major shell problem. Here we consider its extension to multi-major-shells.

The connection of the shell model and the cluster model is especially interesting due to the fact that both models have a complete set of basis states, i.e. any nuclear states can be expanded in both bases. Depending on the simple or complicated nature of these expansions we can distinguish four different cases. i) If it is simple in the shell basis, but complicated in the cluster one, then we can speak about a simple shell state, which is a poor cluster state. ii) If it is simple in the cluster basis, but complicated in the shell basis, one has a good cluster state, which is a poor shell state. (Later on we will refer to these states as rigid molecule-like cluster states.) iii) The expansion may be simple in both bases. We call this situation as a shell-like cluster state. iv) If it is complicated in both bases, than the state is not a simple one.

Of course it is not the name that matters. The important point is that there are three different kinds of simple states when we investigate them from the viewpoint of the cluster-shell competition. Sometimes only the kind ii), i.e. rigid molecule-like states are considered as cluster states. The reason for our vocabulary, i.e. speaking about two kinds of cluster states: shell-like and molecule-like is inspired by their experimentally observable characteristics. Both of them prefer a certain reaction channel, have a large cluster spectroscopic factor, etc. They represent two different quantum phases of the nucleus, as discussed in [4]. By choosing these names we are in line with the general definition of a simple nuclear state in terms of experimental observation: a state is called simple if its wavefunction has a large overlap, or form a large matrix element with the wavefunction of a reaction channel, in which it can be observed [5].

2. Models of different pictures

2.1. Shell models

Elliott's U(3) model bridges the spherical shell model and the collective model, and describes rotational states in the p and sd shell nuclei. Here the U(3) symmetry is that of the space part, while the spin-isospin section is characterized by Wigner's $\mathrm{U}^{ST}(4)$ [6]. Thus the group structure of the model is $\mathrm{U}^{ST}(4) \otimes \mathrm{U}(3)$. In constructing the model space the spin-isospin degrees of freedom are essential, of course. The physical operators, however, often contain simply spin-isospin zero terms, in which case only the generators of the U(3) group contribute. The space part of the basis states is defined by the representation labels of the group chain: $U(3) \supset SU(3) \supset SO(3) \supset SO(2)$.

The quadrupole-quadrupole interaction plays a major role, and it can be written as a linear combination of the quadratic invariant operators ($C^{(2)}$) of the group-chain: SU(3) $\supset$ SO(3): $QQ = \frac{1}{2}C^{(2)}_{SU(3)} - \frac{3}{2}C^{(2)}_{SO(3)}$. The electromagnetic transition rates are obtained by applying effective charge. The many-nucleon states are classified according to their SU(3) symmetry, and in the simplest case the interactions are expressed in terms of its generators.

The single major shell model has an algebraic structure defined by a larger group, which contains the symmetry groups of both the space part U(3), and the spin-isospin part $\mathrm{U}^{ST}(4)$. It is the unitary group of U(4Ω), where Ω denotes the orbital degeneracy: Ω=1, 3, 6,... for the the s, p, sd, ... shells, respectively. U(4Ω) is a dynamical group of the single major shell model in the sense that the physical operators are obtained in terms of its generators, and the whole spectrum is provided by a single irreducible representation (irrep) of it.

The description of the electromagnetic transitions without an effective charge requires the incorporation of the major shell excitations; i.e. a vertical extension of the SU(3) shell model. For this purpose the symplectic group proved to be very useful [7].

The Sp(3,R) group is generated by the position vectors of the nucleons, and their canonically conjugate momenta. An alternative set of its generators is expressed in terms of harmonic oscillator operators, containing the 9 generators of the U(3) group, which preserve the number of oscillator quanta, and in addition 6 creation $B^{\dagger(l)}_m = [\pi^\dagger \times \pi^\dagger]^{(l)}_m$, $l = 0, 2$ and 6 annihilation $B^{(l)}_m = [\pi \times \pi]^{(l)}_m$ operators, which ladder by 2 or -2 quanta. The creation operators are $[2, 0, 0]$ U(3) tensors, therefore, their products also carry U(3) labels: $[n^e_1, n^e_2, n^e_3]$ (e stands for excitation).

The algebraic structure of the simplectic shell model is provided by the Sp(3,R) group in the sense that the physical operators are expressed in terms of its generators, and the collective bands are given by its (infinite dimensional) irreducible representations (which are built on U(3) irreps).

2.2. Collective model

Several algebraic collective models have been developed. In what follows we focus on a model with a simple algebraic structure, which is very illuminative from the viewpoint of the interrelation of the fundamental structure models. It is the large n limit of the Sp(3,R) symplectic model.

The dynamical group in this limit simplifies to $U_s(3) \otimes U_b(6)$, i.e. to a compact group, as opposed to the noncompact Sp(3,R). Technically the simplification is achieved by replacing the creation and annihilation operators of the Sp(3,R) model (in which the description is based on nucleon degrees of freedom) by boson operators. The $U_s(3)$ is Elliott's shell model symmetry of the $0\hbar\omega$ shell, and $U_b(6)$ is the group of the six dimensional oscillator, generated by the bilinear products of the (l = 0 and 2) boson creation and annihilation operators. It is realised in a similar way as the U(6) group of the interacting boson model (IBM) [8], nevertheless physically it is different, because in the case of the contracted symplectic model the bosons are associated to intershell excitations, not to intrashell ones.

This model is called U(3) boson model [9], or contracted symplectic model [10]. The contracted boson operators which generate the intershell excitations do introduce here a simplifying approximation. Therefore, the antisymmetry requirement is not fully incorporated in this description. This model is more easily applicable (compared to the Sp(3,R) model), e.g. it has an orthonormal set of basis states.

The algebraic structure of the contracted simplectic collective model is provided by the $U_s(3) \otimes U_b(6)$ group in a sense (similarly to that of the symplectic model) that the physical operators are expressed in terms of its generators, and the collective bands are given by its (finite dimensional) irreducible representations (which are built on U(3) irreps).

2.3. Cluster models

There are different kinds of cluster models, which all share the general picture of dividing the relevant degrees of freedom into two cathegories: those belonging to the relative motion of the clusters (usually in a large variety) and those of the internal structure of them (usually in a rather limited number).

In searching for the symmetry-based relations to the shell and collective models, two aspects of the models are especially important: i) how much they are microscopic, and ii) to what extent they are equipped with an algebraic structure.

A model is called fully microscopic if the antisymmetrization is completely involved, and the interactions are (effective) nucleon-nucleon forces. It is semimicroscopic, if the exclusion principle is appreciated, but the interactions are (phenomenological) cluster-cluster interactions. It is fully phenomenological, if cluster-cluster interactions are applied in a model space, which is constructed phenomenologically, i.e. without taking into account the Pauli-principle.

The algebraic structure on the other hand reveals the symmetries of the model, and this can provide us with a connection to the shell and collective picture.

For the sake of simplicity we consider here binary cluster configurations. When the SU(3) shell model [1] is applied for the description of the internal structure of the clusters then the spin-isospin degrees of freedom of the clusters are coupled together, while the space-part is characterized by the group-chain:

$$U_{C_1}(3) \otimes U_{C_2}(3) \otimes \ U_R(3) \supset U_C(3) \ \otimes U_R(3) \supset U(3) \supset SU(3) \supset SO(3) \supset SO(2), \qquad (1)$$

here $U_C(3)$ stands for the coupled space symmetry of the two clusters.

This basis is especially useful for treating the exclusion principle, since the U(3) generators commute with those of the permutation group, therefore, all the basis sates of an irrep are either Pauli-allowed, or forbidden [11].

On the fully microscopic level much work has been done in developing an algebraic description. Especially great progress was made in the works [11, 12], in calculating the necessary matrix elements in an U(3) basis.

In the fully microscopic description usually effective nucleon-nucleon forces are applied without specific symmetry character. Therefore, this is not a fully algebraic description (the physical operators are not expressed in terms of the generators of a dynamical group), contrary to the single major shell model of U(4Ω), the symplectic model of Sp(3,R), and the contracted symplectic model of $U_s(3)\otimes U_b(6)$.

The semimicroscopic algebraic cluster model (SACM) [13] is a fully algebraic approach with transparent symmetry-properties. The internal structure of the clusters is described here by the Elliott model [1] too, therefore, this part of the wavefunction has a $U_C^{ST}(4)\otimes U_C(3)$ symmetry. The relative motion of the clusters is accounted for by the modified vibron model [14].

The coupling between the relative motion and internal cluster degrees of freedom for a binary cluster system results in a group structure: $G_{2C} \equiv U_{C_1}^{ST}(4) \otimes U_{C_1}(3) \otimes U_{C_2}^{ST}(4) \otimes U_{C_2}(3) \otimes U_R(4)$. The spin and isospin degrees of freedom are essential in this case, too, from the viewpoint of the construction of the model space. However, if one is interested only in a single supermultiplet [$U_C^{ST}(4)$] symmetry, which is typical in cluster problems, then the relevant group structure simplifies to that of the space part. In particular the U(3) (strong) coupled basis is defined by the group chain:

$$U_C(3) \otimes U_R(4) \supset U_C(3) \otimes U_R(3) \supset U(3) \supset SU(3) \supset SO(3) \supset SO(2) \qquad (2)$$

The exclusion of the Pauli-forbidden states amounts up to a truncation of the coupled U(3) basis from the side of the small number of oscillator quanta. Some major shells are completely missing, and from some other ones parts of the single-nucleon states are excluded. This is the modification [13] with respect to the original vibron model, as it is applied e.g. in molecular physics [15].

The relation of the SACM and fully microscopic description from the viewpoint of the model space is that they contain the same U(3) [and $U^{ST}(4)$] irreps, but the complete antisymmetrization is carried out only in the fully microscopic descriptions. Therefore, the calculation of the cluster spectroscopic amplitude in the semimicroscopic model is being done by the introduction of phenomenologic parameters [16].

The connection between the fully microscopic and the semimicroscopic cluster models is somewhat similar to the relation between the symplectic shell model and the contracted symplectic model. The previous ones are fermionic models, accounting for the antisymmetrization to full details, while the latter ones are their simplifying (bosonic) approximations. Therefore, their wavefunctions are not the same, though the U(3) content is identical in both cases.

3. Intersection

When major shell excitations are incorporated, then both the (symplectic) shell model, and the (contracted symplectic) collective model, as well as the (microscopic or semimicroscopic algebraic) cluster model has a set of basis states characterised by the irreps of the

$$U_x(3) \otimes U_y(3) \supset U(3) \supset SU(3) \supset SO(3) \supset SO(2) \qquad (3)$$

group chain, as seen above. For the shell and collective models x stands for the band-head (valence shell), for the cluster model it refers to the internal cluster structure. y indicates in each case the major shell excitations; in the shell and collective model cases it takes place in steps of $2\hbar\omega$, connecting oscillator shells of the same parity, while in the cluster case it is in steps of $1\hbar\omega$, incorporating all the major shells. For the cluster model it has only completely symmetric (single-row Young-tableux) irreps: $[n, 0, 0]$, while in the case of the shell and collective models it can be more general. As a consequence the model space of the three models have a considerable overlap, but they are not identical.

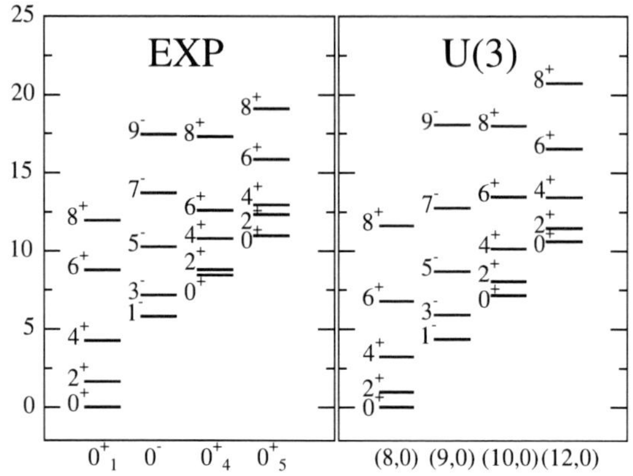

Figure 1. Experimental and U(3) model spectra in the ^{20}Ne nucleus. The experimental bands are labelled by the K^π, and the model states by the (λ, μ) quantum numbers.

As for the relation of the model spaces (wavefunctions) of the three basic structure models (shell, collective and cluster models) is considered the following can be said. When multi major shells are involved then the (quadrupole) collective or (dipole) cluster bands can be picked up from the microscopic shell model basis according to their $U_x(3) \otimes U_y(3) \supset U(3)$ symmetry. Some irreducible representations are present in each of the three models. Having the same SU(3) basis, however, does not necessarily mean 100% overlap of the wavefunction, it can be less, too. This situation is similar to what was found for the single major shell problem in terms of the historical SU(3) connection.

Another interesting question of the intersection of the different structure models is how their spectra compare to each other, when similar interactions applied. In this respect first we should note that the Wildermuth-connection between the shell and cluster model states, which was found originally for harmonic oscillator Hamiltonians, turned out to be valid for more general interactions, too [17], just like the Elliott-connection. If the Hamiltonian is expressed in terms of the invariant operators of the group chain (3) then the energy-eigenvalue has an analytical solution i.e. a (broken) dynamical symmetry is present. This $U_x(3) \otimes U_y(3) \supset U(3)$ dynamical symmetry is the common intersection of the shell, collective and cluster models of multi major shells.

Figure 1. shows a comparison between the experimental and the U(3) dynamically symmetric (3) spectrum of the ^{20}Ne nucleus. The energy (in MeV) was obtained with the formula $E = 13.19\lambda - 0.4579\lambda(\lambda+3) + 0.8389\frac{1}{2\theta}L(L+1)$. The oscillator energy is determined according to the systematics [18], while θ is the moment of inertia calculated classically for the rigid shape determined by the U(3) quantum numbers. The parameters of the quadratic and the rotational terms were fitted to the experimental data. (The ground state energy is taken to be zero.) The experimental data are from [19]. (For more details, including $E2$ transitions, see [20].)

The purpose of this calculation is not to give a detailed description of the experimental data, rather to show how the common intersection of the three basic structure models compares to the experiment. The right panel in Fig. 1. can be considered as a (part of a) shell, collective or cluster spectrum, when the basis states and the operators can be characterized by the SU(3) (and subgroup) symmetries, i.e. for the case of the dynamical symmetry. These circumstances are similar again to those of the single-shell problem.

4. Summary and conclusion

In this paper we have discussed the interrelation of the fundamental nuclear structure models, the shell, collective and cluster models from the viewpoint of symmetries. These models are based on different physical pictures, and their connection was established first in terms of the SU(3) symmetry for a single shell problem [1, 2, 3]. We have considered here the generalization of this relation along the major shell excitations.

Algebraic models have been constructed for the description of this vertical extension in each of the three approaches. The most relevant ones from the viewpoint of the symmetry-based interrelations are the symplectic shell model of Sp(3,R) algebraic structure [7], the contracted symplectic model of $U_b(6) \otimes U_s(3)$, which is the large n limit of the multi major-shell symplectic model [9, 10], and the fully microscopic [11, 12] as well as the semimicroscopic algebraic cluster models [13], with $U_C(3) \otimes U_R(3)$ basis. The common intersection of these models is provided by the $U_x(3) \otimes U_y(3) \supset U(3)$ dynamical symmetry, i.e. for the many major-shell problem this symmetry substitutes the simple SU(3). Figure 1 shows how the schematic spectrum of this dynamical symmetry, which can be considered as a shell, collective or cluster description, compares with the experimental one.

This work was supported by the OTKA (Grant No K106035). Helpful discussions with Professor J.P. Draayer are kindly acknowledged.

References

[1] Elliott J P 1958 *Proc. Roy. Soc. A* **245** 128, 562
[2] Wildermuth K and Kanellopoulos Th 1958 *Nucl. Phys.* **7** 150
[3] Bayman B F and Bohr A 1958/59 *Nucl. Phys.* **9** 596
[4] Cseh J and Darai J 2014 arXiv 1404.3503 [nucl-th]
[5] Mahaux C 1973 *Ann. Rev. Nucl. Sci.* **23** 193
[6] Wigner E P 1937 *Phys. Rev.* **51** 106
[7] Rosensteel G and Rowe D J 1977 *Phys. Rev. Lett.* **38** 10; 1980 *Ann. Phys. (N.Y.)* **126** 343
[8] Iachello F and Arima A, The Interacting Boson Model, Cambridge University Press, Cambridge, 1987.
[9] Rowe D J and Rosensteel G 1982 *Phys. Rev. C* **25** 3236(R)
[10] Castanos O and Draayer J P 1989 *Nucl. Phys. A* **491** 349
[11] Horiuchi H 1977 *Prog. Theor. Phys. Suppl.* **62** 90
[12] Hecht K T Reske E J Seligman T H Zahn W 1981 *Nucl. Phys. A* **356** 146
[13] Cseh J 1992 *Phys. Lett. B* **281** 173; Cseh J and Lévai G 1994 *Ann. Phys. (NY)* **230** 165
[14] Iachello F 1981 *Phys. Rev. C* **23** 2778(R); Iachello F and Levine R D 1982 *J. Chem. Phys.* **77**, 3046
[15] Iachello F Cseh J and Lévai G 1995 *APH N.S. Heavy Ion Phys.* **1** 91
[16] Cseh J Lévai G and Katō K 1991 *Phys Rev. C* **43** 165;
Hess P O Algora A Cseh J and Draayer J P 2004 *Phys Rev. C* **70** 051303(R)
[17] Cseh J and Katō K 2013 *Phys. Rev. C* **87** 067301
[18] Blomqvist J and Molinari A 1968 *Nucl. Phys. A* **106** 545
[19] Tilley D R et. al. 1998 *Nucl. Phys. A* **636** 249
[20] Cseh J 2014 arXiv 1404.3500 [nucl-th]

Quantum phase transitions in odd-A nuclei: The effect of the odd particle from spherical to oblate shapes

M Böyükata[1], C E Alonso[2], J M Arias[2], L Fortunato[3] and A Vitturi[3]
[1] Department of Elementary Science Education, Faculty of Education, Çanakkale Onsekiz Mart University, TR-17100 Çanakkale, Turkey
[2] Departamento de Física Atómica, Molecular y Nuclear, Universidad de Sevilla, ES-41080 Sevilla, Spain
[3] Dipartimento di Fisica e Astronomia "G. Galilei", Università di Padova and INFN, Sezione di Padova, IT-35131 Padova, Italy
E-mail: boyukata@comu.edu.tr

Abstract.
Quantum shape-phase transitions in odd-nuclei are investigated within the framework of the interacting boson-fermion model (IBFM). We consider the case of a single-j fermion coupled to an even-even boson core that performs a transition from spherical to oblate shapes varying a control parameter in the boson Hamiltonian. The aim of this work is to see the effect of the coupling of the unpaired fermion on the transition, to understand how the coupled single particle modifies the geometric shape of the system and how each of the odd states behaves when the boson core shifts along the transitional path.

1. Introduction

The study of phase transitions in finite nuclear quantal systems has been the subject of many investigations in Nuclear Physics. These studies include the Quantum Phase Transitions (QPT) in nuclei within both the Geometric Collective Model (GCM) [1] and also the Interacting Boson Model (IBM) [2], mostly on even-even systems [3, 4]. Moreover the corresponding phase transitions in Bose-Fermi systems [5-16] have been studied within the IBFM [16].

In our previous works [11, 15], we focused on quantum phase transitions for Bose-Fermi systems: the effect of the coupling of a single $j = 9/2$ fermion to an even-even boson core that performs the transition from spherical to γ-unstable shapes [11] and from spherical to axially prolate shapes [15]. These situations are described within the framework of the intrinsic frame formalism for the IBFM [17, 18, 19]. In Ref [15], results of both studies are also compared to see the differences of the odd particle effect along both transitional paths.

In this work, we intend to see the effect of the coupling of the unpaired fermion in a orbit of definite angular momentum $j = 9/2$ on the transition from spherical to oblate shapes, to understand how the coupled single particle modifies the geometric shape of the system and how each of the odd states behaves when the boson core shifts along the transitional region. Moreover, the overall results of the odd particle effect along the spherical to oblate transition are presented for different angular momenta, with j ranging from 3/2 to 13/2.

The structure of this proceeding is organized as follows: we first give the general IBFM Hamiltonian in section 2, then the intrinsic frame formalism for Bose-Fermi systems is described in section 3. Finally, results are presented and conclusions are remarked in the last section.

2. The IBFM Hamiltonian

The interacting boson-fermion (IBFM) Hamiltonian is generally written as

$$H = H_B + H_F + V_{BF} \ , \tag{1}$$

where H_B describes the bosonic part, the H_F term is the fermionic part, and the V_{BF} is the interaction between bosons and fermion.

The boson Hamiltonian in equation (1) is given by

$$H_B = (1-c)n_d - \frac{c}{4N_B} Q_B^\chi \cdot Q_B^\chi \ , \tag{2}$$

where N_B is the total boson number, c is the control parameter, n_d is the the d-boson number operator defined as

$$n_d = \sum_\mu d_\mu^\dagger d_\mu \ , \tag{3}$$

and Q_B^χ is the boson quadrupole operator given by

$$Q_B^\chi = (s^\dagger \times \tilde{d} + d^\dagger \times \tilde{s})^{(2)} + \chi(d^\dagger \times \tilde{d})^{(2)}. \tag{4}$$

Taking into account that the pure fermion part is a constant for the single-j shell case as the one presented here, the boson-fermion Hamiltonian (1) can be written as follows:

$$H = (1-c)n_d - \frac{c}{4N_B} Q_{BF}^\chi \cdot Q_{BF}^\chi , \tag{5}$$

where Q_{BF}^χ is the quadrupole operator for the odd-even system

$$Q_{BF}^\chi = Q_B^\chi + q_F \ , \tag{6}$$

and $\hat{q}_F$ is the fermion quadrupole operator given by

$$q_F = (a_j^\dagger \times \tilde{a}_j)^{(2)} \ . \tag{7}$$

In this case, the boson-fermion interaction V_{BF} included in equation (1) is

$$V_{BF} = -\frac{c}{2N_B} \ Q_B^\chi \cdot q_F \ . \tag{8}$$

3. The Intrinsic Frame Formalism

The concept of intrinsic frame formalism is one of the useful ways to look into quantum shape-phase transitions in atomic nuclei. This concept associates a potential energy surface to a Hamiltonian as (2) in terms of shape variables. Within the IBM, the intrinsic state for the ground state band of an even-even nucleus is written as

$$|\Phi_{gs}(\beta,\gamma)\rangle = \frac{1}{\sqrt{N_B!}} [b_{gs}^\dagger(\beta,\gamma)]^{N_B} |0\rangle \ , \tag{9}$$

where $|0\rangle$ is the boson vacuum, $b_{gs}^\dagger$ is the ground state boson creation operator and the variational parameters, also called deformation parameters β and γ, play a similar role to the one played

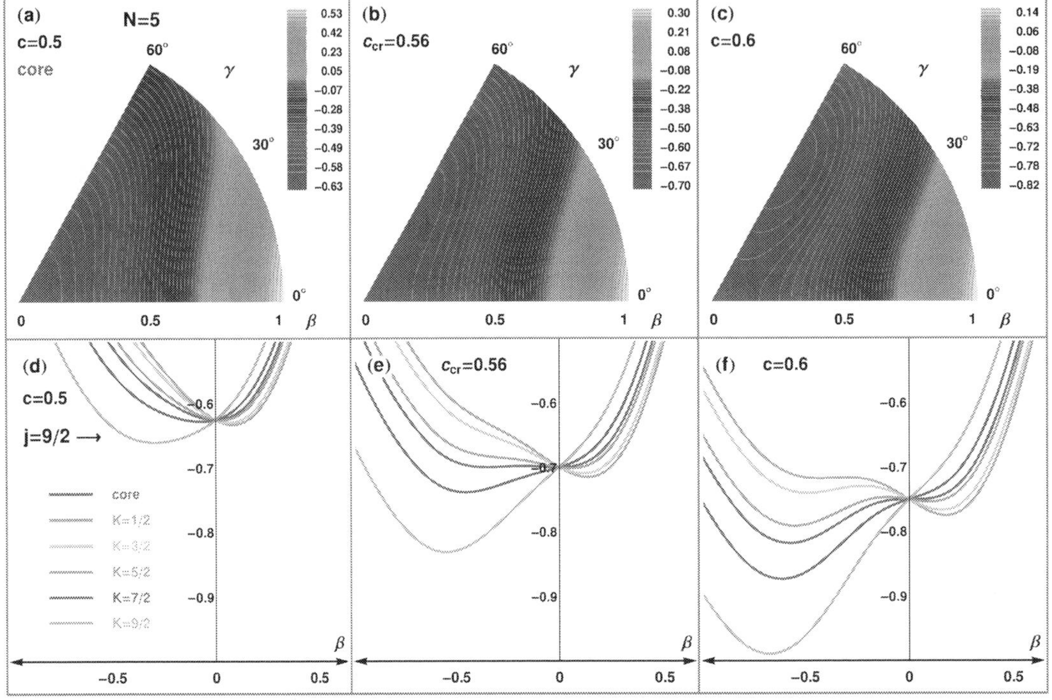

Figure 1. Even-even and odd-even energy surfaces. In panels (a), (b) and (c) energy surfaces for the even-even system in the β-γ plane are shown for different values of c. The selected values for the control parameter c are chosen around the critical value, c_{cr}, in the Hamiltonian describing the transition from spherical to oblate shapes ($\chi = +\sqrt{7}/2$ is fixed). In panels (d), (e) and (f), the five energy surfaces for the different K states coming from $j = 9/2$ in the odd-even system are plotted as a function of β for the same control parameters. In these panels, the energy surface for the even-even case (red color) is also plotted for reference.

by the intrinsic collective shape variables in the Bohr Hamiltonian. The β variable measures the axial deviation from sphericity and the angle variable γ controls the departure from axial deformation.

The ground state boson creation operator is given by

$$b^{\dagger}_{gs}(\beta,\gamma) = \frac{1}{\sqrt{1+\beta^2}} \left[s^{\dagger} + \beta \cos\gamma d^{\dagger}_0 + \frac{\beta}{\sqrt{2}} \sin\gamma \left(d^{\dagger}_2 + d^{\dagger}_{-2} \right) \right] . \tag{10}$$

The ground state energy surface is obtained by calculating the expectation value of the boson Hamiltonian (2) in the intrinsic state (9)

$$E_{gs}(\beta,\gamma) = \langle \Phi_{gs}(\beta,\gamma) | H_B | \Phi_{gs}(\beta,\gamma) \rangle . \tag{11}$$

Intrinsic frame states for the mixed boson-fermion system can be constructed by coupling the odd single-particle states to the intrinsic states of the even core. The lowest intrinsic states of the odd nucleus originate from this coupling to the intrinsic ground-state $\Phi_{gs}(\beta,\gamma)$. To obtain them, we first construct the coupled states

$$|\Psi_{jK}(\beta,\gamma)\rangle = |\Phi_{gs}(\beta,\gamma)\rangle \otimes |jK\rangle , \tag{12}$$

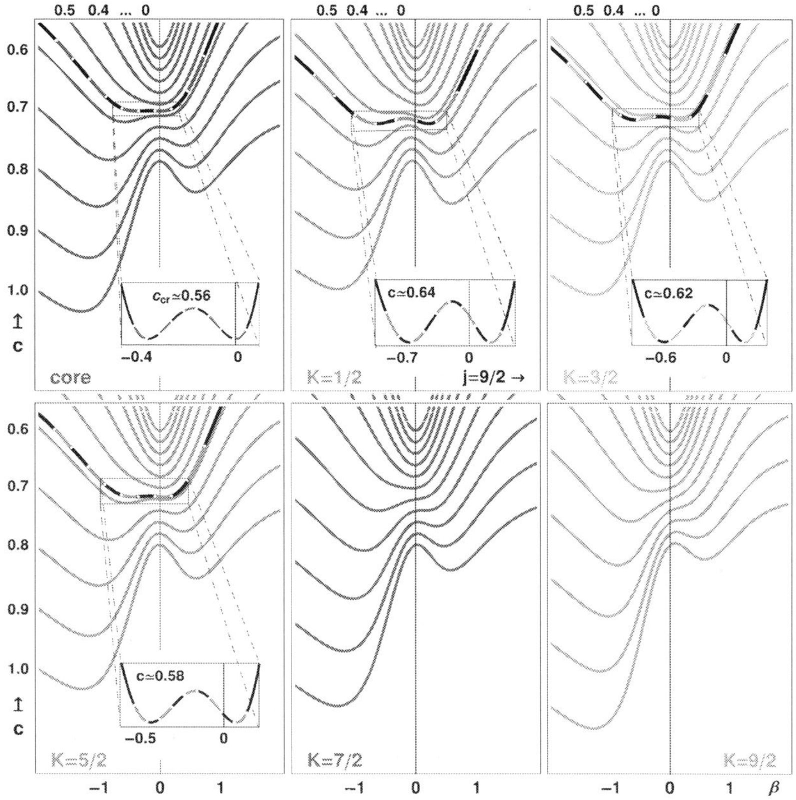

Figure 2. Evolution of the energy surfaces for the bosonic core (red color) and for the different K states (different colors) in the odd-even system with $j = 9/2$ as a function of β. Here, the control parameter c is changed from *0* to *1* for the transition from spherical to oblate shape. Calculations are performed for $N_B = 5$.

and then diagonalize the total boson-fermion Hamiltonian in this basis, yielding a set of energy eigenvalues $E_n(\beta, \gamma)$, where n is an index to count solutions in the odd-even system and K is the projection of the total angular momentum j on the symmetry axis.

4. Results and Conclusions

A single fermion with $j = 9/2$ is coupled to an even-even boson core that undergoes the shape phase transition from spherical to deformed oblate shapes to observe the corresponding shape phase transition in a mixed Bose-Fermi system. The spherical shape is obtained for $c = 0$, while $c = 1$ gives pure oblate shape, fixing $\chi = +\sqrt{7}/2$ in equation (2) and equation (4), unlike Ref. [15] where $\chi = -\sqrt{7}/2$ to obtain pure prolate shape. If it were only for the boson part, the two cases would be completely symmetric with respect to the change of sign in chi, as expected. However, in the case of coupling to a fermion, the behaviours of K-states are quite different as described in the following.

The expectation value of the boson Hamiltonian in the intrinsic state (9), $E_{gs}(\beta, \gamma)$, is calculated to obtain the ground state energy surface. For the bosonic system, a first-order shape phase transition is observed at $c_{cr} = 16N_B/(34N_B - 27)$ in between U(5)-$\overline{SU(3)}$, same as in Ref. [10] for spherical to prolate shapes. Therefore, the critical point of the first-order transition is $c_{cr} = 0.56$ in the case of *five* bosons, as in Ref. [15]. The energy surfaces for the

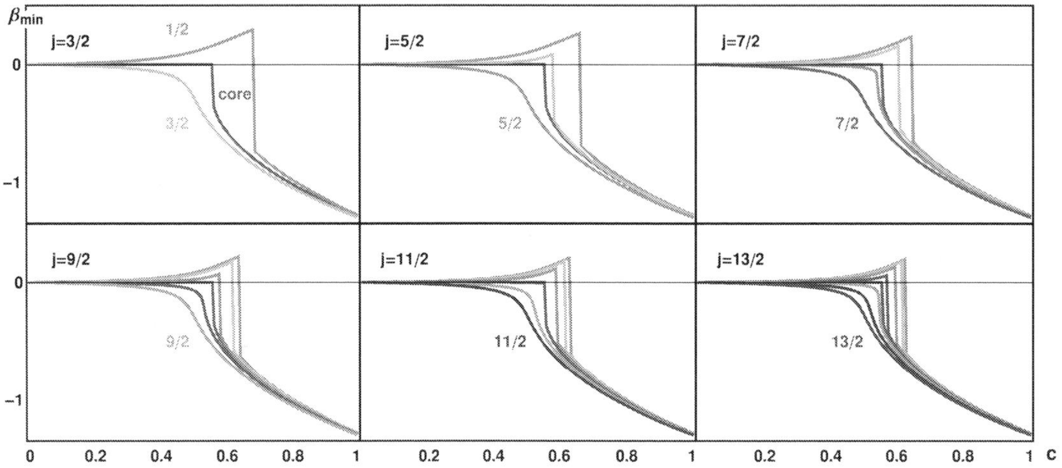

Figure 3. Evolution of the equilibrium deformation parameter in odd-even system as a function of the control parameter c along the U(5) to $\overline{SU(3)}$ transition. Calculation have been done for $N_B = 5$. Each panel represents a different single-j case. Within a panel, different lines correspond to the different possible $K-$states. The red line, which is the same in all panels, gives the even-even value.

boson core as a function of the deformation parameters (β,γ) are plotted around the critical point as shown in panels (a), (b), (c) of figure 1. In the case of $c_{cr} = 0.56$ (middle panels in figure 1) the even-even surface has two degenerate minima (one spherical and one oblate). The even-even system is spherical before this point then it jumps to oblate deformation.

Then the odd $j = 9/2$ particle is coupled. The behaviour of the odd-even energy surfaces as a function of β is illustrated around the critical point in panels (d), (e) and (f) of figure 1. In these panels, the boson core is denoted with red color as a reference. For these cases, around the even-even critical point, we see that independently of the core shape (spherical in figure 1(a), critical figure 1(b), or oblate in figure 1(c)) the intrinsic states of the odd-even system prefer either prolate or oblate shapes depending on their K value (see figures 1(d), 1(e) and 1(f), respectively).

In the case of angular momentum $j = 9/2$, the possible $K-$projections are $K = -9/2, ..., 9/2$. Therefore there are *10* different states that are restricted to *5* because of the symmetry $K \leftrightarrow -K$. The evolution of the energy surfaces along the shape phase transition for the boson core and the odd-even system with five magnetic components K are shown in figure 2. Different lines in a panel correspond to different control parameter values c. The bosonic core (red color) is given in the first panel. For $c = 0$, uppermost line, the core is spherical and it changes to well defined oblate shape for $c = 1$ (lowest line). At the critical point, $c_{cr} = 0.56$, the energy surface presents two degenerate minima as seen in the inset of this panel. The rest of panels correspond to the evolution of the energy surface as a function of β for the different $K-$values allowed for $j = 9/2$ in the odd-even system. States with $K = 7/2, 9/2$ always favor oblate shapes, while states with $K = 1/2, 3/2, 5/2$ are prolate up to $c_{1/2} \simeq 0.64$, $c_{3/2} \simeq 0.62$ and $c_{5/2} \simeq 0.58$, respectively. After these points, they suddenly change to oblate shapes. As seen in the insets for $K = 1/2, 3/2$ and $5/2$, they present for the mentioned values of the control parameter two degenerate minima. Therefore, we can suggest that components with $K = 1/2, 3/2, 5/2$ show first order transitions. This is similar to the case of transitional nuclei along the U(5) $\rightarrow$ SU(3) line but the role and behaviour of high-K and low-K states are interchanged in a non-trivial way: states with $K = 1/2, 3/2, 5/2$ always favor prolate shapes, while states with $K = 7/2, 9/2$

are oblate up to $c_{7/2} \simeq 0.6$ and $c_{9/2} \simeq 0.69$. This is evidently seen when comparing figure 2 of Ref. [15].

Now we study the same transition for a single-j case but considering different angular momenta j: from 3/2 to 13/2. The behaviour of the order parameter β that minimizes the energy surface as a function of the control parameter c for the different j cases is depicted in figure 3. Here, the number of K states changes according to the angular momentum j considered The vertical scale displays the β deformation, with positive values indicating prolate deformation and negative ones for oblate deformed shapes. The bosonic case is plotted as a reference in red. Panels of figure 3 refer to the angular momenta; $j = 3/2, 5/2, 7/2, 9/2, 11/2$ and $13/2$. It is evident from figure 3 that along the transitional path, states with lower K values prefer to be prolate up to a critical point, and then jump to oblate shape at specific c values. On the contrary, the states with higher K values always prefer oblate shapes for all different angular momenta. In the pure oblate deformed region, $c = 1$, the dominant action of the strongly oblate core drives all odd states into the oblate side and they have approximately the same β_{min} as the oblate core. Moreover, notice that on left side of figure 3 of Ref.[15] high-K states jump from oblate to prolate, while in figure 3 of this work, they behave smoothly for $j = 9/2$. Their tendency to the oblate side is not hampered by the core's deformation. The reverse is true for low-K states.

Summarizing, in this work the coupling of a single $j = 9/2$ particle to a boson core that undergoes a transition from spherical to oblate shapes has been investigated and results are compared with our previous work [15]. The overall results of the odd particle effect are also presented for different angular momenta, j from 3/2 to 13/2, along the spherical to oblate transition.

Acknowledgments

One of us (M. Böyükata) thanks the Scientific and Technical Research Council of Turkey (TÜBİTAK) within the program BİDEB-2224A for the support to attend the 11th International Spring Seminar on Nuclear Physics. This work has been partially supported by the Spanish Ministerio de Economía y Competitividad and FEDER funds under Project No. FIS2011-28738-c02-01; by the Spanish Consolider-Ingenio 2010 Programme CPAN (CSD2007-00042); and by Junta de Andalucía (FQM160, P11-FQM-7632).

References

[1] Bohr A and Mottelson B 1975 *Nuclear Structure. II Nuclear Deformations* (Benjamin: New York)
[2] Iachello F and Arima A 1987 *The Interacting Boson Model* (Cambridge: Cambridge University Press)
[3] Casten R F 2009 *Prog. Part. Nucl. Phys.* **62** 183
[4] Cejnar P and Jolie J 2009 *Prog. Part. Nucl. Phys.* **62** 210
[5] Jolie J, Heinze S, Van Isacker P and Casten R F 2004 *Phys. Rev. C* **70** 011305(R)
[6] Iachello F 2005 *Phys. Rev. Lett.* **95** 052503
[7] Alonso C E, Arias J M, Fortunato L and Vitturi A 2005 *Phys. Rev. C* **72** 061302(R)
[8] Alonso C E, Arias J M and Vitturi A 2007 *Phys. Rev. Lett.* **98** 052501
[9] Alonso C E, Arias J M and Vitturi A 2007 *Phys. Rev. C* **75** 064316
[10] Alonso C E, Arias J M, Fortunato L and Vitturi A 2009 *Phys. Rev. C* **79** 014306
[11] Böyükata M, Alonso C E, Arias J M, Fortunato L and Vitturi A 2010 *Phys. Rev. C* **82** 014317
[12] Fortunato L, Alonso C E, Arias J M, Böyükata M and Vitturi A 2011 *Int. J. Mod. Phys. E* **20** 207
[13] Petrellis D, Leviatan A and Iachello F 2011 *Ann. of Phys.* **326** 926
[14] Iachello F, Leviatan A and Petrellis D 2011 *Phys. Lett. B* **705** 379
[15] Böyükata M, Alonso C E, Arias J M, Fortunato L and Vitturi A 2014 *EPJ. Web. of. Conf.* **66** 02014
[16] Iachello F and Van Isacker P 1991 *The Interacting Boson-Fermion Model* (Cambridge: Cambridge University Press)
[17] Leviatan A 1988 *Phys. Lett. B* **209** 415; Leviatan A, Shao B 1989 *Phys. Rev. Lett.* **63** 2204
[18] Leviatan A, Shao B 1989 *Phys. Rev. Lett.* **63** 2204
[19] Alonso C E, Arias J M, Iachello F and Vitturi A 1992 *Nucl. Phys. A* **539** 59

Experimental First Order Pairing Phase Transition in Atomic Nuclei

L. G. Moretto,[1] A. C. Larsen,[2] F. Giacoppo,[2] M. Guttormsen,[2] and S. Siem[2]

[1]Lawrence Berkeley National Laboratory and University of California, Berkeley, CA 94720, USA
[2]Department of Physics, University of Oslo, N-0316 Oslo, Norway

E-mail: lgmoretto@lbl.gov

Abstract. The natural log of experimental nuclear level densities at low energy is linear with energy. This can be interpreted in terms of a nearly 1st order phase transition from a superfluid to an ideal gas of quasi particles. The transition temperature coincides with the BCS critical temperature and yields gap parameters in good agreement with the values extracted from even-odd mass differences from rotational states. This converging evidence supports the relevance of the BCS theory to atomic nuclei.

1. Introduction

For conventional superconductors, the standard Bardeen-Cooper-Schrieffer (BCS) theory [1] predicts a critical temperature/angular momentum at which the superconducting phase reverts to the normal one through a second order phase transition. However, for atomic nuclei, this second-order phase transition has never been truly verified experimentally, in spite of long and intense efforts. On the other hand, first-order phase transitions can also arise from the BCS Hamiltonian, as demonstrated in reference. [2]. In the present paper, we show that a first-order, rather than a second-order, phase transition is dramatically evident in experimental nuclear level densities below neutron threshold, and that this first-order transition is indisputably related to the presence of an energy gap in the quasiparticle spectrum. A large body of high-quality nuclear level-density data exemplified in figure 1. are now available in the literature [3–5]. The stunning, common feature of the level densities, particularly evident for deformed, mid shell nuclei, is the linear dependence of their logarithm with excitation energy. Above $\approx 2\Delta$, where Δ is the pair-gap parameter, and up to about the neutron separation energy, they are well described by the constant temperature expression proposed by Gilbert and Cameron [6]:

$$\rho(E) \approx \exp\left(\frac{E}{T}\right) \qquad (1)$$

where E is the excitation energy and T is the constant nuclear temperature. They found this expression to be in good agreement with the cumulative number of levels at low excitation energy, but did not provide any fundamental, quantitative explanation for this relation.

2. The Phase Transition

This experimental linear dependence of the entropy $S(E) \approx \ln \rho(E)$ as given by equation (1) and

Journal of Physics: Conference Series **580** (2015) 012048 doi:10.1088/1742-6596/580/1/012048

shown in figure 1 is the micro-canonical hallmark of first-order phase transitions. Surprisingly, we may have been staring at the biggest signal yet of such a transition without seeing it. This transition is, at least for nuclei well away from closed shells, clearly related to pairing. If we, provisionally, take the constant temperature of the experimental level density spectrum to be the BCS critical temperature, then, according to the well-known BCS relation

$$T_{cr} = \frac{2\Delta_0}{3.53} \qquad (2)$$

we can extract the gap parameter Δ_0 and compare it directly with that obtained from even-odd mass differences represented in the liquid-drop term as described e.g. by Bohr and Mottelson [7], and viceversa:

$$\Delta_{\mathrm{BM}} \approx 12A^{-1/2}. \qquad (3)$$

For a wide range of mass number A, the resulting relationship between mass number and temperature using equation (2) is shown in figure. 2, where the experimental constant temperatures T_{CT} are taken from references. [8–10]. The close agreement in magnitude and trend is remarkable for $A > 100$ and away from closed shells, although the assimilation of the constant level density temperature characteristic of a first-order transition to a critical temperature associated with a second-order transition remains to be explained. As a consequence of this observation, given the even-odd mass difference, we can predict the low-energy nuclear level densities throughout the nuclear chart for regions away from magic proton/neutron numbers. Before we embark on the explanation of this remarkable feature, let us consider another striking experimental observation: the level densities of neighbouring even-even and odd-A nuclei have nearly identical slopes, as seen in figure 1 showing data from the rare-earth region [11-15], and several actinides [16-17]. Therefore, the level densities of neighbouring isotopes can be made to overlap by means of a horizontal shift along the excitation-energy axis reference [18].

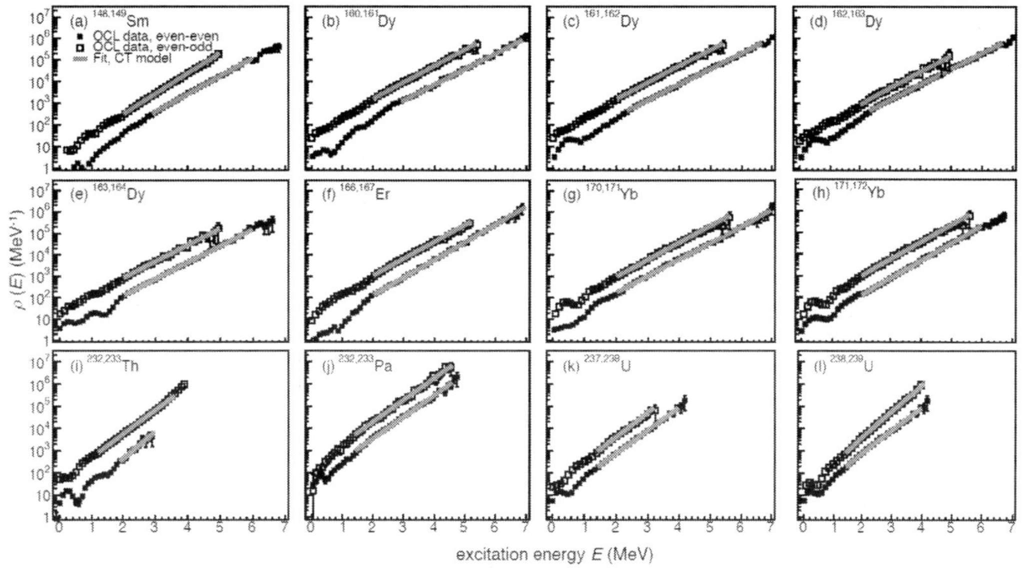

Figure 1. (Color online) Experimental level densities for rare-earth and actinide nuclei measured at the Oslo Cyclotron Laboratory (OCL), with a fit of the constant-temperature model (blue line) for excitation energies above ≈ 2 MeV.

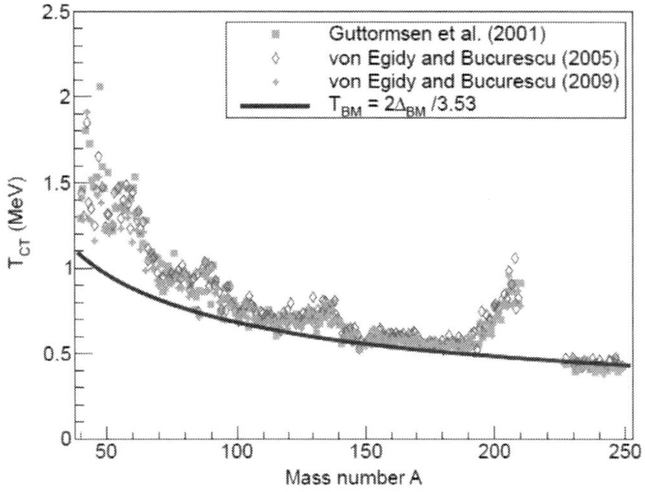

Figure 2. (Color online) Relation between mass number A and constant temperature T_{CT} taken from references [8- 10] for A = 40-250, compared to T_{cr} from equation (3) utilizing equation (2) (dark purple line).

The resulting shift is, not surprisingly, in very good agreement with the even-odd mass difference; see table 1. As a consequence, locally, for any given pair of even-even and odd-A nuclei, we can calculate the common slope of the two level densities directly from the observed excitation-energy shift. Equally intriguing is the vertical shift between the even-even-odd-A nuclear level densities, bringing the lower even-even level density on top of the higher odd-A one (see figure 1). This difference in entropy, approximately constant throughout the energy range [2Δ; S_n-1 MeV], can be interpreted as the entropy carried by the extra quasiparticle. The experimental evidence thus suggests that as the system is excited; quasiparticles are created with a constant energy cost and carrying a constant amount of entropy, see table 1. This is a clear signature of a first-order phase transition, from a superfluid phase to an ideal gas of quasiparticles.

Nuclide	T_{BM} (MeV)	T_{CT} (MeV)	T_{eo} (MeV)	Δ_{BM} (MeV)	Δ_{CT} (MeV)	Δ_{eo} (MeV)	ΔS (k_B)
^{148}Sm	0.56	0.51(1)	–	0.99	0.90(2)	–	–
^{149}Sm	0.56	0.46(1)	0.51(6)	0.98	0.81(2)	0.9(1)	2.0(2)
^{160}Dy	0.54	0.60(1)	–	0.95	1.05(1)	–	–
^{161}Dy	0.54	0.58(2)	0.57(6)	0.95	1.01(3)	1.0(1)	1.9(2)
^{162}Dy	0.54	0.60(1)	–	0.94	1.05(2)	–	–
^{163}Dy	0.53	0.57(2)	0.51(6)	0.94	1.00(4)	0.9(1)	1.9(2)
^{164}Dy	0.53	0.56(1)	–	0.94	0.99(1)	–	–
^{166}Er	0.53	0.52(1)	–	0.93	0.92(1)	–	–
^{167}Er	0.53	0.56(2)	0.51(6)	0.93	0.99(4)	0.9(1)	2.0(2)
^{170}Yb	0.52	0.57(1)	–	0.92	1.00(1)	–	–
^{171}Yb	0.52	0.55(1)	0.51(6)	0.92	0.96(1)	0.9(1)	1.8(2)
^{172}Yb	0.52	0.54(1)	–	0.91	0.95(1)	–	–
^{231}Th	0.45	0.41(1)	0.51(11)	0.79	0.72(2)	0.9(2)	2.4(4)
^{232}Th	0.45	0.34(1)	–	0.79	0.60(2)	–	–
^{233}Th	0.45	0.40(2)	0.51(11)	0.79	0.70(2)	0.9(2)	2.3(4)
^{232}Pa	0.45	0.44(1)	0.40(6)	0.79	0.77(2)	0.7(1)	1.7(2)
^{233}Pa	0.45	0.45(1)	–	0.79	0.79(2)	–	–
^{237}U	0.44	0.40(1)	0.40(6)	0.78	0.70(2)	0.7(1)	1.9(2)
^{238}U	0.44	0.42(1)	–	0.78	0.74(2)	–	–
^{239}U	0.44	0.37(1)	0.37(3)	0.78	0.65(1)	0.65(5)	2.5(5)

Table 1. Extracted temperatures T_{CT} from fitting the CT-model expression to the level-density data of rare-earth and actinide nuclei, and the corresponding pair-gap parameters Δ_{CT} calculated from equation (2). These are compared to the global formula for Δ_{BM} [equation (3)], for which the temperature is deduced using equation (2). Also, the even-odd experimental shift, Δ_{eo} is given, and the corresponding temperature T_{eo} is estimated from this shift for the odd nucleus. The experimental entropy excess ΔS is also given.

3. Level Density with Pairing

The presence of pairing in nuclei away from closed shells dominates the low energy level density and its energy dependence. For the uniform model in the case of an even-even nucleus, the ground state is shifted downward by an amount $E_{cond.} = \frac{1}{2} g\Delta_0^2$, where Δ_0 is the ground state gap parameter and g is the doubly degenerate single particle level density, which is related to the single particle level density parameter a according to $a = \frac{\pi^2}{3} g$. As the temperature increases, quasi particle excitations are produced until their blocking effect leads to a decrease and eventual breakdown of the pairing correlation at the critical temperature $T_{cr} = \frac{2\Delta_0}{3.53}$.

At this temperature the nucleus reverts to a non-interacting Fermi gas with its ground state shifted by an amount equal to the condensation energy. Therefore, at T_{cr} the excitation energy is

$$E_{cr} = \frac{1}{2} g\Delta_0^2 + \frac{\pi^2}{3} gT_{cr}^2 = \frac{1}{2} g\Delta_0^2 \left(1 + \frac{8}{3} \frac{\pi^2}{(3.53)^2}\right). \qquad (4)$$

We can also evaluate the mean number of quasi particles Q_{cr} at the critical point [2]

$$Q_{cr} = 4gT_{cr} \ln 2.$$

We can now calculate the mean energy cost per quasi particle:

$$\frac{E_{cr}}{Q_{cr}} \cong \frac{3.53\pi}{16 \ln 2} \Delta_0 = \Delta_0. \qquad (5)$$

This result is remarkable: it indicates that, if we consider the excitation energy as the independent variable, the energy cost per quasi particle is constant as expected for 1st order transition. This is in contrast with what one observes when the temperature is used as the independent variable, when a distinct 2nd order phase transition is visible. In the same spirit, we note that the heat capacity increases nearly exponentially with temperature up to T_{cr}. This means that most of the energy is absorbed near T_{cr}. From the constant energy cost Δ per quasiparticle, it follows that the entropy per quasiparticle is

$$\frac{\delta S}{\delta Q} = \frac{\Delta_0}{T_{cr}} = \frac{3.53}{2} = 1.77 \qquad (6)$$

to be compared with the empirical, vertical shift as discussed above (see table 1). To summarize, if we use the energy rather than the temperature as the independent variable, we observe the progressive creation of quasiparticles, in number proportional to the energy, like the amount of ice melted is proportional to the absorbed heat, independent of the amount of previously melted ice. This independence, together with the constant entropy per quasiparticle, gives clear evidence of a first-order phase transition.

We can also calculate the entropy at the critical point: $S_{cr} = 2\frac{\pi^2}{3} g\, T_{cr}$ and the overall entropy per quasi particle:

$$\frac{S_{cr}}{Q_{cr}} = \frac{\pi^2}{6 \ln 2} = 2.374. \qquad (7)$$

Correcting for the BCS discontinuity in the specific heat of a factor 2.43

$$\frac{S'_{cr}}{Q_{cr}} = 2.374 - \frac{1}{2} \ln 2.43 = 1.92 \qquad (8)$$

Alternatively

$$\frac{S'_{cr}}{Q_{cr}} = \frac{E/Q}{T_{cr}} = \frac{3.54\Delta}{2\Delta} = 1.77 \qquad (9)$$

in excellent agreement.
The number of the states associated with each quasi particle is then approximately constant and given by $N = e^{\frac{S'}{Q}} = 6.8$.
Given that this transition is "nearly" 1st order, we infer that the level density should be "nearly" exponential: $\rho(E) \approx \exp\frac{E}{T}$ where T should be "about" T_{cr}.

4. Even-Odd effects in level densities

In the pairing picture, an odd nucleus possesses one quasi particle in its ground state, which should control the level density at low energy. Otherwise, the odd-A nucleus should look like an even-even one except for an energy shift which should correspond to the even-odd mass difference Δ. A simple check for this is to verify that the level densities of two adjacent nuclei overlap if a horizontal shift Δ is applied to the odd nucleus. According to the considerations made above, this shift Δ can be related to the level density slope by the expression $T_{\exp} \cong T_{cr} = \frac{2\Delta}{3.53}$.
The next check can be made by overlapping the two level densities by means of a vertical shift. This vertical shift ΔS should be compared with the entropy per quasi particle $\frac{S'}{Q} \cong 1.77$.
These checks are done in table 1 where a substantial consistency is observed. However, somewhat unexpectedly, the linearity of lnϱ with E is also observed at low energy near the magic regions, so for these regions the cause of the linear behaviour must be looked for elsewhere.

5. Spectra with any gap

As discussed above, the origin of the linear dependence is due to the constant energy cost for the production of a quasi particle and a constant entropy per quasi particle. A similar situation occurs for a magic system with a gap in the single particle spectrum. Here, the cost to promote a nucleon and thus create a quasi particle (particle, hole excitation) is constant, at least for a while. This can be illustrated with a simple model.

Let excitations (quasi particles) be created into a state of degeneracy N at the cost δ per excitation.The excitation energy is: $E = n\delta$ and the associated number of states Ω is $\Omega \cong N^n$ for $n \ll N$.
It follows that $S = n \ln N = \frac{E}{\delta} \ln N = \frac{E}{T}$ where $T = \frac{\delta}{\ln N}$ and $\rho(E) = \exp\left(\frac{E}{T}\right)$.
Thus an exponential spectrum is expected if a gap is present irrespective of its origin.

6. Consistency between "Pairing gap" and "any gap"

The entropy per quasi particle in the pairing model is: $\frac{S_{cr}}{Q_{cr}} = \frac{\Delta}{T_{cr}} = 1.77$

For the "any gap" model we have: $\frac{\delta S}{\delta n} = \ln N$

Let us put pairing Δ into δ and equate T with T_{cr} : $T = \frac{\Delta}{\ln N} = \frac{2\Delta}{3.53}$

From this we obtain: ln N =1.77 in exact agreement with equation (6), and in good agreement with the average, experimental values from table 1 giving $\Delta S_{ave} = 2.0(1)$ Further, the number of available states per quasiparticle is exp(ΔS) = exp(1.77) ≈ 6, again agreeing well with the experimental ΔS values in table 1, exp(ΔS_{ave}) = 8(1).

7. Conclusion

In conclusion, we have shown that the low-energy level densities (below the neutron separation energy) portray a strong signature of a first-order phase transition, completely consistent with the BCS framework. The coexistence of a superfluid with a vapor of quasiparticles is easily characterized

Journal of Physics: Conference Series **580** (2015) 012048 doi:10.1088/1742-6596/580/1/012048

thermodynamically, especially through the comparison of even-even and odd-A nuclei. In particular, it is shown that the even-odd mass difference is sufficient to determine the level density in absolute

References

[1] Bardeen J, Cooper L N and Schrieffer J R 1957 *Phys. Rev.* **108** 1175
[2] Moretto L G 1975 *Nucl. Phys. A* **243** 77
[3] Data measured at the Oslo Cyclotron Laboratory with the Oslo method, references and data can be found at http://www.mn.uio.no/fysikk/english/research/about/infrastructure/OCL/nuclear physics-research/compilation/.
[4] Data from particle-evaporation experiments, references and data can be found at http://inpp.ohiou.edu/~voinov/index.html.
[5] Voinov A V, Oginni B M. Grimes S M, Brune C R, Guttormsen M, Larsen A C, Massey T N, Schiller A and Siem S 2009 *Phys. Rev. C* **79** 031301(R)
[6] Gilbert A and Cameron A G W 1965 *Can. J. Phys.* **43** 1446
[7] Bohr A and Mottelson B 1969 *Nuclear Structure*, Benjamin, NewYork, Vol. I.
[8] Von Egidy T and Bucurescu D 2005 *Phys. Rev. C* **72** 044311; 2006 *Phys. Rev. C* **73** 049901(E)
[9] Von Egidy T and Bucurescu D 2009 *Phys. Rev. C* **80** 054310
[10] Siem S et al. 2002 *Phys. Rev. C* **65** 044318
[11] Guttormsen M et al. 2003 *Phys. Rev. C* **68** 064306
[12] Nyhus H T et al. 2012 *Phys. Rev. C* **85** 014323
[13] Melby E et al. 2001 *Phys. Rev. C* **63** 044309
[14] Agvaanluvsan U et al. 2004 *Phys. Rev. C* **70** 054611
[15] Guttormsen M et al. 2013 Phys. Rev. C 88 024307
[16] Guttormsen M et al. 2014 *Phys. Rev. C* **89** 014302
[17] Guttormsen M et al. 2000 *Phys. Rev. C* **61** 067302
[18] Morse R W and Bohm H V 1957 *Phys. Rev.* **108** 1094

Recent Highlights of the ISOLDE Facility

M J G Borge[1]
ISOLDE-PH, CERN, CH-1211-Geneva-23, Switzerland
Instituto de Estructura de la Materia, CSIC, Serrano 113bis, 28006-Madrid, Spain

E-mail: mgb@cern.ch

Abstract. The on-line isotope mass separator ISOLDE is the radioactive beam facility of CERN. Its objective is the production and research of nuclei far from stability. Exotic nuclei of most chemical elements are available for the study of nuclear structure, nuclear astrophysics, fundamental symmetries and atomic physics, and for applications in condensed-matter and life sciences. A description of the facility, some recent results and future plans are presented here.

1. Introduction

The ISOLDE facility is one of the World-leading laboratories for the production and investigation of radioactive nuclei and has the World largest ISOL-production of radioactive nuclei. It is able to produce and deliver one thousand three hundred different nuclei of which seven hundred has been used for physics. This large variety of radioactive ion beams is used for many different experiments in the fields of nuclear and atomic physics, materials science and life sciences. The facility was approved fifty years ago, the 19th of December 1964 by the Council of the European Organization for Nuclear Research, CERN. It started operation three years later. Presently the facility is located at the Proton-Synchrotron Booster, PSB, of the CERN accelerator complex. The radioactive nuclei are produced in reactions of 1.4 GeV protons from PSB into thick targets. The beam delivered by the PSB injector is pulsed and contains up to 3.10^{13} protons/pulse with a minimum spacing of 1.2 s, giving a typical average proton current on target of 2 μA [1]. The use of high-energy protons such as the ones delivered by the CERN PSB injector has been recognized to be optimum for the production of radioactive nuclei. Higher energy of the protons, 2.0 GeV, is expected in 2020 gaining up to a factor of 5 in the deep spallation process, therfore favouring the production of neutron deficient heavy nuclei.

The success of ISOLDE is due to the continuous development of new radioactive ion beams and to the improvement of the experimental conditions. More than 20 different target materials and ionizers are in use. The target material is kept at a temperature between 1000 and 2000 degree Celsius so that the radioactive atoms produced diffuse out of the target into different dedicated ion sources. Ionization can take place in hot plasma, on a hot surface or by laser excitation. Chemical selectivity is obtained by the right combination of target-ion sources giving rise to a selective production of isotopes of, at the moment, 72 of the chemical elements. The knowledge accumulated over decades on how to construct targets and ion sources tailored to release pure beams of specific elements are one of ISOLDE's strong points. The ions are extracted from the ion-source by 30-60 kV acceleration voltages and directed towards an electro-magnet where they are separated according to their mass. ISOLDE has

[1] Correspondence to mgb@cern.ch.

two isotope separators on-line with independent target-ion source systems. The General Purpose Separator (GPS) has a mass resolving power, M/ΔM, of more than 1000 while the resolution of the High Resolution Separator (HRS) exceeds 5000. Figure 1 shows the main elements of the ISOLDE facility.

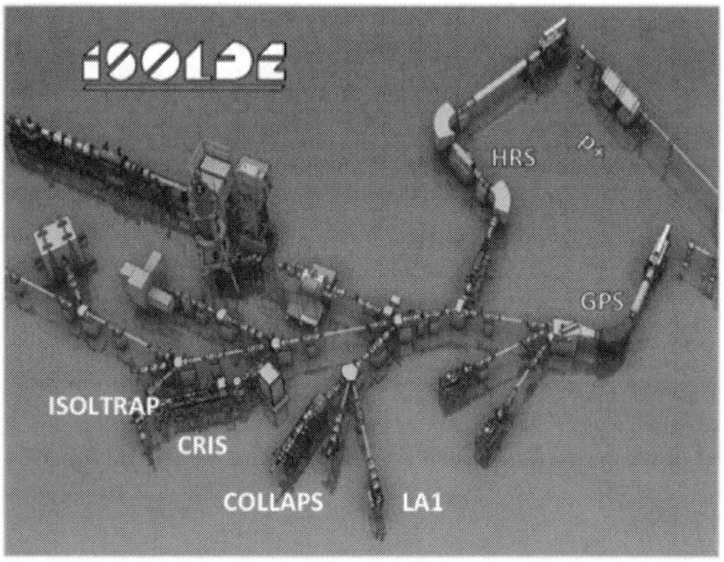

Figure 1. The main elements of the ISOLDE facility are shown. The high intensity (3.10^{13}/pulse) and high energy (1.4 GeV) proton beam from the PS Booster of CERN is directed to one of the two target-ion source stations for the production of radioactive species. The out coming 1^+ ions are mass separated in A/Q either in the HRS or in the GPS magnets and injected into the main beam line. The 60 keV beam is directed to the different low energy experiments such as LA1 for movable experiments, COLLAPS and CRIS for hyperfine structure studies, ISOLTRAP for mass measurements or post-accelerated at HIE-ISOLDE.

The installation of the post-accelerator REX-ISOLDE opened up thirteen years ago new fields of research with radioactive ion beams of higher energies. REX-ISOLDE has made used of the large variety of radionuclides that were extracted from the on-line mass separators GPS and HRS covering the whole mass range from He to U for reaction studies and Coulomb excitation with energies up to 3 MeV/u. REX-ISOLDE used the method of charge-state breeding to enhance the charge state of the ions before injection into a compact linear accelerator. The radioactive singly-charged ions from the separators were first accumulated, bunched and cooled in a Penning trap, REXTRAP. The trap stored the ions during the breeding in the subsequent charge breeder. Bunches of ions were then transferred to an electron beam ion source, REXEBIS, where the ions are charge bred to a mass-to-charge ratio between 2 and 4.5. Finally, the ions were injected into a compact linear accelerator. More details can be found in [2]. The efficiency of the system varied from 5 to 20 %. REX operational since 2001 has accelerated over 100 isotopes of more than 30 different elements.

To increase the energy a superconducting linear accelerator, which is due to start operation in 2015, is under construction downstream the normal conducting accelerator, HIE-ISOLDE project. The first phase of the HIE-ISOLDE project will provide beams accelerated to an energy of 4.3 MeV/u in 2015 and 5.5 MeV/u at the start of 2016. Post-accelerated radioactive beams of 5.5 MeV/u will allow reaching the Coulomb barrier threshold for a wide range of nuclei facilitating multi-step Coulomb excitation in the full mass range. In a second phase of the HIE-ISOLDE project energies up to 10 MeV/u for the radioactive beams will be obtained. This energy will allow to do scattering studies mainly transfer reaction studies in a wide mass range.

2. Recent Highlights from ISOLDE

In the following some of the recent highlights of ISOLDE will be presented covering examples from light to heavy nuclei and concentrating in ground state properties and exotic decay modes.

2.1. Observation of beta-delayed protons in the neutron rich halo nucleus ^{11}Be

Close to the beta-stability line all beta-decays will populate particle bound states, i.e. states that are long lived (stable, beta or gamma decaying) and therefore have narrow widths, less than 1 keV. Moving towards the driplines a larger and larger fraction of beta-decays will feed states that are unbound for the proton emission in the neutron deficient side and neutron emission in the neutron rich side or alpha emission that is possible in both sides of the valley of stability. Further close to the dripline beta-delayed particle emission become often the dominating decay mode. A general overview of the different decay modes and their impact on the knowledge on nuclear structure near the drip line and beyond can be found in recent reviews [3, 4]. At the same time β^- decay and proton emission take a nucleus in almost opposite directions on the nuclear chart, so β^--delayed proton emission, β^-p, is forbidden in all but a few nuclei where is highly suppressed as the available energy is $Q_{\beta\text{-}p}$ = 782 keV – S_n, being S_n the neutron separation energy. This new decay mode only possible for nuclei with very low S_n energy, has been recently observed at ISOLDE in the decay of ^{11}Be. The low decay energy implies that the branching ratio for this decay mode is very low and it is estimated for the most favourable case, ^{11}Be, to be around 10^{-8} [6].

The β^-p decay mode is expected in 1n-halo nuclei due to both energetics and to the more pronounced single-particle behavior of haloes. Two neutron halo nuclei were similarly candidates for β^-d emission and indeed this decay mode has been observed in ^{6}He and ^{11}Li with branching ratios of 10^{-6} and 10^{-4} respectively [4].

To investigate the new decay mode, β^-p, in the more favorable case ^{11}Be one has to face the difficulty of detecting protons of maximum energy of 280.7(3) keV with a very reduced branching ratio. Therefore the idea is to produce a very intense and pure sample and search for the β^-p daughter with a level of sensitivity better than 10^{-8}. Considering that the half-lives of ^{11}Be and the β^-p daughter, ^{10}Be, are very different respectively 13.8 s and $1.5x10^6$ y one can consider to make a sample and sending it to a state of art accelerator mass spectrometry (AMS) device that have the required sensitivity.

A first attempt was made in 2001 but the signal was not strong enough and a β^-p branching ratio of $2.5(25)x10^{-6}$ was obtained [5]. Although compatible with zero this result meant that the branching ratio could be higher than theoretically expected. Due to the improvements in the production of ^{11}Be at ISOLDE and a higher sensitivity in the AMS detection of ^{10}Be, the experiment was tried again in 2012. Samples of ^{11}Be and the possible contaminant ^{11}Li were collected. The activity produced was monitored by the measurements of the gamma activity connected to the respective decays. The ^{10}Be AMS measurements were performed at the Viena Enviromental Research accelerator (VERA). The branching ratio of the β^-p decay mode in ^{11}Be was determined by detecting the number of final nucleus ^{10}Be obtained by the AMS method with respect to the deduced number of ^{11}Be ions collected in the original sample. The branching ratio was of $8.3(9)x10^{-6}$ [6]. This branching ratio is astonishing large but consistent and of the same order of magnitude that the previous measurement [5]. This unexpected high rate can only be understood if one assumes that the decay proceeds through a new very narrow resonance in ^{11}B. The B_{GT} could reach a value of 3, which correspond to a free neutron decay. A natural interpretation will be then in terms of a peripheral beta decay of the halo neutron in ^{11}Be into a single proton state.

2.2. Confirmation of magicity in the calcium isotopes at N=32 and N = 34

A pioneering work done at CERN thirty years ago [7] indicated that the magic numbers of neutrons and protons corresponding to closed shells could loose their special character due to the proton-neutron asymmetry. It is since then a great challenge for theory to predict the new magic numbers emerging far from stability, and closely related, to understand the different components of the strong force that acts between neutrons and protons.

The mass of a nucleus is together with its half-life and decay modes the first quantity to measure and reveals the strength of nuclear binding. The total binding energy of a nucleus contains great physics information and precision mass measurements often provide important test of nuclear models. The ISOLTRAP mass measurements have successfully pioneered other precision mass devices established nowadays worldwide [8]. The beam from ISOLDE is collected and bunched before being sent to the preparation trap where purification takes place. A detailed analysis of the ion motion in the precision trap shows that the cyclotron resonance of the ion can be determined through the measurement of the time of flight of the ion when ejected from the trap.

The study of calcium isotopes with two double magic isotopes is always a good test ground for shell model. At the same time calcium isotopes with close proton shell mark the frontier of applicability of 3-nucleon forces from chiral effective field theory. The magic number character could be partially determined by measuring the mass and deducing the two neutron separation energy in the calcium isotopes around N = 32.

At ISOLDE the objective was to map the masses below and above N= 32 to properly characterize this new magic number for very neutron rich nuclei. The masses of ^{51}Ca and ^{52}Ca were measured previously at TRIUMF [9] by the TITAN trap confirming the expected new magic number far from stability, N=32. At ISOLDE these masses were revisited using ISOLTRAP for $^{51\text{-}52}$Ca getting compatible values but much more accurate due to the higher yield. The Multi-Reflection time of flight device, MR-ToF, is a mass analyser where the ions are separated by multiple reflections between electrostatic mirrors. The flight time is proportional to the square root of charge over mass, so well calibrated it can be used to determine the mass of a nucleus. It fact it was used for first time at ISOLDE to go beyond the double magic nucleus and determine the mass of ^{53}Ca and ^{54}Ca demonstrating the feasibility of precision mass measurement at the production level of 10 atom/s. These measurements allow for the determination of the two-neutron separation energies, $S_{2n} = B(Z,N) - B(Z,N\text{-}2)$ i.e. the difference of binding energy of two isotopes that differ in two neutrons. The S_{2n} is a sensitive probe of the evolution of nuclear structure with neutron number. The pronounced decrease in S_{2n} deduced from the data is well followed by the microscopic calculations using three nucleon forces based on chiral effective theory, and unambiguously establish N=32 as a closed shell neutron number for neutron rich nuclei very far from stability. The results that follow surprising well the predictions of 3-body forces increase our understanding of neutron-rich matter and were published recently in Nature [10].

2.3. Sphericity preserved for the neutron-rich Cd-isotopes

One of the main advantages of an ISOL-type facility is that one can do a systematic study of the ground state properties of a certain element and study how these properties change by adding an extra neutron. Magnetic dipole and electric quadrupole moments are other fundamental properties of nuclei that can provide stringent tests of nuclear models.

The experimental determination of nuclear radii and nuclear moments [11] has been for many years a major activity at ISOLDE [12]. The nuclear properties are deduced from accurate laser spectroscopy measurements of the hyperfine atomic spectrum. The systematic measurement of nuclear moments along an isotopic, isotonic or isobaric chain is a powerful tool to extract nuclear structure changes.

A recent example is the measurement of the individual hyperfine spectra of the cadmium isotopes (Z = 48) spanning over twenty masses, A = 107 to 129. The atomic spectra of the cadmium ions have common atomic features but differ in the nuclear magnetic and quadrupole moments that influence the hyperfine structure of each isotope. Yordanov et al [13] have observed that in spite of the complexity of the nuclei studied, the quadrupole moments of the nuclear state $h_{11/2}$ exhibit a linear behaviour with changing neutron number. The unpaired neutron in the $11/2^-$ state in each of the odd-mass cadmium isotopes behaves nearly the same way. This behaviour was thought to occur only when protons or neutrons were in close shell and therefore inactive. The results of the cadmium isotopes indicate that the simplicity for the neutrons has survived although it has not a magic number of protons. This is one more proof of the validity of nuclear shell model and the pairing force.

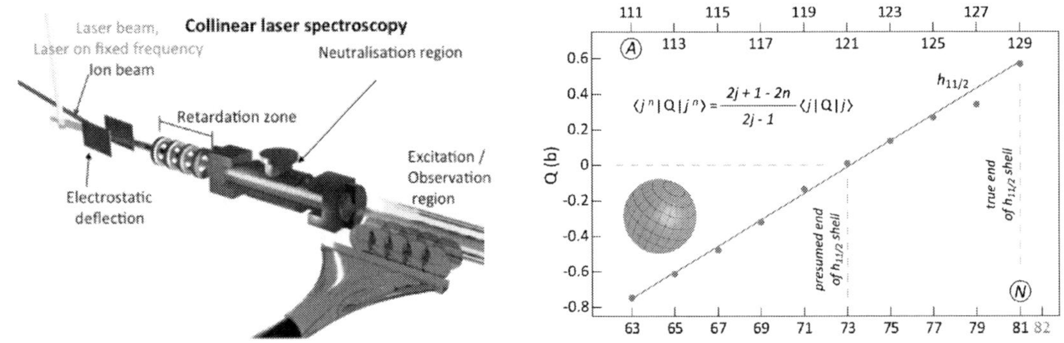

Figure 2. On the left hand side a schematic description of the elements involved in the collinear laser spectroscopy, COLLAPS, setup at ISOLDE. On the right hand side the quadrupole moments of $^{111\text{-}129}$Cd isotopes obtained with COLLAPS [15]. A straight line is fitted through the $h_{11/2}$ quadrupole moments according with the formula given in the graph showing that the spherical character unexpectedly continues until N=82 .

The linear behaviour of the $11/2^-$ quadrupole moments, shown in figure 2, for ten odd-mass isotopes, is the most relevant feature of the cadmium isotopes. This behaviour is most probably due to the unique parity of the $h_{11/2}$ orbit. The formula given in figure 2 follows the de-Shalit and Talmi formalism [14] for seniority $\nu = 1$ where all but one particle are coupled to spin zero. The fitting parameters are Q_{sp} = -667(31) mb and Q_{const} = -85(8) mb, where the offset term, Q_{const} represents a constant quadrupole moment contribution from correlations with the core. The fact that the line crosses zero around the middle of the $h_{11/2}$ shell indicates the spherical shape for the $11/2^-$ states. One should notice that the deviation from the straight line in ^{127}Cd indicates a change in shape consistent with the abnormal energy of the first 2^+ state for the even isotopes ^{126,128}Cd reported in [15]. Further long-lived $11/2^-$ states are identified in ^{127}Cd and ^{129}Cd for first time. This is a recent example of how advanced laser spectroscopy provided access to the properties of very exotic nuclei.

2.4. Collinear Resonance Ionization Spectroscopy of neutron deficient and neutron rich Fracium isotopes

The phenomenon of shape coexistence in the lead region has been studied in the francium isotopes by the determination of the magnetic moments and mean-square charge radius via the measurement of the hyperfine interaction and the determination of the isotope shifts for neutron deficient and neutron rich francium isotopes. It has been suggested the presence of intruder proton orbit in the neutron deficient francium nuclei that could polarize the nucleus creating significant deformation.

Figure 3 shows the hyperfine spectra of the ground state of the neutron deficient $^{202\text{-}207}$Fr isotopes as well as for the ^{218m,219,229,231}Fr isotopes. The measurements of the very exotic francium isotopes have been possible by the realization of a new highly sensitive, high-resolution technique of bunched collinear beam ionization spectroscopy (CRIS). The CRIS setup is designed to realize hyperfine structure studies with sensitivity up to 1 atom/s. The high sensitivity is reached through a combination of excellent overlap of laser and beam and the high efficiency of the ion detection system. The hyperfine structures associated with ground state and isomeric states were identified by the use of a decay station. The spectra were recorded by scanning the laser frequency. The ^{221}Fr was used as reference and allowed to determine the A factor and from it the magnetic moments using previously assigned spins. The charge radii were extracted from the measured isotope shifts determined as the centroid of the hyperfine structure. The extracted $\delta\langle r^2\rangle$ for $^{202-213}$Fr isotopes is compared with predictions from the spherical droplet model. The resulting charged radii [16,17] show a remarkably close odd-even staggering trend

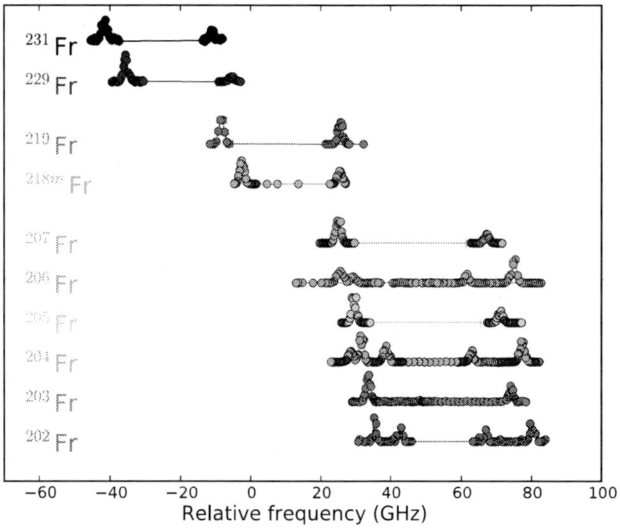

Figure 3 Hyperfine spectra of the ground state of $^{202-207,218m,219,229,231}$Fr measured at ISOLDE using the CRIS device.

to the one observed for the lead isotopes for N=118 to N=126. A departure from the trend is seen at 116, suggesting an earlier onset of deformation in the neutron-deficient francium isotopes [16] than in the lead [18] and polonium[19] chains where this happens below N=114.

3. The HIE-ISOLDE Project

The main features of the HIE ISOLDE project (HIE stands for "High Intensity and Energy) are to boost the energy of the beams, going in steps from currently 3 MeV/u via 5.5 MeV/u to finally 10 MeV/u. A factor of six in intensity is expected from the higher energy, 2 GeV from PSB and higher intensity of the proton injector expected to occur in 2019. In addition improvements in several aspects of the secondary beam properties such as purity, ionization efficiency and optical quality are addressed.

For HIE-ISOLDE phase I where the beam energy is going to be boosted to 5.5 MeV/u twenty-seven experiments have been approved with more than six hundred shifts for day-one physics. The approved physics cases expand over the wide range of post-accelerated beams available at ISOLDE. The increase in energy of the radioactive beams will enhance the cross section and the accessibility to detailed nuclear structure information at higher excitation energy. The proposed studies will be realised with the existing workhorses MINIBALL and T-REX plus new instrumentation for transfer reaction studies such as the active targets MAYA and the future ACTAR, a new general purpose scattering chamber and the two arms CORSET setup from GSI. Plans exist of complementing the facility with a highly performing Storage Ring, TSR presently at Heidelberg in Germany. The addition of the TSR will allow the beams from HIE-ISOLDE to be stored and cooled, providing much greater luminosity and much better beam definition. A HELIOS-type solenoidal spectrometer is envisioned to be located at the exit of the TSR.

4. Summary and Outlook

In summary, the future of ISOLDE is bright. It restarts first of August 2014 for physics with the low energy program. With more than 45 year of operation ISOLDE remains as the pioneer ISOL-installation both at the level of designing new devices and production of frontier Physics.

Post-accelerated beams up to 5.5 MeV/u for the wide range of nuclei produced at ISOLDE will be available very soon. HIE-ISOLDE will be the only next-generation radioactive beam facility (as identified by the NuPECC LRP) available in Europe in 2015, and the most advanced ISOL facility World-wide.

References

[1] Kugler E 2000 *Hyperfine Interactions* **129**, 23
[2] The REX-ISOLDE Facility edited by Cederkall J *et al.* 2005 *CERN Report*, CERN-2005-009
[3] Blank B and Borge M J G 2008 *Prog. Part. Nucl. Phys.* **60** 403
[4] Pfützner M, Karny M, Grigorenko L V and Riisager K 2012 *Rev. Mod. Phys.* **84** 567
[5] Borge M J G *et al.* 2013 *J. Phys.* G *Nucl. Part.* **40** 035109
[6] Riisager K *et al.* 2014 *Phys. Letts.* B **732** 305
[7] Thibaud C *et al.* 1975 *Phys. Rev.* C **12** 644
[8] Blaum K, 2006 *Phys. Rep.* **425** 1
[9] Gallant A T *et al.* 2012 *Phys. Rev. Lett. 109*
[10] Wienholtz F *et al.* 2013 *Nature* **498** 346
[11] Neyens G 2003 *Rep. Prog. Phys.* **66** 633
[12] Otten E W 1993 *Phys. Rep.* **225** 157
[13] Yordanov D T et al. 2013 *Phys. Rev. Lett.* **110** 192501
[14] de-Shalit A and Talmi I 1963 *Nuclear Shell Theory* (Academic Press, New York)
[15] Kautzsch T *et al.* 2000 *Eur. Phys. J.* A **9** 201
[16] Flanagan K T *et al.* 2013 *Phys. Rev. Lett.* **111** 212501
[17] Lynch K M *et al.* 2014 *Phys. Rev. X* **4** 011055
[18] DeWitte H *et al.* 2007 *Phys. Rev. Lett.* **98** 112502
[19] Cocoolios T E *et al.* 2011 *Phys. Rev. Lett.* **106** 052503

Journal of Physics: Conference Series **580** (2015) 012050 doi:10.1088/1742-6596/580/1/012050

Thermal pairing and giant dipole resonance in highly excited nuclei

Nguyen Dinh Dang

1) Theoretical Nuclear Physics Laboratory, Nishina Center for Accelerator-Based Science, RIKEN, 2-1 Hirosawa, Wako city, 351-0198 Saitama, Japan
2) Institute for Nuclear Science and Technology, Hanoi, Vietnam

E-mail: dang@riken.jp

Abstract. Recent results are reported showing the effects of thermal pairing in highly excited nuclei. It is demonstrated that thermal pairing included in the phonon damping model (PDM) is responsible for the nearly constant width of the giant dipole resonance (GDR) at low temperature $T < 1$ MeV. It is also shown that the enhancement observed in the recent experimentally extracted nuclear level densities in ^{104}Pd at low excitation energy and various angular momenta is the first experimental evidence of the pairing reentrance in finite (hot rotating) nuclei. In the study of GDR in highly excited nuclei, the PDM has been extended to include finite angular momentum. The results of calculations within the PDM are found in excellent agreement with the latest experimental data of GDR in the compound nucleus ^{88}Mo. Finally, an exact expression is derived to calculate the shear viscosity η as a function of T in finite nuclei directly from the GDR width and energy at zero and finite T. Based on this result, the values η/s of specific shear viscosity in several medium and heavy nuclei were calculated and found to decrease with increasing T to reach $(1.3-4) \times \hbar/(4\pi k_B)$ at $T = 5$ MeV, that is almost the same value obtained for quark-gluon-plasma at $T > 170$ MeV.

1. Introduction

Superfluid pairing plays an important role in the study of nuclear structure. In infinite systems the pairing gap Δ vanishes at a temperature $T_c \simeq 0.567\Delta(T = 0)$, which is the critical temperature of the superfluid-normal (SN) phase transition. In deformed nuclei, the Coriolis force, which breaks the Cooper pairs, increases with the total angular momentum up to a certain critical angular momentum J_c, where all Cooper pairs are broken and the nucleus undergoes the SN phase transition, as suggested by the Mottelson-Valatin effect [1]. The combination of temperature and angular momentum effects causes the pairing reentrance phenomenon, which was predicted 50 years ago by Kammuri [2]. The physical interpretation of this phenomenon was given by Moretto [3] as follows. In spherical nuclei at $T = 0$, the total angular momentum J is made up by the nucleons from the broken pairs, which occupy the single-particle levels around the Fermi surface and block them against the scattered pairs. As J increases, more levels around the Fermi surface are blocked, which make the pairing correlations decrease until a sufficiently large total angular momentum J_c, where the pairing gap Δ completely vanishes in infinite systems. The increase of T spreads the quasiparticles, which occupy the levels near the Fermi surface, farther away from it. Some levels become partially unoccupied, which are available for scattered pairs. As the result, when T reaches a critical value T_1, the pairing correlations are energetically favored, and the pairing gap reappears. As T goes higher, the increase of a large number of quasiparticles quenches and eventually breaks down the pairing gap at T_2 $(> T_1)$.

It is now well known that, because of thermal fluctuations in finite nuclei, the paring gap does not vanish at $T \geq T_c$ and/or J_c, but moronically decreases with increasing T and/or J. The pairing reentrance in finite nuclei is also modified so that the pairing gap remains finite after its first reappearance at finite T and J. It has been shown within the phonon damping model (PDM) [4] that the presence of a thermal pairing gap at $T > T_c$ makes the width of the giant dipole resonance (GDR) remain nearly unchanged at low temperatures ($T \leq 1$ MeV) in heavy nuclei. This lecture will show the most recent experimental facts that support this mechanism. It also demonstrates that the recently observed enhancement of the nuclear level densities [5, 6] is a clear evidence of pairing reentrance in ^{104}Pd [7]. In the description of the GDR width in hot rotating nuclei, the PDM has been extended to include the effect of angular momentum [8]. It will be shown that the predictions of the PDM agree well with the latest experimental data for the GDR in highly excited ^{88}Mo. Finally, the GDR widths and energies predicted by the PDM and experimentally extracted are also used to calculate the shear viscosity of finite hot nuclei.

2. Pairing reentrance in hot rotating nuclei

The pairing pairing Hamiltonian for a spherical system rotating about the symmetry z axis, where the total angular momentum J is completely determined by its z-projection M alone, has the form $H = \sum_k \epsilon_k(a^\dagger_{+k}a_{+k} + a^\dagger_{-k}a_{-k}) - G\sum_{kk'} a^\dagger_k a^\dagger_{-k} a_{-k'} a_{k'} - \lambda\hat{N} - \omega\hat{M}$, where $a^\dagger_{\pm k}(a_{\pm k})$ are the creation (annihilation) operators of a particle (neutron or proton) in the k-th state, whereas ϵ_k, λ and ω are the single-particle energies, chemical potential and rotation frequency for an axially deformed system, respectively. This Hamiltonian was used to derive the FTBCS1 equations including the effect of angular momentum [9], which consist of a set of FTBCS (finite-temperature BCS) -based equations, corrected by the effects of quasiparticle-number fluctuations. The FTBCS1 equation for the pairing gap is written as a sum of the level-independent term Δ and level-dependent term $\delta\Delta_k$ as $\Delta_k = \Delta + \delta\Delta_k$, where $\Delta = G\sum_{k'} u_{k'}v_{k'}(1 - n^+_{k'} - n^-_{k'})$, $\delta\Delta_k = G\delta\mathcal{N}^2_k u_k v_k/(1 - n^+_k - n^-_k)$, $u^2_k = [1 + (\epsilon_k - Gv^2_k - \lambda)/E_k]/2$, $v^2_k = [1 - (\epsilon_k - Gv^2_k - \lambda)/E_k]/2$, $E_k = \sqrt{(\epsilon_k - Gv^2_k - \lambda)^2 + \Delta^2_k}$, $n^\pm_k = [1 + e^{(E_k \mp \omega m_k)/T}]^{-1}$. with the quasiparticle-number fluctuations $\delta\mathcal{N}^2_k = (\delta\mathcal{N}^+_k)^2 + (\delta\mathcal{N}^-_k)^2 = n^+_k(1 - n^+_k) + n^-_k(1 - n^-_k)$. The equations for the particle number and total angular momentum are $N = 2\sum_k \left[v^2_k(1 - n^+_k - n^-_k) + \frac{1}{2}(n^+_k + n^-_k)\right]$, $M = \sum_k m_k(n^+_k - n^-_k)$, respectively, with $\pm m_k$ being the angular momentum projection quantum number. Although the sums run over all neutron and proton single-particle levels respectively, only the levels close to the Fermi surface give dominant contributions. For the levels far from the Fermi surface, $u_k \simeq 1$ (0) and $v_k \simeq 0$ (1) so that $u_k v_k$ are small. By solving these FTBCS1 equations, one can easily calculate the total energy $\mathcal{E}$ and entropy S of the system, from which the level density $\rho(\mathcal{E}, M)$ is calculated by using the invert Laplace transformation of the grand partition function [10]. The total level density $\rho(\mathcal{E})$ of a system with the total energy $\mathcal{E}$ is a sum of J-dependent level densities (densities of states), namely $\rho(\mathcal{E}) = \sum_J (2J + 1)\rho(\mathcal{E}, J)$. The J-dependent level density $\rho(\mathcal{E}, J)$ is calculated by differentiating $\rho(\mathcal{E}, M)$, that is $\rho(\mathcal{E}, J) = \rho(\mathcal{E}, M = J) - \rho(\mathcal{E}, M = J + 1)$.

Recently, a series of measurements has been conducted at the Bhabha Atomic Research Center for the reaction ^{12}C + ^{93}Nb $\rightarrow$ ^{105}Ag* $\rightarrow$ ^{104}Pd* + p at the projectile energy $E(^{12}C) =$ 40 - 45 MeV [5, 6]. In these experiments the exclusive proton spectra in coincidence with a γ-ray multiplicity detector array reveal broad structures at higher multiplicities and energies. The obtained spectra including these broad structures can be fitted by using the statistical model only after multiplying the phenomenological level densities for ^{104}Pd by an enhancement function, which depends on excitation energy and angular momentum. It has been suggested in

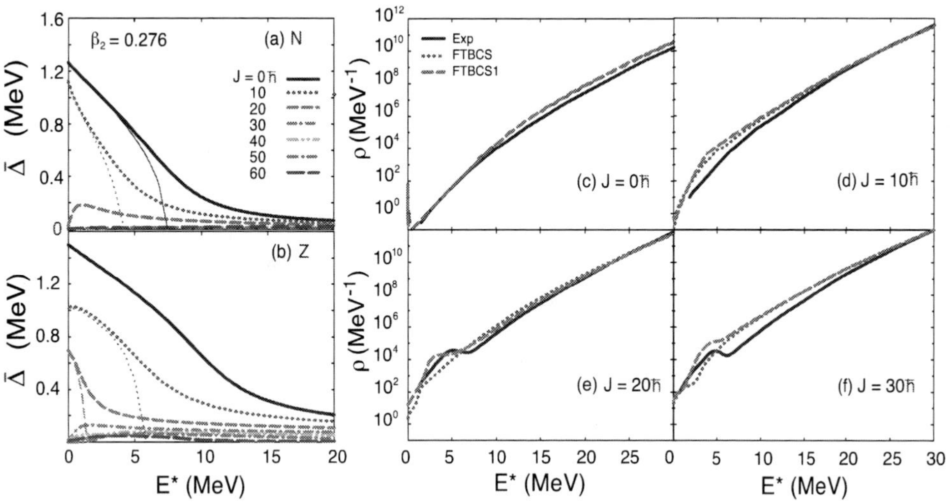

Figure 1. (Color online) Level-weighted pairing gaps $\overline{\Delta}$ for neutrons (a) and protons (b) as well as the level densities (c – f) for ^{104}Pd obtained within the FTBCS and FTBCS1 as functions of excitation energy E^* at deformation parameter $\beta_2 = 0.276$ for different values of angular momentum J. In (a) and (b) the thin and thick lines denote the FTBCS and FTBCS1 pairing gaps, respectively. In (c) – (f), the dotted and dashed lines stand for the level densities $(2J+1)\rho(\mathcal{E}, J)$ predicted by the FTBCS and FTBCS1, respectively, whereas the solid lines are the empirical level densities (See Refs. [5, 6, 7] for details).

Ref. [6] that this enhancement might come from the pairing reentrance.

The level-weighted pairing gaps $\overline{\Delta} \equiv \sum_k \Delta_k/\mathcal{N}$ (with $\mathcal{N}$ being the sum of all levels) obtained within the conventional FTBCS show no pairing reentrance. The FTBCS neutron and proton gaps shown in panels (a) and (b) of Fig. 1 decrease with increasing E^* (or T) and collapses at some E_c^*, whose value decreases as J increases. At the same time, the FTBCS1 gaps do not vanish at $E^* > 0$. At $J < 20\hbar$ they decrease monotonically as E^* increases and remain finite even at $E^* > 15$ MeV. The pairing reentrance shows up at $J \geq 20\hbar$ for neutrons and at $J \geq 30\hbar$ for protons, where the gaps first increase with E^* then decrease as E^* increases further but do not vanish even at high E^*. Since this pairing reentrance takes place at rather high values of J, where the pairing gaps already vanish within the FTBCS, its origin resides in the quasiparticle number fluctuations $\delta\mathcal{N}_k^2$ in the FTBCS1 gap equation.

Our preliminary calculations in Ref. [7] show that the corresponding level densities obtained within the FTBCS are smooth for all values of E^* and J, whereas those predicted by the FTBCS1 have a local enhancement at $2 < E^* < 4.5$ MeV for all values of J under consideration. For $\beta_2 = 0.276$, the FTBCS1 predictions agree fairly well with the empirical level densities obtained at $J \leq 30\hbar$ [See Figs. 1 (c) – 1 (f)]. For $\beta_2 = -0.3$, a better agreement between theoretical predictions and empirical level densities is seen at $J > 30\hbar$ (not shown). The fact that theoretical and empirical level densities agree well for $J \leq 30\hbar$ at $\beta_2 = 0.276$, and for $J > 30\hbar$ at $\beta_2 = -0.3$ is a clear indication of a transition from a prolate shape at $J \leq 30\hbar$ to an oblate shape at higher J.

3. Damping of GDR in highly excited nuclei

The PDM Hamiltonian includes the independent single-particle (quasiparticle) field, GDR phonon field, and the coupling between them [4]. The single-particle energies ϵ_k are obtained

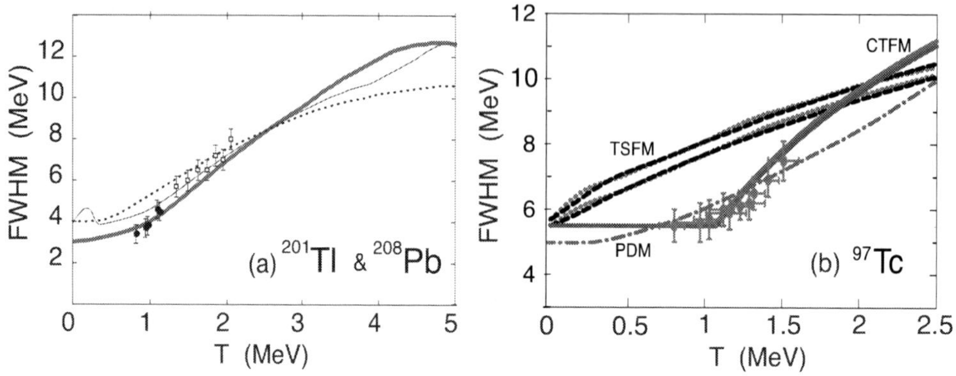

Figure 2. GDR width predicted within the PDM as a function of temperature for ^{201}Tl, ^{208}Pb (a) and ^{97}Tc (b). In (a) the dotted and solid lines denotes the GDR width in ^{208}Pb and ^{201}Tl, respectively. The corresponding experimental data are shown as open boxes and full circles, respectively [12]. In (b) the PDM prediction (dash-dotted line) is compared with the experimental data and predictions of the thermal shape fluctuation model (TSFM) with shell effect (dotted lines) and without shell effect (dashed lines) for J = $0\hbar$ (lower) and J = $30\hbar$ (upper) as well as those of the phenomenological model (continuous lines) for J = $10\hbar$ (lower) and J = $20\hbar$ (upper) (See Ref. [11] for details).

within the Woods-Saxon potentials. The GDR width $\Gamma(T)$ consists of the quantal width, $\Gamma_{\rm Q}$, and thermal width, $\Gamma_{\rm T}$, as $\Gamma(T) = \Gamma_{\rm Q} + \Gamma_{\rm T}$. Including thermal pairing, the quantal and thermal widths are given as $\Gamma_{\rm Q} = 2\gamma_Q(E_{GDR}) = 2\pi F_1^2\sum_{ph}[u_{ph}^{(+)}]^2(1-n_p-n_h)\delta[E_{\rm GDR}-E_p-E_h]$, and $\Gamma_{\rm T} = 2\gamma_T(E_{GDR}) = 2\pi F_2^2\sum_{s>s'}[v_{ss'}^{(-)}]^2(n_{s'}-n_s)\delta[E_{\rm GDR}-E_s+E_{s'}]$, where $u_{ph}^{(+)} = u_pv_h + u_hv_p$, $v_{ss'}^{(-)} = u_su_{s'} - v_sv_{s'}$ $(ss' = pp', hh')$, n_k are quasiparticle occupations numbers, which are well approximated with the Fermi-Dirac distribution for independent quasiparticles, $n_k = [\exp(E_k/T)+1]^{-1}$. The parameter F_1 is chosen so that Γ_Q at $T = 0$ is equal to GDR's width at $T = 0$, whereas the parameter F_2 is chosen so that, with varying T, the GDR energy E_{GDR} does not change significantly. The latter is found as the solution of the equation $E_{GDR} - \omega_q - P_q(E_{GDR}) = 0$, where ω_q is the energy of the GDR phonon before the coupling between the phonon and single-particle mean fields is switched on, and $P_q(\omega)$ is the polarization operator owing to this coupling (Its expression is given in Ref. [4]). The GDR strength function is found as $S_q(\omega) = (1/\pi)[\gamma_Q(\omega)+\gamma_T(\omega)]/\{(\omega-E_{GDR})^2+[\gamma_Q(\omega)+\gamma_T(\omega)]^2\}$. The representation $\delta(x) = \lim_{\varepsilon\to 0}\varepsilon/[\pi(x^2+\varepsilon^2)]$ is used for the δ-function with $\epsilon = 0.5$ MeV in numerical calculations. The GDR widths predicted by the PDM, thermal shape fluctuation model (TSFM), and the phenomenological model called critical temperature included fluctuation model (CTFM) are shown in Fig. 2 in comparison with the experimental systematics for ^{97}Tc [11], ^{201}Tl [12], and ^{208}Pb [13]. The PDM results for open shell nuclei include the effect of non-vanishing thermal pairing gap owing to thermal fluctuations in finite nuclei. The PDM is the only up-to-date semi-microscopic model that is able to describe well the experimental data in the entire temperature region including $T \leq 1$ MeV as well as high T, where the GDR width seems to saturate.

To describe the non-collective rotation of a spherical nucleus, the z-projection M of the total angular momentum J is added into the PDM Hamiltonian as $-\gamma\hat{M}$, where γ is the rotation frequency [8]. The latter and the chemical potential are defined, in the absence of pairing, from the equation $M = \sum_k m_k(f_k^+ - f_k^-)$, and $N = \sum_k(f_k^+ + f_k^-)$, where N is the particle number and $f_k^{\pm}$ are the single-particle occupation numbers, $f_k^{\pm} = 1/[\exp(\beta E_k^{\mp})+1]$, and $E_k^{\mp} = \epsilon_k - \lambda \mp \gamma m_k$.

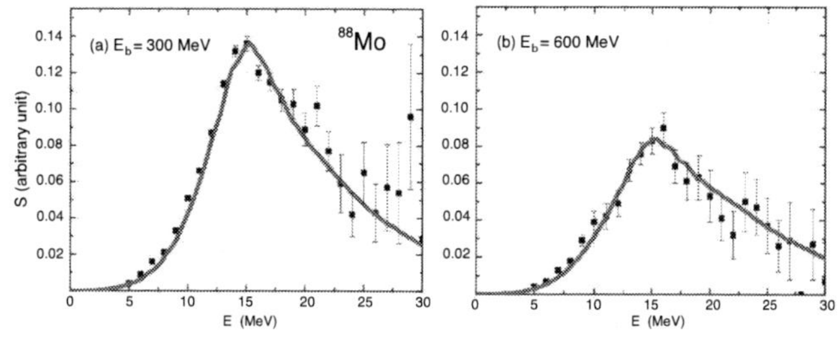

Figure 3. PDM predictions for the GDR line shapes in ^{88}Mo (solid lines) in comparison with the experimental data at the beam energies of 300 MeV (a) and 600 MeV (b), which correspond to the average temperatures of 2.04 and 3.06 MeV, respectively. The average angular momentum is equal to $38\hbar$ (Adapted from Figs. 5.5 and 5.6 of Ref. [14]. Also see Ref. [15] for details of theoretical calculations).

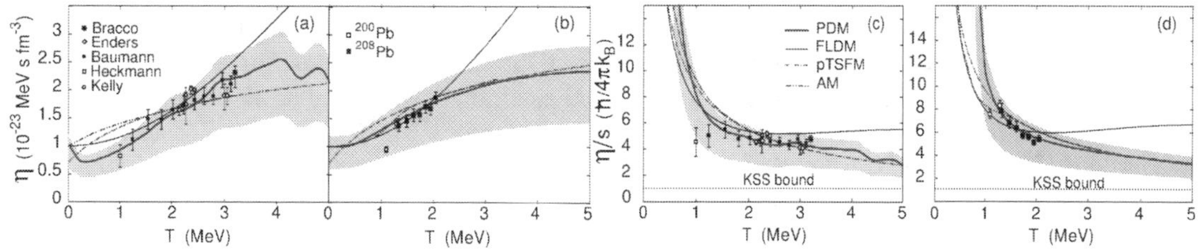

Figure 4. Shear viscosity $\eta(T)$ [(a) and (b)] and ratio η/s [(c) and (d)] as functions of T for nuclei in tin [(a) and (c)], and lead [(b) and (d)] regions. The gray areas are the PDM predictions by using $0.6u \leq \eta(0) \leq 1.2u$ with $u = 10^{-23}$ Mev s fm^{-3}(See Ref. [17] for details).

The prediction of the PDM for the GDR line shape agrees remarkably well with the latest experimental date for ^{88}Mo at high T and J [14], as shown in Fig. 3.

4. Shear viscosity of hot nuclei

For the application of hydrodynamics to nuclear system, the quantum mechanical uncertainty principle requires a finite viscosity for any thermal fluid. Kovtun, Son and Starinets (KSS) [16] conjectured that the specific viscosity, which is the ratio η/s of shear viscosity η to the entropy volume density s, is bounded at the lower end for all fluids by the universal value, which is called the KSS bound (or unit), $\eta/s = \hbar/(4\pi k_B)$.

According to collective theories, one of the fundamental explanations for the giant resonance damping is the friction term (viscosity) of the neutron and proton fluids. By using the Green-Kubo's relation, Ref. [17] expresses the shear viscosity $\eta(T)$ at finite T in terms of the GDR's energies and widths at zero and finite T as

$$\eta(T) = \eta(0)\frac{\Gamma(T)}{\Gamma(0)}\frac{E_{GDR}(0)^2 + [\Gamma(0)/2]^2}{E_{GDR}(T)^2 + [\Gamma(T)/2]^2} \ . \tag{1}$$

The predictions for the shear viscosity η and the specific viscosity η/s by several theoretical models for ^{120}Sn and ^{208}Pb are plotted as functions of T in Fig. 4 in comparison with the empirical results. The latter are extracted from the experimental systematics for GDR in tin and lead regions [13] making use of Eq. (1). It is seen from Fig. 4 that the predictions by the PDM agree best with the empirical results. The ratio η/s decreases sharply with increasing T up to $T \sim 1.5$ MeV, starting from which the decrease gradually slows down to reach (2 - 3) KSS units at $T = 5$ MeV. The fermi-liquid drop model (FLDM) has a similar trend up to $T \sim$ 2 - 3 MeV, but at higher T ($T > 3$ MeV for ^{120}Sn or 2 MeV for ^{208}Pb) both η and η/s with T sharply increase. At $T = 5$ MeV the FLDM model predicts the ratio η/s within (3.7 - 6.5) KSS units, that is roughly $1.5-2$ times larger than the PDM predictions. The TSFM in two versions as adiabatic model (AM) and phenomenological TSFM (pTSFM) show a similar trend for η and η/s, provided $\eta(0)$ in the pTSFM calculations is reduced to $0.72u$. However both versions of TSFM overestimate η at $T < 1.5$ MeV. Based on these results and on a model-independent estimation, it is concluded that the PDM predicts $1.3 \leq \eta/s \leq 4$ KSS units for medium and heavy nuclei at $T = 5$ MeV. This is about (3 - 5) times smaller (and of much less uncertainty) than the value between (4 - 19) KSS units predicted by the FLDM for heavy nuclei [18], where the same lower value $\eta(0)$ =$0.6u$ was used.

5. Conclusions

This lecture has confirmed that thermal pairing included in the phonon damping model (PDM) is indeed responsible for the nearly constant width of the giant dipole resonance (GDR) at low temperature $T < 1$ MeV. It also demonstrates that the enhancement, which has been observed in the recent experimentally extracted nuclear level densities in ^{104}Pd at low excitation energy and various angular momenta, is the first experimental evidence of the pairing reentrance in finite (hot rotating) nuclei. The extension of the PDM to finite angular momentum offers predictions in excellent agreement with the latest experimental data of GDR in the compound nucleus ^{88}Mo. The specific shear viscosity η/s in several medium and heavy nuclei is calculated from the GDR widths and energies at zero and finite temperatures and found to decrease with increasing T to reach $(1.3-4) \times \hbar/(4\pi k_B)$ at $T = 5$ MeV, which is almost the same value obtained for quark-gluon-plasma at $T > 170$ MeV.

The calculations were carried out on the RIKEN Integrated Cluster of Clusters (RICC) system. Thanks are due to M. Ciemala for permission to use the experimental GDR line shapes in ^{88}Mo [14].

References

[1] Mottelson B R and Valatin J G 1960 *Phys. Rev. Lett.* **5** 511.
[2] Kammuri T 1964 *Prog. Theor. Phys.* **31** 595.
[3] Moretto L G 1971 *Phys. Lett. B* **35** 379; Moretto L G 1972 *Nucl. Phys. A* **185** 145.
[4] Dinh Dang N and Arima A 1998 *Phys. Rev. Lett* **80** 4145; Dinh Dang N and Arima A 2003 *Phys. Rev. C* **68** 044303.
[5] Mitra A *et al.* 2002 *Nucl. Phys. A* **707** 343; Mitra A *et al.* 2006 *Nucl. Phys. A* **765** 277; Mitra A *et al.* 2009 *J. Phys. G* **36** 095103; Mitra A *et al.* 2010 *EPJ Web of Converences* **2** 04004.
[6] Datar V M, Mitra A, and Chakrbarty D R, *unpublished.*
[7] Dinh Dang N, Quang Hung N, Agrawal B K, Datar V M, Mitra A, and Chakrabarty D R, *in preparation.*
[8] Dinh Dang N 2012 *Phys. Rev. C* **85** 064323.
[9] Quang Hung N and Dinh Dang N 2008 *Phys. Rev. C* **78** 064315.
[10] Quang Hung N and Dinh Dang N 2011 *Phys. Rev. C* **84** 054324.
[11] Dey B *et al.* 2014 *Phys. Lett. B* **731** 92.
[12] Dinh Dang N and Quang Hung N 2012 *Phys. Rev. C* **86** 044333.
[13] Schiller A and Thoennessen A 2007 *Atomic Data and Nuclear Data Tables* **93** 548.
[14] Ciemala M *Ph.D. thesis, Niewodniczanski Institute of Nuclear Physics PAN, Krakow* 2013 **Report No. 2062/PL** (http://www.ifj.edu.pl/publ/reports/2013/2062.pdf)
[15] Dinh Dang N, Ciemala M, Kmiecik M, and Maj A 2013 *Phys. Rev. C* **87** 054313.
[16] Kovtun P K, Son D T, and Starinets A O 2005 *Phys. Rev. Lett.* **94** 111601.
[17] Dinh Dang N 2011 *Phys. Rev. C* **84** 034309.
[18] Auerbach N and Shlomo S 2009 *Phys. Rev. Lett.* **103** 172501.

Ischia14 IOP Publishing
Journal of Physics: Conference Series **580** (2015) 012051 doi:10.1088/1742-6596/580/1/012051

Complex configuration effects on β-decay rates

A. P. Severyukhin[1], V. V. Voronov[1], I. N. Borzov[1], N. N. Arsenyev[1] and Nguyen Van Giai[2]

[1]Bogoliubov Laboratory of Theoretical Physics, Joint Institute for Nuclear Research, 141980 Dubna, Moscow region, Russia
[2]Institut de Physique Nucléaire, CNRS-IN2P3, Université Paris-Sud, F-91406 Orsay Cedex, France

Abstract. The effects of the phonon-phonon coupling on the β-decay rates of neutron-rich nuclei are studied in a microscopic model based on Skyrme-type interactions. Making use of the finite rank separable approach for the quasiparticle random phase approximation enables one to perform nuclear structure calculations in very large configurational spaces. The inclusion of the tensor correlations and the $2p-2h$ configurations results in a redistribution of the Gamow-Teller strength and a consequent reduction of the β-decay half-life. Such an effect is discussed for some neutron-rich $N = 82$ isotones below the doubly magic nucleus ^{132}Sn. The calculations are performed by using the SGII Skyrme interactions with the tensor terms in the particle-hole channel and the surface-peaked interaction in the particle-particle channel. Predictions are given for ^{126}Ru and ^{128}Pd in comparison to ^{130}Cd which is the r-process waiting-point nucleus.

The study of spin-isospin excitations in neutron-rich nuclei is presently an important problem not only from the nuclear structure point of view but also because of the special role they play in many astrophysical processes. Many fundamental issues depend on our quantitative understanding of the β-decay of atomic nuclei. It is desirable to have theoretical models which can describe the data wherever they can be measured and which can predict the properties related to spin-isospin excitations in systems too short-lived to allow for experimental studies. One of the successful tools for studying charge-exchange nuclear modes is the quasiparticle random phase approximation (QRPA) with the self-consistent mean-field derived from a Skyrme-type energy-density functional (EDF), see e.g., [1, 2, 3, 4]. These QRPA calculations enable one to describe the properties of the ground state and excited charge-exchange states using the same EDF.

A comparison of such calculations with recent experimental data [5] demonstrates that QRPA approach cannot reproduce correctly the strength distribution of the spin-isospin resonances. It is necessary to take into account a coupling with more complex configurations that results in shifting some strength upward in excitation energy [6, 7, 8]. Using the Skyrme EDF and the RPA, such attempts in the past [9, 10] have allowed one to understand the damping of charge-exchange resonances and their particle decay. Recently, the damping of the Gamow-Teller (GT) mode was investigated using the Skyrme-RPA plus particle-vibration coupling (PVC) [11]. However, the size of the configuration space increases very rapidly and one has to work within limited number spaces.

It would be helpful to study the $2p-2h$ configuration effect on the β^--decay rates of neutron-rich nuclei. It is somewhat simpler to include the PVC in QRPA calculations if one uses separable forces [7, 12]. Our tool is the QRPA with Skyrme interactions in the finite rank separable approximation (FRSA) [13, 14, 15, 16], allowing one to perform calculations in large configuration spaces. Successful applications of the method to study the electric low-lying states and giant resonances within and beyond the QRPA can be found in Refs [14, 15, 16, 17]. We use an extension of this approach proposed for the charge-exchange nuclear excitations [18] and also for accommodating the tensor correlations which mimic the Skyrme-type tensor interactions [19]. Taking into account the basic ideas of the quasiparticle-phonon model (QPM) [12], the Hamiltonian is then diagonalized in a space spanned by states composed of one and two QRPA phonons [20],

$$\Psi_\nu(JM) = \left(\sum_i R_i(J\nu)Q^+_{JMi} + \sum_{\lambda_1 i_1 \lambda_2 i_2} P^{\lambda_1 i_1}_{\lambda_2 i_2}(J\nu)\left[Q^+_{\lambda_1\mu_1 i_1}\bar{Q}^+_{\lambda_2\mu_2 i_2}\right]_{JM}\right)|0\rangle\ , \qquad (1)$$

where $Q^+_{\lambda\mu i}|0\rangle$ ($\bar{Q}^+_{\lambda\mu i}|0\rangle$) is the GT (electric) excitation having the QRPA energy $\omega_{\lambda i}$ ($\bar{\omega}_{\lambda i}$). The normalization condition for the wave functions (1) is

$$\sum_i R_i^2(J\nu) + \sum_{\lambda_1 i_1 \lambda_2 i_2} (P^{\lambda_1 i_1}_{\lambda_2 i_2}(J\nu))^2 = 1. \qquad (2)$$

The amplitudes $R_i(J\nu)$ and $P^{\lambda_1 i_1}_{\lambda_2 i_2}(J\nu)$ are determined from the variational principle which leads to a set of linear equations

$$(\omega_{\lambda i} - \Omega_\nu)R_i(J\nu) + \sum_{\lambda_1 i_1 \lambda_2 i_2} U^{\lambda_1 i_1}_{\lambda_2 i_2}(Ji)P^{\lambda_1 i_1}_{\lambda_2 i_2}(J\nu) = 0, \qquad (3)$$

$$(\omega_{\lambda_1 i_1} + \bar{\omega}_{\lambda_2 i_2} - \Omega_\nu)P^{\lambda_1 i_1}_{\lambda_2 i_2}(J\nu) + \sum_i U^{\lambda_1 i_1}_{\lambda_2 i_2}(Ji)R_i(J\nu) = 0. \qquad (4)$$

The rank of the set of linear equations (3) and (4) is equal to the number of one- and two-phonon configurations included in the wave function (1). Its solution requires to compute the matrix elements coupling one- and two-phonon configurations

$$U^{\lambda_1 i_1}_{\lambda_2 i_2}(Ji) = \langle 0|Q_{Ji}H\left[Q^+_{\lambda_1 i_1}\bar{Q}^+_{\lambda_2 i_2}\right]_J|0\rangle. \qquad (5)$$

Equations (3) and (4) have the same form as the QPM equations [7, 12], but the single-particle spectrum and the residual interaction are derived from the same Skyrme EDF [20].

To calculate the half-lives, the same ansatz as Sec. II of Ref. [21] with the ratio of the weak axial-vector and vector coupling constants G_A/G_V=1.25 [22] is used. In the allowed GT approximation, the β^--decay rate is expressed by summing the probabilities of the energetically allowed GT transitions (in units of $G_A^2/4\pi$) weighted with the integrated Fermi function

$$T^{-1}_{1/2} = \sum_m \lambda^m_{if} = D^{-1}\left(\frac{G_A}{G_V}\right)^2 \sum_m f_0(Z, A, E_i - E_{1^+_m})B(GT)_m, \qquad (6)$$

$$E_i - E_{1^+_m} \approx \Delta M_{n-H} + \mu_n - \mu_p - E_m, \qquad (7)$$

where λ^m_{if} is the partial β^--decay rate. D=6147 [22] is a constant, $\Delta M_{n-H} = 0.782$ MeV is the mass difference between the neutron and the hydrogen atom, μ_n and μ_p are the neutron and proton chemical potentials respectively, E_i is the ground state energy of the parent nucleus, and $E_{1^+_m}$ denotes a state of the daughter nucleus (Z, A). E_m are the eigenvalues of the QRPA equations or Eqs. (3) and (4) taking into account the two-phonon configurations. The wave functions allow us to determine $B(GT)_m$.

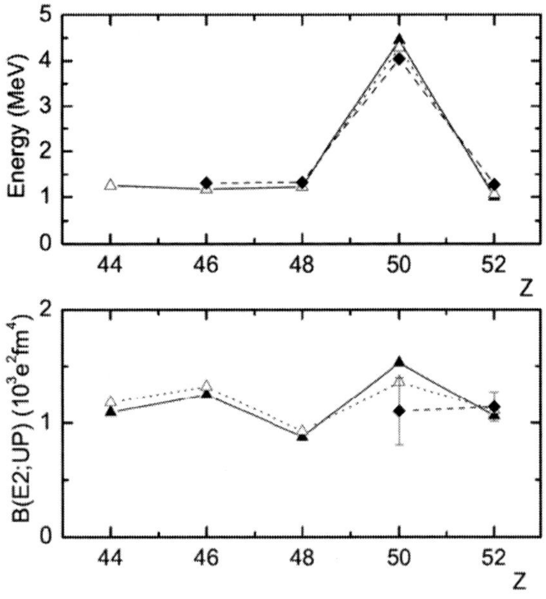

Figure 1. Energies and $B(E2)$ values for up-transitions to the $[2_1^+]_{QRPA}$ states in the neutron-rich N=82 isotones. Results of the calculations without the tensor interaction (open triangles) and with the tensor interaction (filled triangles) are shown. Experimental data (filled diamonds) are taken from Ref. [31].

There is a relation between the $N = 82$ shell closure and the $A \approx 130$ peak of the solar r-process abundance distribution [23]. The $N = 82$ isotones below ^{132}Sn are important for stellar nucleosynthesis, see, e.g., Refs. [24, 25, 26, 27, 28]. It is interesting to study the influence of the coupling between one- and two-phonon terms in the wave functions and the tensor force effects on the strength distributions of the GT states in the neutron-rich $N = 82$ isotones, namely, ^{126}Ru, ^{128}Pd and ^{130}Cd. As the parameter set in the particle-hole channel, we use the central Skyrme interaction SGII [29] and the same zero-range tensor interaction as that in the paper [30]. Since the SGII parametrization gives reasonable values for the Landau parameters $F_0' = 0.73$ and $G_0' = 0.50$, one obtains a successful description of the spin-dependent properties and, in particular, experimental energies of the GT resonances of ^{90}Zr [29]. For the studied region of nuclei we use the surface-peaked pairing force fixed in Refs. [18, 19]. The single-particle continuum is discretized by diagonalizing the HF hamiltonian on a basis of 12 harmonic oscillator shells and cutting off the single-particle spectra at the energy of 100 MeV. This is sufficient to exhaust the Ikeda sum rule $3(N - Z)$ [18, 19] as well as the sum rule for the electromagnetic excitation modes [16]. Having taking into account the tensor correlation effects within the $1p - 1h$ and $2p - 2h$ configurational spaces, we do not need any quenching factor [6].

To construct the wave functions (1) of the low-lying 1^+ states we use only the $[1_i^+ \otimes \lambda_{i'}^+]_{QRPA}$ terms and all electric phonons with $\lambda > 2$ vanish. All one- and two-phonon configurations with the transition energies $|E_{1_m^+} - E_i|$ up to 10 MeV are included. We have checked that the inclusion of the high-energy configurations leads to minor effects on the half-life values. Since there is a clear influence of the 2_1^+ phonon on the half-lives [20, 32], we examine the energies and transition probabilities of the 2_1^+ QRPA states of the neutron-rich $N = 82$ isotones (see Fig.1). There is a remarkable increase of the 2_1^+ energy of ^{132}Sn in comparison with those in ^{134}Te and ^{130}Cd. It corresponds to a standard evolution of the 2_1^+ energy near closed shells [16]. The

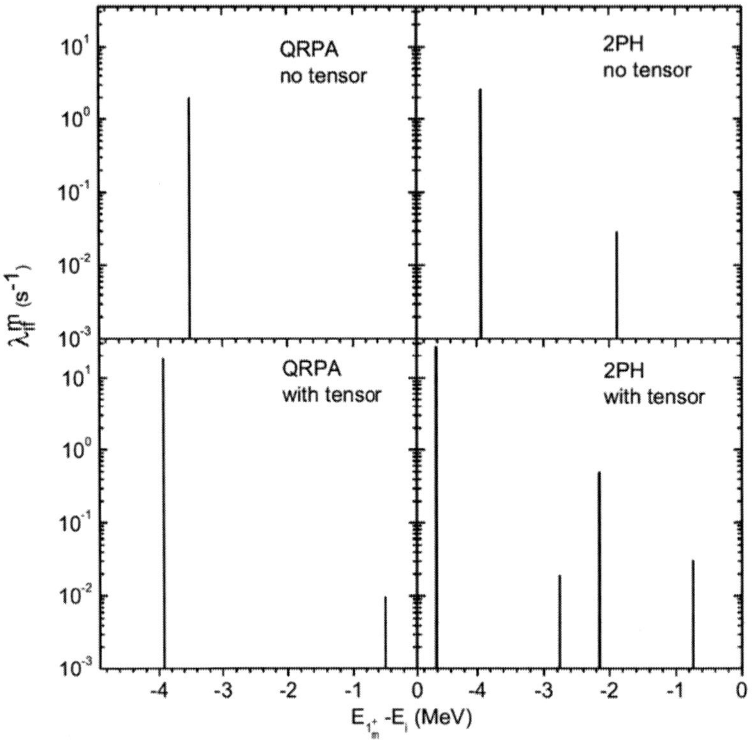

Figure 2. The phonon-phonon coupling effect on the β-transition rates in ^{128}Pd. The left and right panels correspond to the calculations within the QRPA and taking into account the $2p-2h$ configurations, respectively. Results of the calculations without (resp. with) the tensor interaction are shown in the upper (resp. lower) panels.

2^+_1 states in ^{126}Ru, ^{128}Pd and ^{130}Cd have a non collective structure with a domination of the proton configuration $\{1g_{9/2}, 1g_{9/2}\}$. As shown in [16], the neutron transition density is shifted outwards as compared to the proton transition density due to the presence of the neutron skin. As can be seen from Fig.1, the FRSA model with the SGII+tensor interaction reproduces the experimental data [31] very well. Also, one can see that the inclusion of the tensor interaction does not change the energies and the transition probabilities.

The QRPA analysis within the one-phonon approximation can help to clarify the tensor correlation effect on the GT strength distributions. When the tensor terms are included, for the region of nuclei near ^{132}Sn, the GT strength distributions calculated with the same parameter set are rather well studied in Refs. [19, 33]. As a general trend, there is a redistribution of the GT strengths by the inclusion of the tensor interaction. In particular, for ^{130}Cd, the tensor correlations induces 3.6 MeV downward shift of the main peak of GT resonance and the part of the main peak strength is fragmented in the low-energy peaks and also the high-energy tail is formed [33]. Let us now focus on the properties of the low-energy spectrum of the 1^+ states of the daughter nuclei, which influence β-decay half-lives. As expected, the largest contribution in the calculated β^--decay half-life comes from the $[1^+_1]_{QRPA}$ state. To illustrate it, the β-transition rates λ^m_{if} of ^{128}Pd are shown in Fig. 2. The transition energies $|E_{1^+_m} - E_i|$ refer to the ground state of the parent nucleus. As pointed out in Ref. [25, 26], we find that the correct description of the two-quasiparticle configuration $\{\pi 1g\frac{9}{2}\nu 1g\frac{7}{2}\}$ is important for the half-life description in

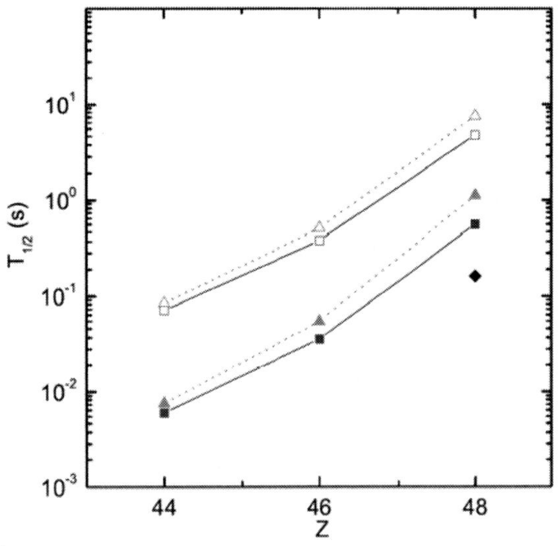

Figure 3. The phonon-phonon coupling effect on β^--decay half-lives of ^{126}Ru, ^{128}Pd and ^{130}Cd. Results of the calculations without the tensor interaction (open triangles, squares) and with the tensor interaction (filled triangles, squares) are shown. The squares correspond to the half-lives calculated with inclusion of the phonon-phonon coupling, the triangles are the QRPA calculations. Experimental data (filled diamonds) are from Refs. [34].

all the cases considered. As can be seen from Fig. 3, the half-lives calculated with the tensor force are about 10 times shorter than those calculated without. It is worth to mention that the first discussion of the strong impact of the tensor correlations on the β^--decay half-lives based on QRPA calculations with Skyrme forces has been done in Ref. [4].

The inclusion of $2p - 2h$ configurations results a further increase of the transition energies $|E_{1^+_m} - E_i|$ and partial rates of the main GT transitions (see, e.g., right bottom panel of Fig. 2). Fig. 3 shows that this effect produces an impact on the β-decay half-life which is reduced by a factor 2. The main two-phonon components of the 1^+_1 wave function are the $[1^+_1 \otimes 2^+_1]_{QRPA}$ and $[1^+_1 \otimes 0^+_2]_{QRPA}$ configurations. The calculated value in the case of ^{130}Cd overestimates the experimental data [34]. Moreover, for ^{132}Sn, the $[1^+_1]_{QRPA}$ state of the daughter nucleus is above the parent ground state, i.e., this calculation gives stable ^{132}Sn and one should seek for further improvements of the effective interaction. Finally, we predict the half-lives: 36 ms for ^{128}Pd and 6 ms for ^{126}Ru.

In summary, starting from the Skyrme mean-field calculations the GT strength in the Q_β-window has been studied within the extended FRSA model including both the tensor interaction effect and 2p-2h configurations. The suggested approach enables one to perform the calculations in very large configurational spaces. Choosing as an example the nuclei ^{126}Ru, ^{128}Pd and ^{130}Cd, the β^--decay rates are calculated. The inclusion of the tensor interaction leads to a redistribution of the GT strength. The low-energy GT strength is fragmented and the 1^+_1 state is moved downwards in energy. Taking into account these effects results in a dramatic reduction of β-decay half-lives. At a qualitative level, we observe a improvement of the half-life description of ^{130}Cd by the tensor force. The half-lives of $N = 82$ isotones are predicted.

Acknowledgments
We thank H. Sagawa for useful discussions. This work is partly supported by the IN2P3-JINR agreement.

References

[1] Bender M, Dobaczewski J, Engel J and Nazarewicz W 2002 *Phys. Rev.* C **65** 054322
[2] Fracasso S and Colò G 2007 *Phys. Rev.* C **76** 044307
[3] Bai C L, Zhang H Q, Sagawa H, Zhang X Z, Colò G and Xu F R 2011 *Phys. Rev.* C **83** 054316
[4] Minato F and Bai C L 2013 *Phys. Rev. Lett.* **110** 122501
[5] Ichimura M, Sakai H and Wakasa T 2006 *Prog. Part. Nucl. Phys.* **56** 446
[6] Bertsch G F and Hamamoto I 1982 *Phys. Rev.* C **26** 1323
[7] Kuzmin V A and Soloviev V G 1984 *J. Phys.* G **10** 1507
[8] Drożdż S, Osterfeld F, Speth J and Wambach J 1987 *Phys. Lett.* B **189** 271
[9] Colò G, Nguyen Van Giai, Bortignon P F and Broglia R A 1994 *Phys. Rev.* C **50** 1496
[10] Colò G, Sagawa H, Nguyen Van Giai, Bortignon P F and Suzuki T 1998 *Phys. Rev.* C **57** 3049
[11] Niu Y F, Colò G, Brenna M, Bortignon P F and Meng J 2012 *Phys. Rev.* C **85** 034314
[12] Soloviev V G 1992 *Theory of Atomic Nuclei: Quasiparticles and Phonons* (Bristol and Philadelphia: Institute of Physics)
[13] Nguyen Van Giai, Stoyanov Ch and Voronov V V 1998 *Phys. Rev.* C **57** 1204
[14] Severyukhin A P, Stoyanov Ch, Voronov V V and Nguyen Van Giai 2002 *Phys. Rev.* C **66** 034304
[15] Severyukhin A P, Voronov V V and Nguyen Van Giai 2004 *Eur. Phys. J.* A **22** 397
[16] Severyukhin A P, Voronov V V and Nguyen Van Giai 2008 *Phys. Rev.* C **77** 024322
[17] Severyukhin A P, Arsenyev N N and Pietralla N 2012 *Phys. Rev.* C **86** 024311
[18] Severyukhin A P, Voronov V V and Nguyen Van Giai 2012 *Prog. Theor. Phys.* **128** 489
[19] Severyukhin A P and Sagawa H 2013 *Prog. Theor. Exp. Phys.* **2013** 103D03
[20] Severyukhin A P, Voronov V V, Borzov I N, Arsenyev N N and Nguyen Van Giai *arXiv* 1410.0614v1
[21] Engel J, Bender M, Dobaczewski J, Nazarewicz W and Surman R 1999 *Phys. Rev.* C **60** 014302
[22] Suhonen J 2007 *From Nucleons to Nucleus* (Berlin: Springer-Verlag)
[23] Burbidge E M, Burbidge G R, Fowler W A and Hoyle F 1957 *Rev. Mod. Phys.* **29** 547
[24] Martínez-Pinedo G and Langanke K 1999 *Phys. Rev. Lett.* **83** 4502
[25] Borzov I N 2003 *Phys. Rev.* C **67** 025802
[26] Dillmann I *et al.* (ISOLDE collaboration) 2003 *Phys. Rev. Lett.* **91** 162503
[27] Jungclaus A *et al.* 2007 *Phys. Rev. Lett.* **99** 132501
[28] Cuenca-García J J, Martínez-Pinedo G, Langanke K, Nowacki F and Borzov I N 2007 *Eur. Phys. J.* A **34** 99
[29] Nguyen Van Giai and Sagawa H 1981 *Phys. Lett.* B **106** 379
[30] Bai C L, Zhang H Q, Zhang X Z, Xu F R, Sagawa H and Colò G 2009 *Phys. Rev.* C **79** 041301(R)
[31] Pritychenko B, Birch M, Singh B and Horoi M 2014 *Nuclear Data Sheets* **120** (2014) 112
[32] Severyukhin A P, Voronov V V, Borzov I N and Nguyen Van Giai 2013 *Rom. J. Phys.* **58** 1048
[33] Severyukhin A P and Sagawa H 2012 *Eur. Phys. J. Web of Conf.* **38** 06001
[34] Hannawald M *et al.* (ISOLDE collaboration) 2001 *Nucl. Phys.* A **688** 578c

The Pygmy Dipole Resonance – status and new developments

A Zilges[1], V Derya[1], D Savran[2,3], A Hennig[1], SG Pickstone[1] and M Spieker[1]
[1]Institute for Nuclear Physics, University of Cologne, 50937 Köln, Germany
[2]EMMI and Research Division GSI Helmholtzzentrum, 64291 Darmstadt, Germany
[3]Frankfurt Institute for Advanced Studies FIAS, 60438 Frankfurt am Main, Germany

E-mail: zilges@ikp.uni-koeln.de

Abstract. At energies well below the isovector electric Giant Dipole Resonance, a concentration of electric dipole strength is observed in many stable and radioactive nuclei. This low-lying excitation mode is usually denoted as Pygmy Dipole Resonance (PDR). Different theoretical approaches can reproduce the gross features, like the energetic position and partly the excitation strength of the PDR, but they differ considerably in its underlying structural description. More sophisticated experiments looking for the isospin character, the detailed decay pattern, and the single-particle structure of the low-lying E1 excitations are therefore mandatory to gain a deeper understanding. This manuscript will give an overview about the most recent experimental results and present new experimental tools for the future.

Introduction

Atomic nuclei irradiated by photons in the MeV range can be excited to various electric and magnetic dipole modes (see figure 1). The electric dipole strength is dominated by the Giant Dipole Resonance (GDR) which was discovered already in 1937 by Bothe and Gentner [1]. Baldwin and Klaiber continued with systematic studies on various nuclei which enabled to derive the main parameters of the GDR [2].

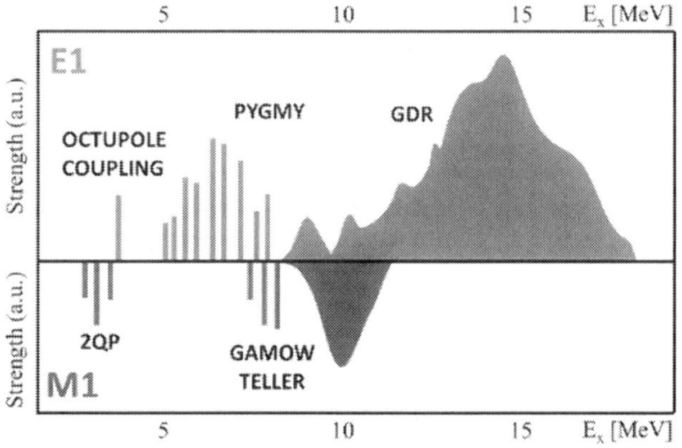

Figure 1: Schematic distribution of dipole strength in an atomic nucleus showing electric dipole modes (E1) in the upper part and magnetic dipole modes (M1) in the lower part. The two-quasi-particle (2QP) strength is often denoted "scissors mode" in deformed nuclei.

The GDR exhausts about 100% of the energy-weighted sum rule (EWSR) for isovector E1 strength. However, the last decades showed that an additional concentration of E1 strength exhausting about one to a few percent of the EWSR (i.e., a few percent of the strength found in the GDR) can be detected at lower energies. These excitations are commonly denoted as Pygmy Dipole Resonance (PDR); its properties are summarized in a recent review [3]. Different experimental probes have been used to determine the distribution of B(E1) strength including Nuclear Resonance Fluorescence (NRF) using photons from bremsstrahlung, see, e.g., refs. [4,5,6], Coulomb excitation of stable isotopes in (p,p') experiments at proton energies of around 400 MeV [7], and Coulomb excitation using relativistic beams in inverse kinematics [8]. The precise knowledge of the E1 strength distribution in atomic nuclei can help to get a better understanding of the neutron skin of nuclei and of the slope of the symmetry energy in the Equation of State [9-13].

Systematics of E1 strength distribution

An overview of most available data on E1 strength in the region below and around the particle threshold derived from electromagnetic excitation experiments is displayed in figure 2. Here, the summed E1 strengths are given as the exhaustion of the isovector E1 EWSR. The strength is plotted versus the Coulomb corrected Fermi energy Δ_{CCF} which is a good measure for the exoticity of a nucleus [3]. Studies on radioactive nuclei usually result in larger error bars. In an often discussed simplified picture of the PDR as an oscillation of excess neutrons against an isospin saturated core one would expect a correlation between the PDR strength and Δ_{CCF}. The figure shows quite obviously that the existing data are far from being consistent, different methods give different values for the summed B(E1) strength [3]. One of the reasons is different approaches how to treat the unresolved background in the region of the PDR. Another important reason is the difference in the definitions, which energy range and excitations have to be included in the summed strength, i.e., which states represent the PDR. We note that the data for radioactive nuclei usually stem from excitations above the neutron threshold (because the neutron is observed in the exit channel). This is in contrast to the situation for stable nuclei. The typical observables derived from electromagnetic excitation make it difficult to distinguish the structural differences in E1 excitation modes. Therefore, alternative probes are necessary to learn more about the E1 excitations.

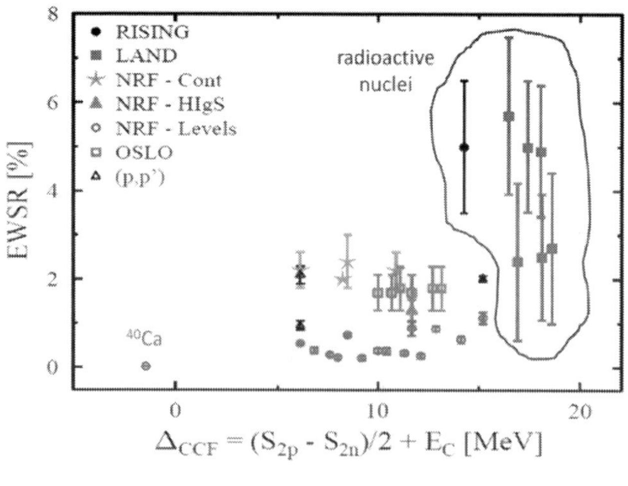

Figure 2: Summed electric dipole strength in the region of the PDR versus the Coulomb corrected Fermi energy Δ_{CCF}.This parameter is a good measure for the exoticity of a nucleus. (adapted from [3].)

Isospin splitting of the E1 strength

The scattering of α particles at energies of around 35 MeV/u under forward angles favors isoscalar ΔT=0 excitations at the surface of a nucleus. Coulomb excitation is strongly suppressed under these kinematical conditions. Therefore, they represent a complementary probe to photons which interact only with the protons in the nucleus and induce ΔT=1 E1 transitions. Figure 3 exhibits as an example the situation in the nucleus ^{124}Sn: α scattering can only populate states with the same isospin T=12 as the groundstate, whereas the excitation by isovector photons can populate T=12 as well as T=13 states [14].

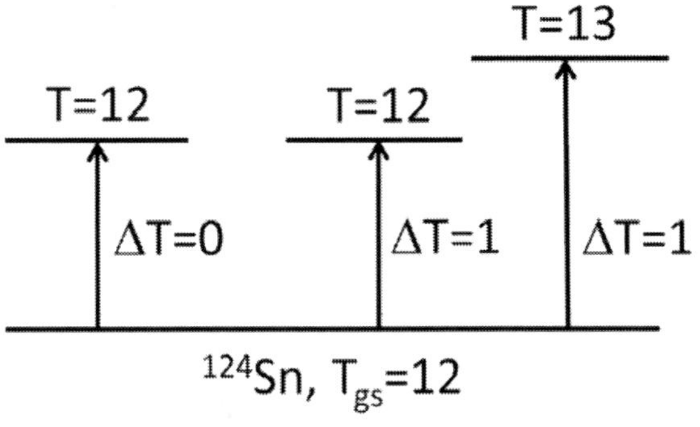

Figure 3: From the T_{gs}=12 ground state in ^{124}Sn, an excited state with T=12 can either be reached by an isoscalar ΔT=0 transition or by an isovector ΔT=1 transition. In addition, the ΔT=1 transition can populate a T=13 state (e.g. belonging to the GDR) which is at higher energies.

In pioneering experiments at the KVI Groningen, a group around M.N. Harakeh combined medium resolution spectroscopy of scattered α particles at the QMG/2 spectrometer with NaI detectors to measure the subsequent γ decay in coincidence [15]. This technique allows to select low-multipolarity groundstate transitions (as, e.g., E1 transitions) from the data. One could identify isoscalar E1 strength around 5-7 MeV in four nuclei with rather low level densities, namely ^{40}Ca, ^{58}Ni, ^{90}Zr, and ^{208}Pb. Low level density was a prerequisite due to the limited energy resolution of the NaI detectors. About one decade later, an improved setup, using the Big-Bite Spectrometer (BBS) for α spectroscopy and an array of HPGe detectors for high-resolution γ spectroscopy allowed to extend the studies to nuclei with higher level densities as well [16]. Systematic studies on various isotopes followed, which revealed a surprising structural splitting of the E1 strength below the particle threshold. Figure 4 shows as an example the results for the Z=50 nucleus ^{124}Sn. The lower part gives the B(E1) strength distribution measured in real-photon scattering experiments, the upper part the cross sections measured in the α scattering experiment. Below about 7 MeV, the excitation pattern seems to be quite similar in (α,α'γ) and (γ,γ') whereas nearly all excitations above about 7 MeV could not be excited with the isoscalar probe. Similar results have been obtained for all other heavy nuclei investigated in (α,α'γ) and in a very recent study using ^{17}O as a projectile by the Milano group [17]. Therefore, one can conclude that the low-lying E1 strength splits into a part below about 7 MeV which can be excited with α particles and γ rays and a higher lying part which can only be excited by γ rays. This observation was reproduced qualitatively by a number of theoretical calculations [see, e.g., 18-21]. A systematic comparison of PDR strength should therefore take into account possible structural differences of the E1 excitations.

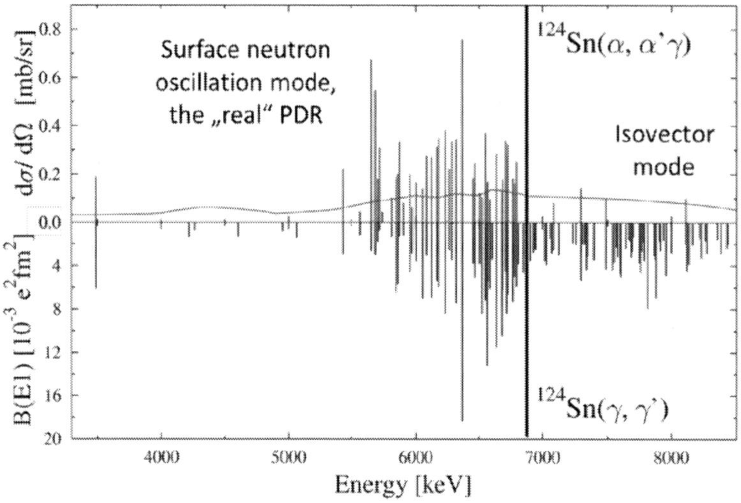

Figure 4: Cross section for E1 excitations observed in the (α,α'γ) experiment (upper part) in comparison with the E1 strength observed in (γ,γ'). Whereas all low lying states are populated in both experiments, the α particles do not populate the majority of the higher lying states (data taken from ref. [19]).

Outlook

The main challenges that need to be addressed to to obtain a better understanding of the PDR can be summarized as follows:

- What is the systematics of the E1 strength in light nuclei, in nuclei away from shell closures and in exotic nuclei?
- What are the decay properties of the E1 excitations into other low-lying states and do the theoretical models predict the decay pattern correctly?
- Is the observed difference in the excitation pattern using isoscalar and isovector probes a general phenomenon?
- What is the single-particle structure of the E1 excitations?

Ambitious experimental programs have been initiated and partly started at several stable and radioactive beam facilities and photon beam facilities worldwide to answer the questions above. An incomplete list includes experiments using photons at HIγS, Duke University; at the S-DALINAC at Darmstadt University; ELBE at FZ Dresden; and in the future at the ELI-NP facility in Bukarest. The PDR in exotic nuclei will be studied at GSI Darmstadt using the virtual photon bath at relativistic beam energies; and at RIKEN, Tokyo using a He target to study the response to an isoscalar probe. Finally, stable ion beams are used, e.g., at INFN Legnaro, at iThembaLABS Cape Town, at RCNP Osaka, at MLL Munich, and at the IKP Cologne. One very recent result from the latter facility is shown in figure 5. A (d,pγ) reaction on ^{119}Sn has been performed to populate states in ^{120}Sn. Due to the coincident spectroscopy of the proton and the emitted γ rays, this represents a very efficient tool for selecting the desired reaction and decay channels in the data analysis. The first preliminary results show that a strong excitation of the PDR in this one-neutron transfer reaction, tentative parities have been assigned by comparison with existing data. A more detailed analysis may allow an insight into the single-particle structure of the excitations.

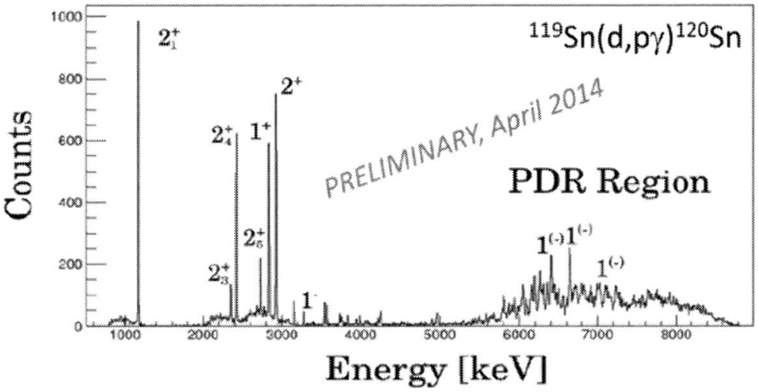

Figure 5: γ spectrum from a (d,pγ) reaction at E_d=8.5 MeV measured at the Cologne Tandem accelerator using the particle detector array SONIC and the HORUS γ spectroscopy array.

Acknowledgements

Our work on the Pygmy Dipole Resonance was made possible and advanced by the enthusiastic help and discussions with numerous colleagues. We are especially grateful to P. von Brentano, J. Endres, F. Iachello, M.N. Harakeh, and H.J. Wörtche for their contributions. It is always a pleasure to discuss with Aldo Covello to whom this article is devoted; and I wish him many more years of scientific work.

This research was supported by the Deutsche Forschungsgemeinschaft (ZI 510/4-2) and by the Alliance Program of the Helmholtz Association (HA216/EMMI). S.G.P. and M.S. are members of the honors branch of the BCGS within the German federal excellence initiative.

[1] Bothe W and Gentner W 1937 *Z. Phys.***71** 236
[2] Baldwin CG and Klaiber GS 1947 *Phys. Rev.* **C 71** 3
[3] Savran D, Aumann T, Zilges A 2013 *Prog. Part. Nucl. Phys.* **C 70** 210
[4] Govaert K et al. 1998 *Phys. Rev.* **C 57** 2229
[5] Zilges A, Volz S et al. 2002 *Phys. Lett.* **B 542** 43
[6] Schwengner R, Rusev G, Tsoneva N et al., 2008 *Phys. Rev.* **C 78** 064314
[7] Tamii A, Poltoratska I, von Neumann-Cosel P, Fujita Y et al., 2011 *Phys. Rev. Lett.* **107** 062502
[8] Wieland O et al, 2009 *Phys. Rev. Lett.* **102** 024310
[9] Reinhard PG and Nazarewicz W 2010 *Phys. Rev.* **C 81** 051303(R)
[10] Piekarewicz J 2012 *Phys. Rev.* **C 85** 041302(R)
[11] Erler J 2013 *Phys. Rev.* **C 87** 044320
[12] Carbone A et al., 2010 *Phys. Rev.* **C 81** 041301(R)
[13] Brown BA and Schwenk A 2014 *Phys. Rev.* **C 89** 011307(R)
[14] Satchler GR 1987 *Nucl. Phys.* **A 472** 215
[15] Poelhekken TD at al. 1992 *Phys. Lett.* **B 278** 423
[16] Savran D et al. 2006 *Phys. Rev. Lett.* **97** 172502
[17] Crespi FCL et al., 2014 *Phys. Rev. Lett.***,** in press
[18] Paar N, Niu YF, Vretenar D, Meng J, 2009 *Phys. Rev. Lett.* **103** 032502
[19] Endres J, Litvinova E, Savran D et al, 2010 *Phys. Rev.Lett.* **105** 044325
[20] Bianco D, Knapp F, Lo Iudice N et al., 2012 *Phys. Rev.* **C 86** 044327
[21] Lanza EG, Vitturi A, Litvinova E, Savran D, 2014 *Phys. Rev.* **C 89** 041601(R)

Deformation effects in Giant Monopole Resonance

J Kvasil[1], V O Nesterenko[2], A Repko[1], D Bozik[1], W Kleinig[2,3] and P -G Reinhard[4]

[1] Institute of Particle and Nuclear Physics MFF UK, Charles University, CZ-18000 Prague 8, Czech Republic
[2] Laboratory of Theoretical Physics, Joint Institute for Nuclear Research, Dubna, Moscow region, 141980, Russia
[3] Technical University of Dresden, Institute for Analysis, D-01062, Dresden, Germany
[4] Institute of Theoretical Physics II, University of Erlangen, D-91058, Erlangen, Germany

E-mail: kvasil@ipnp.troja.mff.cuni.cz

Abstract. The isoscalar giant monopole resonance (GMR) in Samarium isotopes (from spherical ^{144}Sm to deformed $^{148-154}$Sm) is investigated within the Skyrme random-phase-approximation (RPA) for a variety of Skyrme forces. The exact RPA and its separable version (SRPA) are used for spherical and deformed nuclei, respectively. Consistently with earlier investigations, the quadrupole deformation is shown to yield two effects: the GMR broadens and attains a two-peak structure due to the coupling with the quadrupole giant resonance.

1. Introduction

During last decades, the GMR remains to be a subject on intense studies (see [1, 2, 3] for recent reviews and discussions) as it provides a valuable information on the nuclear incompressibility [4]. Unlike early explorations, the recent theoretical investigations are mainly done within the self-consistent mean field models (SC-MFM) (see [5, 6] for a review), in particular with Skyrme forces [5, 7, 8, 9]. A variety of experimental data is also now available, see e.g. [10, 11, 12, 13, 14, 15].

Despite an impressive theoretical and experimental effort, some GMR problems are not yet resolved. For example, the GMR experimental data in spherical nuclei, Sn/Cd isotopes [11, 12] from one side and Pb/Sm isotopes [13, 14] from another side, cannot be simultaneously reproduced by any SC-MFM [1, 2, 16, 17, 18, 19]. It is not yet clear if this is caused by a poor theoretical description (e.g., too rough treatment of the pairing impact [16, 17, 18, 19]) or by a lack of accuracy of experimental data [10, 11, 12, 13, 14, 15]. For example, the TAMU (Texas A&M University) [13, 15] and RCNP (Research Center for Nuclear Physics at Osaka University) [10, 11, 14] (α, α') experiments produce noticeably different responses.

The situation with GMR in deformed nuclei is even more vague. Though there is an evident progress in experimental studies, e.g. for soft Mo [15] and deformed Sm [10, 13] isotopes, the SC-MFM calculations are yet at the very beginning, which is explained by a need to deal with a huge configuration space. So the theoretical results on the GMR in deformed nuclei are now mainly from early non self-consistent studies based on phenomenological mean fields [20, 21]. More than three decades ago, two deformation effects have been predicted [20, 21] and observed [22, 23] in GMR: i) broadening the resonance and ii) onset of two-peak structure due to a coupling with the $\mu = 0$ branch of the giant quadrupole resonance (GQR). Obviously it is worth to revisit these results by using a modern theoretical framework, e.g. the Skyrme SC-MFM. This is just the aim of the present study.

We consider GMR in a chain of Sm isotopes, from spherical ^{144}Sm to deformed ^{154}Sm, within the Skyrme random-phase approximation (RPA) approach. For spherical nuclei, the exact RPA method [24] is used. For deformed nuclei, the separable RPA (SRPA) [25, 26] is implemented. This method exploits the self-consistent factorization of the residual interaction, which drastically decreases the computational effort while keeping high accuracy of the calculations. SRPA was shown to be a very reliable and effective theoretical tool in investigation of various electric [27, 28, 29] and magnetic [30] giant resonances in both axially deformed and spherical nuclei. In particular, SRPA was used for exploration of the impact of the time-odd current density on the properties of GMR in ^{208}Pb [29]. In spherical nuclei, SRPA and exact RPA results are about identical. Both methods are fully self-consistent in the sense that their mean fields and residual interactions are derived from the initial Skyrme functionals and all the obtained terms are taken into account.

The calculations employ a wide set of Skyrme forces with various isoscalar effective masses m_0/m: SkT6 [31], SVbas [32], SkM*[33], SGII [34], SLy6 [35] and SkI3 [36].

2. Theoretical framework

The calculations are performed within exact RPA [24] for spherical ^{144}Sm and SRPA [25, 26] for soft and axially deformed ^{148,150,152,154}Sm. For every nucleus, the equilibrium deformation β_2 is determined. Then, for the isoscalar monopole transition operator $\hat{M} = \sum_i^A (r^2 Y_{00})_i$, the strength function

$$S(E0; E) = \sum_{\nu} E_{\nu} \mid \langle \nu | \hat{M} | 0 \rangle |^2 \, \xi_{\Delta}(E - E_{\nu}) \tag{1}$$

with the Lorentz weight $\xi_{\Delta}(E - E_{\nu}) = \frac{1}{2\pi} \frac{\Delta}{(E-E_{\nu})^2 - \Delta^2/4}$ is computed. Here, $|0\rangle$ is the ground state wave function, $|\nu\rangle$ and E_{ν} are RPA states and energies, Δ is the averaging parameter. The Lorentz weight roughly simulates smoothing effects beyond RPA (coupling to complex configurations and escape widths) and allows a convenient comparison of the calculated and experimental strengths. In the present study, the averaging Δ= 2 MeV is found optimal. Note that the same averaging was used for the giant dipole resonance [27].

In SRPA calculations for deformed nuclei, the input (doorway) operators $r^2 Y_{00}$, $r^4 Y_{00}$, $j_0(qr) Y_{00}$ (with q=0.4, 0.6) and $r^2 Y_{20}$ are used to construct time-even $\hat{X}^{\tau}_{k,k'}$ and time-odd $\hat{Y}^{\tau}_{k,k'}$ operators and corresponding strength constants of the separable residual interaction (where τ marks protons and neutrons and k, k' numerate the input operators) [25, 26]. The residual interaction is self-consistent since separable operators $\hat{X}^{\tau}_{k,k'}, \hat{Y}^{\tau}_{k,k'}$ and strength constants involve by construction (through the second density derivatives of the initial Skyrme functional) all the contributions of the Skyrme forces [25, 26]. The quadrupole doorway operator $r^2 Y_{20}$ is included to generate through $\hat{X}^{\tau}_{k,k'}, \hat{Y}^{\tau}_{k,k'}$ the quadrupole excitation branch $\lambda\mu$ =20 and thus cause the quadrupole-monopole coupling pertinent for deformed nuclei. For the set of 5 input operators, the rank of the SRPA matrix is 5×4, which is much less than a huge rank of conventional RPA matrices. At the same time, this set is enough for a good convergence of results.

In SRPA calculations for $^{148-152}$Sm, the pairing δ-force

$$V_{pair}(\vec{r}, \vec{r}\,') = V_0 \left[1 - \eta \left(\frac{\rho(\vec{r})}{\rho_0} \right)^{\gamma} \right] \delta(\vec{r} - \vec{r}\,') \tag{2}$$

is used at the BCS level [37]. For SVbas, both volume ($\eta = 0$) and density-dependent surface (η =1) pairing options are used. Other Skyrme forces exploit the standard volume pairing.

The calculations use a large configuration space with particle-hole (two-quasiparticle) energies up to 70-75 MeV. For the force SVbas, the monopole strength summed at the relevant energy interval 9-45 MeV exhausts the energy weighted sum rule EWSR $= \frac{\hbar^2}{2\pi m}$ A $\langle r^2 \rangle_0$ by 100-105%, depending on the isotope. Similar results hold for the other forces. The spurious mode lies at 2-7 MeV, i.e. safely below the GMR concentrated at 10-20 MeV.

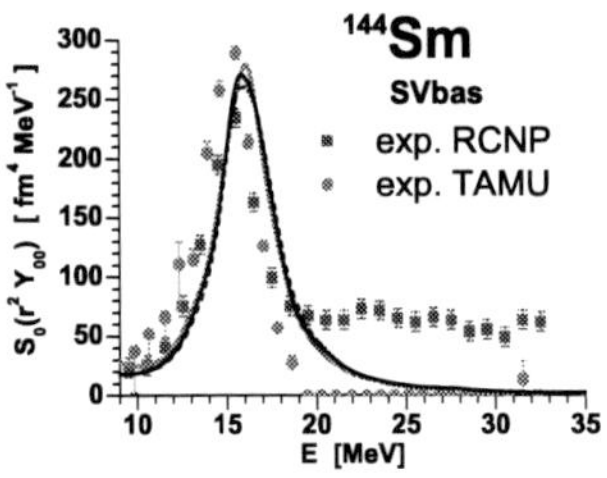

Figure 1. E0 strength function in the spherical ^{144}Sm calculated within full RPA (thick red line) and SRPA (black lines). In SRPA, the input operators $f(r)Y_{00}$ with radial dependence r^2 (dotted line), r^2, r^4 (dashed line) and $r^2, r^4, j_0(0.4r), j_0(0.6r)$ (thin line) are used. The RCNP [10] and TAMU [13] experimental data are depicted.

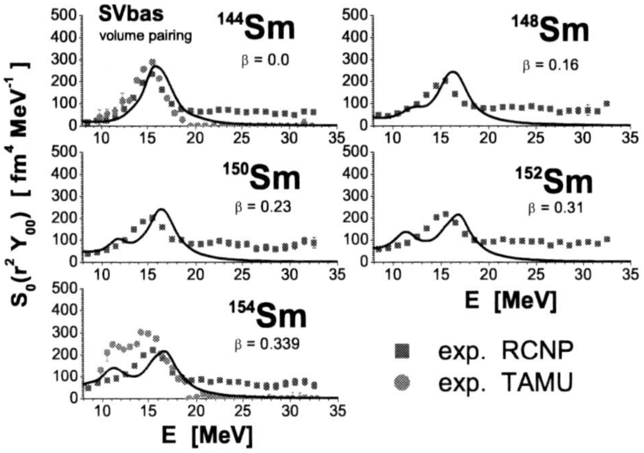

Figure 2. E0 strength functions in Sm isotopes, calculated with SVbas force within full RPA (^{144}Sm) and SRPA ($^{148-154}$Sm), as compared to RCNP [10] and TAMU [13] experimental data. For every isotope, the deformation parameter β_2 is shown.

3. Results and discussion

In Figure 1, the E0 strengths calculated within full RPA and SRPA (with different sets of input operators) are compared. It is seen that RPA and SRPA strengths are about identical already for one input operator $r^2 Y_{00}$, which indicates a high accuracy of SRPA. This is partly caused by a simple one-peak form of the GMR. Next three SRPA input operators are almost irrelevant here. However, following our calculations, they are necessary for description of a more complicated GMR form in deformed nuclei.

Figure 1, as well as next figures, exhibit the RCNP [10] and TAMU [13] experimental data. For our comparative analysis, we find convenient to convert the TAMU data, given in EWSR/E in [13], to the units used in RCNP (fm^4 MeV^{-1}). Following Fig. 1, the RCNP and TAMU data give about the same GMR peak energy but considerably deviate at energies above the GMR location, i.e. at $E > 19$ MeV. Namely, RCNP indicates a large and about uniform tail of E0 strength at 19 MeV$< E <$ 33 MeV while TAMU gives here a vanishing strength (see critical discussion of this discrepancy in [13]). This difference may affect determination of experimental GMR energy centroids which are usually estimated through evaluation of sum rules with different energy weights and thus depend on the chosen energy interval.

As seen from Fig. 1, both full RPA and SRPA results well reproduce the peak energy and width of the GMR (though the later is mainly attributed to the proper choice of the average parameter Δ=2 MeV). For $E > 19$ MeV, our calculations, in agreement with TAMU data [13], do not give any significant tail.

In Figure 2, the full RPA and SRPA results obtained with the force SVbas are compared to TAMU [13] and RCNP [10] data for the set of Sm isotopes. Parameters of the equilibrium quadrupole deformation β_2 determined from the minimum of the nuclear energy show that only ^{144}Sm is spherical while the other isotopes are soft (^{148}Sm) or well deformed ($^{150-154}$Sm). The E0 strength is calculated within full RPA in ^{144}Sm and SRPA in $^{148-154}$Sm (with 5 input operators as discussed above). It is seen that our calculations clearly demonstrate the broadening of the GMR and the onset of the GMR two-peak structure in going from the spherical ^{144}Sm to the deformed ^{154}Sm. The latter effect is caused by the coupling of E0 and E2 modes in deformed nuclei [20, 21, 22, 23]. Note that the two-peak structure in ^{154}Sm is observed in TAMU [13] but not in RCNP [10] experiments, which once more signals on the essential discrepancy between the two experiments. Further, in contrast to our predictions, the RNCP data do not show any double-peak structure in deformed ^{150,152}Sm. Taking into account the RCNP/TAMU disagreement for ^{154}Sm, the corroboration of the RNCP data for ^{150,152}Sm is desirable.

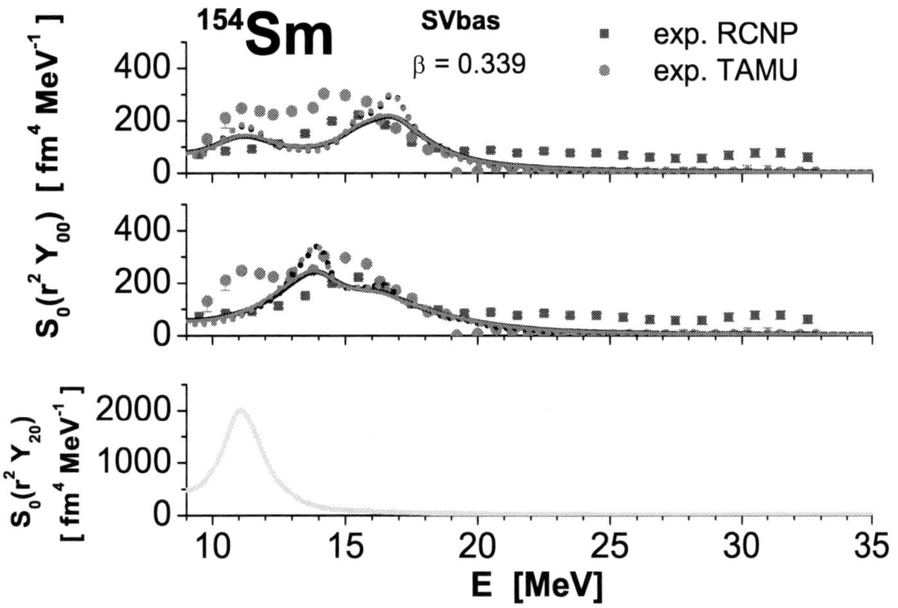

Figure 3. Isoscalar E0 (two upper panels) and E2 (bottom panel) strength functions in deformed ^{154}Sm, calculated within SRPA with the force SVbas. The E0 strength is determined: i)) with (upper panel) and without (middle panel) coupling to the quadrupole excitations, ii) with the volume (black curves) and surface (red curves) pairing, ii) with the averaging parameter Δ=1 MeV (dotted curves) and 2 MeV (solid curves). The RCNP [10] and TAMU [13] experimental data are shown.

The origin of the two-peak structure is demonstrated in Fig. 3 for ^{154}Sm. It is seen that the first peak takes place if E0-E2 coupling is included through the input operator r^2Y_{20} (upper plot) but is absent if the coupling is switched off (middle plot). Moreover, the position of the first peak in E0 strength coincides with position of $\mu = 0$ branch of the quadrupole resonance, exhibited at the bottom plot. The upper and middle plots also show sensitivity of the results to the choice of pairing (volume vs surface) and average parameter Δ (1 or 2 MeV). Though these factors somewhat change the results, the qualitative picture remains the same.

Finally in Fig.4, the GMR calculated in ^{154}Sm with different Skyrme forces is shown. It is seen that the first GMR peak is generally upshifted as the isoscalar effective mass m_0/m decreases. The better agrement is obtained for the forces with a large m_0/m, from SkT6 to SGII. All the Skyrme forces give the two-peak GMR structure. Our results for E0 strength at the GMR region

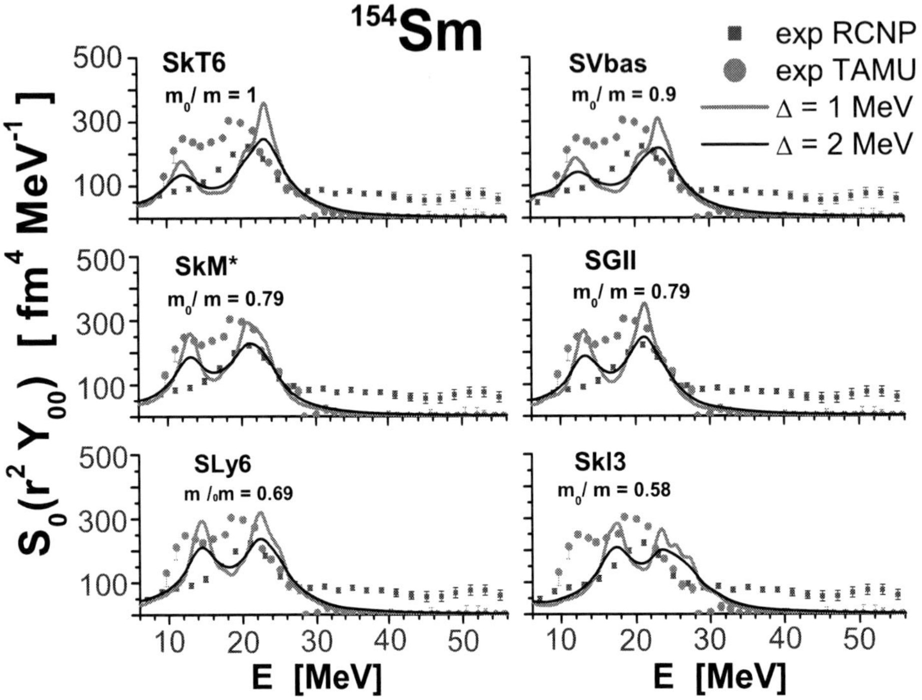

Figure 4. Isoscalar E0 strengths in ^{154}Sm, calculated for different Skyrme parametrizations (SkT6, SVbas, SkM*, SGII, SLy6 and SkI3) and two values of the Lorents averaging parameter, $\Delta = 1$ and 2 MeV, as compared to TAMU [13] and RCNP [10] experimental data. For each force, the isoscalar effective mass m_0/m is shown.

and above are in a general agrement with TAMU [13] data and do not correspond to RCNP [10] distributions. The results with Δ=1 and 2 MeV look qualitatively similar though the larger averaging is more convenient for the comparison to experiments. It is remarkable that changing Δ from 1 to 2 MeV practically does not affect the description of the GMR width in deformed nuclei. The same was found for the giant dipole resonance [27].

4. Conclusions

The GMR in Sm isotopes, from spherical ^{144}Sm to deformed $^{148-154}$Sm, was explored. We used exact RPA [24] for spherical ^{144}Sm and separable RPA (SRPA) [25, 26] for deformed isotopes. Various Skyrme forces (SkT6, SVbas, SkM, SGII, SLy6 and SkI3) were involved. In accordance to earlier studies [20, 21, 22, 23], the SRPA calculations have shown distinct deformation effects: i) broadening the GMR and ii) splitting of the resonance into two peaks. The latter effect was shown to be caused by the coupling between GMR and μ=0 branch of the quadrupole giant resonance. The obtained results are in a good agreement with TAMU experimental data [13] which support the two-peak GMR structure in ^{154}Sm. At the same time, they deviate from RCNP [10] data which exhibit in all Sm isotopes a one-bump GMR structure and an impressive high-energy distribution of E0 strength. Certainly, further exploration of GMR in Sm isotopes needs a harmonization of available experimental data.

The calculations with different Skyrme forces give rather close results though the forces with a large isoscalar effective mass m_0/m (from SkT6 to SGII) look more promising. In accordance to previous studies, the volume and surface pairings give similar effects in GMR.

5. Acknowledgments

The work was partly supported by the DFG grant RE 322/14-1, Heisenberg-Landau (Germany-BLTP JINR), and Votruba-Blokhincev (Czech Republic-BLTP JINR) grants. P.-G.R. is grateful for the BMBF support under Contracts No, 06 DD 9052D and 06 ER 9063. The support of the Czech Science Foundation (P203-13-07117S) is appreciated.

References

[1] Coló G 2008 *Phys. Part. Nucl.* **39** 286
[2] Avogadro P and Bertulani C A 2013 *Phys. Rev.* **C 88** 044319
[3] Stone J R, Stone N J and Mozskowski S A 2014 *Phys. Rev.* **C 89** 044316
[4] Blaizot J 1980 *Phys. Rep.* **64** 171
[5] Bender M, Heenen P-H and Reinhard P-G, 2003 *Rev. Mod. Phys.* **75** 121
[6] Vretenar D, Afanasjev A V, Lalazissis G A and Ring P 2005 *Phys. Rep.* **409** 101
[7] Skyrme T H R 1956 *Phil. Mag.* **1** 1043
[8] Vauterin D and Brink D M 1972 *Phys. Rev.* **C 5** 626
[9] Engel Y M et al. 1975 *Nucl. Phys.* **A 249** 215
[10] Itoh M et al. 2003 *Phys. Rev.* **C 68** 064602
[11] Li T et al. 2007 *Phys. Rev. Lett.* **99** 162503
[12] Patel D et al. 2012 *Phys. Letters* **B 718** 447
[13] Youngblood D H et al. 2004 *Phys. Rev.* **C 69** 034315
[14] Uchida M, et al. 2004 *Phys. Rev.* **C 69** 051301(R)
[15] Youngblood D H et al. 2013 *Phys. Rev.* **C 88** 021301(R)
[16] Khan E 2009 *Phys. Rev.* **C 80** 011307(R); ibid **80** 057302
[17] Vesely P, Toivanen J, Carlsson B C, Dobaczewski J, Michel M and Pastore D 2012 *Phys. Rev.* **C 86** 024303
[18] Cao L, Sagawa H and Colo G 2012 *Phys. Rev.* **C 86** 054313
[19] Avogadro P and Nakatsukasa T 2013 *Phys. Rev.* **C 87** 014331
[20] Zawischa D, Speth J and Pal D 1978 *Nucl. Phys.* **A 311** 445
[21] Abgrall Y, Morand B, Caurier E and Grammaticos N 1980 *Nucl. Phys.* **A 346** 431
[22] Buenerd M, Lebrun D, Martin Ph, de Saintignon P and Perrin C 1980 *Phys. Rev. Lett.* **45** 1667
[23] Garg U et al 1984 *Phys. Rev.* **C 29** 93
[24] Repko A, Kvasil J, Nesterenko V O and Reinhard P-G 2014 (to be submitted for publication)
[25] Nesterenko V O, Kvasil J, Reinhard P -G (2002) *Phys. Rev.* **C 66** 044307
[26] Nesterenko V O, Kleinig W, Kvasil J, Vesely P, Reinhard P-G and Dolci D S 2006 *Phys. Rev.* **C 74** 064306
[27] Kleinig W, Nesterenko V O, Kvasil J, Reinhard P-G and Vesely P 2008 *Phys. Rev.* **C 78** 044313
[28] Kvasil J, Nesterenko V O, Kleinig W and Reinhard P-G 2014 *Phys. Scr. 89* 054023
[29] Nesterenko V O, Kleinig W, Kvasil J, Vesely P and Reinhard P-G 2007 *Proceed. of 26th Intern. Workshop on Nucl. Theory (Rila, Bulgaria, 2007)* ed. S. Dimitrova (INRNE BAS, 2007) p. 273
[30] Vesely P, Kvasil J, Nesterenko V O, Kleinig W, Reinhard P-G and Ponomarev V Yu 2009 *Phys. Rev.* **C 80** 031302
[31] Tondeur F, Brack M, Farine M and Pearson J M 1984 *Nucl. Phys.* **A 420** 297
[32] Kluepfel P, Reinhard P-G, Buervenich T J and Maruhn J A 2009 *Phys. Rev.* **C 79** 034310
[33] Bartel J, Quentin P, Brack M, Guet C and Haakansson H-B 1982 *Nucl. Phys.* **A 386** 79
[34] Van Giai N and Sagawa H 1981 *Phys. Lett.* **B 106** 379
[35] Chabanat E, Bonche P, Haensel P, Meyer J and Schaeffer R 1997 *Nucl. Phys.* **A 627** 710
[36] Reinhard P-G and Flocard F 1995 *Nucl. Phys.* **A 584** 467
[37] Bender M, Rutz K, Reinhard P-G and Maruhn J A 2000 *Eur. Phys. J.* **A 8** 59

Ischia14 IOP Publishing
Journal of Physics: Conference Series **580** (2015) 012054 doi:10.1088/1742-6596/580/1/012054

Nuclear pairing correlations within and beyond HFB-BCS models

Danilo Gambacurta
GANIL, CEA/DSM and CNRS/IN2P3, Boîte Postale 55027, 14076 Caen Cedex, France
E-mail: gambacurta@ganil.fr

Denis Lacroix
Institut de Physique Nucléaire, IN2P3-CNRS, Universite Paris-Sud, F-91406 Orsay Cedex, France

Abstract. Pairing correlations in nuclear systems play a crucial role in several aspects, i.e. binding energies and odd-even effects, superfluid phenomena and pair transfer mechanisms, just to quote few of them. On the theoretical side, the standard description of these features is done by using BCS or HFB models which allow to describe in a simple way pairing effects. However, due to the explicit breaking of the particle number, these theories present some limitations which can be cured by using particle number projection techniques. In this paper, we will show the merits of these techniques and their accuracy in treating pairing correlations. First, a beyond BCS-HFB method is introduced where the effect of four quasi-particle states is included perturbatively and subsequently the particle number is restored. We will then show the need of restoring the good particle number also for excited states that are essential in the pair transfer process between superfluid systems. Applications to the Richardson model are shown and discussed.

1. Introduction

The mean-field description of pairing correlations in superconducting systems is typically described within the Bardeen-Cooper-Schrieffer (BCS) or Hartree-Fock-Bogoliubov (HFB) approximations, eventually augmented by Quasi-particle Random Phase Approximation (QRPA) for the description of excited states. Both levels of approximation violate the particle number symmetry. These methods suffer thus from particle number fluctuations that have to be cured in the case of finite systems, (for a recent review see refs [1]). This goal can be accomplished by using projection techniques, usually performed in the projection after variation (PAV) scheme, where unphysical components are "a posteriori" projected out. Only few calculations have been done at the variation after projection scheme, which allows a fully self-consistent treatment of the symmetries but is numerically highly demanding. In the last decades, several approaches have been used to overcome these difficulties. When the number of particle is small enough, the exact solution of the pairing problem is accessible by direct diagonalization of the Hamiltonian [2, 3, 4], Quantum Monte-Carlo technique (see for instance [5, 6]) or extending quasi-particle theories using Variation After Projection approaches [7, 8] (for recent applications see [9, 10, 11]), or the generalized seniority scheme [12]. All of these methods are however rather involved and demanding in terms of computational power. In this paper we discuss two accurate methods where the particle number symmetry is restored by using projection techniques in a PAV like spirit. In Section 2 the treatment of the ground state correlation are discussed [13] while in Section 3 a method for the description of excited states is proposed [14]. Applications are done in the Richardson model [15].

2. Quasi-particle perturbative approach for g.s. correlations

Let's consider a two-body pairing Hamiltonian given by:

$$H = \sum_{i=1}^{\Omega} \varepsilon_i (a_i^\dagger a_i + a_{\bar{i}}^\dagger a_{\bar{i}}) + \sum_{i \neq j}^{\Omega} v_{ij} a_i^\dagger a_{\bar{i}}^\dagger a_{\bar{j}} a_j \equiv H_0 + V. \quad (1)$$

Here, $(i, \bar{i})$ denotes time-reversed pair and Ω denotes the number of levels. In the following, we will consider the specific case of equidistant levels with $\varepsilon_i = (i-1)\Delta\varepsilon$ with $\Delta\varepsilon = 1$ MeV and a number of particles $N = \Omega$.

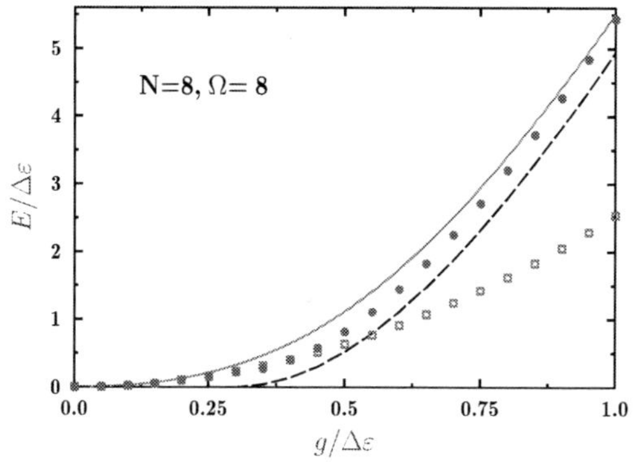

Figure 1. (color online) Illustration of the correlation energy as a function of the coupling strength for the case of N=8 particles and $\Omega = 8$. The exact result solution (red solid curve), BCS (black dashed line), standard perturbation theory (green open squares), QP^2T theory with second order correction equation (6) (blue filled circles), are displayed.

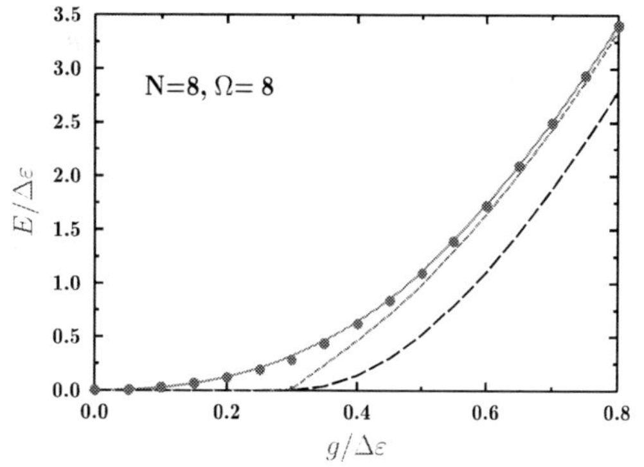

Figure 2. (color online) Illustration of the correlation energy as a function of the coupling strength for the case of N=8 particles and $\Omega = 8$. The exact result solution (red solid line), PAV result (green short-dashed line) and the combination of projection and perturbation theories (blue filled circles) are shown. The original BCS (black long dashed line) is also shown as a reference.

In figure 1, an example of standard perturbation theory (SPT) is presented for the $N = 8$ particles and constant coupling case, i.e. $v_{ij} = -g$. The correlation energy defined as the difference between the Hartree-Fock energy E_0 and the ground state energy $\mathbf{E}_0$ obtained within the SPT are compared to the exact solution and BCS result. In the latter case, pairing correlation is non-zero only above the threshold value $g/\Delta\varepsilon \simeq 0.3$. As illustrated from Fig. 1, standard perturbation theory matches with the exact result below the threshold but significantly underestimates the correlation for larger g value. This aspect underlines the highly non-perturbative nature of the pairing quantum phase-transition. On the opposite, one of the advantage a theory like BCS is the possibility to incorporate non-perturbative physics.

In order to obtain a better description we can combine theories based on quasi-particles and perturbative approaches. Let's consider then the BCS/HFB state

$$|\Phi_0\rangle = = \prod_{i>0}\left(U_i + V_i a_i^\dagger a_{\bar{i}}^\dagger\right)|-\rangle, \tag{2}$$

and the effective mean field BCS/HFB Hamiltonian

$$H_0 = E_0 + \sum_i E_i\left(\beta_i^\dagger \beta_i + \beta_{\bar{i}}^\dagger \beta_{\bar{i}}\right) \tag{3}$$

where E_0 is the BCS/HFB ground state energy, while E_i corresponds to the quasi-particle energy given by. It is easy to see that

$$H|\Phi_0\rangle = \left(H_0 - \sum_{i\neq j} v_{ij} U_i^2 V_j^2 \beta_i^\dagger \beta_{\bar{i}}^\dagger \beta_j^\dagger \beta_{\bar{j}}^\dagger\right)|\Phi_0\rangle, \tag{4}$$

showing that that the main effect of the neglected residual interaction is due to the coupling of Φ_0 with the 4QP states. We consider thus the perturbation V:

$$V = -\sum_{i\neq j} v_{ij} U_i^2 V_j^2 \beta_i^\dagger \beta_{\bar{i}}^\dagger \beta_j^\dagger \beta_{\bar{j}}^\dagger. \tag{5}$$

coupling the ground state with the 4QP states. The present approach, that is a direct extension of standard perturbation theory, is called hereafter quasi-particle perturbation theory (QP^2T). The second order correction to the ground state energy is equal to:

$$E_0^{(2)} = -\frac{1}{2}\sum_{i>j} v_{ij}^2 \frac{(U_i^2V_j^2 + U_j^2V_i^2)^2}{E_i + E_j}. \tag{6}$$

The result obtained with the QP^2T approach at second order in perturbation (6) are displayed in figure 1 with filled circles. Similarly to the original quasi-particle theory, the energy deduced from the QP^2T contains spurious contribution coming from the fact that the perturbed state does not preserves particle number.The most direct way to remove spurious contributions due to the mixing of different particle number is to introduce the operator P^N

$$P^N = \frac{1}{2\pi}\int_0^{2\pi} d\varphi \; e^{i\varphi(\hat{N}-N)}, \tag{7}$$

that projects onto particle number N. The most straightforward way to combine projection with quasi-particle perturbation theory is to directly take the expectation value

$$\mathbf{E}_0 = \frac{\langle \Psi_0^N | H | \Psi_0^N \rangle}{\langle \Psi_0^N | \Psi_0^N \rangle} \tag{8}$$

with $|\Psi_0^N\rangle = P^N|\Psi_0\rangle$ and where $|\Psi_0\rangle$ is truncated at a given order in perturbation. This approach will be referred as the projected quasi-particle perturbation theory (QP^3T) in the following. The correlation energies obtained for $N = \Omega = 8$ using the second order QP^3T are shown in figure 2. In each case, the original BCS, the PAV and the exact solution are also shown. The result of QP^3T almost superimposed with the exact solution.

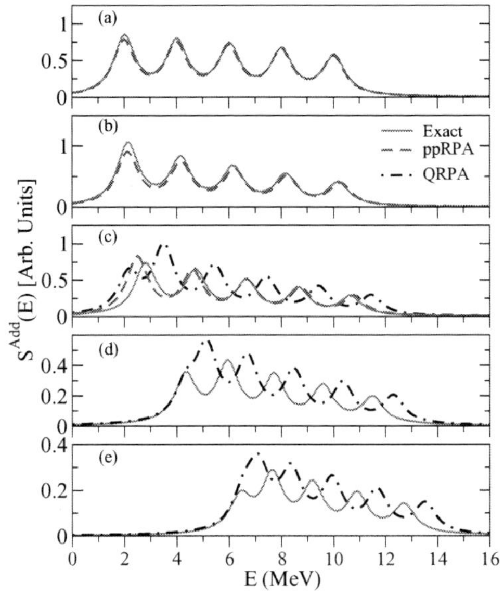

Figure 3. Addition strength function as a function of the excitation energy E for different pairing strength g obtained for a system of $N = 8$ particles to $N = 10$ in the case of $\Omega = 10$ equally spaced levels. From top to bottom: (a) $g/\Delta\varepsilon = 0.1$, (b) 0.3, (c) 0.5, (d) 0.7 and (e) 0.9 are shown. The results of the ppRPA (blue dashed line) and QRPA (black dot-dashed line) are presented and compared with the exact results (red solid line).

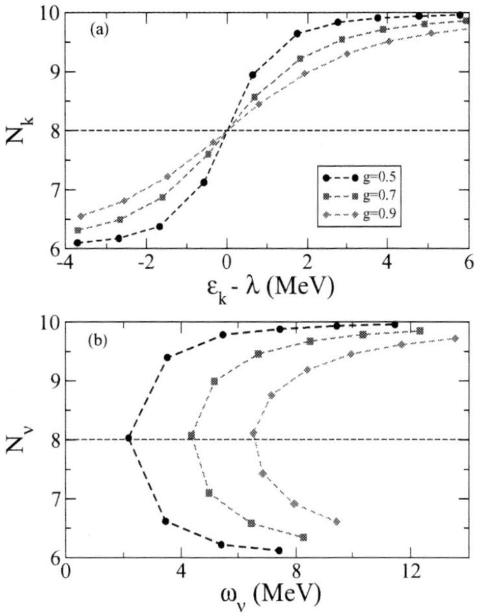

Figure 4. (color online) Bottom: mean number of particles N_ν of the QRPA states as a function of their excitation energy $\hbar\omega_\nu$. Top: mean number of particles N_k of 2QP states as a function of $(\epsilon_k - \lambda)$, where λ is the Fermi energy.

3. Pair transfer and excited states

Due to the absence of residual coupling between quasi-particle excitations, it is known that HFB alone cannot properly describe excited states. Then, linear response theory including possible particle-particle (pp), hole-hole (hh), or particle-hole (ph) excitations is applied to describe excited states, and then transfer probabilities, within the QRPA approach. In QRPA, the excited states, denoted by $|\nu\rangle$, are obtained by considering coherent superposition of 2 quasi-particle (2QP) excitations through the operators

$$Q^\dagger_\nu = \sum_i (X^\nu_j \alpha^\dagger_i \alpha^\dagger_{\bar{i}} - Y^\nu_j \alpha_{\bar{i}} \alpha_i). \tag{9}$$

In QRPA, the addition transition probability is given by

$$P^{\rm Add}_\nu = |\langle 0|\hat{T}^{\rm Add}|\nu\rangle|^2 = |\sum_i (V_i^2 X_i^{(\nu)} - U_i^2 Y_i^{(\nu)})|^2. \tag{10}$$

where $\hat{T}^{\rm Add} = \sum_i T_{i\bar{i}} a^\dagger_i a^\dagger_{\bar{i}}$ is non-Hermitian addition transition operator.

In the weak coupling limit, below a certain threshold value of g denoted by $g_{\rm cr}$, the minimization of the energy in HFB identifies to the Hartree-Fock approach with no pairing and the excited states are described by using the particle-particle RPA (ppRPA) where the phonon creation operators and the corresponding addition probability simply write as

$$Q_\nu^\dagger = \sum_p X_p^\nu a_p^\dagger a_{\bar{p}}^\dagger + \sum_h Y_h^\nu a_h^\dagger a_{\bar{h}}^\dagger, \qquad P_\nu^{\rm Add} = |\sum_p X_p^\nu - \sum_h Y_h^\nu|^2. \tag{11}$$

Contrary to the QRPA, the $U(1)$ symmetry associated to particle number conservation is not broken in ppRPA.

In Fig 3 the ppRPA (dashed line) and QRPA (dot-dashed line) are compared to the exact results. Above a given threshold g_{cr}^{RPA}, ppRPA collapses and leads to imaginary energies making not possible a direct comparison with the exact and QRPA results. However, in Fig. 3 we show in the panel (c) corresponding to a pairing strength $g = 0.5 MeV$ the ppRPA results obtained at the collapse point, i.e. $g_{cr}^{RPA} = 0.48$, in order to show that its description is still reasonable even in the superfluid phase. From these comparisons we can also see that although the QRPA provides a global reproduction of the exact results, in general QRPA leads to peaks in the strength that are at slightly higher energies compared to the exact solution while probabilities are slightly overestimated. These discrepancies can be, at least partially, traced back to the violation of the number of particle in QRPA.

As an illustration, the mean number of particles $N_\nu = \langle \nu | \hat{N} | \nu \rangle$ is displayed in the bottom panel of Fig. 4 as a function of the excitation energy ω_ν. N_ν has been estimated using the quasi-boson approximation leading to:

$$N_\nu = \sum_i 2(U_i^2 - V_i^2)(X_i^{\nu 2} + Y_i^{\nu 2}) + \langle \hat{N} \rangle, \tag{12}$$

where $\langle \hat{N} \rangle$ is the number of particle in the QP vacuum.

In the same figure, the mean-particle number of the two quasi-particle (2QP) excited states $|k\rangle$ as a function of $(\epsilon_k - \lambda)$ is also shown, where λ is the Fermi energy. The 2QP states and the corresponding mean particle number are given by:

$$|k\rangle = \alpha_k^\dagger \alpha_{\bar{k}}^\dagger |0, QP\rangle, \quad N_k = \langle k | \hat{N} | k \rangle = \langle \hat{N} \rangle + 2U_k^2 - 2V_k^2. \tag{13}$$

The Figure 4 clearly shows that the 2QP states will be close to a state with $N+2$ (resp. $N-2$) particles only if $U_k \to 1$, i.e. well above the Fermi energy λ (resp. $U_k \to 0$, i.e. well below the Fermi energy), but will be a bad approximation if the 2QP state involves single-particle state in the vicinity of λ. Consequently, QRPA states will also suffer from the same problems if the state is constructed from 2QP states that are close to the Fermi energy. In order to improve on that, we follow a Tamm-Dancoff approximation spirit, introducing a set of states $|\Phi_k\rangle \hat{P}_{N+2} \alpha_k^\dagger \alpha_{\bar{k}}^\dagger |0, QP\rangle$. The excited states are then obtained diagonalizing the Hamiltonian in this new space which preserves the particle number symmetry , (for more details see[14]), and the corresponding results are labeled as P-QTDA$^{(GS)}$.

To illustrate the predictive power of the P-QTDA$^{(GS)}$ method, some ratios of pair transfer probabilities are shown in Figure 5. In this figure, $P_{\rm gs}$, $P_{\rm ex}$ and $P_{\rm tot}$ correspond respectively to the probability to transfer to the ground state, the sum of probabilities to transfer to any excited states, while $P_{\rm tot} = P_{\rm gs} + P_{\rm ex}$. Below the pairing threshold, the P-QTDA$^{(GS)}$ reduces to a ppTDA$^{(GS)}$, where the diagonalization is made in a reduced space of Slater determinant. In that case, the result are of the same quality as for the ppRPA. This figure clearly confirms that while QRPA is rather far from the exact results in the superfluid phase, especially in the vicinity of the BCS threshold, the projected theory provides a much better reproduction of ground state to excited state pair transfer probabilities. Moreover,theoretical predictions based on the QRPA [16] might significantly overestimate the Giant Pairnig Vibration cross section, as suggested also in [17], especially in the vicinity of the normal-superfluid transition.

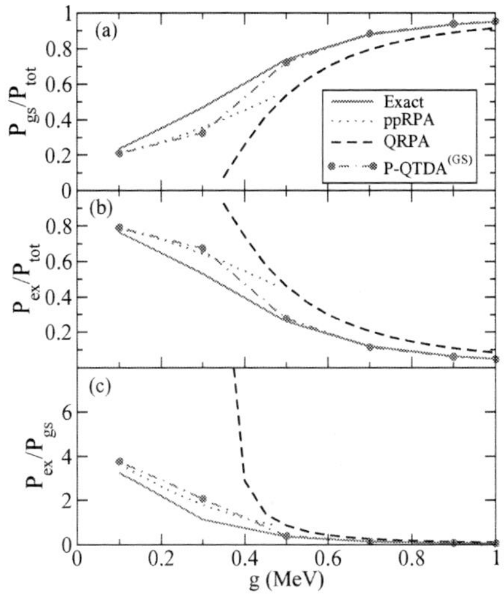

Figure 5. (color online) Ratios of probabilities estimated with different theories for the equidistant level case. The exact (red solid line), QRPA (black dashed line) and P-QTDA (green filled circles) are shown. The ppRPA is also shown (dotted line) up to $g = 0.48$ MeV.

4. Conclusions

In this paper, two methods, preserve the particle number symmetry, for the treatment of pairing correlations in ground state and excited states of finite systems which have been presented. Applications to the Richardson model are shown and discussed. The comparison with exact results allow to show clearly the quality of the presented methods and their improvement with respect to other approaches that do not preserve particle number.

References

[1] Von Delft J and Ralph D C 2001 Phys. Rep. **61**,345; Birman J L, Nazmitdinov R G, and Yukalov V I 2013 Phys. Rep. **526** 1
[2] Volya A, Brown B A and Zelevinsky V 2001 Phys. Lett. B **509** 37
[3] Zelevinsky V and Volya A 2003 Phys. of Atomic Nuclei, **66** 1781
[4] Sumaryada T and Volya A 2007 Phys. Rev. C **76** 024319
[5] Capote R, Mainegra E, Ventura A 1998 J. Phys. G **24** 1113
[6] Mukherjee A, Alhassid Y, and Bertsch G F 2011 Phys. Rev. C **83** 014319
[7] Dietrich K, Mang H J and Pradal J 1964 Phys. Rev. **135** B22
[8] Hupin G and Lacroix D 2011 Phys. Rev. C **83** 024317
[9] Rodriguez T R and Egido J L 2007 Phys. Rev. Lett. **99** 062501
[10] Rodriguez T R and Egido J L 2010 Phys. Rev. C **81** 064323
[11] Hupin G and Lacroix D 2012 Phys. Rev. C **86** 024309
[12] Talmi I 2001, Nuclear Physics A **686** 217
[13] Lacroix D and Gambacurta D 2012 Phys. Rev. C **86** 014306
[14] Gambacurta D and Lacroix D 2012 Phys. Rev. C **86** 064320
[15] Richardson R.W. and Sherman N 1964 Nucl. Phys. **52** 221
[16] Fortunato L, von Oertzen W, Sofia H M and Vitturi A 2002, Eur. Phys. J. **A14**, 37.
[17] Mouginot B *et al* 2011 Phys. Rev. C **83**, 037302.

Self-consistent studies of the dipole response in neutron rich nuclei using realistic potentials

F Knapp[1], N Lo Iudice[2,3], P Veselý[4], F.Andreozzi[2,3], G De Gregorio[2,3], A Porrino[2,3]

[1] Faculty of Mathematics and Physics, Charles University in Prague, Czech Republic
[2] Dipartimento di Fisica, Università di Napoli Federico II, Napoli, Italy
[3] Istituto Nazionale di Fisica Nucleare (INFN), Sezione di Napoli, Napoli, Italy
[4] Institute of Nuclear Physics, Academy of Sciences of the Czech Republic, 250 68 Řež, Czech Republic

E-mail: frantisek.knapp@mff.cuni.cz

Abstract. The dipole response in neutron rich nuclei is investigated within self-consistent approaches which make direct use of a nucleon-nucleon optimized chiral potential complemented with a density dependent term simulating a three-body force. Hartree-Fock-Bogoliubov plus Tamm-Dancoff and random-phase approximations show that such a potential improves the description of the dipole modes with respect to other realistic interactions. The inclusion of the two-phonon states within an equation of motion method induces a pronounced fragmentation of both giant and pygmy resonances in agreement with recent experiments.

1. Introduction

Collective modes, and, more specifically, giant (GDR) and pygmy (PDR) dipole resonances in neutron rich nuclei are mostly investigated in phenomenological mean field approaches embedded within energy density functionals derived from Skyrme forces [1] or from relativistic meson-nucleon Lagrangians [2].

More recently, we have followed an alternative route proposed in [3], which consists in carrying out self-consistent calculations using directly realistic potentials derived from nucleon-nucleon (NN) forces, and adopted a V_{lowk} potential derived from the CD-Bonn NN interaction. Using such a potential, we have generated a Hartree-Fock-Bogoliubov (HFB) basis and, then, adopted both (quasi-particle) Tamm-Dancoff ((Q)TDA) and random-phase approximation ((Q)RPA) approaches to carry out a systematic study of the dipole response in several chains of neutron rich isotopes [4]. As in [3], however, we found necessary to add a corrective density dependent piece in order to get single-particle spectra sufficiently compressed so as to reproduce the main peak of the experimental GDR. Using such a modified Hamiltonian, we obtained spectra qualitatively consistent with experiments for all light, medium mass and heavy isotopes [4].

We have now performed analogous self-consistent (Q)TDA and (Q)RPA studies by making direct use of a recently determined chiral potential, dubbed NNLO_{opt}, with its parameters optimized so as to minimize the effects of the three-body forces [5]. This potential was adopted for ^{132}Sn in [6] and found to yield a much more compact single particle spectrum compared to V_{lowk} and, consequently, to produce a dipole strength distribution much closer to the region of observation of the GDR peak. It was, nonetheless, still necessary to add the same density dependent term, though with a weaker coupling constant, in order to reproduce the centroid of the resonance. The potential, so modified, reproduces satisfactorily the gross features of both GDR and PDR. This is illustrated here for a chain of Ca isotopes.

(Q)TDA and (Q)RPA, however, were unable to describe the fine structure of the two resonances. So, we went beyond the mean field approximation and adopted the equation of motion phonon method (EMPM) [7–9] to study the dipole response within a space spanned by one plus two-phonon basis states [6]. The two phonons enhanced enormously the fragmentation of the dipole strength in both GDR a PDR regions, in better agreement with experiments. We report here also on the findings of such an investigation.

2. Brief outline of the EMPM method

The primary goal of the EMPM method is to generate a basis of n-phonon states $|n;\beta>$ of energies E_β having the form

$$|n;\beta>=\sum_{\lambda\alpha} C^{\beta}_{\lambda\alpha}\left\{O^{\dagger}_{\lambda}\times\mid n-1,\alpha>\right\}^{\beta}, \qquad (1)$$

where the TDA particle-hole (p-h) phonon operator

$$O^{\dagger}_{\lambda}=\sum_{ph} c^{\lambda}_{ph}(a^{\dagger}_{p}\times b_h)^{\lambda} \qquad (2)$$

of energy E_λ acts on the $(n-1)$-phonon states $\mid n-1,\alpha>$ of energies E_α. The procedure starts with the equations of motion

$$< n,\beta \parallel [H,O^{\dagger}_{\lambda}] \parallel n-1,\alpha >= \Big(E_\beta - E_\alpha\Big) < n,\beta \parallel O^{\dagger}_{\lambda} \parallel n-1,\alpha > . \qquad (3)$$

We, then, expand the commutator and invert equation (2) in order to express the p-h operators, present in the expanded commutator, in terms of the phonon operators $O^{\dagger}_{\lambda}$. The outcome of this action is [9] the generalized eigenvalue equation

$$\sum_{\lambda'\alpha'\lambda_1\alpha_1}\Big[\mathcal{A}^{\beta}(\lambda\alpha,\lambda_1\alpha_1)-E_\beta\delta_{\lambda_1\lambda}\delta_{\alpha_1\alpha}\Big]\mathcal{D}^{\beta}(\lambda_1\alpha_1,\lambda'\alpha')C^{\beta}_{\lambda'\alpha'}=0. \qquad (4)$$

Here

$$\mathcal{D}^{\beta}(\lambda_1\alpha_1;\lambda'\alpha')=\Big[< n-1,\alpha_1\mid \times O_{\lambda_1}\Big]_{\beta}\Big[O^{\dagger}_{\lambda'}\times\mid n-1,\alpha'>\Big]_{\beta} \qquad (5)$$

is the metric matrix and $\mathcal{A}^{\beta}$ is a matrix of the simple structure

$$\mathcal{A}^{\beta}(\lambda\alpha,\lambda'\gamma)=(E_\lambda+E_\alpha)\delta_{\lambda\lambda'}\delta_{\alpha\gamma}+\sum_{\sigma}W(\beta\lambda'\alpha\sigma;\gamma\lambda)\mathcal{V}^{\sigma}_{\lambda\alpha,\lambda'\gamma}, \qquad (6)$$

where W is a Racah coefficient and $\mathcal{V}^{\sigma}$ a phonon-phonon potential given by

$$\mathcal{V}^{\sigma}_{\lambda\alpha,\lambda'\gamma}=\sum_{rstq}\rho_{\lambda\lambda'}([q\times t]^{\sigma})F^{\sigma}_{qtrs}\rho^{(n)}_{\alpha\gamma}([r\times s]^{\sigma}). \qquad (7)$$

We have denoted by $\rho_{\lambda\lambda'}$ and $\rho^{(n)}_{\alpha\alpha'}$, respectively, the $n=1$ (TDA) and the n-phonon $(n>1)$ density matrices

$$\rho_{\lambda\lambda'}([r\times s]^{\sigma})=<\lambda'\parallel\Big[a^{\dagger}_{r}\times b_s\Big]^{\sigma}\parallel\lambda> \qquad (8)$$

$$\rho^{(n)}_{\alpha\alpha'}([r\times s]^{\sigma})=\langle n;\alpha'\parallel\Big[a^{\dagger}_{r}\times b_s\Big]^{\sigma}\parallel n;\alpha\rangle. \qquad (9)$$

The formal analogy between the structure of the phonon matrix $\mathcal{A}^{\beta}(\lambda\alpha, \lambda'\alpha')$ and the form of the TDA matrix $A^{\lambda}(ph; p'h')$ was pointed out [9]. The first is deduced from the second by replacing the TDA p-h energies with the sum of phonon energies $(E_\lambda + E_\alpha)$ and the TDA p-h interaction with the phonon-phonon interaction $\mathcal{V}^{\sigma}$.

Equation (4) represents the eigenvalue equation in the overcomplete basis $\left\{O_\lambda^\dagger \times \mid n-1, \alpha\rangle\right\}^{\beta}$. The redundant states are eliminated by the procedure outlined in [7, 8], based on the Cholesky decomposition method.

Since recursive formulas hold for all quantities entering $\mathcal{A}$ and $\mathcal{D}$, the eigenvalue equations are solved iteratively starting from the TDA phonons and, thereby, yield a set of orthonormal multiphonon states $\{|0>, |1, \lambda>, \ldots |n, \alpha> \ldots\}$.

In such a basis, the Hamiltonian matrix is composed of a sequence of diagonal blocks, one for each n, mutually coupled by off-diagonal terms $\langle n' \mid H \mid n\rangle$ which are non vanishing only for $n' = n \pm 1, n \pm 2$ and are computed by means of recursive formulas. A matrix of such a simple structure can be easily diagonalized yielding eigenfunctions of the form

$$\mid \Psi_\nu\rangle = \sum_{n\alpha} C_\alpha^{(\nu)} \mid n; \alpha\rangle. \tag{10}$$

3. Calculations and results

We adopted an intrinsic Hamiltonian obtained by subtracting the center of mass kinetic energy from the shell model kinetic operator and used the two-body potential

$$V_2 = NNLO_{opt} + V_\rho. \tag{11}$$

where $NNLO_{opt}$ is the NN chiral potential optimized so as to minimize the effects of the three-nucleon force [5], while V_ρ is a corrective repulsive, density dependent, two-body potential simulating a three-body contact force [10] which improves the description of bulk properties in closed shell nuclei [11] and yields more realistic single particle spectra and multipole nuclear responses [3, 4].

V_2 is adopted to generate a HFB basis in a configuration space which includes 13 harmonic oscillator major shells, up to the principal quantum number $N_{max} = 12$. Such a space is sufficient for reaching a good convergence. The self-consistent basis is used to compute the dipole strength distribution. We performed such a calculation in both (Q)TDA and (Q)RPA and found that the two approaches yield almost identical dipole spectra in all chains of isotopes [4, 6].

As illustrated in [6], the $NNLO_{opt}$ yields proton and neutron single particle spectra considerably more compressed than the ones produced by V_{lowk} derived from the CD-Bonn potential [4]. This compression is more consistent with experiments and improves the (Q)TDA and (Q)RPA descriptions of the dipole response. In ^{132}Sn, indeed, the peak of the strength distribution gets shifted from ~45 MeV, obtained when V_{lowk} is used, to ~23 MeV.

Such a strong shift, however, is not sufficient to reach the strongest experimental peak, observed at ~16 MeV. Hence the need of adding V_ρ, though with a weaker coupling constant. Such a potential pushes the dipole strength distribution in the region of observation as shown in figure 1 for a chain of Ca isotopes. Analogous results are obtained for O and Sn isotopes. (Q)RPA and (Q)TDA cross sections, computed by using a width $\Delta = 2.0$ MeV, follow closely the measured GDR cross sections. Such a good agreement is illustrated in the plots of figure 2 for ^{40}Ca and ^{48}Ca and holds for all the other nuclei.

Figure 1 shows that a peak appears around 10 MeV and becomes more and more pronounced as the excess neutrons increase. This trend, consistent with experiments, points toward the pygmy nature of these transitions, a suggestion supported by the analysis of the transition densities.

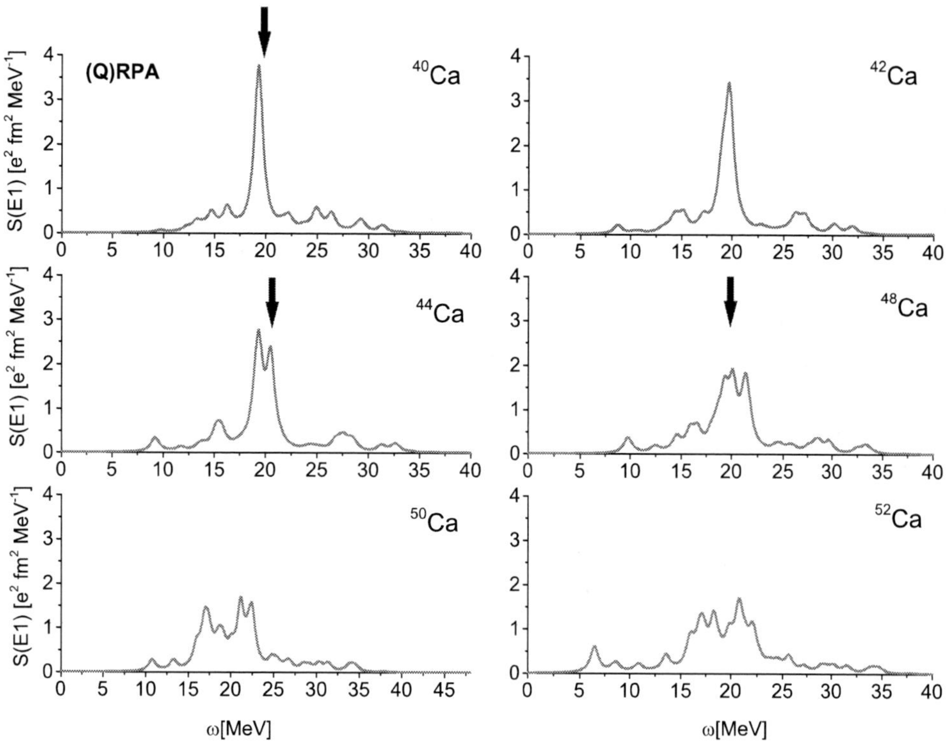

Figure 1. (Q)RPA $E1$-strength distribution in a chain of Ca isotopes obtained using $NNLO_{opt} + V_\rho$.

Both (Q)TDA and (Q)RPA describe properly the gross features of GDR and PDR. To reproduce their fine structure it is necessary to go beyond the mean field approximation. This was done for ^{132}Sn by using the EMPM [6].

In the EMPM, the 1^- states are obtained by diagonalizing the Hamiltonian in a space spanned by one- ($| \lambda, 1^-\rangle$) plus two-phonon ($| (n = 2)\beta, 1^-\rangle$) basis states. The latter states are generated in a subspace which includes the states $| (\lambda_1 \times \lambda_2)^{1^-}\rangle \equiv \left\{O^\dagger_{\lambda_1} \times | \lambda_2 >\right\}^{1^-}$ of energies $E_{\lambda_1} + E_{\lambda_2} \leq 30$ MeV,

The inclusion of the two-phonon states has a strong damping effect on the response. As shown in figure 3, compared to TDA, the EMPM cross section is severely quenched and reshaped due to the one- to two-phonon coupling. It has a smoother behavior and follows more closely the experimental points [12].

Both TDA and EMPM calculations yield a small peak in the cross section around ~ 10 MeV, fairly close in position and height to the one at ~ 9.8 MeV observed experimentally [12]. The EMPM strength collected by the low-lying states up to ~ 11 MeV exhausts $\sim 5.8\%$ of the TRK sum rule. This value is within the error of the measured fraction 4(3)% [12].

The phonon coupling enhances greatly the fragmentation of the strength. This effect, partly hidden in the cross section due to the smoothing action of the Lorentzian, is clearly visible in the $E1$ spectra shown in [6]. As compared to TDA, the EMPM spectrum is much more dense and is composed of peaks of considerably shorter height in both GDR and PDR regions.

The low-lying spectrum is composed of a large number of levels excited by both isoscalar and isovector probes (figure 4), in analogy with the experimental spectrum produced by (γ, γ') and

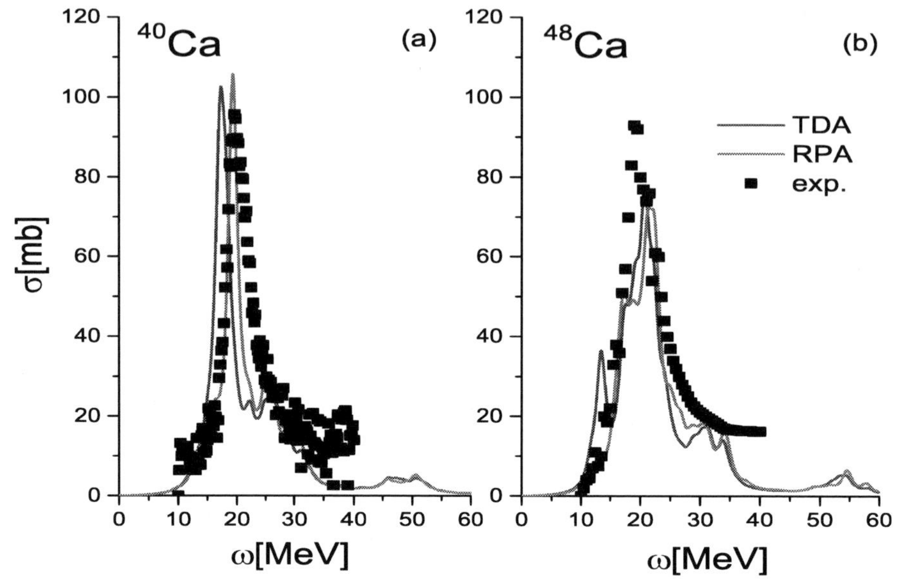

Figure 2. TDA and RPA versus experimental [17] cross sections in ^{40}Ca and ^{48}Ca

$(\alpha, \alpha'\gamma)$ experiments in the open shell ^{124}Sn [13–15]. Our calculation, however, does not predict for ^{132}Sn a splitting between isoscalar and isovector dipole modes at low-energy, at variance with the conclusions drawn from the analysis of the observed spectra in ^{124}Sn [16].

The mechanism of excitation suggests the pygmy nature of the low-lying states, a suggestion supported by the analysis of the wavefunctions and the transition densities [6].

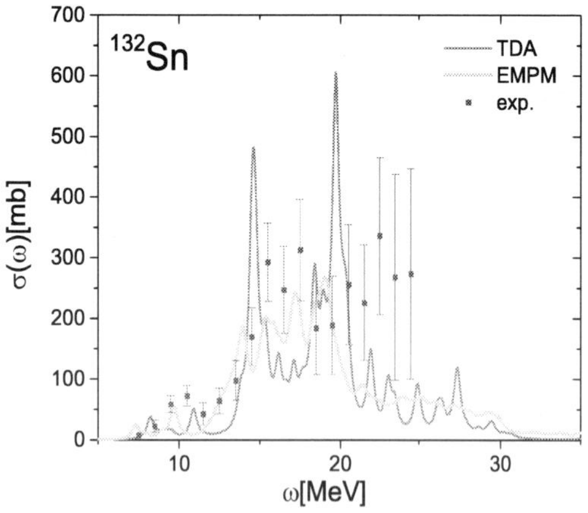

Figure 3. EMPM and TDA versus experimental cross sections in ^{132}Sn. A Lorentzian width $\Delta = 0.5$ MeV was used to compute the cross sections.

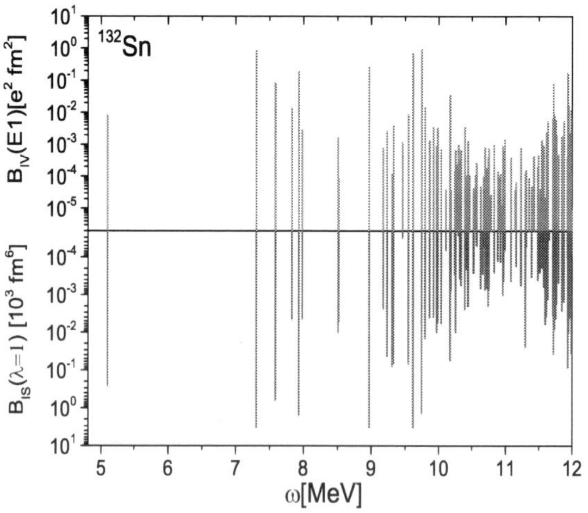

Figure 4. Low-lying isoscalar and isovector dipole transition strengths in ^{132} Sn

4. Conclusions

The first conclusion to be drawn from the above analysis is that the direct use of the optimized chiral potential $NNLO_{opt}$ yields much more realistic HF spectra compared to other realistic potentials and improves drastically the description of the dipole response. Such a potential, however, is not able to fill completely the gap between theory and experiments. This gap was filled here by adding a phenomenological density dependent repulsive term.

Another aspect to be pointed out is the high sensitivity of the response to the NN interaction. This property may be exploited to test different NN forces and to suggest a way of improving their parametrization.

Last but not least, the present study has pointed out the crucial role of complex configurations, two-phonon states in our case, in damping the GDR and in enhancing the density of low-lying levels as recent experiments require.

Acknowledgments

This work was partially supported by the Czech Science Foundation (Project No. P203-13-07117S)

References

[1] Bender M, Heenen P H and Reinhard P G 2003 *Rev. Mod. Phys.* **75** 121
[2] Vretenar D, Afanasjev A V, Lalazissis G A and Ring P 2005 *Phys. Rep.* **409** 101
[3] Hergert H, Papakonstantinou P and Roth R 2011 *Phys. Rev. C* **83**(6) 064317
[4] Bianco D, Knapp F, Lo Iudice N, Veselý P, Andreozzi F, De Gregorio G and Porrino A 2014 *Journal of Physics G: Nuclear and Particle Physics* **41** 025109
[5] Ekström A, Baardsen G, Forssén C, Hagen G, Hjorth-Jensen M, Jansen G R, Machleidt R, Nazarewicz W, Papenbrock T, Sarich J and Wild S M 2013 *Phys. Rev. Lett.* **110**(19) 192502
[6] Knapp F, Lo Iudice N, Vesely P, Andreozzi F, De Gregorio G and Porrino A 2014 *Phys. Rev. C* **90** 014310
[7] Andreozzi F, Knapp F Lo Iudice N, Porrino A and Kvasil J 2007 *Phys. Rev. C* **75**(4) 044312
[8] Andreozzi F, Knapp F, Lo Iudice N, Porrino A and Kvasil J 2008 *Phys. Rev. C* **78**(5) 054308
[9] Bianco D, Knapp F, Lo Iudice N, Andreozzi F and Porrino A 2012 *Phys. Rev. C* **85**(1) 014313
[10] Waroquier M, Heyde K and Vincx H 1976 *Phys. Rev. C* **13**(4) 1664–1673
[11] Günther A, Roth R, Hergert H and Reinhardt S 2010 *Phys. Rev. C* **82**(2) 024319
[12] Adrich P, Klimkiewicz A, Fallot M, Boretzky K, Aumann T, Cortina-Gil D, Pramanik U D, Elze T W, Emling H, Geissel H, Hellström M, Jones K L, Kratz J V, Kulessa R, Leifels Y, Nociforo C, Palit R, Simon H, Surówka G, Sümmerer K and Waluś W (LAND-FRS Collaboration) 2005 *Phys. Rev. Lett.* **95**(13) 132501
[13] Govaert K, Bauwens F, Bryssinck J, De Frenne D, Jacobs E, Mondelaers W, Govor L and Ponomarev V Y 1998 *Phys. Rev. C* **57**(5) 2229–2249
[14] Endres J, Litvinova E, Savran D, Butler P A, Harakeh M N, Harissopulos S, Herzberg R D, Krücken R, Lagoyannis A, Pietralla N, Ponomarev V Y, Popescu L, Ring P, Scheck M, Sonnabend K, Stoica V I, Wörtche H J and Zilges A 2010 *Phys. Rev. Lett.* **105**(21) 212503
[15] Endres J, Savran D, Butler P A, Harakeh M N, Harissopulos S, Herzberg R D, Krücken R, Lagoyannis A, Litvinova E, Pietralla N, Ponomarev V, Popescu L, Ring P, Scheck M, Schlüter F, Sonnabend K, Stoica V I, Wörtche H J and Zilges A 2012 *Phys. Rev. C* **85**(6) 064331
[16] Savran D, Babilon M, van den Berg A M, Harakeh M N, Hasper J, Matic A, Wörtche H J and Zilges A 2006 *Phys. Rev. Lett.* **97**(17) 172502
[17] Erokhova V A, Elkin M A, Izotova A V, Ishkhanov B S, Kapitonov I M, Lileeva E I and Shirokov E V 2003 *Izv. Rossiiskoi Akademii Nauk, Ser.Fiz.* **67** 1479 (EXFOR M0653.002 and M0653.005)

Journal of Physics: Conference Series **580** (2015) 012056 doi:10.1088/1742-6596/580/1/012056

Study of the Pygmy Dipole Resonance in ^{124}Sn with AGATA

L. Pellegri[1,2], A. Bracco[1,2], F.C.L. Crespi[1,2]
and the AGATA collaboration
[1] INFN Sezione di Milano, I-20133 Milano, Italy
[2] Università degli Studi di Milano, I-20133 Milano, Italy

E-mail: luna.pellegri@mi.infn.it

Abstract. The γ decay of the pygmy dipole resonance of ^{124}Sn was measured with the AGATA demonstrator coupled to an array of 9 large volume $LaBr_3$:Ce scintillators. This resonance was populated by the inelastic scattering reaction ^{17}O+^{124}Sn at 20 MeV/u. With AGATA, the γ decay up to the neutron separation energy was measured with high resolution. The angular distribution was measured both for the γ rays and the scattered ^{17}O ions. The present results are presented in comparison with the previous findings for (γ, γ') and (α,α'γ).

1. Introduction

In atomic nuclei the electric-dipole strength is exhausted to a large extent by the Giant Dipole Resonance (GDR). However a few percent of the E1 strength is taken by the so-called Pygmy Dipole Resonance (PDR), a concentration of 1^- states around the neutron separation threshold [1]. The study of the Pygmy Dipole Resonance (PDR) is particularly relevant to investigate the nuclear structure and also in connection with photo-disintegration reaction rates in astrophysical scenarios [2, 3, 4, 5, 6, 7]. Its description within the hydrodynamical model corresponds to a vibration of the neutron skin against a N=Z core [8]. In recent years, the study of the PDR has attracted particular attention and moreover its microscopic structure is presently under discussion. Efforts in the direction of understanding its nature require its excitation using different probes. Indeed, recent works comparing results of photon and α-scattering experiments show the presence of a different behaviour in the population of these states [9, 10, 11]. While a set of states at lower energy is excited with both types of reactions, the other set at higher energies is not populated by α scattering. This interesting finding has motivated further work based on the use of another probe with strong isoscalar character as ^{17}O [12] used in this work.

2. Experimental setup

The experiment was performed at the Legnaro National Laboratories (LNL), in Italy. The Giant Resonance modes were excited by inelastic scattering of ^{17}O ions at 20 MeV/u, in the laboratory frame, on a ^{124}Sn target (3 mg/cm^2 thick, enriched to 99%). ^{17}O has been chosen since it is loosely bound (being the neutron separation energy 4.1 MeV), and thus it allows to have a γ-spectrum mainly containing target de-excitation, particularly in the region of interest $E^* >$ 4.5 MeV. The detection of the scattered ions was performed with two ΔE-E silicon telescopes mounted inside the scattering chamber. The detectors used in this experiment are the prototypes

Figure 1. Left panel: Schematic view of the experimental setup, comprising five triple clusters of AGATA, an array of 9 $LaBr_3$:Ce detectors named HECTOR$^+$ and two E-ΔE Silicon telescope of the TRACE projects. Right panel: Picture of the experimental setup.

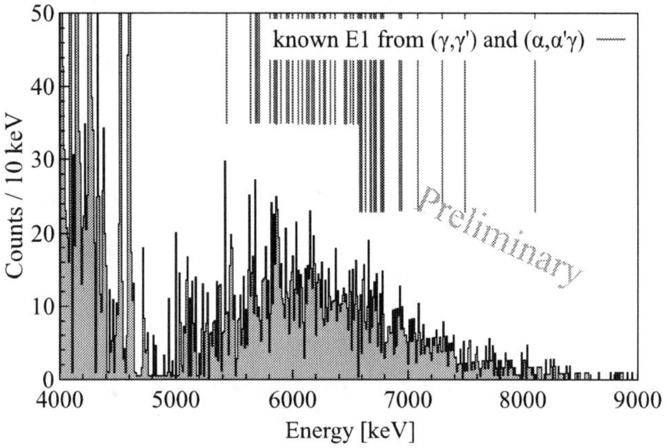

Figure 2. AGATA energy spectrum of the ^{124}Sn in the PDR region with the gating condition that the energy of the scattered ^{1}7O is equal to the γ-ray energy of the de-exciting nucleus as described in the text.

for the TRACE array [13, 14]. The main feature of these detectors is the identification in charge and mass of the scattered ions. Thanks to the use of these detectors, the excitation energy transferred to the target nucleus was measured with medium resolution (1.5 MeV). The γ rays produced during the de-excitation of the reaction participants were measured using an array of HPGe segmented detectors (the AGATA demonstrator [14, 15, 16]) coupled to an array of 9 large volume $LaBr_3$:Ce scintillators, (the HECTOR$^+$ array [17]) used to increase the detection efficiency for high-energy γ rays. A picture of the experimental setup is shown in the right panel of Fig. 1; the AGATA Demonstrator detectors are in the centre surrounded by the HECTOR$^+$ array. A schematic view of the complete experimental set-up is shown in the left panel of Fig. 1.

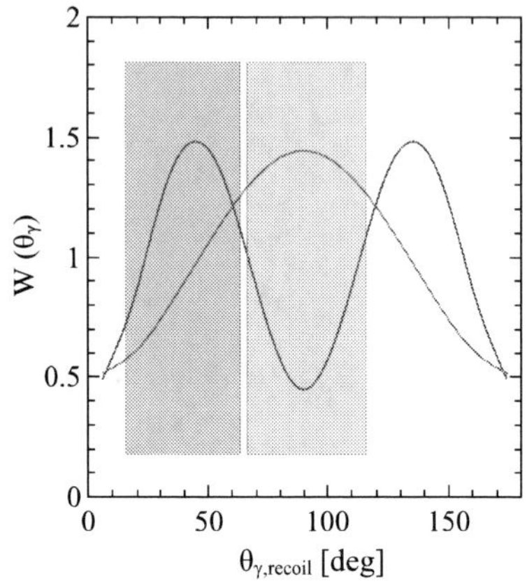

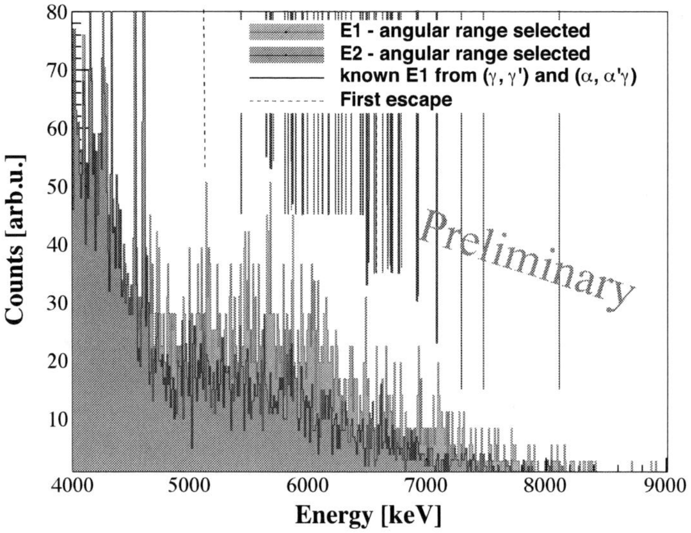

Figure 3. Top panel: expected angular distributions for an E1, red solid line, and E2, blue solid line, transition. The highlighted blue and red region represent the angular gate used to enhanced the E2 and E1 transitions respectively. Bottom panel: γ-rays spectra of ^{124}Sn in the PDR energy region measured with the AGATA demonstrator. The red spectrum is gated on 65°-115° angular range (red region of the top panel) in order to enhanced the E1 transitions while the blue spectrum is gate on 15°-65° angular range (blue region of the top panel) to enhanced the E2 transitions.

3. Preliminary results

Fig. 2 shows the γ-ray spectrum measured with the AGATA demonstrator array in the energy region above 4 MeV where the γ decay of the pygmy states is expected to be. This spectrum is obtained after selecting the inelastically scattered ^{17}O events and with the additional requirement that the energy of the γ rays equals the Total Kinetic Energy Loss (TKEL) values within a window ± 1 MeV wide. This requirement permits the selection of the ground-state transitions which are the dominant γ decay of the the pygmy dipole resonance. In addition, since these states had a life-time of the order of the femtoseconds a Doppler correction for the recoil motion was also needed. Finally, to reject the accidental background and the feeding from higher-lying

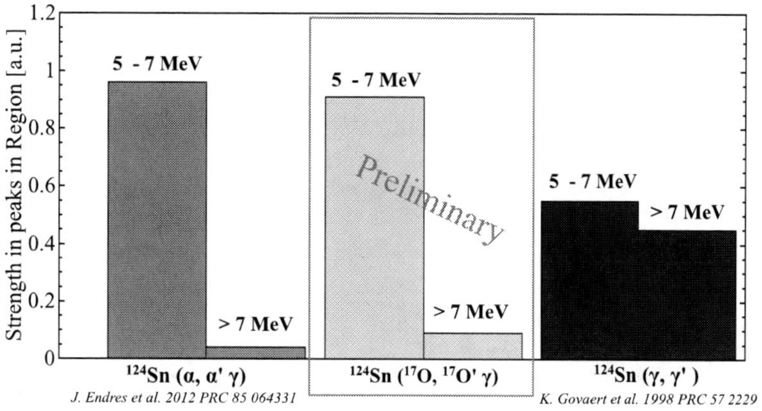

Figure 4. Ratio of the strength in peaks evaluated in the two energy region, below and above 7 MeV, for the three experiments.

states a background subtraction was applied. In Fig. 2 several E1 transitions known from (γ, γ') [9] and (α,α'γ) [11] measurements were identified and are indicated with the lines at the top of the figure.

Exploiting the position sensitivity of AGATA ($\Delta\theta \sim 1\,°$) and silicon detectors ($\Delta\theta \sim 3\,°$), it was possible to infer the character of the PDR. The multipolarity of the pygmy states, indeed, can be deduced by comparing the γ-ray energy spectrum gated on the angular distribution of the emitted γ rays. If we set a gate on the angular distribution, it is possible to identify the E1 and E2 components in the γ-ray spectra. The bottom panel of Fig. 3 shows the comparison between the γ-ray spectra of ^{124}Sn in the PDR energy region. The red spectrum is gated on the 65 °-115 ° angular range (red region in the top panel of Fig. 3) in order to enhanced the E1 transitions while the blue spectrum is gate on the 15 °-65 ° (blue region in the top panel of Fig. 3) angular range to enhanced the E2 transitions. The PDR transitions are clearly enhanced in the red spectrum, showing as expected an E1 behaviour.

An interesting question is the comparison of the pygmy states population with this reaction with those induced by the (α, α'γ) and (γ, γ'). Since the (α, α'γ) results indicated that the PDR region, seen in (γ, γ'), is split into two part, one below 7 MeV populated by both reactions and the other above 7 MeV populated only by the photon scattering, we integrated our data in these two specific regions. The comparison of the present results with the existing photon [9] and α-scattering [11] data are shown in Fig. 4. It is evident from this figure that the strength measured in the (γ, γ') experiment are almost equal in both energy ranges, which is not the case for the (^{17}O, ^{17}O'γ) and (α, α'γ) experiments. From these interesting results we can therefore say that the inelastic scattering of ^{17}O at 20 MeV/u was successfully used to measure the gamma decay from the PDR states of ^{124}Sn, and that the results confirm very well the presence of a splitting of the PDR similar to that observed with the (α, α'γ) technique. In order to be more quantitative, the isoscalar strength associated to this resonance has to be extracted and this work is in progress.

4. Conclusions

The low-lying E1 states in the ^{124}Sn nucleus were populated via inelastic scattering of a ^{17}O beam at bombarding energy of 20 MeV/u. Their subsequent γ decay was measured with the AGATA demonstrator array coupled to an array of large volume $LaBr_3$:Ce scintillators. The multipolarity of the observed gamma transitions was determined with remarkable sensitivity thanks to angular distribution measurements. The interesting result obtained from the preliminary data analysis concerns the nature of pygmy dipole resonance. Similarly to what was found using the (α, $\alpha'\gamma$) reaction, also in this case the results seem to indicate a selectivity in the population of specific pygmy states as compared to photon-scattering results. This finding points out the more isoscalar nature of the low-energy part of this resonance. A more complete analysis can be found in [18].

References

[1] Savran D *et al.* 2013 *Prog. Part. Nucl. Phys.* **70** 210
[2] Klimkiewicz A *et al.* 2007 *Phys. Rev. C* **76** 051603
[3] Carbone A *et al.* 2010 *Phys. Rev. C* **81** 041301(R)
[4] Roca-Maza X *et al.* 2011 *Phys. Rev. Lett.* **106** 252501
[5] Tamii A *et al.* 2011 *Phys. Rev. Lett.* **107** 062502
[6] Wieland O *et al.* 2009 *Phys. Rev. Lett.* **102** 092502
[7] Goriely S *et al.* 2004 *Nucl. Phys. A* **739** 331
[8] Mohan R *et al.* 1971 *Phys. Rev. C* **3** 1740
[9] Govaert K *et al.* 1998 *Phys. Rev. C* **57** 2229
[10] Endres J *et al.* 2010 *Phys. Rev. Lett.* **105** 212503
[11] Endres J *et al.* 2012 *Phys. Rev. C* **85** 064331
[12] Crespi F C L *et al.* 2014 *Phys. Rev. Lett.* **113** 012501
[13] Gadea A *et al.* 2011 *Nucl. Instr. and Meth. A* **654** 88
[14] Akkoyun S *et al.* 2012 *Nucl. Instr. and Meth. A* **668** 26
[15] Crespi F C L *et al.* 2013 *Nucl. Instr. and Meth. A* **705** 47
[16] Farnea E *et al.* 2010 *Nucl. Instr. and Meth. A* **621** 331
[17] Giaz A *et al.* 2013 *Nucl. Instr. and Meth. A* **729** 910
[18] Pellegri L *et al.* *Phys. Lett. B* URL `http://dx.doi.org/10.1016/j.physletb.2014.08.029`

Journal of Physics: Conference Series **580** (2015) 012057 doi:10.1088/1742-6596/580/1/012057

Shape isomers and their clusterization

J. Darai,[1] J. Cseh[2]

[1]Department of Experimental Physics, University of Debrecen, Debrecen, Pf. 105, Hungary-4010

[2]Institute for Nuclear Research, Hungarian Academy of Sciences, Debrecen, Pf. 51, Hungary-4001

E-mail: [1]darai@atomki.mta.hu, [2]cseh@atomki.mta.hu

Abstract. The role of the symmetries, phases and deformation is analysed in the interrelation of the cluster, shell and collective models. Based on that a symmetry-adapted method is presented for the determination of the shape isomers, and for the study of their possible fragmentation. Comparison with experimental data is discussed both for the cluster-shell competition, and for the existence and fragmentation of shape isomeres.

1. Introduction

The search for shape isomers, e.g. superdeformed (SD), or hyperdeformed (HD) states is an exciting topic of the present-day nuclear structure studies. Especially interesting is their existence in the $N = Z$ nuclei, in which the role of pairing, quartering, clustering, etc. can be studied.

Finding the possible clusterization of a specific nuclear state is important from two aspects. First it contributes to our understanding of the nuclear structure. Second it gives valuable information on the nuclear reactions, which can populate the state in question. In particular, a binary cluster configuration with the clusters in their ground intrinsic state is uniquely related to a reaction channel, like target and projectile in their ground state. Actually, one of the possible definitions (probably the best one) of a cluster state is provided by a reaction channel.

Here we discuss the relation of the collective, shell and cluster models for large deformation, and based on that we present a new method for the determination of stable nuclear shapes, as well as for finding their binary fragmentations. In doing so different kinds of symmetries are applied, and they allow us to study the phases of the shell and cluster models, too.

2. Collective, shell and cluster models

2.1. Symmetries

The connection of the shell, collective [1] and cluster models [2, 3], found in terms of the SU(3) symmetry in 1958, was based on a single-shell problem, described in spherical basis, related to simple symmetries. In a recent paper [4] the extension to multi-major shells was addressed, and the $U(3) \otimes U(3) \supset U(3)$ dynamical symmetry was found as the common intersection of the three models. Here we consider the role of more general symmetries and the case of large deformation.

Historically two kinds of the SU(3) symmetry were applied in this respect. i) The exact symmetry, and ii) the dynamical SU(3) symmetry (sometimes called dynamically broken symmetry). The latter kind of special breaking is achieved by incorporating an interaction

Table 1. Different SU(3) symmetries in nuclear stucture models. The signs '+' and '-' indicate if the Hamiltonian and its eigenvectors are symmetric or not. The operator is said to be symmetric, if it is a scalar, while a set of eigenvectors is symmetric if they transform according to an irreducible representation.

symmetry	Hamiltonian	eigenvectors	model
exact	+	+	harmonic oscillator
dynamical(ly broken)	-	+	Elliott, symplectic, IBM, vibron, SACM
quasi-dynamical	-	-	shell, quantum phase, IBM, SACM

which is expressed in terms of the invariant operator of the SO(3) subgroup, in addition to the SU(3) scalar part.

The relation between the shell and collective models, established by Elliott [1] was based on the SU(3) $\supset$ SO(3) dynamical symmetry. The Wildermuth-connection between the shell and cluster models was originally based on harmonic oscillator Hamiltonians of exact SU(3) symmetry, but it turns out to be valid also for the dynamical symmetry, as discussed e.g. in [5].

In studying the case of large deformation the more general quasi-dynamical SU(3) symmetry [6] turns out to be important, too. It is a generalization of the concept of the U(3) dynamical symmetry in the following sense. The Hamiltonian breaks the symmetry in such a way that the U(3) quantum numbers are not valid for its eigenvectors (contrary to the case of the real U(3) dynamical symmetry). In other words neither the operator is symmetric, nor its eigenvectors [7]. (For comparison with the exact and dynamical symmetries see Table I.) Yet, the symmetry is present in some sense, and it may survive even for strong symmetry-breaking interactions [6].

An asymptotic Nilsson-state serves as an intrinsic state for the quasi-dynamical SU(3) representation. Thus the effective quantum numbers are determined by the Nilsson-states in the regime of large deformation [8]. When the deformation is not large enough, then we can expand the Nilsson-states in the asymptotic basis, and calculate the effective quantum numbers based on this expansion [9].

The concept of effective U(3) symmetry is applicable also for the case when the simple leading representation approximation is valid, and then the real and effective U(3) quantum numbers usually coincide [9]. The quasi-dynamical symmetry appears in the investigation of both quantum phases and the large deformation.

Some interesting extensions of Elliott's SU(3) symmetry is illustrated in Figure 1.

2.2. Phases

Symmetry-adapted models are especially suitable for the studies of the phases and phase-transitions in finite quantum systems [10].

In the limit of large particle number phase-transitions are seen in the sense that the derivative of the energy-minimum, as a function of the control-parameter is discontinuous. The order of the derivative, showing the discontinuity, gives the order of the phase-transition. A phase is defined as a region of the phase diagram between the endpoint of the dynamical symmetry and the transition point. It is also conjectured [11] that such a quantum phase is characterised by a quasi-dynamical symmetry. Therefore, although the real dynamical symmetry is valid only at a single point of the phase-diagram, the more general quasi-dynamical symmetry may survive, and in several cases does survive [11, 12], in a finite volume of the phase diagram. If this conjecture really turns out to be true, then the situation is similar to Landau's theory: different phases are determined by different (quasi-dynamical) symmetries, and phase transitions correspond to a change of the symmetry.

In the case of the finite particle number the discontinuities are smoothed out, as the

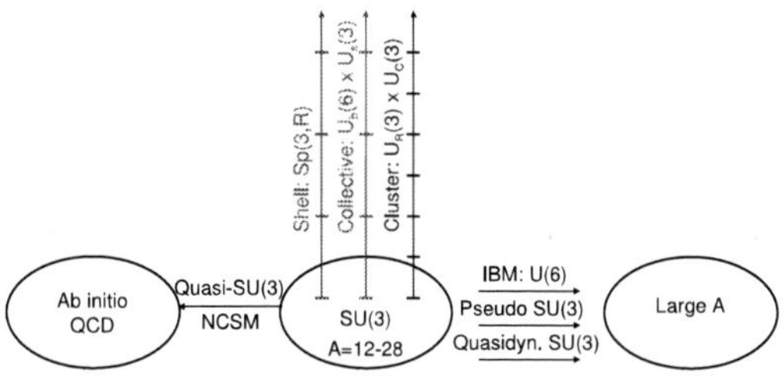

Figure 1. Extension of Elliott's SU(3) symmetry in different directions.

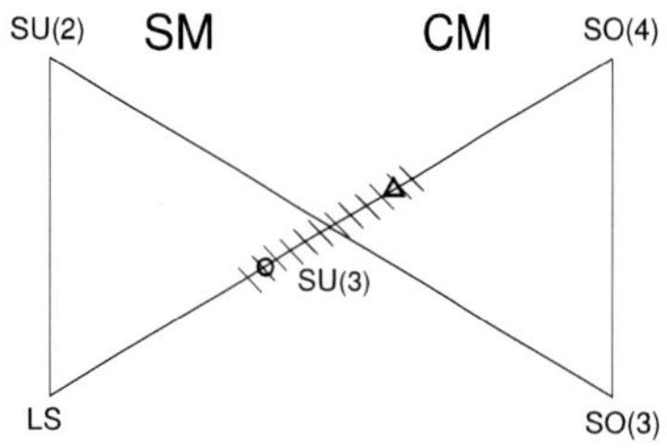

Figure 2. Joint phase diagram of the shell and cluster models, and the location of the ground-band of the ^{20}Ne nucleus, from shell (circle), cluster (triangle) and quasicluster (diffuse) model calculations.

consequence of the finite size effect, but still remarkable changes can be detected in the behaviour of the corresponding functions.

The semimicroscopic algebraic cluster model (SACM) [13] of a binary cluster system has three dynamical symmetries. Two of them come from the vibron model of the relative motion: $U_R(3)$ corresponds to shell-like clusterization, or in the language of the collective motion to a soft vibrator, while $O_R(4)$ represents a rigid molecule-like rotator. The third symmetry corresponds to a situation, when the coupling between the relative motion and internal degrees of freedom of the clusters is weak. Therefore, the phase diagram of a binary system is two-dimensional, and can be illustrated by a triangle [14].

Schematic calculations show that there is a second order phase transition (in between the shell-like ($U_R(3)$) and rigid molecule-like ($O_R(4)$) clusterizations) when interaction terms up to second order are included, but also a first order transition shows up, if third order interactions are involved [15].

A triangle-like phase diagram has been proposed for the shell model, too [16], which in addition to the SU(3) and SU(2) symmetries has the independent-particle model as the third corner. The limiting cases correspond to the situations, in which the quadrupole-quadrupole, pairing (both of them are two-body) interactions, and the single-particle energies dominate, respectively. (When a single shell calculation is performed then the spin-orbit interaction may be the most relevant part of the single-particle contribution.) The two phase diagrams match each other at the SU(3) corner, as shown in Figure 2. The real nuclear systems can be allocated to this diagram.

Here we refer to the results of three different calculations on the ground-band of the ^{20}Ne nucleus. Vargas et al [17] performed a full shell calculation for the 4 nucleons outside the core. Yepez-Martinez et al have carried out a similar calculation on the cluster-side of the phase diagram [18], by applying the semimicroscopic algebraic cluster model [13].

Itagaki et al used the antisymmetrized quasi-cluster model [19] for the description of the ground band. This model can take a direct route from the rigid molecule-like (SO(4)) clusterization via the shell-like cluster limit (SU(3)) to the jj-coupled shell model dominated by a strong LS interaction. Thus, it is especially illuminative from the viewpoint of the cluster-shell competition. It does not have, however, a well-defined algebraic structure, therefore, it does not determine a well-defined position.

It is remarkable that three different model calculations have similar conclusion on the closeness of the ground-band of ^{20}Ne to the matching point between the shell and cluster models.

2.3. Deformation

In [20] it was shown that the symmetry algebra of the anisotropic harmonic oscillator is SU(3), whenever its frequencies are commensurate. In considering realistic interactions the Nilsson model plays an important role. Soon after the experimental discovery of the superdeformation, it was realised that the $L-S$ coupling recovers for the SD shape [21].

We also apply the Nilsson-model in order to investigate the survival (or appearance) of the SU(3) symmetry systematically in a large range of the quadrupole deformation. We obtain the shape isomers from a selfconsistent calculation concerning the quadrupole deformation. In particular: one varies systematically the parameters $(\beta_{in}, \gamma_{in})$, as an input for the Nilsson-model. The calculations provide us with the effective U(3) quantum numbers, which can be translated into the $(\beta_{out}, \gamma_{out})$ quadrupole deformation, since they are uniquely related to each other [22]. Then one can check if the selfconsistency is satisfied, as well as the question whether or not the result is stable with respect to the (small) changes of the input values. This method for the determination of the shape isomers is an alternative to the standard energy-minimum calculation and has been shown to be effective for a range of light nuclei [23, 24, 25].

The result of the Nilsson model + quasi-dynamical SU(3) calculation for ^{36}Ar is shown in Figure 3.

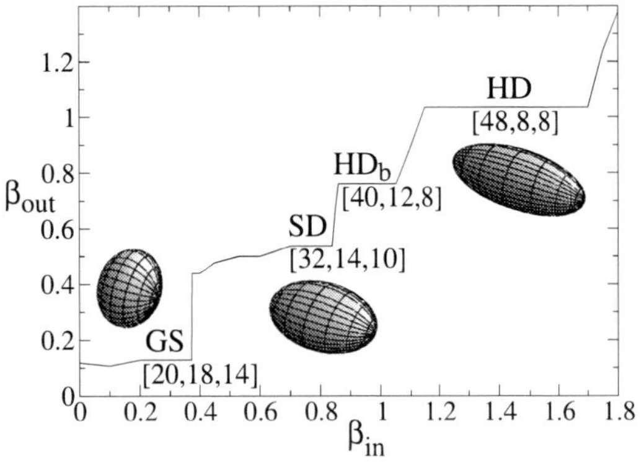

Figure 3. Shape isomers of the ^{36}Ar nucleus from the deformation self-consistency calculation, with effective U(3) quantum numbers and the schematic illustration of the shape at the plateaus.

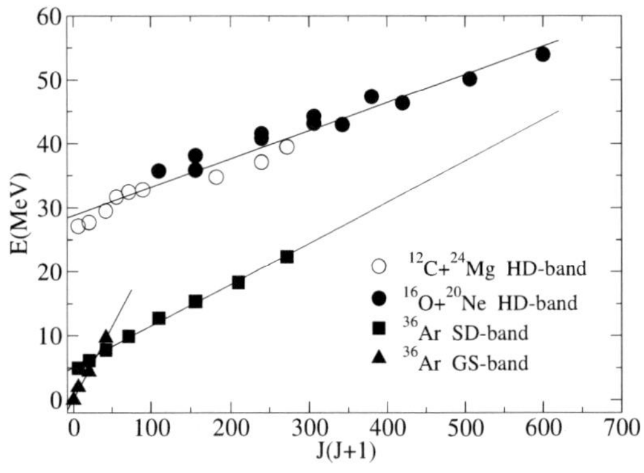

Figure 4. Excited states in ^{36}Ar, observed as resonances in the ^{12}C+^{24}Mg (open circles) and ^{16}O+^{20}Ne (full circles) reactions. The SD band observed in ^{36}Ar and the ground-state band are also included.

2.4. Fragmentation

The U(3) connection [1, 3] between the collective, shell and cluster models works well in case i) the U(3) symmetry is approximatelly valid, and ii) the relation between the cluster and shell model wavefunctions is simple. This latter condition means that the expansion of the cluster U(3) state in terms of shell basis reduces to a few terms.

As discussed in the previous subsection the U(3) symmetry recovers for the superdeformed, hyperdeformed, etc shapes, in spite of the important role of the symmetry-breaking spin-orbit interaction. The second condition (on the simple shell model expansion) turns out to be valid also for several shape isomers. E.g. in case of the ^{20}Ne nucleus each of the 4 shape isomers (of the U(3) symmetry: [12,4,4], [16,8,0], [24,4,0], [40,0,0]) has single multiplicity in the shell model basis [26]. Therefore, if a cluster state with the same U(3) symmetry is allowed it is identical with the shell state.

In addition to the U(3) selection, there is another simple prescription by Harvey [27] for the determination of the allowed clusterizations. It also applies harmonic oscillator basis, so the two requirements are somewhat similar. Nevertheless, their physical content is not the same; in some sense they are complementary to each other. Therefore, the best way is to apply them in a combined way [28, 29, 30]. (Their relation is discussed more in detail in [31].) When a cluster configuration is forbidden, one can characterize its forbiddenness quantitatively [32].

The energetic preference represents a complementary viewpoint for the selection of clusterization. We usually incorporate it in two different ways: i) by applying simple binding-energy arguments [33], and ii) by performing double-folding calculations, according to the dinuclear system model [34, 35].

3. Application

We have applied the above methods for the determination of the shape isomers from the ground state up to the linear alpha-chain states in several light $N = Z$ nuclei: ^{20}Ne [26], ^{28}Si [25], ^{36}Ar [23], ^{56}Ni [24]. Several isomers have been found, and their possible binary fragmentations have also been determined. In our description the clusters are supposed to be in their intrinsic ground state, like the free nucleus, without any simplifying assumption. Their relative orientation is also free to change, without any constraint. Some general features of these results are as follows. The same binary cluster-configuration can be present in different shape isomers of a nucleus due to the different orientation of the deformed clusters with respect to the molecular axis. The ground state usually prefers asymmetric fragmentations, while in the hyperdeformed state more symmetric clusterizations appear. The superdeformed state represents a situation in between. This feature is a consequence of the similar shapes and volumes of close-lying nuclei in their ground state.

As a specific example we cite here the case of the ^{36}Ar nucleus. In [28] the possible binary clusterizations of the ground, the (experimentally known) SD, and the (theoretically predicted [36]) HD states was studied systematically. It was found that the HD state could be populated in the ^{24}Mg+^{12}C and ^{20}Ne+^{16}O reactions. A recent analysis of the ^{24}Mg+^{12}C elastic scattering [37] revealed the existence of resonances, as shown in the upper part of Fig. 4. The moment of inertia is in a very good agreement with that of the HD shape predicted from alpha-cluster model [36]. Nilsson-model+quasi-dynamical SU(3) calculation gave the same HD state [23], as the alpha-cluster model.

The similarity of the (predicted and observed) moments of inertia, and the fact that the resonances were seen in exactly those reactions, which define the preferred cluster-configurations of the HD shape suggest that the recently observed band in Fig. 4 is a good candidate for the hyperdeformed shape isomer of the ^{36}Ar nucleus. Thus ^{36}Ar may be a nucleus to have experimental evidence for the existence of ground, superdeformed, as well as hyperdeformed bands. (The question of the ternary clusterization for these shape isomers have been addressed in [29].)

4. Summary and conclusion

In this contribution we have considered the interrelation of the fundamental structure models of atomic nuclei for the case of general symmetries and large deformation.

The cluster-shell competition or coexistence has been interpreted in terms of the joint phase diagram of the shell and cluster models. Three different (shell, cluster and quasi-cluster) calculations indicate the position of the ground-band of ^{20}Ne very close to the SU(3) matching point of the two models.

Concerning the large deformations the quasi-dynamical SU(3) symmetry is found to be stable for Nilsson-type interactions for several shapes with commensurable ratios of the main axes. This symmetry-oriented method allows the determination of the possible clusterizations, as well, which reveals the connection to reaction channels.

All these considerations, as well as many others, are based on the extension of Elliott's SU(3) symmetry. In Figure 1 we summarized some of its generalization along different directions.

This work was supported by the OTKA (Grant No K106035).

References

[1] Elliott J P 1958 *Proc. Roy. Soc. A* **245** 128, 562
[2] Wildermuth K and Kanellopoulos Th 1958 *Nucl. Phys.* **7** 150
[3] Bayman B F and Bohr A 1958/59 *Nucl. Phys.* **9** 596
[4] Cseh J 2014 arXiv 1404.3500 [nucl-th]; Cseh J contribution to the present volume
[5] Cseh J and Katō K 2013 *Phys. Rev. C* **87** 067301
[6] Rochford P and Rowe D J 1988 *Phys. Lett. B* **210**, 5;
Rochford P Rowe D J Repka J 1988 *J. Math. Phys.* **29**, 572
[7] Cseh J 2006 *Proc. IV Int. Symp. on Quantum Theory and Symmetries (Varna)* (Sofia: Heron Press) p. 918
[8] Jarrio M Wood J L and Rowe D J 1991 *Nucl. Phys. A* **528**, 409
[9] Hess P O Algora A Hunyadi M and Cseh J 2002 *Eur. Phys. J. A* **15**, 449
[10] Cejnar P and Iachello F 2007 *J. Phys. A* **40** 581; Cseh J et al. 2008 *Int. J. Mod. Phys. E* **17** 2296;
Cejnar P and Jolie J 2009 *Progr. Part. Nucl. Phys.* **62** 210; Iachello F 2010 *J. Phys. Conf. Ser.* **237** 012014
[11] Rowe D J 2004 *Nucl. Phys. A* **745** 47
[12] Yepez-Martinez H Cseh J and Hess P O 2006 *Phys. Rev. C* **74** 024319
[13] Cseh J 1992 *Phys. Lett. B* **281** 173; Cseh J and Lévai G 1994 *Ann. Phys. (NY)* **230** 165
[14] Cseh J 2010 *J. Phys. Conf. Ser.* **205** 012021
[15] Fraser P R Yepez-Martinez H Hess P O and Lévai G 2012 *Phys. Rev. C* **85** 014317
[16] Van Isacker P 2002 Nucl. Phys. A **704** 232c
[17] Vargas C E Hirsch J and Draayer J P 2001 *Nucl. Phys. A* **690** 409
[18] Yepez-Martinez H Ermamatov M J Fraser P R and Hess P O 2012 *Phys. Rev. C* **86** 034309
[19] Itagaki N Cseh J and Ploszajczak J M 2011 *Phys. Rev. C* **83** 014302
[20] Rosensteel G and Draayer J P 1989 *J. Phys. A* **22** 1323
[21] Sugawara-Tanabe K and Arima A 1993 *Phys. Lett. B* **317** 1;
Sugawara-Tanabe K Arima A and Yoshida N 1995 *Phys. Rev. C* **51** 1809
[22] Rowe D J 1985 Rep. Prog. Phys. **48**, 1419
[23] Cseh J et al. 2009 *Phys. Rev. C* **80**, 034320
[24] Darai J et al. 2011 *Phys. Rev. C* **84**, 024302
[25] Darai J Cseh J and Jenkins D G 2012 *Phys. Rev. C* **86**, 064309
[26] Cseh J and Darai J 2014 arXiv 1404.3503 [nucl-th]
[27] Harvey M 1975 *Proc. 2nd Int. Conf. on Clustering Phenomena in Nuclei, (College Park) USDERA report ORO-4856-26* p. 549
[28] Cseh J Algora A Darai J and Hess P O 2004 *Phys. Rev. C* **70**, 034311
[29] Algora A Cseh J Darai J Hess P O 2006 *Phys. Lett. B* **639**, 451
[30] Darai J Cseh J Adamian G and Antonenko N 2012 *EPJ Web of Conferences* **38**, 16001
[31] Cseh J and Darai J 2009 *AIP Conf. Proc.* **1098**: *Fusion08* 225
[32] Algora A and Cseh J 1996 *J. Phys. G* **22**, L39
[33] Buck B Merchant A C and Perez S M 2000 *Few-Body Systems* **29**, 53
[34] Volkov V V 1978 Phys. Rep. **44**, 93; 1982 *Deep inelastic nuclei reactions* Energoizdat, Moscow
[35] Shneidman T M *et al.* 2002 *Phys. Lett.* **B526**, 322
[36] Rae W D M and Merchant A C 1992 *Phys. Lett. B* **279**, 207
[37] Sciani W et al. 2009 *Phys. Rev. C* **80**, 034319

Journal of Physics: Conference Series **580** (2015) 012058 doi:10.1088/1742-6596/580/1/012058

Search for E1 strength in ^{62,64}Fe around the threshold

O. Wieland[1], A. Bracco[1,2], F. Camera[1,2] and R. Avigo[1,2] for the PRESPEC and AGATA collaborations

[1]INFN sez. di MILANO
[2]Università degli studi di MILANO

E-mail: wieland@mi.infn.it

Abstract. The structure and nature of the pygmy dipole resonance (PDR) states below and above the neutron threshold is a recent open problem, particularly in exotic, neutron rich nuclei. Present experimental observations give only limited information on this subject. New experiments using different methods are needed. A recent measurement at the GSI laboratories on ^{62,64}Fe with the PRESPEC (2014) setup, following a past experiment with the RISING (2005) setup on ^{68}Ni, will contribute to solve the open questions. The setup located at GSI consists of the segmented HPGe detector array AGATA, scintillators (HECTOR), an Energy Loss / Total Energy time of flight measuring detector system called LYCCA and the fragment separator (FRS) apparatus. The experiment is based on relativistic Coulomb excitation together with the detection of the incoming and outgoing particles event by event. The detection of the produced γ-rays in the reactions, provides insight into the problem of the electric dipole response and E1 strength distribution around particle separation threshold.

1. Introduction

Giant resonances are among the basic building blocks of nuclear structure. Since they dominate the nuclear response, particularly above the particle binding energies, investigations of their main features such as centroid energy and resonance width have been carried on for many years in experimental nuclear physics. In the case of the simplest mode, the giant dipole resonance (GDR), a reasonable knowledge of its systematic features has been achieved. However, its fine structure, which carries unique information on the underlying nature of the mode and on the decay mechanisms, is still not known. A very interesting long-standing question is the nature of the so-called electric pygmy dipole resonance close to the neutron threshold in medium-to-heavy nuclei. When the pygmy strength is rather collective, it is also predicted that it might result from a neutron excess density vibration relative to the $N \approx Z$ core. A good understanding of the properties of soft E1 (or pygmy) modes is important, particularly in connection with exotic neutron-rich nuclei, because in that case the corresponding strength is expected to be higher than in stable nuclei. Furthermore, the presence of a resonance of E1 character close to particle threshold has important astrophysical implications, because it considerably modifies the equilibrium of the (γ, n) and (n, γ) reactions occurring in explosive nucleosynthesis scenarios. This knowledge is crucial for statistical reaction codes applied in network calculations and plays therefore an important role in the astrophysical r-process [1,2,3].

The E1 strength distribution around the separation energy depends mainly on the neutron excess [4]. At present, there's an ongoing discussion on the possibility of extracting information on the neutron skin from the pygmy resonance structure [5,6,7]. This quantity is related to the isospin-

Journal of Physics: Conference Series **580** (2015) 012058 doi:10.1088/1742-6596/580/1/012058

dependent part of the nuclear equation of state (EOS), which, in turn, has relevant implications for the description of neutron stars. For this reason, there is an enormous effort aimed at determining the parameters that govern the asymmetric matter EOS, using both experimental and theoretical tools.
The first experiment searching pygmy states in the ^{68}Ni exotic nucleus, measuring γ-ray emission with the RISING setup [8], shows interesting results [9]. They allowed to deduce the neutron skin thickness [6] in a model dependent way. That work has motivated further investigations and thus new experiments regarding the search of the pygmy resonance in exotic ^{64}Fe and ^{62}Fe nuclei, using relativistic Coulomb excitation and the γ-tracking array AGATA, have been recently made.

2. Experimental set up and measurements

The experimental PRESPEC set up consisted in the combination of the GSI Fragment Separator (FRS) with the AGATA and HECTOR array systems [10] (figure 1). The segmented HPGe detectors of AGATA are equipped with electronics allowing the pulse shape analysis procedure which provides the position information with high resolution. To increase the efficiency and timing properties, the LaBr3:Ce scintillator detectors were mounted around the target position. The setup used for γ-ray detection provided an excellent Doppler correction, good background reduction and good efficiency.

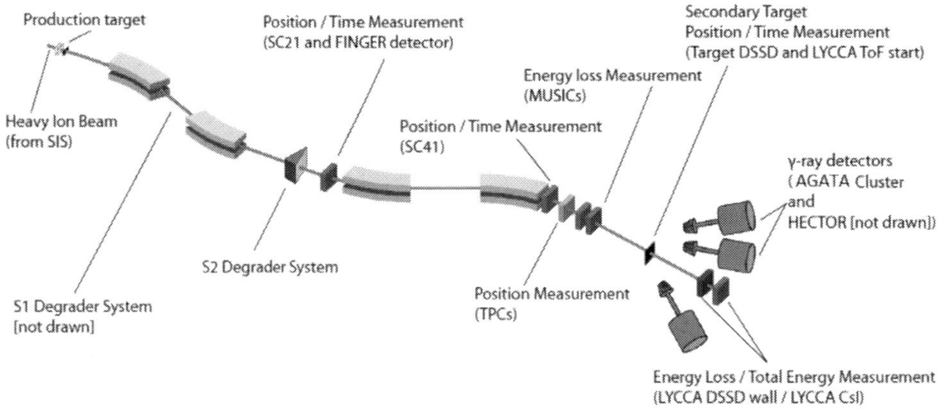

Figure 1. Schematic view of the FRS (sequence of 4 magnets on the left) together with the PRESPEC detectors (on the right). FRS is used for selecting the isotopes of interest. PRESPEC consist mainly in γ-ray detectors as HPGe segmented detectors (AGATA, mounted under forward angles), BaF_2 and $LaBr_3$:Ce scintillators (HECTOR, not shown and mounted additionally around the secondary target, mainly at backward angles) and an Energy Loss / Total Energy time of flight measuring detector system called LYCCA which detects and identifies the outgoing particles after the target.

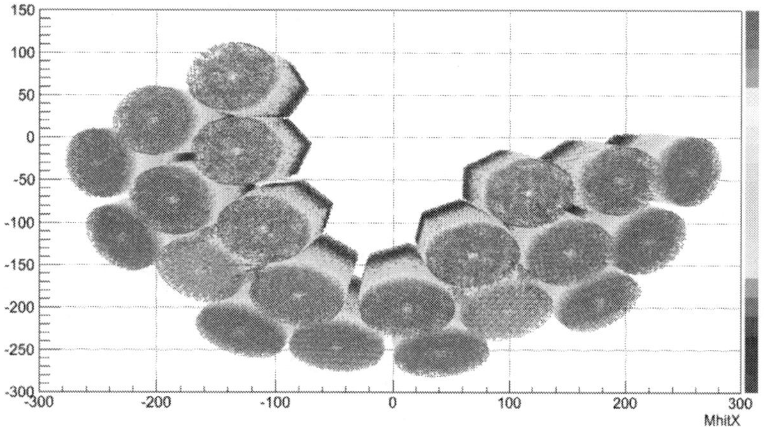

Figure 2. Perspective (x,y,z) plot of the measured γ-ray interaction points in the HPGe segmented detectors of AGATA. One can see the shape of the single detectors. The germanium capsules are mounted around the target position (in the centre at 0,0 coordinates). Every single interaction point, which contains additionally time and energy information, is coloured respectively to the local intensity. The spatial resolution around 5mm allows the reconstruction of the γ-ray interaction path with the help of tracking algorithms.

The radioactive $^{62,64}Fe$ beams were produced by fragmentation of the primary ^{86}Kr beam delivered by the SIS synchrotron at GSI at 700 or 900 AMeV and focused on a Be target. The ions of interest were selected and transported with the fragment separator FRS, (figure 1). The settings of the FRS were chosen to accept secondary fragments with a magnetic rigidity corresponding to a certain mass-over-charge ratio. This provided a beam cocktail containing the ions of interest in large fraction. The different nuclei contained in the secondary beam were identified uniquely according to their nuclear charge and mass number on an event-by-event basis. The beam cocktail was then impinging on the Au target (2 g/cm^2 thick) or Pb target (1 g/cm^2 thick), which were surrounded by the γ-ray detectors (see figure 1). The incoming isotopes of interest were selected using a reconstructed ID, mass versus charge (AoQ-Z) plot, while the outgoing ions were selected using the LYCCA-array placed at zero degree [10 and references therein] γ-ray spectra were measured. On an event-by event basis the incoming and outgoing beam particles were identified and very small deflection angles far below the grazing angle were selected. This guarantees pure Coulomb excitation mechanism. A selection of an impact parameter that is larger than 13 fm strongly suppresses the nuclear contribution to the reaction [8]. After Doppler correction for the radiation emitted from the projectile at 70% of the speed of light only the projectile γ-ray emission mainly produces detectable structures. The information from the pulse shape of the signals from the AGATA detectors leads to a very precise information of the position (figure 2), which is crucial if one wants to minimize the Doppler broadening of the measured spectral lines. The effect in the γ-ray spectra when one uses or not the PSA analysis is shown in figure 3, one sees that the γ-ray tracking helps to increase the peak to total ratio in the spectrum and lowers the background. Especially at lower energies where a high atomic background and statistical gamma decay from beam induced excitations is present.

With AGATA we measured the first 2^+ state in ^{64}Fe at 746 keV. The energy and strength of this state is known, so it can be used as a benchmark to estimate the strength of possible higher lying E1 states in the same nucleus.

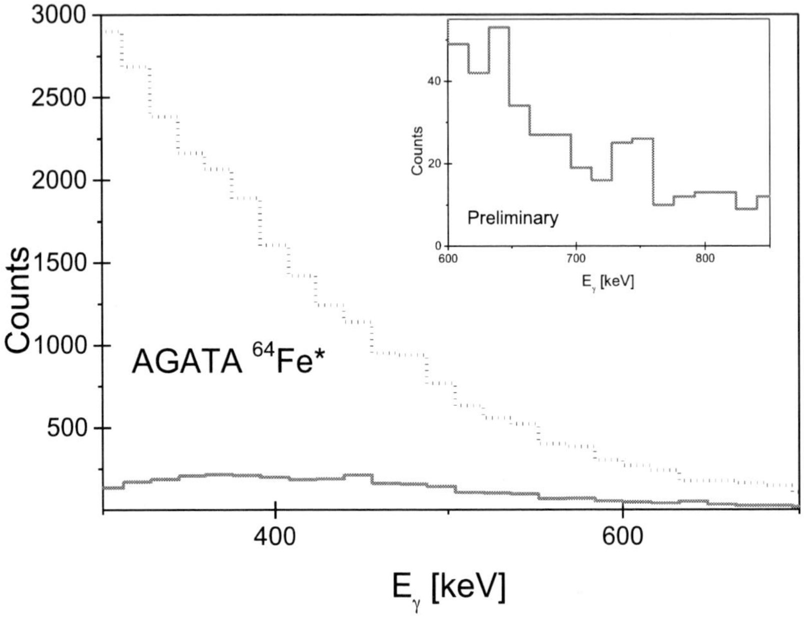

Figure 3. Doppler corrected γ-ray spectra measured in AGATA from Coulomb excited ^{64}Fe beam (E_{beam}=400AMeV and v/c=0.7) at very small scattering angles. The dotted line is the result obtained using only the position information whereas the continuous line is obtained using tracking algorithms. The peak structure between 700 and 800 keV (in the inset) corresponds to the decay of the first 2^+ state in ^{64}Fe.

3. First results

A first result of the measurement is the proof that highly segmented HPGe detectors can perform the pulse shape analysis and tracking of the γ-rays in a relativistic Coulomb excitation experiment in the presence of a high background rate.

A preliminary analysis of the AGATA (figure 4) and LaBr3:Ce data (not presented in this paper as the analysis has still to be finalized) showed in both different detectors an accumulation of strength in the γ-ray energy spectra (figure 4) below threshold. This seems to happen not only in the ^{64}Fe nucleus but also in ^{62}Fe at slightly different energies. The strength distribution shows the typical shape and strength as expected [11,12]. On the basis of the angular distribution of these structures the E1 character may be demonstrated. Preliminary simulations with the GEANT Monte Carlo code and calculations of the cross section with an eikonal, Coulomb excitation approach together with branching ratio simulations and past measurements [9,13] appear to support the thesis of the E1 character of these structures. However, confirmations with higher statistics need to be obtained. Further analysis work is in progress.

4. Conclusion

After obtaining interesting results in 2005 with the RISING set up at GSI laboratories on the pygmy dipole state in ^{68}Ni, new measurements were made again at GSI in 2012 and 2014 with the AGATA-PRESPEC set up to search for pygmy states and to measure the low lying E1 strength distribution in the exotic $^{64,62}Fe$ nuclei.

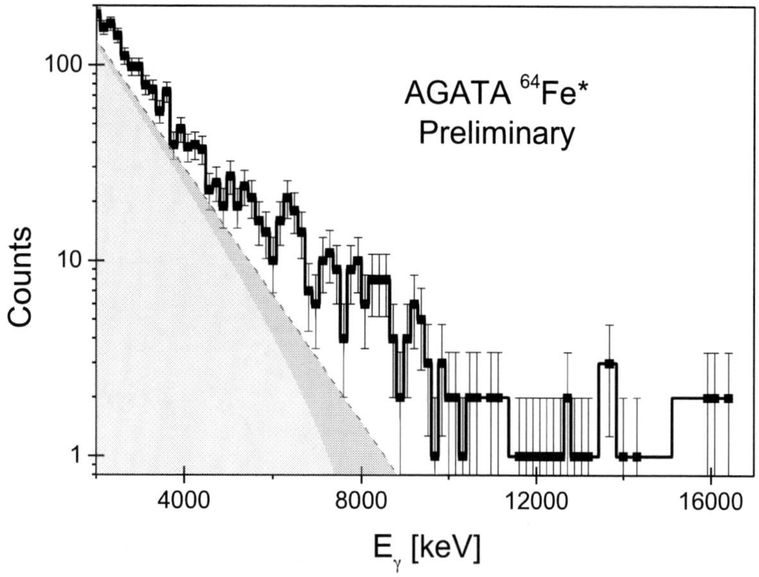

Figure 4. Preliminary Doppler corrected high energy γ-ray spectra for ^{64}Fe measured with AGATA using pulse shape analysis and tracking techniques. One can note a number of peaks sitting on an exponential background, which mainly comes from target excitation [9, ref. therein]. Width and shape of the peaks are similar to typical E1 strength distribution below the threshold [11,12].

The preliminary results obtained in these measurements contain interesting signals for the investigation of the pygmy states. Structures are present in the region 5-8 MeV in the measured γ-ray spectra and these structures seem to have an E1 character as this is the dominant excitation mode at these high beam energies. The structure shows similar strength distributions as seen in stable Fe isotopes [13] and a comparison with the yield obtained for the first 2^+ state shows that these E1 states would follow approximately the strength rule given in [12]. We also plan to extract from the data information for the region above the neutron threshold of 7.4 MeV up to the GDR region. This experiment is expected to provide the first and unique information for the E1 strength in the exotic $^{62,64}Fe$ nuclei, important for the modelling of the r-process path.

References

[1] Goriely S 1998 *Phys. Lett.* B **436** 10
[2] Goriely S and Khan E 2002 *Nucl. Phys. A* **706** 217
[3] Arnould M *et al.* 2007 *Phys. Rep.* **450** 97
[4] Litvinova E *et al.* 2010 *Phys. Rev. Lett.* **105** 022502
[5] Klimkiewicz A *et al.* 2007 *Phys. Rev. C* **76** 051603
[6] Carbone A *et al.* 2010 *Phys. Rev. C* **81** 041301(R)
[7] Reinhard P G and Nazarewicz W 2010 *Phys. Rev. C* **81** 051303
[8] Wollersheim H J *et al.* 2005 *NIM* A **537** Issue 3 637
[9] Wieland O *et al.* 2009 *Phys. Rev. Lett.* **102** 092502
[10] Pietralla N *et al.* 2014 *EPJ* Web of Conferences **66** 02083
[11] Bauwens F *et al.* 2000 *Phys. Rev.* C **62** 024302
[12] Massarczyk R *et al.* 2014 *Phys. Rev. Lett.* **112** 072501
[13] Rossi D M *et al.* 2013 *Phys. Rev. Lett.* **111** 242501
[14] Aumann T and T Nakamura T 2013 *Phys. Scr.* **T152** 014012 and references therein